Intelligent Infrastructures

International Series on
INTELLIGENT SYSTEMS, CONTROL, AND AUTOMATION:
SCIENCE AND ENGINEERING

VOLUME 42

Editor

Professor S. G. Tzafestas, National Technical University of Athens, Greece

For other titles published in this series, go to
www.springer.com/series/6259

Rudy R. Negenborn • Zofia Lukszo
Hans Hellendoorn

Editors

Intelligent Infrastructures

 Springer

Editors

Rudy R. Negenborn
Faculty of Mechanical, Maritime
& Materials Engineering
Delft University of Technology
Mekelweg 2
2628 CD Delft
The Netherlands

Zofia Lukszo
Faculty of Technology, Policy
& Management
Delft University of Technology
2628 BX Delft
Jaffalaan 5
The Netherlands

Hans Hellendoorn
Faculty of Mechanical, Maritime
& Materials Engineering
Delft University of Technology
Mekelweg 2
2628 CD Delft
The Netherlands

ISBN 978-90-481-3597-4   e-ISBN 978-90-481-3598-1
DOI 10.1007/978-90-481-3598-1
Springer Dordrecht Heidelberg London New York

Library of Congress Control Number: 2009940709

Printed on acid-free paper

9 8 7 6 5 4 3 2 1

Springer is part of Springer Science+Business Media (www.springer.com)

Foreword

Our society and economy have come to rely on services that depend on networked infrastructure systems, like highway and railway systems, electricity, water and gas supply systems, and telecommunication networks. A sustainable, efficient, safe, and reliable operation of these infrastructure systems is of crucial importance for the functioning of today's society. Currently, the operation of infrastructure systems is pushed to its limits. Infrastructure operators must survive in a competitive and regulated world where they have to maintain the operation whilst maximizing capacity utilization and revenue. Moreover, the variety of involved owners, operators, suppliers, and users has created enormously complex systems.

At the same time significant changes take place in the organization and the environment of various infrastructure sectors: the restructuring of the energy sector due to liberalization, the introduction of distributed electricity generation, the global temperature rise due to the climate change, the increasing traffic demand on railway, highway, waterway, and in the air, the pollution of the environment due to particles, NO_x and CO_2, and the development of a Pan-European or global market with well defined legislation.

The increasing complexity as well as the significant changes going on in various infrastructure sectors demands the definition of new ways to manage such infrastructures. The difficulty in the management of these infrastructure systems is understood by considering the vast amount of actors acting in them: energy producers, operators, and consumers all work on one large international electricity network, ship brokers, ship owners, district water boards, and lock keepers all work on one large international waterway network. Achieving an infrastructure that from an overall point of view operates well, requires in some way or another managing or regulating the behavior of all these individual actors. This is clear from a European point of view, where regulations are determined for the various countries with the aim of achieving a Pan-European desired performance.

In this book, the authors show how passive networks can be transformed into intelligent infrastructures; they have succeeded to bring together a wide range of approaches to solve the complex task of infrastructure operation. In particular, the authors focus on the analogies and differences between different types of infrastructures, stimulating cross-infrastructural discussion. Research in supranational networks is important for a sustainable, prosperous, and reliable future.

Rotterdam, September 2009   Prof.Dr. Ruud Lubbers

Former Dutch Prime Minister,
Chairman of the Supervisory Board of the
Energy Research Centre of The Netherlands

Foreword

Infrastructure systems constitute one of the prototype instances of a complex system in the sense of general systems theory. This theory has been introduced in the 1930s by Ludwig von Bertalanffy, a biologist, who was concerned with the accelerating differentiation of the scientific disciplines, the evolving diversity of conceptualizations and terminologies in these disciplines, and the emerging communication barriers between the disciplines. Von Bertalanffy was convinced that the trap of scientific silos can only be overcome if a meta-science is established which provides a common set of fundamental scientific concepts and problem solving strategies. This meta-science should – at least in principle – be applicable in all scientific fields and thus could contribute to bridge the gap between the sciences.

Inspired by the analysis of biological organisms and ecosystems, von Bertalanffy identified a general theory of systems as the core of this new meta-science. Such a theory should provide the scientific basis and the methods and tools for the representation, the analysis and understanding as well as the synthesis of any kind of system. A system is considered as a set of interacting, often hierarchically organized subsystems, with defined interfaces to its environment. The behavior of such systems is typically not only determined by the nature of the individual subsystems but in particular by the emergent behavior resulting from their competitive or cooperative interaction. Obviously, this characterization not only holds for any biological entity, a cell, an organ, or a complete organism, but also for any infrastructure entity which could be as diverse as the transportation systems in an urban region, the water supply and disposal network of a city, the telecommunication system on a continent, or the hydrogen production and distribution facilities of an industrial complex in the chemical industries. Infrastructure systems are not only composed of technological entities but also comprise decision making and disturbing human actors. Their use and evolution is strongly determined by the political, economical, and societal habits and norms. Though man-made and hence artificial, infrastructure systems share another feature with biological systems: they typically evolve as a consequence of a kind of selection pressure resulting from the apparent system behavior in its environment.

Since infrastructure systems are typically mission-critical entities in society and industry, the evolutionary development is not fully satisfactory. The complexity and the increasing interrelation of the infrastructure systems in the various sectors like industry, water, energy, communication, or transportation, calls for scientifically sound procedures for infrastructure systems design and operation which account for their socio-economic-technological nature. General systems theory provides a perfect foundation from which a theory of infrastructure systems can be deduced. Such theory has to bridge between existing scientific disciplines covering not only all the engineering sciences but also economics and sociology. This challenging mission can only be successful if a dedicated scientific community with a super-critical number of scientific groups around the world representing the essential scientific disciplines can be established. This is a complex systems problem in itself! One instrument to help make this happen is the publication of a series of monographs which have to both, raise awareness for the subject and introduce key areas of the emerging discipline for reference.

This book constitutes such an effort. It presents a collection of well-selected contributions associated with three selected types of infrastructure systems, namely electricity, road traffic, and water infrastructures. These three topical areas are introduced by two contributions which are of a more general nature. They discuss modeling and predictive control issues for infrastructure systems in general. The selection of contributions focuses on highly relevant topical areas, not only for geographical regions with an already well-developed but continuously reengineered set of infrastructures, but also for developing geographical regions where sustainable infrastructure systems are crucial for economical development. The topics of the individual contributions are nicely spread between infrastructure design and infrastructure control and operations. They also show the different perspective and scientific background of the authors and the diversity of methodologies employed.

This book will hopefully not only serve the purpose of disseminating research results but also of raising the awareness for the interesting, timely, and relevant research topics in the area of infrastructure systems. Together with the editors and authors, I hope that this book will find attention in the diverse systems engineering communities and thus contributes to the probably overdue formation of infrastructure systems engineering as a well-defined research field – with its own and distinct identity, but still well-connected with the established systems engineering communities.

Aachen, September 2009   Prof.Dr.-Ing. Wolfgang Marquardt
                                                          Member of the Scientific Advisory Board
                                                          Next Generation Infrastructures

Table of Contents

Part II Electricity Infrastructures

Preface

Modern societies heavily depend on infrastructure systems such as road-traffic networks, water networks, and electricity networks. Nowadays infrastructure systems are large-scale, complex, networked, socio-technical systems, that almost everybody uses on a daily basis, and that have enabled us to live closely together in large cities. Infrastructure systems are so vital that their incapacity or destruction would have a fatal effect on the functioning of our society. The complexity of these systems is defined by their multi-agent and multi-actor character, their multi-level structure, their multi-objective optimization challenges, and by the adaptivity of their agents and actors to changes in their environment. The operation and control of existing infrastructures are fallible: too often people are confronted with capacity problems, dangerous situations, unreliability, and inefficiency.

To address these issues, infrastructures have to be made more intelligent, i.e., they should be equipped with senses, ICT programs, and actuators. They should be able to determine autonomously how to operate the infrastructure, taking into account the most up-to-date state of the infrastructure and the existence of several decision makers, such that ultimately the infrastructure is operated in a pro-active way and issues are resolved quickly. This book focuses on how to make infrastructures intelligent.

In this book a wide range of problems about how infrastructures are functioning today is discussed and novel advanced methods and tools for the operation and control of existing infrastructures are proposed. Different points of view on intelligent control of infrastructure systems (such as power networks, road traffic networks, water networks) are brought together. In particular, the question of how intelligence can contribute to solving the following problems that are common to many types of infra-structure systems will be addressed:

- How to more efficiently use available transport capacity?
- How to improve the reliability of service?
- How to make infrastructures more environmentally sustainable?
- How to enhance infrastructure security?

The book is divided into four parts: one generic infrastructure modeling and control part, and three parts considering particular infrastructures in detail. Each of the chapters addresses one or more of the above mentioned problem areas, therewith stimulating cross-infrastructural discussion on how to address the mentioned issues.

The contributions in this book are made by a large number of international scientists, some of which are also active within the subprogram Intelligent Infrastructures of the Next Generation Infrastructures Foundation (http://www.nginfra.nl/), which aims at developing theory and applications concerning novel modes for control and management of existing infrastructures. Together the contributions create a broad and profound coverage of the main issues related to the intelligent operation of infrastructures intended for researches, graduate students, policy makers, system operators, and managers working on organization, modeling, optimization, or control of infrastructures.

We wish to thank all authors for their high-quality contributions and the reviewers for their constructive remarks and suggestions; without them this book would not exist. Furthermore, we thank Ms. Hardy and Ms. Sherwood for their careful proof reading work, Ms. Jacobs and Ms. Pot for their guidance in the publication process at Springer, and the Next Generation Infrastructures Foundation for its financial support.

Thanks to the efforts of all people involved, we are convinced that the methods and tools presented in this book will give an invaluable support to a better understanding of the analysis and control of multi-actor multi-level infrastructural systems.

Delft, September 2009   Rudy R. Negenborn
                                                                                                      Zofia Lukszo
                                                                                                      Hans Hellendoorn

List of Contributors

Göran Andersson
Power Systems Laboratory, ETH Zürich,
Zürich, Switzerland, e-mail: andersson@eeh.ee.ethz.ch

Michèle Arnold
Power Systems Laboratory, ETH Zürich,
Zürich, Switzerland, e-mail: arnold@eeh.ee.ethz.ch

Toni Barjas Blanco
Department of Electrotechnical Engineering, Katholieke Universiteit Leuven,
Leuven, Belgium, e-mail: tonibarjasmartinez@gmail.com

Jean Berlamont
Department of Civil Engineering, Katholieke Universiteit Leuven,
Leuven, Belgium, e-mail: Jean.Berlamont@bwk.kuleuven.be

Michiel C.J. Bliemer
Faculty of Civil Engineering and Geosciences, Delft University of Technology,
Delft, The Netherlands, e-mail: m.c.j.bliemer@tudelft.nl

Ettore Bompard
Dipartimento di Ingegneria Elettrica, Politecnico di Torino,
Torino, Italy, e-mail: ettore.bompard@polito.it

Piet H.L. Bovy
Faculty of Civil Engineering and Geosciences, Delft University of Technology,
Delft, The Netherlands, e-mail: p.h.l.bovy@tudelft.nl

Kris Cauwenberghs
Division Water, Flemish Environment Agency,
Brussels, Belgium, e-mail: k.cauwenberghs@vmm.be

Po-Kuan Chiang
Department of Civil Engineering, Katholieke Universiteit Leuven,
Leuven, Belgium, e-mail: `pokuan.chiang@student.kuleuven.be`

Francesco Corman
Department of Transport and Planning, Delft University of Technology,
Delft, The Netherlands, e-mail: `f.corman@tudelft.nl`

Andrea D'Areano
Dipartimento di Informatica e Automazione, Università degli Studi Roma Tre,
Roma, Italy, e-mail: `a.dariano@dia.uniroma3.it`

Geert Deconinck
Department of Electrical Engineering, Katholieke Universiteit,
Leuven, Belgium, e-mail: `Geert.Deconinck@esat.kuleuven.ac.be`

Bart De Moor
Department of Electrotechnical Engineering, Katholieke Universiteit Leuven,
Leuven, Belgium, e-mail: `bart.demoor@esat.kuleuven.be`

Bart De Schutter
Delft Center for Systems and Control & Marine and Transport Technology,
Delft University of Technology, Delft, The Netherlands,
e-mail: `b@deschutter.info`

Mathijs M. de Weerdt
Department of Software Technology, Delft University of Technology,
Delft, The Netherlands, e-mail: `m.m.deweerdt@tudelft.nl`

Michel dos Santos Soares
Faculty of Technology, Policy, and Management, Delft University of Technology,
Delft, The Netherlands, e-mail: `m.dossantossoares@tudelft.nl`

Rui Duan
Department of Electrical Engineering, Katholieke Universiteit,
Leuven, Belgium, e-mail: `Rui.Duan@esat.kuleuven.ac.be`

Jinliang Gao
School of Municipal and Environment Engineering, Harbin Institute of Technology,
Harbin, China, e-mail: `gjl@hit.edu.cn`

Andreas Hegyi
Faculty of Civil Engineering and Geosciences, Delft University of Technology,
Delft, The Netherlands, e-mail: `a.hegyi@tudelft.nl`

Hans Hellendoorn
Delft Center for Systems and Control, Delft University of Technology,
Delft, The Netherlands, e-mail: `j.hellendoorn@tudelft.nl`

Paul Hines
School of Engineering, University of Vermont,
Burlington, Vermont, e-mail: `paul.hines@uvm.edu`

Marija D. Ilić
Electrical & Computer Engineering Department, Carnegie Mellon University,
Pittsburgh, Pennsylvania, and Faculty of Technology, Policy, and Management,
Delft University of Technology, Delft, The Netherlands,
e-mail: `milic@ece.cmu.edu`

Andrej Jokic
Department of Electrical Engineering, Eindhoven University of Technology,
Eindhoven, The Netherlands, e-mail: `a.jokic@tue.nl`

René Kamphuis
Power Systems and Information Technology, Energy Research Centre of The
Netherlands, Petten, The Netherlands, e-mail: `kamphuis@ecn.nl`

Koen Kok
Power Systems and Information Technology, Energy Research Centre of The
Netherlands, Petten, The Netherlands, e-mail: `j.kok@ecn.nl`

Mircea Lazar
Department of Electrical Engineering, Eindhoven University of Technology,
Eindhoven, The Netherlands, e-mail: `m.lazar@tue.nl`

Sylvain Leirens
Departamento de Ingeniería Eléctrica y Electrónica, Universidad de Los Andes,
Bogotá, Colombia, e-mail: `sleirens@uniandes.edu.co`

Hao Li
Faculty of Civil Engineering and Geosciences, Delft University of Technology,
Delft, The Netherlands, e-mail: `h.li@tudelft.nl`

Zofia Lukszo
Faculty of Technology, Policy, and Management, Delft University of Technology,
Delft, The Netherlands, e-mail: `Z.Lukszo@tudelft.nl`

Manfred Morari
Automatic Control Laboratory, ETH Zürich,
Zürich, Switzerland, e-mail: `morari@control.ee.ethz.ch`

Roberto Napoli
Dipartimento di Ingegneria Elettrica, Politecnico di Torino,
Torino, Italy, e-mail: `roberto.napoli@polito.it`

Rudy R. Negenborn
Delft Center for Systems and Control, Delft University of Technology,
Delft, The Netherlands, e-mail: `r.r.negenborn@tudelft.nl`

Dario Pacciarelli
Dipartimento di Informatica e Automazione, Università degli Studi Roma Tre,
Roma, Italy, e-mail: `pacciarelli@dia.uniroma3.it`

Marco Pranzo
Dipartimento di Ingegneria dell'Informazione, Università degli Studi di Siena,
Siena, Italy, e-mail: `pranzo@dii.unisi.it`

Akın Şahin
Automatic Control Laboratory, ETH Zürich,
Zürich, Switzerland, e-mail: `sahin@control.ee.ethz.ch`

Martin J.J. Scheepers
Power Systems and Information Technology, Energy Research Centre of The
Netherlands, Petten, The Netherlands, e-mail: `scheepers@ecn.nl`

Koen H. van Dam
Faculty of Technology, Policy, and Management, Delft University of Technology,
Delft, The Netherlands, e-mail: `K.H.vanDam@tudelft.nl`

Nick van de Giesen
Department of Water Management, Delft University of Technology,
Delft, The Netherlands, e-mail: `N.C.vandeGiesen@tudelft.nl`

Monique van den Berg
Delft Center for Systems and Control, Delft University of Technology,
Delft, The Netherlands, e-mail: `monique.vandenberg@tudelft.nl`

Paul P.J. van den Bosch
Department of Electrical Engineering, Eindhoven University of Technology,
Eindhoven, The Netherlands, e-mail: `p.p.j.v.d.bosch@tue.nl`

Rutger van der Brugge
Dutch Research Institute for Transitions, Erasmus University Rotterdam,
Rotterdam, The Netherlands, e-mail: `VanderBrugge@fsw.eur.nl`

Josee van Eijndhoven
Dutch Research Institute for Transitions, Erasmus University Rotterdam,
Rotterdam, The Netherlands, e-mail: `VanEijndhoven@fsw.eur.nl`

Arjan J.C. van Gemund
Department of Software Technology, Delft University of Technology,
Delft, The Netherlands, e-mail: `a.j.c.vangemund@tudelft.nl`

Peter-Jules van Overloop
Department of Water Management, Delft University of Technology,
Delft, The Netherlands, e-mail: `P.J.A.T.M.vanOverloop@tudelft.nl`

Jos Vrancken
Faculty of Technology, Policy, and Management,
Delft University of Technology, Delft, The Netherlands, e-mail:
`j.l.m.vrancken@tudelft.nl`

Margot Weijnen
Faculty of Technology, Policy, and Management, Delft University of Technology,
Delft, The Netherlands, e-mail: `M.P.C.Weijnen@tudelft.nl`

Patrick Willems
Department of Civil Engineering, Katholieke Universiteit Leuven,
Leuven, Belgium, e-mail: `patrick.willems@bwk.kuleuven.be`

Cees Witteveen
Department of Software Technology, Delft University of Technology,
Delft, The Netherlands, e-mail: `c.witteveen@tudelft.nl`

Wenyan Wu
Faculty of Computing, Engineering and Technology, Staffordshire University,
Stafford, UK, e-mail: `w.wu@staffs.ac.uk`

Le Xie
Electrical & Computer Engineering Department, Carnegie Mellon University,
Pittsburgh, Pennsylvania, e-mail: `lx@ece.cmu.edu`

Fei Xue
Dipartimento di Ingegneria Elettrica, Politecnico di Torino,
Torino, Italy, e-mail: `fei.xue@polito.it`

Solomon Zegeye
Delft Center for Systems and Control, Delft University of Technology,
Delft, The Netherlands, e-mail: `s.k.zegeye@tudelft.nl`

Jonne Zutt
Department of Software Technology, Delft University of Technology,
Delft, The Netherlands, e-mail: `j.zutt@tudelft.nl`

Introduction

Z. Lukszo and M.P.C. Weijnen

Service economy

Infrastructures are of vital importance for the economy and the society. The current proliferation of telecommunication and information infrastructures has a profound influence on the structure of the economy and on our social life, in analogy with historic events of infrastructure innovation, which have boosted economic growth, contributed to public health, changed our mobility patterns, and brought comfort to our homes. However, most of the present day infrastructures were laid out to service the development of an industrial economy, and seem inadequate as a backbone for a new economic structure that relies on information and knowledge intensive services more than manufacturing activities. The emerging service economy requires high reliability, flexibility, and quality of service rather than cheap utilities and commodity services [3].

There is evident public dissatisfaction with the performance of many infrastructures. Users asking for tailor made quality-of-service and service-on-demand are confronted instead with problems of road congestion, power outages, inadequate public transport, virus attacks on the internet and insecure financial services. In many of these instances it is not clear which actor should be held responsible and which actor or actors should be taking action for improvement of the infrastructure performance. Since most former infrastructure monopolies have been unbundled, followed by the introduction of competition in segments of the value chain, the ownership of the system and the operational responsibilities are often distributed. The multi-actor complexity of the system is further increased by the entrance of competitors on the market and the emergence of new players in new roles, such as traders and brokers. At the same time, the complexity of vital infra-structures is further increased by the event of new information and telecommunication infrastructure

Z. Lukszo, M.P.C. Weijnen
Delft University of Technology, Faculty of Technology, Policy, and Management, Delft, The Netherlands, e-mail: {z.lukszo,m.p.c.weijnen}@tudelft.nl

pervading in all traditional infrastructure sectors, holding tremendous promise for, e.g., smarter infrastructure capacity management, the emergence of new value added services and even active participation of consumers in infrastructure bound markets, such as the electricity market. As infrastructure operators are thus confronted with entirely new challenges, they are struggling with how to employ innovative information, communication and control systems to satisfy social demands and adequately serve the end-users, within constraints of affordability, safety, security, etc. Last but not least, the increasing interconnectedness between infrastructures which creates new interactions and interdependencies that may entail new and unknown risks, adds an extra dimension to this complexity [1].

These specific infrastructure related innovation challenges are reflected in the research programme "Next Generation Infrastructures" (run by the Next Generation Infrastructures Foundation, http://www.nginfra.nl/) initiated and coordinated by the Delft University of Technology.

Next Generation Infrastructures

Next Generation Infrastructures represents an international consortium of public and private organizations working in a concerted research and development effort to improve the performance of critical infrastructures in the face of disturbances, changing economic conditions, emerging technologies, changing societal preferences, and end-user demands. The programme involves a variety of disciplines from the engineering sciences and the social sciences, so that it can effectively address the interdisciplinary challenges of infrastructure innovation and governance by technical, organizational, and institutional means. The programme is instigated by the growing public concern on the future reliability and quality of infrastructure related services. New approaches to the design, control, management, and regulation of infrastructure systems (more flexible, reliable, intelligent, and leaner) are required to counter new risks and vulnerabilities and avoid massive social costs in case of infrastructure malfunctioning. Rather than focussing on specific system components, the Next Generation Infrastructures research effort is aimed at understanding the structure and steering the overall behavior of infrastructure systems to ensure the satisfaction of societal demands and consumer interests. The predominant challenge at the overall programme level is to characterize the typical technological and social complexities of infrastructure systems, and to develop innovative ways for handling these complexities in the design, operation and governance of infrastructure systems. Therefore the mission of the pro-gramme is formulated as follows:

> "To stimulate and support the development of *more flexible*, *more reliable*, and *more intelligent infrastructures and services*, with respect for public values and consumer interests, and serve therewith a sustainable development of the economy, society and its natural environment".

The programme takes a cross-sectoral approach rather than just a sector or infrastructure specific approach, for two main reasons. In all public utilities and network industries the infrastructure network systems are going through similar processes of technological, economic and institutional change, while at the same time they are becoming more and more interconnected and interdependent, both within and across infrastructure sectors. For adequate management of these inter-dependencies it is vital to bridge the knowledge gap between experts in different infrastructure sectors and stimulate a process of cross-sectoral learning.

One of the main themes in the Next Generation Infrastructures research programme is Intelligent Infrastructures. It is aimed at developing intelligent methods and tools for the operation and control of existing infrastructures, deploying advanced information, communication and control technologies and systems. The central question it addresses is how to make better use of scarce infrastructure capacity, while at the same time improving reliability, safety, efficiency and environmental performance. For this aim an adequate balance between individual actor or subsystem interests and overall system performance should be found.

Intelligent Infrastructures

The operation and control of existing infrastructures is often insufficient or inefficient: too often we are confronted with capacity problems and a lack of safety, reliability, and efficiency. Nowadays, infrastructure operators are in need of new (traffic) routing protocols, path selection and capacity allocation algorithms and online monitoring methods. The huge multi-actor complexity of today's infrastructure networks entailing conflicting interests and demands, the emergence of new technologies, regulations and societal demands, the increasing interconnectedness and interdependencies between infrastructures within and across infrastructure sectors, all hamper the effective and efficient operation and control of these systems. The urgent need for improving the performance of critical infrastructures creates a large demand for innovative optimization and control methods and challenges the scientific research community in [5]:

- Acquiring a deeper understanding of the physical and social network complexity of infrastructure systems and their socio-technical interactions;
- Dealing with more and more deeply distributed autonomous control and its effects on overall network behavior;
- Coping with new needs for flexibility (in time and functionality) in combination with more stringent demands on capacity utilization, reliability and quality of service, health, safety and environment;
- Dealing with the need for a well-defined decision-making process to guarantee the efficiency and effectiveness of decision making in the shorter and longer term for multi-actor, multi-level, multi-objective and dynamic problems.

The problems of different infrastructure sectors are similar: How to maximize the use of the available capacity? How to do this in the most efficient way? How to prevent congestion, without neglecting the proper safety precautions? Therefore,

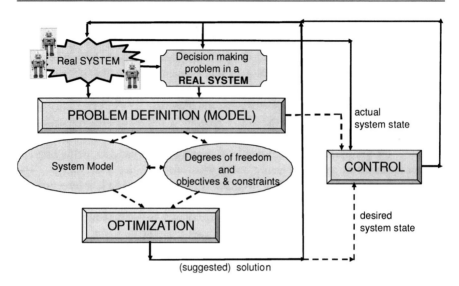

Figure 1: Modeling, optimization, and control scheme.

it is a challenging task to develop generic modeling approaches and control methods that can actually be implemented (e.g., by a regulatory authority) to influence local decision-making by a multitude of actors in respect of societal (overall) interests. In such a distributed approach a leader can coordinate the decision-making of distributed subsystems, but it is also possible that the interaction between the actors takes place without any coordinator. In these settings the decision makers are authorized to make decisions regarding their domain while agreeing to operate according to a common overall goal, or they interact in a non-cooperative way according to their own goals anticipating the responses of other actors and subsystems to their actions.

As large infrastructure systems are multi-level systems, composed of millions of physical components, involving different owners, operators and users, all primarily pursuing their own local performance objectives, there are no easy solutions to these problems. The Intelligent Infrastructures research programme accepted the challenge in its endeavor to answer the following core research question:

> How to influence the physical, organizational, economic and institutional configuration of an infrastructure system in such a way that predefined (dynamic) objectives are satisfied taking into account market and system constraints?

The schematic representation of the modeling, optimization and control problem as depicted in Figure 1 may at first sight look like a standard control theory problem formulation. After a proper problem definition and a modeling step, an optimization procedure results in an optimal system state and the desired values and profiles for the controller, which compares them with the actual system state and if needed performs an appropriate control action. However, as such a (dynamic) decision making problem in infrastructure systems is generally characterized by multi-actor, multi-criteria and multi-level complexity, it is extremely challenging to solve. For example, when looking at transport infrastructures, decision making can take place at many levels in the management and operation of the infrastructure, and it can range from the capacity utilization of national complex transport infrastructures down to urban traffic management and further down to decisions by individual drivers or travelers.

In such a system the objectives and constraints of any decision-maker may be determined in part by variables controlled by other authorities. In some situations, one decision maker may control all variables that permit him to influence the behavior of other decision-makers as in traditional hierarchical control. In essence, each actor is a decision-maker responding to external information: using this information as input he determines his own response. The extent of the interaction may vary depending on the particular environment and time dimension: in some cases, decision makers might be tightly linked, in other, they may have little effect on each other, if at all. *Intelligent* modeling, optimization, and control methods to cope with dynamic multi-actor, multi-level and multi-criteria problems are called for.

The term *intelligent methods* needs further explanation. Intelligence is an ill-defined concept, so depending on a purpose or domain different definitions can be given. The word comes from the Latin verb *intellegere*, which means "to understand". Mostly, intelligence is used as an umbrella term to describe a property of the mind that encompasses many related abilities as the capability to reason, to think abstractly, to behave socially, to learn, to apply knowledge and experience to solve problems, to plan, etc. It may include traits as cleverness, knowledge, wisdom, creativity or autonomy.

In this book the term intelligent infrastructures refers to the application of advanced (intelligent) modeling, optimization and control concepts to cope with distributed multi-actor, multi-criteria and multi-level decision problems.

Agent-based modeling of infrastructures

The common denominator of complex decision problems in the domain of public utilities and network industries is their socio-technical complexity. The physical network of an infrastructure system and the social network of actors involved in its operation collectively form an interconnected complex network where the multi-actor network determines the development and operation of the physical network, and the physical network structure and behavior affect the behavior of the actors. Unpredictable dynamic behavior of the system is often the result of the multi-agent and multi-actor character of the operation, the multi-level structure of the infrastructure system, the multi-objective optimization challenge, and the adaptivity of agents and actors to changes in their environment, together with non-linear response functions. To support the search for innovative control concepts aimed at improving the performance of existing infrastructures during normal as well as abnormal operation appropriate models should be developed. To model infrastructures as socio-technical

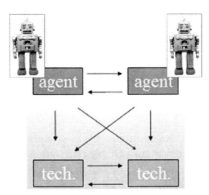

Figure 2: Infrastructure as interconnected system of autonomous agents interacting with technical subsystems.

systems the concept of agent-based systems composed of multiple interacting agents and physical elements is a promising approach [8], see Figure 2. The term *agent* can represent actors in the social network (e.g., travelers taking autonomous decisions which route to follow to avoid road congestion, or companies involved in the production of gas or the generation of power) as well as a control mechanism of a component or a sub-system in the physical network (e.g., controllers for irrigation channels aimed at guaranteeing the adequate delivery of water at minimal water spillage).

In general, agent-based concepts are applicable for (conceptual) modeling of complex systems if the following conditions are satisfied [7]:

- The problem has a distributed character;
- The subsystems (agents) operate in a highly dynamic environment;
- Subsystems (agents) have to interact in a flexible way, where flexibility means reactivity, pro-activeness, cooperativeness, and social ability.

Such models have already successfully been designed and implemented for various infrastructures, including transport, energy and industrial networks, as will be demonstrated in this book.

Multi-level decision making

Capacity management at the operational level addresses day-to-day and hour-to-hour capacity allocation issues, which relate to how the flows (of goods, gas) are directed over the network. In the gas sector, international trade flows through the national grid should be controlled so as not to hamper an adequate supply of gas to national users by excessive use of transport capacity or quality conversion capacity. In the road transport sector, intelligent road capacity allocation principles are designed to achieve more balanced capacity utilization in time and space, i.e., to minimize congestion. In dynamic road pricing schemes price levels for tolls are

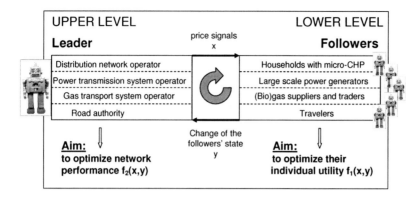

Figure 3: Schematic representation of bi-level decision problem examples found in energy and transport infrastructures.

dynamically varied over space and time depending on the traffic conditions in the network and the policy objectives of the road authority. A challenging question is what kind of operational models are needed to accommodate optimal distributed dynamic pricing schemes. In such a system price levels for tolls can vary over space and time dynamically depending on the traffic conditions in the network and the policy objectives of the road authority. A challenging question is what kind of operational models are needed to generate optimal distributed dynamic pricing schemes. Let us consider a network design problem in which the aim is to determine a set of (time-varying) road-pricing levels on road segments in a transportation network.

The aim of the road authority is to optimize system performance by choosing the optimal tolls for a subset of links within realistic constraints and subject to the dynamic route and departure time choice of the users, that is, the travel behavioral part. For a more detailed problem definition, see [4]. In this problem which can be represented as a bi-level programming problem we can distinguish two different submodels: the dynamic road pricing model of the road authority and the dynamic route and departure time choice model of travelers.

The problem of distributed dynamic pricing is not unique for the road infrastructure, see Figure 3. Similar issues are found in the operation of next generation electric power systems with many small-scale distributed generating units, such as gas turbines, photovoltaics, wind turbines, fuel cells, or micro combined heat and power (micro-CHP) units. These distributed technologies have many advantages, e.g., high fuel efficiency, modular installation, low capital investment, and relatively short construction time [2]. However, distributed generation in a competitive electricity market creates major uncertainties to the operation of the system: as (millions of) power users can switch to the role of power producers, the amount and quality of power produced in such a distributed system can vary enormously. Similarly, wind power fluctuations can pose management problems related to the frequency stability and the desired voltage profile. As a consequence of distributed power generation new control techniques need to be developed and implemented in order to guarantee

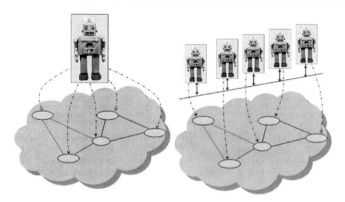

Figure 4: Single-agent versus multi-agent control of a complex network at a single level.

power availability and quality of service (such as frequency, bounds on deviations, stability, elimination of transients for electricity networks, and so on), so as to meet the demands and requirements of the users.

Distributed control

Typically infrastructures are not controlled by a single agent [6]. Reasons for this may be found in technical constraints related to, e.g., communication delays and computational requirements, but also originate from more practical organizational and institutional issues like distributed ownership, unavailability of information from one subsystem to another, and restricted control access. The associated dynamic control problem should be broken up into a number of smaller problems, see Figure 4.

Final remarks

Many new concepts aimed at improving operation and control, mostly relaying on distributed multi-agent strategies for multi-actor, multi-level and multi-control problems, are described in this book. In all presented cases the overall multi-agent system has its own overall objective, while the agents have their own individual objectives. To safeguard adequate functioning of the infrastructure the actions of the individual agents must be steered towards an acceptable overall performance of the system in terms of, e.g., availability, reliability, affordability, and quality of service.

This book presents a variety of approaches rendering vital infrastructure systems more intelligent. Evidently, advanced modeling, optimization and control strategies hold great promise to improve the efficiency, reliability, and resilience of infrastructure systems, without violating the sustainability and safety requirements which they are subject to, in fact, even improving their environmental and safety performance.

The techniques developed in the research contributions brought together in this book need to be made generic so as to be applicable to other types of infrastructure such as telecommunication, railway, water and gas networks, and to interconnected infrastructure systems across sectors. Despite the challenges waiting in the future, we are convinced that the methods and tools we can present today will turn out to be of invaluable support in understanding and solving the multi-level, multi-actor and multi-objective control problems which are characteristic for the operation of modern infrastructure systems.

Infrastructures offer many examples of settings in which multi-actor, multi-objective and multi-level decision making is daily practice and in which well-defined and thoroughly proven intelligent methods for modeling, optimization and control are called for. Virtually all of the methods assembled in this book are applicable for operations management in infrastructure systems. What we observe in practice is that different methods are being applied in combination, that computer-based simulation plays an increasing role, and that automation and computerization have strengthened the role of the lowest and most distributed level of actors in the hierarchy of decision makers, the individual consumers.

References

1. I. Bouwmans, M. P. C. Weijnen, and A. Gheorge. Infrastructures at risk. In A. V. Gheorge, M. Masera, M. P. C. Weijnen, and L. J. de Vries, editors, *Critical Infrastructures at Risk, Securing the European Electric Power System*, pages 19–36. Springer, 2006.
2. J. B. Cardell and M. Ilić. The control and operation of distributed generation in a competitive electric market. In *Electric Power Systems Restructuring*, pages 453–518. Kluwer Academic Publishers, 1998.
3. P. H. Herder and Z. Verwater-Lukszo. Towards Next Generation Infrastructures. An introduction to the contribution in this issue. *International Journal of Critical Infrastructures*, 2(2/3), 2006.
4. D. Joksimovic. *Dynamic bi-level optimal toll design approach for dynamic traffic networks*. PhD thesis, Delft University of Technology, Delft, The Netherlands, 2007.
5. Z. Lukszo, M. P. C. Weijnen, R. R. Negenborn, and B. De Schutter. Tackling challenges in infrastructure operation and control – cross-sectoral learning for process and infrastructure engineers. *International Journal of Critical Infrastructures*, 9, 2009.
6. R. R. Negenborn. *Multi-Agent Model Predictive Control with Applications to Power Networks*. PhD thesis, Delft University of Technology, Delft, The Netherlands, December 2007.
7. K. H. van Dam, A. Adhitya, R. Srinivasan, and Z. Lukszo. Critical evaluation of paradigms for modelling integrated supply chains. *Computers and Chemical Engineering*, 2009.
8. G. Weiss. *Multiagent Systems: A Modern Approach to Distributed Artificial Intelligence*. MIT Press, Cambridge, Massachusetts, 2000.

Part I
Generic Infrastructures

Chapter 1
Intelligence in Transportation Infrastructures via Model-Based Predictive Control

R.R. Negenborn and H. Hellendoorn

Abstract In this chapter we discuss similarities and differences between transportation infrastructures like power, road traffic, and water infrastructures, and present such infrastructures in a generic framework. We discuss from a generic point of view what type of control structures can be used to control such generic infrastructures, and explain what in particular makes *intelligent* infrastructures intelligent. We hereby especially focus on the conceptual ideas of model predictive control, both in centralized, single-agent control structures, and in distributed, multi-agent control structures. The need for more intelligence in infrastructures is then illustrated for three types of infrastructures: power, road, and water infrastructures.

R.R. Negenborn, H. Hellendoorn
Delft University of Technology, Delft Center for Systems and Control, Delft, The Netherlands,
e-mail: {r.r.negenborn, j.hellendoorn}@tudelft.nl

R.R. Negenborn et al. (eds.), *Intelligent Infrastructures*, Intelligent Systems, Control and
Automation: Science and Engineering 42, DOI 10.1007/978-90-481-3598-1_1,
© Springer Science+Business Media B.V. 2010

1.1 Transportation infrastructures

Transportation infrastructures, like power distribution networks [27], traffic and transportation systems [12], water distribution networks [7], logistic operations networks [28], etc., (see Figures 1.1–1.3) are the corner stones of our modern society. A smooth, efficient, reliable, and safe operation of these systems is of huge importance for the economic growth, the environment, and the quality of life, not only when the systems are pressed to the limits of their performance, but also under regular operating conditions. Recent examples illustrate this. E.g., the problems in the USA and Canada [44], Italy [42], Denmark and Sweden [16], The Netherlands, Germany, Belgium, and France [43], and many other countries [36] due to power outages have shown that as power network operation gets closer to its limits, small disturbances in heavily loaded lines can lead to large black-outs causing not only huge economic losses, but also mobility problems as trains and metros may not be able to operate. Also, as road traffic operation gets closer to its limits, unexpected situations in road traffic networks can lead to heavy congestion. Not only the huge traffic congestion after incidents such as bomb alerts are examples of this, also the almost daily road-traffic jams due to accidents illustrate this convincingly.

Expanding the physical infrastructure of these networks could help to relieve the issues in transportation networks, although at extremely high costs. As alternative to spending this money on building new infrastructure, it is worth spending effort on investigating improved use of the existing infrastructure by employing intelligent control techniques that combine state-of-the-art techniques from fields like systems and control engineering [4], optimization [6], and multi-agent systems [47], with domain-specific knowledge.

The examples of networks just mentioned are only some particular types of networks within the much larger class of transportation networks. Common to transportation networks is that at a generic level they can be seen as a set of nodes, representing the components or elements of the network, and interconnections between these nodes. In addition, transportation networks have some sort of commodity, that is brought into the network at source nodes, that flows over links to sink nodes, and that is influenced in its way of flowing over the network by elements inside the network, as illustrated in Figure 1.4. Other characteristics that are common to transportation networks are:

- they typically span a large geographical area;
- they have a modular structure consisting of many subsystems;
- they have many actuators and sensors;
- they have dynamics evolving over different time scales.

In addition to this, transportation networks often contain both continuous (e.g., flow evolution) and discrete dynamics (e.g., on and off switching), and can therefore also be referred to as hybrid systems [45]. This mixture of characteristics makes that transportation networks can show extremely complex dynamics.

Figure 1.1: The water network of The Netherlands.

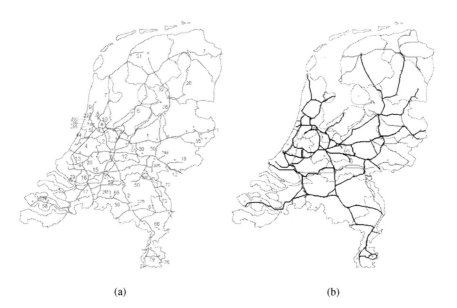

(a)   (b)

Figure 1.2: (a) The national road network of The Netherlands; (b) The national train network of The Netherlands.

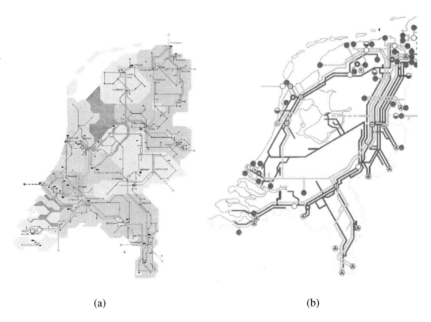

(a)   (b)

Figure 1.3: (a) The national electricity grid of The Netherlands (Illustration courtesy of TenneT); (b) The national gas network of The Netherlands (Illustration courtesy of Gasunie).

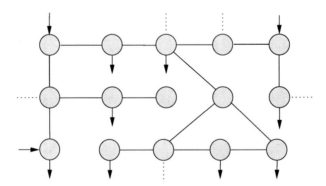

Figure 1.4: Networks of a generic transportation infrastructure. Commodity enters the network at sources (circles with an arrow pointing towards them), flows over links to other elements in the network that alter the flows (at each circle), and leaves the network at sinks (circles with an arrow pointing outward). Dotted lines represent connections with other parts of the network.

All the networks described in this book generically consist of sinks, sources, transition nodes, and interconnections. The commodity that is being transported over the network, however, determines how these components function. In a water network, the water always flows downwards unless a pump is deployed, when two waterways come together in a node the flow will increase, there may be intermediary sinks, ultimately there is one sink, the sea. The source may be distant mountains or rainfall, the amount of water can fairly precisely be predicted. In a traffic network, the sources in the morning are residential areas, the sinks are business districts, in the afternoon sources and sinks change roles. When two roads come together in a node the traffic flow will slow down or stagnate. Traffic lights will give the network a start-and-stop character. Highway and city traffic are very different in character. Electricity networks are determined by Kirchhoff's circuit laws that state that in any node in an electrical circuit the sum of currents flowing into a node is equal to the sum of the currents flowing out of that node. Electricity will always flow and can hardly be stored or stopped, sinks are industries and families, sources are power plants and increasingly often wind turbine parks and solar panels. Gas networks are determined by pressure with usually one source and many sinks. In railway networks it is hard to overtake, which gives these networks a strong sequential character, a chain which is very dependent on the weakest links.

Even though transportation networks differ in the details of commodity, sources, sinks, etc., it is worth to consider them in a generic setting. On the one hand, methods developed for generic transportation networks can be applied to a wide range of specific domains, perhaps using additional fine-tuning and domain-specific enhancements to improve the performance. On the other hand, approaches specifically developed for a particular domain can be applied to other domains after having transferred them to the generic framework.

1.2 Towards intelligent transportation infrastructures

There are many users, controllers, players, actors, and operators involved in the evolution of transportation networks. Each of these concepts refers to entities that directly or indirectly change the way commodity is flowing. Different users may have different objectives, and these objectives may be conflicting. Depending on their objectives, the users choose different actions, resulting in a different operation of the network.

1.2.1 The system versus its control structure

In order to formalize the operation of transportation networks, consider Figure 1.5. The figure illustrates the overall picture of a *system* on the one hand and a *control structure* on the other. The system is the entity that is under control, and the control structure is the entity that controls the system. Hence, the control structure is the concept used to indicate the structure that produces actuator settings. The control

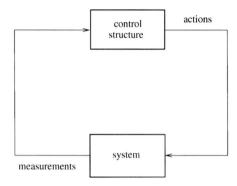

Figure 1.5: The relation between a general system and the control structure that controls the system.

structure monitors the system by making measurements and based on these chooses control actions that are implemented on the system. The system evolves subject to these actions to a new state, which is again measured by the control structure. The control structure consists of one or more components, called *control agents*. These control agents try to determine settings for the actuators inside the system in such a way that their own objectives are met as closely as possible and any constraints are satisfied. In our case, the system consists of the transportation network, and the components of the control structure consist of all the users, controllers, operators, players, etc., from now on only referred to as the control agents.

1.2.2 Intelligent infrastructures

Traditionally, operators play an important role in infrastructural networks. They monitor the system in large local or regional decision rooms, they open valves or pumps, they start up generators or shift transformer taps, and they can remotely close road lanes. More and more operators are replaced by ICT systems. In road traffic and railway control this development is well developed, in waste and drinking water networks this development has hardly begun. ICT solutions are in the beginning focused on local optimization and do not regard the whole network. In the case of, e.g., traffic lights this is not so much a problem, but in railway or electricity networks local optimization is of minor importance.

In the beginning operators observe the ICT solutions in decision rooms, but when confidence in the ICT system increases local control actions will be functioning fully automatically. Simultaneously with the development of locally more optimal control, it becomes apparent that coordination with other parts of the network becomes necessary. Green waves in road traffic networks are simple examples of this that show well how far more complicated this larger system is in a city road network with many junctions and crossing lanes. Communication between local controllers is necessary and intelligent systems have to be built that take into account all the flows and that have the capability of forecasting the near future. As confidence in these larger systems increases, coordination and negotiation between nodes will be a regular practice and transportation infrastructures have become *intelligent transportation infrastructures.*

1.2.3 Control structures

The control structure is a very general concept and can have many different shapes. A first important distinguishing feature between control structures is the number of control agents that constitute the control structure. E.g., the control structure can consist of a single control agent or multiple control agents. Some other properties in which control structures can differ are:

- the access that the control agents have to the sensors and actuators,
- the communication that the control agents have among one another,
- the way in which the control agents process sensor data to obtain actions,
- the authority relations between the control agents,
- the beliefs, desires, and intentions of the control agents.

Defining different types of control structures is difficult due to the large amount of properties that they can have. However, some general types of control structures can be identified, that have increasing complexity, that are commonly encountered in theory and practice, and that will also be of particular interest in the subsequent chapters of this book:

- When it is assumed that there is only one control agent, that has access to all actuators and sensors of the network and thus directly controls the physical network, then this control structure is referred to as an *single-agent* control structure, as illustrated in Figure 1.6(a). The control structure is in a sense ideal structure, since in principle such a control structure can determine actions that give optimal performance, although this may be at significant computational costs.
- When there are multiple control agents, each of them considering only its own part of the network and being able to access only sensors and actuators in that particular part of the network, then the control structure is referred to as a *multi-agent single-layer* control structure, as illustrated in Figure 1.6(b). If in addition the agents in the control structure do not communicate with each other, the control structure is *decentralized*. If the agents do communicate with each other, the control structure is *distributed*.
- When there are multiple control agents, and some of these control agents have authority over other control agents, in the sense that they can force or direct other control agents, then the control structure is a *multi-layer* control structure, as illustrated in Figure 1.6(c) A multi-layer control structure typically is present

when one control agent determines set-points to a group of other control agents, that work in a decentralized or distributed way. Due to the authority relationship between agents or groups of agents, the multi-layer control structure can also be referred to as a supervisory control structure, or a hierarchical control structure.

1.2.4 Control structure design

Suppose that a particular network is given and that any control structure can be implemented on it. The question that then arises is the question of how it can be determined what the best control structure is. Unfortunately, theories for determining general control structures are lacking. However, motivations for preferring one type of control structure over another can be given.

Advantages of single-agent control structures are in general that they can deliver the best performance possible, and that they have been studied extensively in the literature, in particular for small-scale systems. However, there are several issues that complicate the use of single-agent control structures for large-scale transportation networks such as:

- undesirable properties with respect to robustness, reliability, scalability, and responsiveness;
- technical issues related to communication delays and computational requirements;
- commercial, legal, and political issues related to unavailability of information and restricted control access.

These reasons motivate the use of multi-agent control structures [40, 41, 47], which are expected to be able to deal or at least relieve these issues. Multi-agent control structures can in principle:

- improve robustness and reliability, since if one control agents fails, another can take over, and improve responsiveness, since the control agents typically use only local measurements and therefore can react quicker to changing situations;
- reduce communication delays, since the control agents operate locally and therefore solve problems that may be smaller, and since communication typically takes place among nearby control agents;
- deal with unavailability of information and restricted control access, since the control agents only require information of their own part of the network and since they determine actions only for their own part of the network.

However, multi-agent control structures typically have a lower performance than the performance of ideal single-agent control structures; implementing schemes that give desired performance is far from trivial.

An advantage of the decentralized over the distributed multi-agent single-layer control structures is that there is no communication between the controllers, resulting in lower computational requirements and faster control. However, this advantage will typically be at the price of decreased overall performance. The advantage

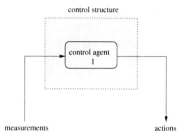

(a) Single-agent control structure. The single control agent makes measurements of the system and provides actions to the network.

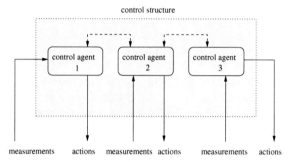

(b) Multi-agent single-layer control structure. Multiple control agents make measurements and provide actions to the network. Communication between the control agents is optionally present (dashed line).

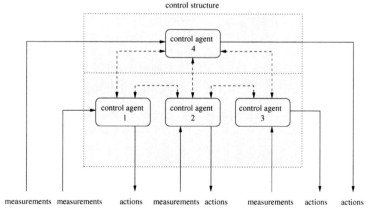

(c) Multi-layer control structure. A higher-layer control agent can make measurements and provide actions to the network and can in addition direct or steer a lower control layer.

Figure 1.6: Some important types of control structures.

of a distributed multi-agent single-layer control structure is therefore that improved performance can be obtained, although at the price of increased computation time due to cooperation, communication, and perhaps negotiation among control agents. However, even though improved performance can be obtained, the performance will still typically be lower than the performance of an ideal single-agent control structure.

The multi-agent multi-layer control structure provides the possibility to obtain a trade-off between system performance and computational complexity. A higher layer considers a larger part of the system and can therefore direct the lower control layer to obtain coordination. Such a multi-layer control structure can thus combine the advantages of the single-agent control structure with the multi-agent single-layer control structure, i.e., overall system performance with tractability. It is noted, however, that communication in a multi-agent multi-layer control structure is typically more complex than in a single-agent control structure and a multi-agent single-layer control structure.

Note that in practice often a particular control structure is already in place, and that the control structure cannot be redesigned from scratch. The question in this case is not so much the question of what control structure is best, but of how the currently existing control structure can be changed, such that the performance is improved. Of course, here it has to be defined what the performance is, and in a control structure with control agents with conflicting objectives it may not be possible to reach consensus on this.

1.2.5 Assumptions for design and analysis

In this book control strategies for several control structures are developed. Due to the complexity of transportation networks, the scope of control problems that is considered is narrowed down. The focus will mostly be on the most fundamental of transportation network control problems: the operational control of transportation networks, in which amounts of commodity to be transported over the network are given, and controllers have to ensure that transport over the network can take place at acceptable service levels, while satisfying any constraints, both under normal and emergency operating conditions.

In order to make the analysis and the design of the control structures more tractable, assumptions have to be made, both on the network and the control structure. Assumptions relating to the network are made on the dynamics of the network, i.e., the way in which the components in the network function. E.g., the dynamics can be assumed to evolve over continuous time or in discrete-time, they can be assumed to involve only continuous dynamics, or both continuous and discrete dynamics, and they can be assumed to be instantaneous or not. In each chapter we explicitly point out which particular assumptions are made on the network. With respect to the control structure, assumptions are made on the following:

- the control agents are already present (however, it is not known yet how they should behave);

- the control agents control fixed parts of the network, and they can access actuators and sensors in these parts of the network;
- the control agents know what qualitative behavior is desired for the parts of the network they control;
- the control agents strive for the best possible overall performance of the network;
- the control agents can measure the state of the parts of the network that they control.

Under such assumptions it remains to be decided how the agents in the control structure derive actuator settings from their measurements, i.e., what protocols, computations, and information exchanges take place inside the control structure. Assumptions on these are made in other chapters of this book. In the following section we discuss a promising approach for use by the control agents in a multi-agent control structure for transportation network control: model predictive control.

1.3 Model predictive control

To find the actions that meet the control objectives as well as possible, the control agents have to make a trade-off between the different available actions. In order to make the best decision and hence find the best actions, all relevant information about the consequences of choosing actions should be taken into account. For power networks, typical information that is available consists of forecasts on power consumption and exchanges [20], capacity limits on transmission lines, dynamics of components like generators, capacitor banks, transformers, and loads [27]. Furthermore, typically area-wide measurements of voltage magnitude and angles across the network can be made to provide an up-to-date status of the situation of the network. A particularly useful form of control for transportation network that in principle can use all information available is model predictive control (MPC) [10, 29].

1.3.1 Single-agent MPC

Over the last decades MPC (also knowns as receding horizon control or moving horizon control) has become an important methodology for finding control policies for complex, dynamic systems. MPC in a single-agent control structure has shown successful application in the process industry [29, 31], and is now gaining increasing attention in fields like, amongst others, power networks [18, 35], road traffic networks [21], railway networks [13], steam networks [30], multi-carrier systems [3], greenhouse control [38], and drug delivery [8].

1.3.1.1 Concept

MPC is a control methodology that is typically used in a discrete-time control context, i.e., control actions are determined in discrete control cycles of a particular

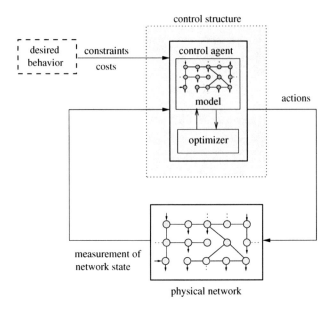

Figure 1.7: Single-agent MPC.

duration which in itself is expressed in continuous time units. From the beginning of one control cycle until the beginning of the next control cycle, the control actions stay fixed, i.e., a zero-order hold strategy is employed.

In each control cycle the MPC control agent uses the following information, as illustrated in Figure 1.7:

- an *objective function* expressing which system behavior and actions are desired;
- a *prediction model* describing the behavior of the system subject to actions;
- possibly *constraints* on the states, the inputs, and the outputs of the system (where the inputs and the outputs of the system correspond to the actions and the measurements of the control agent, respectively);
- possibly known information about future disturbances;
- a *measurement* of the state of the system at beginning of the current control cycle.

The objective of the control agent is to determine those actions that optimize the behavior of the system and minimize costs as specified through the objective function. In order to find the actions that lead to the best performance, the control agent uses the prediction model to predict the behavior of the system under various actions over a certain prediction horizon, starting from the state at the beginning of the control cycle. Once the control agent has determined the actions that optimize the system performance over the prediction horizon, it implements the actions until the beginning of the next control cycle, at which point the control agent determines new actions over the prediction horizon starting at that point, using updated information. Hence, the control agent operates in a receding or rolling horizon fashion to determine its actions.

In general it is preferable to have a longer prediction horizon, since by considering a longer prediction horizon, the control agent can better oversee the consequences of its actions. At some length, however, increasing the length of the prediction horizon may not improve the performance, if transients in the dynamics may have become negligible. For computational reasons, determining the actions over a very long horizon typically is not tractable, and in addition due to potential uncertainty in the prediction model and in predictions of future disturbances, a smaller prediction horizon is usually considered. Hence, in practice, the prediction horizon should be long enough to cover the most important dynamics, i.e., those dynamics dominating the performance, and short enough to give tractable computations. It should hereby also be noted that if a prediction horizon is used that is too short, the system could arrive in states from which it cannot continue due to the presence of constraints, e.g., on the actions. The prediction horizon should thus have such a length that arriving in such states can be avoided.

1.3.1.2 MPC Algorithm

Summarizing, a control agent in a single-agent control structure using MPC to determine its actions performs at each control cycle the following:

1. Measure the current state of the system.
2. Determine which actions optimize the performance over the prediction horizon by solving the following optimization problem:

   minimize   the objective function in terms of actions over the prediction horizon

   subject to   the dynamics of the whole network over the prediction horizon,

                 the constraints on, e.g., ranges of actuator inputs and link capacities,

                 the measurement of the initial state of the network at the beginning
                       of the current control cycle.

3. Implement the actions until the next control cycle, and return to step 1.

1.3.1.3 Advantages and issues

Advantages of MPC are that in principle it can take into account all available information and that it can therefore anticipate undesirable situations in the future at an early stage. Additional advantages of MPC are [29]:

- its explicit way of handling constraints on actions, states, and outputs;
- its ability to operate without intervention for long periods;
- its ability to adapt to slow changes in the system parameters;
- its ability to control systems with multiple inputs and multiple outputs;
- its relatively easy tuning procedure;

- its built-in robustness properties.

However, there are also some issues that have to be addressed before a control agent using an MPC methodology can be implemented successfully:

- the control goals have to be specified;
- the prediction model has to be constructed;
- the measurement of the system state has to be available;
- a solution approach (optimization method) has to be available that can solve the MPC optimization problem;
- the solution approach has to be tractable and efficient.

Basic issues, e.g., stability and robustness, have extensively been studied for MPC in single-agent control structures [31], in particular for linear time-invariant systems. For other classes of systems there are still many open issues. E.g., tractability issues of MPC for nonlinear and discrete-event systems, and for systems in which variables take on discrete values, still deserve attention. E.g., in [34] an approach is proposed to make the MPC problem for a system modeled as a Markov decision process more tractable and to deal with changing system dynamics by including experience using reinforcement learning. Another class of systems for which there are still many open questions are hybrid systems, i.e., systems including both continuous and discrete dynamics.

1.3.2 Multi-agent MPC

As mentioned in the previous section, in a multi-agent control structure, there are multiple control agents, each of them controlling only its own subnetwork, i.e., a part of the overall network. Multi-agent MPC issues have been investigated since the 90s, such as in [1, 2, 5, 15, 17, 19, 22, 26, 37, 39], and more recently in [9, 11, 14, 24, 25, 33, 46] and [32].

In multi-agent MPC, multiple control agents in the control structure use MPC, but now they first measure the subnetwork state, then they determine the best actions over the predicted subnetwork evolution, and then they implement actions. Although this may seem like a straightforward extension of single-agent MPC at first sight, when considering the details it is not.

The actions that an agent in a multi-agent control structure takes influence both the evolution of the subnetwork it controls, and the evolution of the subnetworks connected to its subnetwork. Since the agents in a multi-agent control structure usually have no global overview and can only access a relatively small number of sensors and actuators, predicting the evolution of a subnetwork over a horizon involves even more uncertainty than when a single agent is employed. In addition, when a control agent in a multi-layer control structure provides set-points to another agent, this supervisory control changes the way in which the other agent chooses its actions, and thus the higher-layer control agent changes the performance of the system.

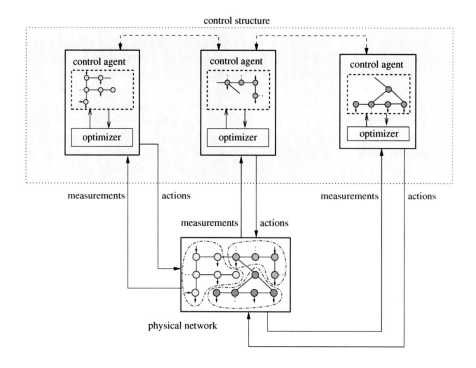

Figure 1.8: Multi-agent single-layer MPC.

Under the assumption that the control agents strive for an optimal overall network performance, the challenge in implementing such a multi-agent MPC methodology comes from ensuring that the actions that the individual agents choose result in a performance that is comparable to the situation when a hypothetical single-agent control structure in which all information is available would be used.

1.3.2.1 Multi-agent single-layer MPC

In the multi-agent single-layer control structure each control agent only has information gathering and action capabilities that are restricted to that part of the network that a particular control agent controls, as illustrated in Figure 1.8. The challenge in implementing multi-agent single-layer MPC comes from predicting the dynamics of the subnetwork, since, as mentioned, its evolution is influenced by the other agents. The underlying problem of MPC for multi-agent control structures can therefore be seen as optimization over a distributed simulation.

Issues

To make accurate predictions of the evolution of the subnetwork, a control agent requires the current state of its subnetwork, a sequence of actions over the prediction horizon, and predictions of the evolution of the interconnections with other subnetworks. The predictions of the evolution of the interconnections with other subnetworks are based on the information communicated with the neighboring control agents. One particular class of methods aims at achieving cooperation among control agents in an iterative way in which in each control cycle control agents perform several iterations consisting of local problem solving and communication. In each iteration agents obtain information about what the plans of neighboring agents are. Ideally at the end of the iterations the agents have found actions that lead to overall optimal performance.

As is the case with MPC for single-agent control structures, having both continuous and discrete dynamics causes computational problems. In transportation networks this combination is commonly encountered, and it is therefore relevant to study models that take this into account. A further complicating issue arises when the subnetworks that the agents control are overlapping. Existing strategies assume that the subnetworks that the control agents control are non-overlapping. However, in some applications the subnetworks considered by the control agents are overlapping. This has to be taken into account explicitly.

1.3.2.2 Multi-agent multi-layer MPC

In the multi-layer multi-agent MPC case there are multiple control layers in the control structure, i.e., there are authority relationships between the agents in the sense that some agents provide set-points or directions to other agents. The agents at higher layers typically consider a larger region of the network and consider slower time scales than agents in lower layers. Figure 1.9 illustrates this.

MPC can also be used by a control agent in a higher layer of the control structure. This higher-layer control agent can then coordinate the lower layer, which may consist of control agents using multi-agent single-layer MPC, or of control agents that use alternative control strategies. The higher-layer control agent then coordinates the lower control layer by enforcing penalty terms, providing additional constraints, or providing set-points. The advantage of the higher-layer control agent is in particular clear when the control agents of the lower layer are working decentralized, i.e., not communicating with one another.

Issues

An important issue to be addressed when designing MPC for multi-agent multi-layer control structures is the choice of the prediction model that the higher-layer control agent uses. A higher-layer control agent has to be able to make relevant predictions of the physical system, but since the physical system is under control of the lower-control layer, the lower-control layer has to be taken into account by the higher-layer control agent as well. In addition, the prediction model that the higher-layer control

control structure

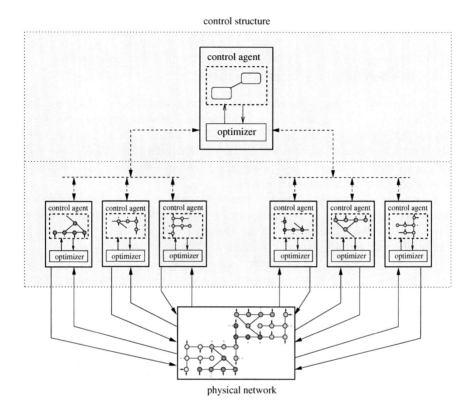

Figure 1.9: Multi-agent multi-layer MPC.

agent uses will typically involve both continuous and discrete elements, since it has to consider a larger part of the network than lower-layer agents. This makes the resulting MPC control problem more complex, and efficient ways have to be found to solve it efficiently.

1.4 MPC for intelligent infrastructures

As stated before, the commodity to be transported in a network determines the meaning of nodes and interconnections. Commodities are electrons in power networks, cars in road networks, water molecules in water networks, trains in railway networks, gas molecules in gas networks, etc. It is important to understand the origin and destination of commodities in networks and the operations that can be performed on them. Recent technological developments, in particular in communication technology, have increased the number of possible operations in various networks considerably. Below, intelligent power, road, and water networks are described as they are the main examples of intelligent infrastructures in this book.

1.4.1 Intelligent power infrastructures

Power networks [27] are large transportation networks consisting of a large number of components. The generators produce power that is injected into the network on the one side, while the loads consume power from the network on the other side. The distribution of the power in the network is dictated by Kirchhoff's laws and influenced by the settings of the generators, loads, transformers, and potentially also by capacitor banks and FACTS devices.

Conventionally, in power networks, power was generated in several large power generators. This power was then transported through the transmission and distribution network to the location where it was consumed, e.g., households and industry. The number of control agents was relatively low. Due to the ongoing deregulation in the power generation and distribution sector in the U.S. and Europe, the number of players involved in the generation and distribution of power has increased significantly. In the near future the number of source nodes of the power distribution network will even further increase as also large-scale industrial suppliers and small-scale individual households will start to feed electricity into the network [23].

As a consequence, the structure of the power distribution network is changing from a hierarchical top-down structure into a much more decentralized system with many generating sources and distributing agencies. This multi-player structure thus results in a system with many interactions and interdependencies. To still guarantee basic requirements and service levels, such as voltage levels, frequency, bounds on deviations, stability, elimination of transients, etc., and to meet the demands and requirements of the users, new control techniques are being developed and implemented resulting in an intelligent power infrastructure. The chapters in Part II of this book focus on this.

1.4.2 Intelligent road infrastructures

Electric traffic lights have now existed for almost a hundred years and are the best well-known control measures in road traffic networks. In the beginning, traffic lights were fixed-time stand-alone systems, later on they evolved into intelligent control systems using sensor information and communication with neighboring junctions. In the last decades a large number of traffic control measures have been added: dynamic speed signs, dynamic route information panels, ramp metering, and parking guidance systems. Furthermore, due to the increased possibilities of traffic measuring traffic information has become available via radio, internet, and route navigation systems. At the same time the amount of traffic has increased enormously, congested inner cities and traffic jams have become a regular phenomenon that cannot be solved by single operating traffic measures anymore. In traffic management the combination of information, prohibition, limitation, and coordination is apparently needed. Model-based predictive control does already play a large role in road networks, but the need for more intelligence and cooperation is huge. Future developments like intelligent (partly) autonomous vehicles in combination with roadside control will open up new possibilities for traffic management and control. The chapters of Part III of this book focus on this.

1.4.3 Intelligent water infrastructures

In the near future the importance of an efficient and reliable flood and water management system will keep on increasing, among others due to the effects of global warming (higher sea levels, more heavy rain during the spring season, but possibly also drier summers). Due to the large scale of water networks, control of such networks in general cannot be done in a centralized way, in which from a single location measurements from the whole system are collected and actions for the whole system are determined. Instead, control is typically decentralized over several local control bodies, each controlling a particular part of the network [41, 47]. Local control actions include activation of pumps or locks, filling or draining of water reservoirs, or controlled flooding of water meadows or of emergency water storage areas. MPC and distributed MPC are suitable techniques for intelligently determining how these local control actions should be implemented. The chapters of Part IV of this book focus on this.

1.5 Conclusions and future research

In this chapter we have discussed various infrastructures and have illustrated that they are similar in structure with regard to their nodes and interconnections, which is in particular visualized in Figures 1.1–1.3, but very different in character due to the very different commodities that are transported over the network. It was made apparent that transporting gas or water molecules is very different from transporting cars and trains, and that this is again different from transporting power. Some commodities are passive, like gas and water, others are active, in particular cars, where drivers play an important role and continuously interact with the control measures. Furthermore, we have discussed several control constellations: single-agent and multi-agent control structures, single-layer and multi-layer control structures, hierarchical and distributed systems, centralized and decentralized systems, and combinations of all of these. We have discussed the pros and cons of each structure and the role of computational power and optimality in each of them. The status quo, the vulnerability of the network, and the technical possibilities determine which control structure is most appropriate to meet the challenges in the networks and the user demands.

We have introduced several kinds of Model Predictive Control (MPC) and explained why this methodology is useful for control of large networks. We have illustrated how MPC and multi-agent techniques can be combined, how this combination can tackle problems in subnetworks and how agents in subnetworks can cooperate and negotiate with each other to find an optimal solution for the whole network.

Finally, we have introduced in somewhat more detail recent technological and communications developments in power networks, road networks, and water networks.

Acknowledgements This research is supported by the BSIK project "Next Generation Infrastructures (NGI)" and the Delft Research Center Next Generation Infrastructures.

References

1. L. Acar. Some examples for the decentralized receding horizon control. In *Proceedings of the 31st IEEE Conference on Decision and Control*, pages 1356–1359, Tucson, Arizona, 1992.
2. M. Aicardi, G. Casalino, R. Minciardi, and R. Zoppoli. On the existence of stationary optimal receding-horizon strategies for dynamic teams with common past information structures. *IEEE Transactions on Automatic Control*, 37:1767–1771, November 1992.
3. M. Arnold, R. R. Negenborn, G. Andersson, and B. De Schutter. Model-based predictive control applied to multi-carrier systems. In *Proceedings of the IEEE PES General Meeting 2009*, Calgary, Canada, June 2009.
4. K. J. Åström and B. Wittenmark. *Computer-Controlled Systems*. Prentice-Hall, Upper Saddle River, New Jersey, 1997.
5. M. Baglietto, T. Parisini, and R. Zoppoli. Neural approximators and team theory for dynamic routing: A receding-horizon approach. In *Proceedings of the 38th IEEE Conference on Decision and Control*, pages 3283–3288, Phoenix, Arizona, 1999.
6. D. P. Bertsekas. *Nonlinear Programming*. Athena Scientific, Beltmore, Massachusetts, 2003.
7. P. R. Bhave and R. Gupta. *Analysis of Water Distribution Networks*. Alpha Science International, Oxford, UK, 2006.
8. L. G. Bleris, P. D. Vouzis, J. G. Garcia, M. G. Arnold, and M. V. Kothare. Pathways for optimization-based drug delivery. *Control Engineering Practice*, 15(10):1280–1291, October 2007.
9. M. W. Braun, D. E. Rivera, M. E. Flores, W. M. Carlyle, and K. G. Kempf. A model predictive control framework for robust management of multi-product, multi-echelon demand networks. *Annual Reviews in Control*, 27:229–245, 2003.
10. E. F. Camacho and C. Bordons. *Model Predictive Control in the Process Industry*. Springer-Verlag, Berlin, Germany, 1995.
11. E. Camponogara, D. Jia, B. H. Krogh, and S. Talukdar. Distributed model predictive control. *IEEE Control Systems Magazine*, 1:44–52, February 2002.
12. C. F. Daganzo. *Fundamentals of Transportation and Traffic Operations*. Pergamon Press, New York, New York, 1997.
13. B. De Schutter, T. van den Boom, and A. Hegyi. A model predictive control approach for recovery from delays in railway systems. *Transportation Research Record*, (1793):15–20, 2002.
14. W. B. Dunbar and R. M. Murray. Distributed receding horizon control for multi-vehicle formation stabilization. *Automatica*, 42(4):549–558, April 2006.
15. H. El Fawal, D. Georges, and G. Bornard. Optimal control of complex irrigation systems via decomposition-coordination and the use of augmented Lagrangian. In *Proceedings of the 1998 International Conference on Systems, Man, and Cybernetics*, pages 3874–3879, San Diego, California, 1998.

16. Elkraft Systems. Power failure in Eastern Denmark and Southern Sweden on 23 September 2003 – preliminary report on the course of events. Technical report, Elkraft Systems, Holte, Denmark, 2003.
17. G. Georges. Decentralized adaptive control for a water distribution system. In *Proceedings of the 3rd IEEE Conference on Control Applications*, pages 1411–1416, Glasgow, UK, 1994.
18. T. Geyer, M. Larsson, and M. Morari. Hybrid emergency voltage control in power systems. In *Proceedings of the European Control Conference 2003*, Cambridge, UK, September 2003. Paper 322.
19. M. Gomez, J. Rodellar, F. Vea, J. Mantecon, and J. Cardona. Decentralized predictive control of multi-reach canals. In *Proceedings of the 1998 IEEE International Conference on Systems, Man, and Cybernetics*, pages 3885–3890, San Diego, California, 1998.
20. E. González-Romera, M. Á. Jaramillo-Morán, and D. Carmona-Fernández. Forecasting of the electric energy demand trend and monthly fluctuation with neural networks. *Computers and Industrial Engineering*, 52:336–343, April 2007.
21. A. Hegyi, B. De Schutter, and J. Hellendoorn. Optimal coordination of variable speed limits to suppress shock waves. *IEEE Transactions on Intelligent Transportation Systems*, 6(1):102–112, March 2005.
22. R. Irizarry-Rivera and W. D. Seider. Model-predictive control of the Czochralski crystallization process. Part I. Conduction-dominated melt. *Journal of Crystal Growth*, 178(4):593–611, 1997.
23. N. Jenkins, R. Allan, P. Crossley, D. Kirschen, and G. Strbac. *Embedded Generation*. TJ International, Padstow, UK, 2000.
24. D. Jia and B. Krogh. Min-max feedback model predictive control for distributed control with communication. In *Proceedings of the 2002 American Control Conference*, pages 4507–4512, Anchorage, Alaska, May 2002.
25. D. Jia and B. H. Krogh. Distributed model predictive control. In *Proceedings of the 2001 American Control Conference*, pages 2767–2772, Arlington, Virginia, June 2001.
26. M. R. Katebi and M. A. Johnson. Predictive control design for large-scale systems. *Automatica*, 33(3):421–425, 1997.
27. P. Kundur. *Power System Stability and Control*. McGraw-Hill, New York, New York, 1994.
28. G. Lodewijks. *Dynamics of Belt Systems*. PhD thesis, Delft University of Technology, The Netherlands, 1996.
29. J. M. Maciejowski. *Predictive Control with Constraints*. Prentice-Hall, Harlow, UK, 2002.
30. Y. Majanne. Model predictive pressure control of steam networks. *Control Engineering Practice*, 13(12):1499–1505, December 2005.
31. M. Morari and J. H. Lee. Model predictive control: past, present and future. *Computers and Chemical Engineering*, 23(4):667–682, 1999.
32. R. R. Negenborn. *Multi-Agent Model Predictive Control with Applications to Power Networks*. PhD thesis, Delft University of Technology, Delft, The Netherlands, December 2007.
33. R. R. Negenborn, B. De Schutter, and J. Hellendoorn. Multi-agent model predictive control for transportation networks: Serial versus parallel schemes. *Engineering Applications of Artificial Intelligence*, 21(3):353–366, April 2008.
34. R. R. Negenborn, B. De Schutter, M.A. Wiering, and H. Hellendoorn. Learning-based model predictive control for Markov decision processes. In *Proceedings of the 16th IFAC World Congress*, Prague, Czech Republic, July 2005. Paper 2106 / We-M16-TO/2.
35. R. R. Negenborn, M. Houwing, B. De Schutter, and J. Hellendoorn. Model predictive control for residential energy resources usin g a mixed-logical dynamic model. In *Proceedings of the 2009 IEEE International Conference on Netwo rking, Sensing and Control*, pages 702–707, Okayama, Japan, March 2009.
36. P. G. Neumann. Widespread network failures. *Communications of the ACM*, 50(2):112, 2007.
37. S. Ochs, S. Engell, and A. Draeger. Decentralized vs. model predictive control of an industrial glass tube manufacturing process. In *Proceedings of the 1998 IEEE Conference on Control Applications*, pages 16–20, Trieste, Italy, 1998.

38. S. Piñón, E. F. Camacho, B. Kuchen, and M. Peña. Constrained predictive control of a greenhouse. *Computers and Electronics in Agriculture*, 49(3):317–329, December 2005.
39. S. Sawadogo, R. M. Faye, P. O. Malaterre, and F. Mora-Camino. Decentralized predictive controller for delivery canals. In *Proceedings of the 1998 IEEE International Conference on Systems, Man, and Cybernetics*, pages 3380–3884, San Diego, California, 1998.
40. K. P. Sycara. Multiagent systems. *AI Magazine*, 2(19):79–92, 1998.
41. D. D. Šiljak. *Decentralized Control of Complex Systems*. Academic Press, Boston, Massachusetts, 1991.
42. UCTE. Final report of the investigation committee on the 28 September 2003 blackout in Italy. Technical report, Union for the Coordination of Transmission of Electricity (UCTE), Brussels, Belgium, 2003.
43. UCTE. Final report system disturbance on 4 November 2006. Technical report, Union for the Coordination of Transmission of Electricity (UCTE), Brussels, Belgium, 2006.
44. U.S.-Canada Power System Outage Task Force. Final report on the August 14, 2003 blackout in the United States and Canada: causes and recommendations. Technical report, April 2004.
45. A. J. van der Schaft and J. M. Schumacher. *An Introduction to Hybrid Dynamical Systems*, volume 251 of *Lecture Notes in Control and Information Sciences*. Springer-Verlag, London, 2000.
46. A. N. Venkat, I. A. Hiskens, J. B. Rawlings, and S. J. Wright. Distributed output feedback MPC for power system control. In *Proceedings of the 45th IEEE Conference on Decision and Control*, San Diego, California, December 2006.
47. G. Weiss. *Multiagent Systems: A Modern Approach to Distributed Artificial Intelligence*. MIT Press, Cambridge, Massachusetts, 2000.

Chapter 2
Model Factory for Socio-Technical Infrastructure Systems

K.H. van Dam and Z. Lukszo

Abstract Decision makers often rely on models and simulations for support in the decision making process. Using insights gained this way, better-informed decisions can be made. Model makers design and build models that can be used to test different scenarios and to gain insight in the possible consequences and results of possible actions using simulations. This chapter presents a "model factory" for socio-technical infrastructure systems, which can be used to set up new models of infrastructures by following a number of modeling steps and re-using already existing building blocks from other models. The approach is demonstrated here through application to two case studies: a decision problem for the location of an intermodal freight hub and disturbance management in an oil refinery supply chain, both inspired by the real-life problems. Application of the framework by others confirms that it is widely applicable in various infrastructure domains. Based on these results and the literature study it can be concluded that the presented "model factory" is clearly rewarding for new application for Intelligent Infrastructures problems and strategic decision making.

K.H. van Dam, Z. Lukszo
Delft University of Technology, Faculty of Technology, Policy, and Management, Delft, The Netherlands, e-mail: {K.H.vanDam, Z.Lukszo}@tudelft.nl

R.R. Negenborn et al. (eds.), *Intelligent Infrastructures*, Intelligent Systems, Control and Automation: Science and Engineering 42, DOI 10.1007/978-90-481-3598-1_2,
© Springer Science+Business Media B.V. 2010

2.1 Introduction

Strategic decision makers in the *infrastructure* domain, which includes the electricity sector, transport networks, industrial clusters and supply chains, have always had to deal with capacity limitations, unexpected disruptions, maintenance and investment decisions, and other challenges. This increased complexity firstly arises from a world that is more and more connected: infrastructure systems are not independent from one another but have significant dependencies and interactions. For example, electricity networks and telecommunication networks are dependent; one does not function without the other. Secondly, these systems have grown over time from small, often local, systems to nationwide and even international and intercontinental networks. Challenges that have always been hard to deal with have now become even more difficult at a larger scale.

A third – perhaps even more important – new challenge is the result of these two developments, but adds an additional layer of complexity: not only the physical system has become more complex, but – mainly due to the increased interconnectivity and growth in scale – also the social network that is directly connected with it. The social network includes *actors* such as the users, network operators, maintenance companies, governmental ministries, regulators, etc. The network is not static anymore: new actors (e.g., regulators) can be installed to manage the interaction among other actors to ensure smooth operation and fair market behavior in an interconnected world.

Infrastructure systems now have to be considered as *social-technical systems* [8, 17, 25]. Having the extra social network layer on top of the physical network layer means that decisions have to be made about both networks including interactions between them. Therefore, the practical challenges that have to be dealt with when designing and managing infrastructure systems have become harder than ever before. Before these problems were *complicated*, but now they have become *complex* and consequently decision makers call for novel solutions to approach the problems of socio-technical systems.

Decision makers often rely on *models and simulations* for support in the decision process to make better-informed decisions. Model makers design and build models that can be used to test different scenarios and to gain insight in the possible consequences and results of many actions, using simulations. In other words: model makers build a *decision support tool*. New challenges for the development of models arise when trying to incorporate both the technical and the social systems in one model. Modelers have to cope with new assignments and requirements for the models they build.

There are existing tools to deal with either the physical (e.g., simulation of industrial processes) or the social network (e.g., economic market models), but these worlds have yet to be brought together in an integrated modeling approach. Since recently this subject is receiving increasing attention. For example, Ortega-Vazquez and Kirschen look at a toolbox for modeling generation expansion in the energy sector [16]; Dignum et al. propose a framework for testing responses of society to various policies [3]; Keirstead is building a model of a large urban area to find a systematic and integrated approach to the design and operation of urban energy systems [11] and Hodge et al. work on a systems modeling perspective to experiment with how new technologies can be incorporated into the existing electricity grid [7].

What is missing in these approaches, however – even though they are valuable for their own goal – is a fully flexible way to make changes in either the social network for a given physical network (e.g., vertical unbundling of energy sector) or the physical network for a given social network (e.g., development of high speed rail) or a combination of these two (e.g., carbon dioxide capture and storage). The goal of decision support models is not to find a system *optimum* (which is a common goal for many existing models), but to improve *understanding* of the dynamics of the whole system and subsystems, explore possible futures, find states that have to be avoided or that are desirable, and most of all to provide a tool for policy makers to experiment with *what-if* scenarios. Which degrees of freedom are there? What are the possible consequences of certain decisions? What are successful configurations of either physical or social networks? It should be stressed that this requires a wider view than traditional engineering. The systems under research are considered as part of a larger system; the view is one of a *system of systems* [6].

This chapter aims at contributing to finding a novel modeling approach for socio-technical systems and to help modelers build better models[1] and thereby supporting decision makers. The intention is to develop a generic approach which can deal with all domains of socio-technical infrastructure systems in which mass, energy, or information is literally *transported* through a physical network. It is an engineered system and the organizational structure is in place to support this transfer or directly use it. Existing models do not succeed in offering the flexibility that is needed to perform simulations in case either the social system or the physical system changes. A new approach to modeling socio-technical systems is required to observe and explain the effect of decisions on the system behavior.

In this chapter, a new approach – the "model factory" for socio-technical infrastructure systems – is presented (Section 2.2) and it is applied to a number of case studies (Section 2.3). Afterwards these models are used as decision support tool (Section 2.4). To conclude, final remarks, including the conditions of use and the benefits compared to alternative modeling approaches and directions for future research are presented (Section 2.5).

2.2 Model factory for socio-technical systems

Here a structured approach to modeling socio-technical systems, with a focus on infrastructure systems, is presented. The approach can be used to set up new models of infrastructures by following a number of steps and re-using already existing building blocks from other models. Hence it can be seen as a "model factory". It aims at offering support to modelers in building and connecting models, but also

---

[1] Note the emphasis on building better *models*, which is not necessarily the same as getting better (e.g., more reliable) *results* from models.

to help other parties (including the problem owner) to be involved in the modeling process and to better understand the results. The functional requirements for the approach can be summed up as follows:

- Support a wide range of socio-technical infrastructure systems including petro-chemical clusters, energy networks, freight transport, and supply chains.
- Flexibility for experiments with varying configurations of the social and technical networks, either one or both.
- Full modularity, which results from the requirement of flexibility, but also offers re-usability.
- Easy to use by modelers, including those not involved in the development of the framework.
- Extendability without losing backward compatibility, so that case specific aspects can be added without breaking other models.
- Easy to explain to new modelers, but especially to the problem owner and other stakeholders in the case studies.

The agent-based paradigm appears most suitable for modeling the socio-technical complexity of infrastructures. A literature study confirmed the hypothesis that the agent-based paradigm (discussed in more detail in Section 2.2.1) is most suitable for modeling the socio-technical complexity of infrastructures.

To be able to bring together different worlds and different models and to allow model components, developed by various people, to interact with one another and be re-used, a shared language to define the key concepts in the model is needed. This shared language can act as an interface between model components, models and modelers. Such a formal definition of concepts is offered by so-called *ontologies* (discussed in more detail in Section 2.2.2). A generic ontology has been developed to describe socio-technical infrastructure systems and it can be used to define model elements independently from a specific model itself.

An agent-based approach, founded on a generic ontology for socio-technical systems and source-code implementations of these concepts together with a set of building blocks, can be used to create flexible, interconnectable and re-usable models. The generic ontology can be extended and specified for specific problems or cases. The result is a "model factory" that can be used to quickly set up new models and to connect models of different infrastructures.

2.2.1 Agent-based modeling

When representing a complex socio-technical system as an agent-based model, one starts with modeling the actors in the system. A model of an actor, or a group of actors (e.g., a community), is called an agent. An agent can be seen as a software entity that is autonomous, reactive, pro-active, and capable of social interaction [9]. The behavior of an actor is formalized using algorithms with, for example, *if-then* rules: the so-called behavioral rules. The key distinguishing element, that sets agent-based

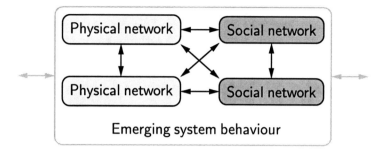

Figure 2.1: The relationship between physical networks and social networks, causing emerging system behavior.

models apart from other models, is a focus on modeling *individuals* who can make decisions.

Agents model the social part of the system, but they can interact in one way or another with one or more physical nodes. For example, an agent is owner of a technology, operates it, or maintains it. This means that interaction between an agent and a technology is needed, such as control actions from the agent to the technology and information flows back. Physical links, such as mass or energy flows, can only occur between two physical nodes while social constructs, such as contracts, can only occur between two social nodes (i.e., two agents). Adding a number of individual agents (including technical components and their links) to create a multi-agent system leads to a model of a system without having to make system level behavior explicit; it emerges from the individual agents' actions (See Figure 2.1).

By modeling components rather than the entire system, the structure of the system is not predefined. Because agents can communicate and link with other agents, different networks can be created by changing the behavioral rules without explicitly defining which relationships are to be made. This way different set-ups of a control system (e.g., hierarchical or coordinated, see [24]) can be tested in a simulated environment and the agents' response to the emergent system behavior can be monitored. Furthermore, both technical and social changes (that are needed for the real challenges in the infrastructure domain, as discussed in Section 2.1) can easily be implemented in such flexible simulation models.

If the system behavior is modeled explicitly, as is common in numerical approaches for example, making changes in the model would require the modeler to adapt the system structure. That way it is possible to compare different configurations of the system, but it is not clear how the most desirable situation can be obtained by influencing lower levels of the system. Agents (like the actors they represent) can exist in several levels of hierarchy, for example if one actor supervises the activities of one or more other actors creating subsystems [22]. To summarize, the main advantage of building models in such a bottom-up approach is that it creates a very flexible environment.

In general, in many of today's infrastructure systems it might not be clear for the problem owner how to directly influence the whole system or even part of it. For the model to be an effective decision support tool, it needs to give insight in how changes at lower system levels impact the emerging system behavior. This way of modeling is close to how it works in the real world: the collective decisions made by more or less autonomous actors at various levels of a hierarchy together result in an overall system behavior.

Agent-based models, due to their bottom-up nature, are suitable for simulating dynamic systems where the structure can or should change during a run, or where experiments with different configurations have to be done. By making a strict distinction between social and physical components and formalizing this domain knowledge in an ontology (see Section 2.2.2) parts of the model can be re-used (e.g., re-using the model of a certain technology with a different agent, or re-using behavioral rules of one agent in another one, or even copying complete agents with their physical nodes into another model) (See Section 2.2.3).

2.2.2 Ontology for socio-technical systems

In the artificial intelligence community ontologies have been developed as a useful means of knowledge representation. Ontologies are formal descriptions of entities and their properties, relationships, constraints and behavior, that are not only machine-readable but also machine-understandable. When two agents communicate about certain concepts, one wants to be sure that they give the same interpretation to the meaning and use of these concepts. Therefore it is of the utmost importance to unambiguously specify each concept and its meaning.

The meaning is stored not only in subclass relationships (*is-a*, e.g., apple is a fruit, red is a color) but also in property relationships (*has-a*, e.g., fruit has a color). An ontology contains explicit formal specifications of the terms in the domain and relations among them. In other words: it is a formal specification of a conceptualization [5]. In this view a *class* is nothing else but a generalization of a number of instances that the modeler chose to put together. An *instance* is a single identifiable object within the limits of the scope of the model, belonging to a class that is formalized in the ontology. To use the fruit example, one can say that this one specific apple has a red color (and two agents communicating about this instances know they both talk about the very same piece of fruit). The class is abstract, where the instance is concrete. Key is describing concepts using already defined concepts, such as in the example above: because the concept of a color was defined, it can be used to define one of the properties of fruit.

When it comes to implementation of a model, an ontology forms the basis of the class structure for object-oriented software implementation. The *is-a* relationship is coded as the subclass relationship in class descriptions and the *has-a* provides information on the properties of the class and the possible values. This is similar to, for example, the Java programming language. The ontology provides an *interface* definition between any two objects in the model implementation. When properties defined in the ontology for a specific class are known, it means this information can be exchanged. The knowledge rules (i.e., the decision making rules of agents) to implement the behavior of the agents can then also be expressed in these formalized concepts.

Ontologies are not only useful for communication between agents, but also for sharing knowledge between modelers, domain experts and users. Which concepts should play a role, depends on the goals of the research and the type of questions that should be answered by the simulation. To build an ontology, a knowledge engineer has to talk with domain experts to analyze the system and to make it explicit.

Ontologies are even more powerful when they can be re-used so it is important to use a generic description as much as possible. This does not only make it possible to re-use domain and expert knowledge, but also to re-use source code. As presented in the next section, by specifying a case study in previously formalized generic concepts previously implemented building blocks can be re-used. In other words: ontologies facilitate re-use, sharing, and interoperability of agent-based models. Section 2.2.3 deals with this aspect of the "model factory".

Figure 2.2 shows a small fraction the ontology for socio-technical infrastructure systems. In this fragment agents (social nodes) and physical systems (physical nodes) are both considered as nodes, with different properties. It shows, for example, that an Agent *is-a* Social Node and that it *has-a* Technology (one or more, as indicated with the asterisk). The ontology contains many formalized concepts, including different types of edges, properties, configurations, labels, etc. Some key concepts from the ontology that are generic for use in socio-technical systems modeling – but that form only a small part of the full rich and detailed ontology – are introduced briefly below[2]:

- **Agent**: A SocialNode is a Node that is capable of making decisions about PhysicalNodes. SocialNode inherits all properties of Node, of which it is a subclass. This inheritance includes, for example, the property PhysicalProperties, which is generic for a Node because a SocialNode can also have a Location (which is a PhysicalProperty). SocialNode has a subclass Agent, representing an actor in the system. This can be a single person (e.g., a truck driver), a group of people (e.g., the operations department) or a whole organization (e.g., the government).

- **Technology**: A PhysicalNode represents an element in the physical world such as an engineered system. The definition of a Technology, a subclass of PhysicalNode, is based on the input-output abstraction and can either be a small unit (e.g., a battery) or a very large system (e.g., a power plant).

- **PhysicalEdge**: PhysicalEdges are Edges between PhysicalNodes. Two different PhysicalEdges are considered: PhysicalConnection and PhysicalFlow. PhysicalConnection is the "hardware" and they represent real links of the infrastructure connecting two PhysicalNodes. One could think of a pipeline, a power cable, or a road connecting two Nodes, making transport of mass or energy possible. A PhysicalFlow is the actual flow of mass or energy between

---

[2] All classes are written with a capital first letter and in so-called *CamelBack* notation.

two PhysicalNodes. PhysicalEdges cannot connect two Agents directly, only via their Technologies.

- **SocialEdge**: Where PhysicalEdges model real connections and mass and energy flows, a SocialEdge is a social construct. It forms the social network of a socio-technical system as well as establishes the link between the physical and social networks. There are two subclasses for SocialEdge in the generic ontology: Contract (again further specified to, among others, PhysicalFlowContract, and ShippingContract) and Ownership.

- **Property**: Different Properties can be defined, such as EconomicProperty, PhysicalProperty or DesignProperty. There is a range of properties in the ontology already, including Voltage, MaximumCapacity, and Price, which can use various well-defined UnitNames. This makes it possible to define properties of Agents (e.g., the amount of money the agent has) and Technologies (e.g., the location of a factory), but also of PhysicalFlows (e.g., the pressure of a gas).

- **OperationalConfiguration**: An OperationalConfiguration is the specification of the connection between input and output of a technical system, precisely defining which GoodNames are at the input (OperationalInputs) and how they are transformed to the output (OperationalOutputs). See Figure 2.3.

- **ComponentTuple**: ComponentTuples are used for the OperationalConfigurations, providing a detailed description of inputs or outputs of the Technologies. They mainly consist of a GoodName and a relative amount and a unit in which the amount is expressed.

- **GoodName**: GoodNames are used to indicate a class of goods, for example crude oil, ethylene, electricity or a container. Specific properties, such as the volume, weight, voltage, and size can be added to actual PhysicalFlows of these goods, or to a ComponentTuple, for example.

Agent and Technology are both defined as Nodes and this way social and physical networks can be built that have relationships between them as defined with the PhysicalEdges and SocialEdges. The elements from Figure 2.2 therefore construct the networks from Figure 2.1.

2.2.3 Library of components of socio-technical systems

To give an impression of what the current[3] contents of the knowledge base are, below is an overview of some often used classes in the ontology and the number of instances they have at this stage:

- **Agent**: 69 Agents have been created, including re-usable ones such as the world market or the environment as well as owners of specific technical installations. Using CaseLabels it is possible to not use all Technologies that are associated

---

[3] The knowledge base is in constant use and new instances are added on a regular basis for new case studies. The data given here is from June 2009

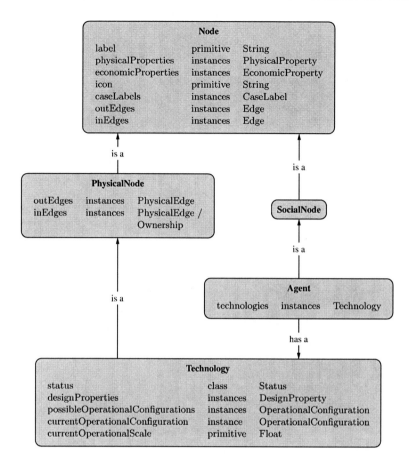

Figure 2.2: The relationship between physical networks and social networks in the ontology (cf. 2.1).

with a specific Agent, but limit it to those important for a specific case.

- **Technology**: over 200 different Technologies are defined in the knowledge base, each providing a rich and detailed definition of the inputs and outputs of the process. Examples range from oil refineries to wind turbines, and from container terminals to storage tanks for diesel.

- **GoodName**: 309 different Goods have been included so far, including many products from the chemical process industry (based on work done in cooperation with industry partners). It is important to re-use these as much as possible between models, as it will make it possible for trading clusters to be formed of Agents using these definitions. Furthermore, GoodNames can be labeled with one or more classes, such as BioMass or EnergyCarrier, enabling reasoning about the products. For example, an Agent who is the owner of a bio-mass

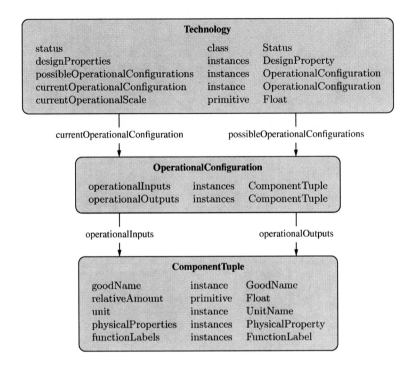

Figure 2.3: The OperationalConfiguration of a Technology as defined with Component-Tuples.

power plant can search for other Agents who can supply any Biomass, regardless of what type. The energy content and other properties of these goods can also be defined (e.g., in a ComponentTuple for the OperationalInputs of the power plant, see Figure 2.3) so more flexible descriptions of technologies are possible.

Any new model can re-use these elements and, if needed, add new ones to quickly create new models. Elements in the physical network are easily re-used. For agents this is less straightforward, because actors often have very specific behavior that needs to be modeled. Still, through the use of the ontology and shared building blocks, also behavioral rules of the agents can be shared between models. One way of doing this is by *extending* already defined agents classes thereby inheriting methods of the super class, or by simply copying the code into a new model.

As an example the trading and production behavior of agents can be given. This has been implemented in a generic way: an agent looks at its technologies to find out which raw materials are needed to run production. Agents will then contact others who are potentially able to supply them with this product (e.g., because a physical connection between two Technologies exists). These potential suppliers again decide, based on the possible outputs of their production technologies, if they can make an offer and send a trade contract. Because the agents use the same ontology, it is well defined that such communication and the exchange of contracts is interoperable. Functions such as paying for maintenance costs, again as defined in the ontology, and arranging shipping contracts, for example, can easily be re-used.

The re-use of building blocks allows new models to be set-up quickly so one can focus either on modeling case specific behavior, or on experimenting with new scenarios.

2.2.4 Procedures and modeling steps

With the classes defined above and (optionally) by re-using instances that have been included in the knowledge base, a socio-technical system definition can be created. Looking back at Figure 2.2, both the social and the physical networks can be defined as well as the links between them. The ontology is stored in a Protégé [4] knowledge base, which can be changed without having to adjust the model source code. The model code works independently from the definition in the ontology. A *knowledge base reader* has been created, which can read the instances in the knowledge base to create instances in the Java model so they can be used in the model. Also, all classes defined in the ontology become objects which can be instantiated at run-time (e.g., when an Agent makes an investment and buys a new Technology). For some classes, such as TransportContract and PhysicalFlow, they are predominantly created during the model run based on actions of the Agents.

The following model building tasks should done to develop an agent-based model using the "model factory" [12]:

M-1. Conceptualize the problem in terms of actors and physical systems, including their relations and properties. Distinguish the set of properties that will act as variable in the model (i.e., the model parameters) and possibly visualize the interaction between model parameters in an influence diagram.
M-2. Extend the generic ontology with new abstract classes applicable for this case.
M-3. Make the model specification by creating concrete instances of the abstract classes from the ontology.
M-4. Implement the behavior of the agents, making use of generic components (e.g., searching for suppliers, determining a price, accepting contracts) and add new components if needed.
M-5. Verify the model.
M-6. Validate the model.

After performing these tasks, the model is ready for simulation and the following tasks have to be executed:

S-1. Formulate experiments: decide which model parameters to vary and what performance indicators to measure.
S-2. Decide on the values for the simulation parameters, (e.g., the number of simulation steps, random seed, etc.) and execute the experiments.
S-3. Analyze the results.

2.2.5 Conclusions

In this section a brief introduction of the "model factory" for socio-technical infrastructure systems was given. The "model factory" consists of three elements:

- **Interface**: An interface definition between any two components, between models, between developers and between developers and problem owners, expressed and formalized in an ontology.

- **Library**: A shared library of source code that can be re-used, including agents (e.g., shipping agent) and specific behavior (e.g., procurement behavior). The components of the library can be seen as "building blocks" to create new models using the "model factory".

- **Procedures**: Procedures on how to use the library and interface to define and build models of socio-technical infrastructure systems.

This way the framework enables developers to set up new models for new challenges decision makers cope with.

2.3 Illustrative case studies

In this section two illustrative case studies are presented: an oil refinery supply chain and an intermodal freight hub. These case show how the "model factory" is used to develop models in different infrastructure domains.

The two cases have been selected from different infrastructure domains, to show the wide applicability of the approach and to highlight that at higher level of abstraction the various domains can be considered as similar. As the first case (Section 2.3.1) an inter-modal transport infrastructure is investigated. The system consists of different actors with their own interests and control over (part of) the physical transport network. In the second case study (Section 2.3.2) a supply chain for an oil refinery, including production, storage and transportation of raw materials and products, is modeled. This supply chain, which goes beyond the control of one actor, can be considered as an infrastructure because it is a typical example of a socio-technical system.

Furthermore, these cases are illustrative of modeling of Intelligent Infrastructures because the main challenge is how to make best use of the already available infrastructure through smart and innovative policies and the focus is on the relationship between the distributed individuals and the effect their decisions have on the overall system performance. Together these two cases illustrate the applicability of the "model factory". Additional cases are mentioned in Section 2.3.3.

2.3.1 An intermodal freight hub

Intermodal freight transportation is defined as a system that carries freight from origin to destination by using two or more transportation modes. In this system, hubs

Table 2.1: Agents, PhysicalNodes, and their relationships for the freight hub case.

Agent	Relationship	Physical Node
World market	owns	Delivery installation
Container terminal operator	owns	Container terminal
Intermodal freight hub operator	owns	Intermodal freight hub
Hub user	owns	Toy factory
Consumer	owns	Consumer installation
Shipper	–	not applicable

are one of the key elements: they function as transferring points of freight between different modes. The success of an intermodal freight hub depends on four major factors [13], namely: location, efficiency, financial sustainability, and level of service (e.g., price, punctuality, reliability, or transit time). The location of hubs is a critical success factor in intermodal freight transportation and needs to be considered very carefully as it has direct and indirect impact on different stakeholders including investors, policy makers, infrastructure providers, hub operators, hub users, and the community [19].

Intermodal freight transportation becomes an attractive alternative to road transportation as the latter no longer assures a reliable and sustainable service delivery owing to traffic congestion, rising fuel price, and air pollution problems. However, the increasing demand of intermodal freight transportation has posed a new challenge on how to provide sufficient infrastructure that will meet that demand and maintain a satisfactory level of services. A comprehensive review of intermodal rail-truck freight transport literature is given by Bontekoning et al. in [1].

The study presented in this chapter relates to making models to support the decision making process for choosing a location and realizing a new freight hub. It is a real policy problem in the Queensland region in Australia [19].

Conceptualization of the problem in terms of actors and physical systems

As an initial model for this case study, a simplified version of the complex realistic transport network is used. The actors that are modeled in the social system are shown in Table 2.1 alongside the physical nodes. Hub users are companies that use the hub for their transport demand. Aggregated demand is modeled as one actor, called the consumer. Some other stakeholders, such as the community, are left out of this version of the model for now (but, because the model is set-up in a bottom-up approach, other actors can easily be added later).

The ownership relationship between the SocialNodes and PhysicalNodes are shown in Table 2.1, but the system contains many other relationships including PhysicalFlowContracts (between Agents) and PhysicalFlows (between Technolo-

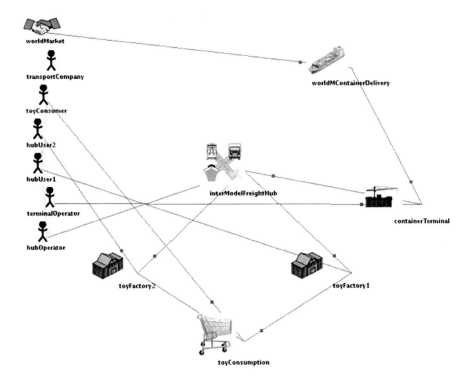

Figure 2.4: Geographical representation of the social and physical nodes in the model. Edges between Agents (on the left) are SocialEdges (e.g., PhysicalFlowContracts) as are those between Agents and Technologies (e.g., Ownership). Edges between the Technologies are PhysicalEdges (e.g., PhysicalFlows).

gies). Most of these Edges are created on the fly in the model (except the PhysicalConnection between the Technologies, which are fixed) and are the result of the behavior of the agents. Figure 2.4 shows a screen capture of the model during a simulation run, to illustrate these other relationships. See [12] for a more detailed description.

Extension of the generic ontology with new abstract classes

The generic ontology for socio-technical systems was extended for the freight hub model. While many concepts could be reused, new abstract concepts such as GIS[4] location (a Property), TransportContract (an Edge), or new transport modalities for rail and truck (a Class) were created to enable the specification of the model in the ontology. These new concepts are shared with other models so they can be re-

---

[4] Geographical Information System, a set of coordinates used to determine a position on Earth.

used. The concept of a TransportContract, for example, was later re-used for the oil refinery case study as presented in Section 2.3.2.

Creation of concrete instances

Instances for the actors, physical nodes, and all fixed relations (recall Table 2.1 for the fixed SocialEdges, but in the instances also the PhysicalConnections are defined) were added to the shared knowledge base too. Instances of agents that are also used in other models, in this case the world market, did not have to be created but can be re-used. The instance of the world market was updated by adding an ownership relationship with the container delivery installation, so in addition to the goods traded on the world market in other models (e.g., petro-chemicals and natural resources) the world market can supply containers with toys.

Implementation of the behavior of the agents

The behavior of all actors is modeled as searching for other agents that can offer the desired goods. This means that agents look for other agents that have an output that matches with their input. This behavior is generic and not dependent on the goods (i.e., containers or toys for this model, and this is the reason why the world market agent needs a container supply technology so it "knows" it can supply them). All behavior is demand driven and agents act in the order of the supply chain (consumer starts first, ending with the world market). Agents collect a number of (unsigned) PhysicalFlowContracts from other agents and then choose the best contract to sign, thereby committing to a transaction.

All agents pay maintenance and operational costs for the physical system they own (if applicable). Again, this is also generic behavior. The transport agent, not owning a physical node, does not need to buy any goods on the market. It does pay maintenance and operational costs for its fleet of vehicles. Agents that arrange to buy goods also need to arrange transport for this, by asking the transport agent for a transport contract.

Specific behavior for the transport agent was needed, as such an agent had not been used in other case studies built using the "model factory". The transport agent is contacted by other agents who have signed a PhysicalFlowContract and the shipping agent makes an offer for the transportation of this flow. This behavior was later, for example, re-used for the oil refinery case study as presented in Section 2.3.2.

Verification and validation of the model

The model presented here is a proof-of-concept model and is designed to illustrate the applicability of the framework. In a series of batch-runs with different parameters the model was verified. However, it is not based on a real-life case and real data, therefore, no validation had to take place. Expert judgement by people in the transport modeling domain and experts in intermodal freight systems was used to determine the usefulness of the model and the added value to solving a real complex problem [20].

2.3.2 An oil refinery supply chain

In this section a model of an oil refinery supply chain is presented. A hierarchy of decisions has to be made in managing the supply chain: strategic (e.g., capacity investments, adding units, upgrading technology, supply chain reconfiguration), tactical (e.g., production planning, policy evaluation, disruption management) and operational (e.g., procurement, storage, scheduling, throughput level). These motivate the development of simulation models of the supply chain, which could reflect the dynamic behavior of the entities in the face of the various uncertainties. This model enables decision making for supply chain management by allowing the user to evaluate the impact of a particular decision on the supply chain performance, analyze different supply chain policies, and identify the consequences of a disruption, through simulation. The model is based on the system description from [18] and is described in more detail in [21].

Conceptualization of the problem in terms of actors and physical systems

An *oil refinery supply chain* begins from the oil reservoirs, both onshore and offshore. Crude oil is tapped from these sites and then transported to various refineries around the world mostly by pipelines or large ships called very large crude carriers (VLCCs). Transportation times of crude are relatively long; it takes four to six weeks for a VLCC carrying crude oil from the Middle East to reach refineries in Asia, for example. The crudes are then processed in crude distillation units (CDUs) and separated into fractions based on their boiling points. These fractions are processed further in different downstream refining units such as reformer, cracker, and blending pool to get the various products. A single crude mix may yield numerous products and their variants through a suitable alteration of processing conditions. Hence, refineries must adapt their operations to the different crude batches to meet the required product specifications from their customers.

The refinery occupies a pivotal position in the supply chain with its functional departments initiating and controlling the interactions with the external entities, which are oil suppliers, 3rd party logistics (3PL) providers, shippers, jetty operators, and customers. The operation of the refinery supply chain requires various decisions in every cycle of 7 days – what mix of products to make, which crudes to purchase and in what quantities, which mix to process and in which processing mode, etc. Different actors are responsible for the different decisions [10]. These actors communicate with one another through information flows in order to control the material flows. The refinery physical units may be further sub-divided into storage units such as crude and product tanks and processing units such as CDU, reformer, cracker, and blend tanks. The functioning of these units and other supply chain activities are overseen by the functional departments: the storage department and

Table 2.2: Agents, PhysicalNodes, and their relationships for the refinery supply chain case.

Agent	Relationship	Physical Node
Refinery company	owns	Refinery units (incl. CDU)
Refinery company	owns	Raw materials storage tanks
Refinery company	owns	End product storage tanks
Operations dept. (Refinery)	controls	Refinery units (incl. CDU)
Storage dept.(Refinery)	controls	Raw materials storage tanks
Storage dept.(Refinery)	controls	End product storage tanks
Sales dept.(Refinery)	–	not applicable
Procurement dept.(Refinery)	–	not applicable
Logistics dept.(Refinery)	–	not applicable
3rd party logistics provider	–	not applicable
Shipper	–	not applicable
Supplier	owns	Oil wells
Jetty owner	owns	Jetty
Consumer	owns	Consumer installation

the operations department. The actors and physical systems are shown in Table 2.2 (note that some actors, for example the logistics department, do not own or control a physical system, but they do have their own specific tasks and communicate with the other actors).

Extension of the generic ontology with new abstract classes

For the development of the oil refinery supply chain model no major changes to the generic ontology were needed. All the key classes needed to define the system were already in place, based on earlier case studies. Only minor additions were needed, such as adding properties to the TransportContract for more detailed registration of transport delays and payment.

Creation of concrete instances

The Agents and Technologies as defined in Table 2.2 were added to the knowledge base. The values of the properties are based on [18]. New additions to the ontology were needed in the form of instances for the technical elements (e.g., the refinery units and storage tanks, delivery installations and fixed infrastructure connections between them) with their properties (e.g., production recipes for the refinery for the various mixes of crudes, maximum capacities of the storage tanks, distances for the shipping routes). Furthermore, the initials conditions of the system (e.g., current stock levels in the storage tanks and current capital of the agents) were defined.

Implementation of the behavior of the agents

Each entity acts based on its policies and the combined actions of the entities determine the overall performance and economics of the supply chain. For example, the procurement department decides on the type and amount of crude to buy, the logistics department oversees transportation of the crude, and the storage department manages the crude unloading from the ship to the storage tanks. The combined actions from these three departments determine crude arrival at the refinery. The complex maze of flows among the entities could lead to unforeseen domino effects. Furthermore, the refinery has to contend with various uncertainties such as prices, supply availability, production yields, and demand variations.

Following the agent paradigm, the tasks are clearly distributed between the agents. Some tasks therefore have to be split into several subtasks (requiring communication between the agents). A schedule is made so that some processes (e.g., procurement) only occur at certain intervals while others (e.g., production) happen at each time step of the simulation. Events such as the arrival of a VLCC at the jetty are monitored each time step. For the purpose of this model one day of 24 hours is 1 simulation time tick, but it should be stressed that for other problems different settings can be chosen (e.g., splitting a day in 100 ticks allows studying certain operational aspects in more detail).

Modeling the behavior of the agents was based on existing behavioral rules for trading (as also used in the freight hub model from Section 2.3.1), but additional rules had to be implemented for various procurement policies (e.g., forecasting of demand deciding on procurement), scheduling which OperationalConfiguration to use and for the activities of the jetty (which had not been used in earlier models), for example [21]. The new elements can now be shared with other modelers or be re-used in new case studies and as such the new modeling effort has contributed to a larger set of building blocks.

Verification and validation of the model

The oil refinery model built using the "model factory" has been tested extensively. A benchmarking study has been executed in which the agent-based model was compared with an equation-based model of the same supply chain [21]. The aim of the benchmarking study was to compare the modeling paradigms (i.e., the agent-based and equation-based paradigms) and to learn about the advantages and disadvantages of the different approaches. For benchmarking modeling paradigms, it was necessary to demonstrate that the models under study are comparable. A numerical analysis was performed, establishing that the two models show the same behavior. This concludes the verification and the validation phase of the agent-based model as successful, the more so because the equation-based model had been validated against the real system [18].

2.3.3 Conclusions

In this section two models built using the "model factory" were presented. The steps from Section 2.2.4 lead to setting up new models through using the ontology and re-using building blocks from other models. The modeling approach has been widely used by others. Representative examples include the development of a model of the Dutch electricity sector to experiment with CO_2 emission trading [2], and a number of models to investigate the evolution of industrial clusters [15].

It should be stressed, that a generic framework for modeling cannot be created from scratch, but has to be developed through gradual changes and iterations. Each iteration follows an application of the framework to a new domain and the lessons learnt from this feed back into the generic framework. This is repeated several times before a state can be reached in which the framework can be called generic (to a certain extent). Each new case study that is executed provides the opportunity to add new re-usable components to the library. After numerous case studies, some of which were highlighted here, this has now lead to a framework that is widely applicable, proved to be re-usable, and can be used as a "model factory" to support decision makers.

2.4 Decision support using the model factory

The objective of the research is to provide decision makers with support when dealing with complex problems in various infrastructure domains. The models created with the "model factory", including the two cases described in Section 2.3, can be used as decision support tools. Here two examples are given, revisiting the intermodal freight hub model and the oil refinery supply chain model.

For the intermodal freight hub study, decision makers have to choose a good location for a new hub and the model is used to experiment with different ways to encourage stakeholders to agree with this proposal (Section 2.4.1). Abnormal situations can occur in the supply chain (e.g., disruption in ship arrival, production not achieving desired quality target and planned corrective maintenance of plant equipment) and simulation models support decision makers on how to best deal with them (Section 2.4.2). It is demonstrated how the models, built using the "model factory", can be applied to these cases as decision support tools to help solve real problems.

2.4.1 Decision support with the intermodal freight hub model

First the model described in Section 2.3.1 is considered and it is demonstrated how it can be used to support the governmental transport agency in deciding on the location of the freight hub.

Formulate experiments

As said before, in Section 2.3.1, a key variable in the development of an intermodal freight transport system is the location for the freight hub. In this case study, three conceptually difficult issues have been identified [19]:

1. Investments in the infrastructure cause dynamic effects, for example on transport demand;

2. Actors can have conflicting objectives that prevent reaching a global optimum;

3. The problem owner cannot enforce other actors to make certain decisions.

The problem owner, in this case the governmental transport agency, has indicated that it needs more insight into the relationships between the stakeholders to make a decision. Three steps can be identified on the path towards the development of a new freight hub in which tools can assist the problem owner:

1. Generating possible solutions for the location of the freight hub;

2. Evaluating the solutions for each stakeholder and the overall system level;

3. Experimenting with different policies and measures to influence the decisions of the individual stakeholders

Steps two and three form an iterative process. A location can only be chosen after experimenting with different instruments to influence the actors. For each potential location the situation should be analyzed using the model and different scenarios be played out, before a judgement on the suitability of the hub location can be given. During the simulation runs the capital of each agent is measured and plotted in a graph. The main model parameter to vary is the location of the intermodal freight hub. Three different locations for the hub are tested, with different effects on the actors (See Figure 2.4 for one such location scenario). When one of the actors loses money, compared to the initial situation, experiments are done to identify measures that can reduce the losses and generate adequate incentives for all actors to agree with a certain proposed location.

Execute the experiments

By changing the location of the hub, it can be shown that some actors benefit while others loose (See Figure 2.5). Also it is illustrated that simple measures to support actors that lose money can encourage them to still support a hub location. Experiments to discover an appropriate tax deduction were conducted. Giving a subsidy for the variable costs of the transport agent can help prevent it from losing money. This influences actors to still accept a hub location that was initially not ideal for them. With these experiments it is demonstrated that a subsidy of 0.05 euro/unit volume/km for the transport agent is enough to compensate for the losses that arise from one of the potential hub locations.

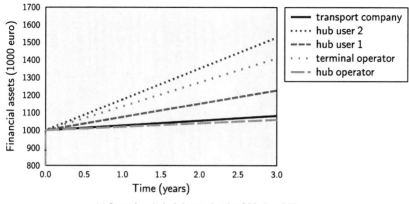

(a) Location 1: hub located at lat:250, lon:250.

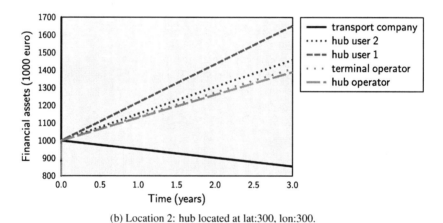

(b) Location 2: hub located at lat:300, lon:300.

Figure 2.5: Illustrative results from the intermodal freight hub case study.

Analyze the results

The results show that not all possible hub locations are suitable for all actors, but that small subsidies can be used to encourage them to still accept this choice. To make realistic policy recommendations, a more realistic network representation is needed together with more dynamic behavior of the actors. Future experiments can be defined to provide an additional layer for simulations that can be explored without having to make many changes to the modeled behavior. This means that only further extension of the ontology and the model specification in the knowledge base (and subsequent implementation of these additional components) is needed. No conceptual changes will have to be made to the model.

2.4.2 Decision support with the oil refinery supply chain model

Here a decision problem for the oil refinery supply chain is presented. To determine nominal process conditions, the model optimally chooses which crudes should be bought for normal operation, how much crude is needed and from which supplier they should be ordered. Furthermore, the mode of operation is scheduled based on predicted demands and the throughput for operation of the refinery is set based on actual demand from the consumer. For an abnormal situation, when a disturbance is manifesting itself, this normal approach is not adequate anymore because the complexity is too big and there are too many interdependent variables. There are more options to choose from, making it more difficult to decide what is the best decision. A model-based decision-support tool is therefore called for.

Formulate experiments

Although the approach proposed in this work is general, here the scope is limited to disturbances dealing with the supply of crude oil to the crude distillation units. A disruption in the supply can be caused by a delay in the shipment of crudes from the supplier (at a large distance from the refinery) or problems in the tank farm, for example. In both cases the operations department runs the risk of not having access to enough crude to perform the scheduled operations.

These disturbances are then defined as follows:

$$\bar{d} = (ShipDelay, StorageProblem) \tag{2.1}$$

with

$$ShipDelay \in \{0, 1, 2, \ldots, n\} \quad \text{in days, for 1 ship for 1 cycle} \tag{2.2}$$

$$StorageProblem \in \{0, 1\}_m \quad \text{for each of the } m = 5 \text{ crude storage tanks.} \tag{2.3}$$

For instance, $\bar{d} = (2, 0, 0, 0, 0, 0)$ indicates that a ship at sea is delayed by 2 days, and there are no problems with the storage installation. Note that the value of n for a ship delay can be larger than the time horizon used, effectively sinking the ship.

For simplicity, it is assumed that the magnitude of the disturbance is known as soon as the disturbance occurs. In reality, this may involve uncertainty. Furthermore, the granularity could be adjusted so that a delay could be expressed in parts of a day (e.g., hours) instead of full days as used here.

When faced with such a disturbance, the problem owner has a number of choices. Firstly, he has to determine if the disturbance has a significant effect on the operation of the supply chain. If the effect is deemed minor, no action may be necessary. One example of a case where corrective action would be required include inability to execute the previously planned schedules due to inadequate crude. This may be addressed by changing the operating mode, the throughput or by emergency crude procurement. Often a combination of these actions may be needed.

For the Emergency Procurement (*EmPr*), the procurement department can contact a local supplier to buy crude at a much higher price but with a short lead time.

The degree of freedom for each of the five crudes is between 0 kbbl[5] to the amount that could reasonably available on short notice, which is assumed 600 kbbl here. The procurement department has to ask logistics department for the expected delay to be able to make this decision. Note that when there is an error in a storage installation ($StorageInstallationError = 1$), emergency procurement will not solve a disturbance, as all crudes have to rest in storage to allow the brine to settle and cannot be transferred directly from the vessel to the crude distillation unit but have to pass storage tanks.

Furthermore, the operations department can choose to Change the Operational Configuration (COC), meaning that a different recipe (one out of four) is selected using crudes that are still in stock, but possibly resulting in yields that are not ideal compared to the scheduled operation. Finally, the operations department can Change the Operational Scale (COS), to run the refinery at a lower throughput (from 40% to 100%, with a minimum to keep plant running), producing less end products but avoiding having to shutdown the plant when crude runs out (or postponing plant shutdown, for example to allow emergency procurement crudes to arrive).

The degrees of freedom are defined as follows:

$$\bar{x} = (EmPr_1, EmPr_2, EmPr_3, EmPr_4, EmPr_5, COC, COS) \tag{2.4}$$

with

$$0 \leq EmPr_i \leq 600 \quad \text{in kbbl, for each of the } i = 5 \text{ crudes} \tag{2.5}$$
$$COC \in \{R1, R2, R3, R4\} \quad \text{discrete choice between operating modes} \tag{2.6}$$
$$40 \leq COS \leq 100 \quad \text{percentage of CDU throughput capacity.} \tag{2.7}$$

There are different options for the criteria with which to choose the best alternative. As examples, one can look at the overall profit of the refinery (for a certain time frame), profit during the production cycle effected by the disturbance, other financial measures, but also non-economical criteria such as customer satisfaction. Here the profit P of refinery, 14 ticks after a disruption took place, was chosen. This means that the effect of a disturbance at $t = 22$ will be simulated over the next two cycles of operation, during which new raw material deliveries and product dispatch. We expect that the impact of the disturbance would have worn off by then.

The objective function is defined as follows:

$$\max \ P(\bar{x}, \bar{d}) = \sum_{t=1}^{50} (Income_{sales}^t(\bar{x}, \bar{d}) - Cost_{procurement}^t(\bar{x}, \bar{d})$$
$$- Cost_{transp}^t(\bar{x}, \bar{d}) - Cost_{maint}^t(\bar{x}, \bar{d}))$$
$$+ Value_{product \ stock}^{t=50}(\bar{x}, \bar{d}) + Value_{raw \ materials}^{t=50}(\bar{x}, \bar{d}). \tag{2.8}$$

---

[5] 1 kbbl is 1000 standard oil barrels.

In this equation the value of the product inventories and raw materials at the end of the simulation run have been included. The consequences of the disruption on future cycles are included in the cost function (e.g., if the response is to switch to another mode of operation without any emergency procurement, it is possible that during a later cycle the planned operation cannot be met) but no new decisions following to any such new disturbances are assumed; a single response is formulated.

The performance function is discontinuous due to the transportation costs which are a function of the amount of crude procured, the capacity of the ships (either a VLCC for long distance shipping or a general purpose tanker with much smaller capacity for short haul in the case of emergency procurement) and the travel time. The transport price is calculated per vessel and the cost for procurement plus transport therefore follows a saw-tooth pattern making it more difficult to determine the right amount to buy, especially in combination with other measures such as switching to another recipe.

Execute the experiments

Figure 2.6(a) illustrates the refinery behavior under normal operation and Figure 2.6(b) after disturbance $\bar{d} = (30,0,0,0,0,0)$ which occurs on day $t = 22$, resulting in a loss of 24 k dollar because of loss of production.

Next, the agent-based simulation model is used to support the choice on which response is the most appropriate, given the many degrees of freedom. COC is set to the planned recipe and not considered as a variable to better illustrate the approach with continuous variables. The Nelder-Mead simplex algorithm [14] is used to search for optimal settings for $\bar{x}$ over a number of iterations [23]. After 20 iterations (each time requiring a new run of the simulation), little change occurs and the population of possible solutions is homogeneous, so the algorithm is stopped. The best outcome of the optimization prevents a shut-down of the refinery by buying crudes from an emergency supplier to make up for the delayed ship and by slightly reducing the throughput. This way the loss caused by the disruption is reduced by 14.7k dollar (not keeping in mind penalties to be paid by the shipper for delays).

Analyze the results

These experiments show that the agent-based model can be used as a decision support tool and to help decide which response to a disturbance is appropriate and it does a good job finding good values for the many degrees of freedom. Determining the right combination of options (e.g., a switch of operational configuration and emergency procurement for the crudes used for that recipe instead of the ones in the delayed tanker) is difficult, and becomes extremely hard especially when responses at different times are allowed (e.g., a small emergency procurement now, switching recipe a few days later, switching back when the delayed ship arrives, etc) and when responses to new disturbances (e.g., the long-term effects caused by not always following planned operation) are also included. The model can be used to evaluate these options and come with a recommendation for a response.

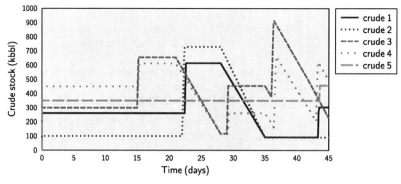

(a) Crude stocks under normal operation.

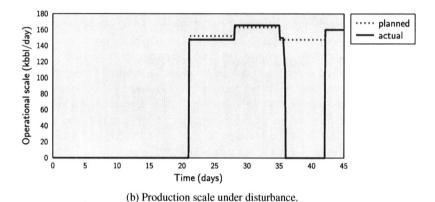

(b) Production scale under disturbance.

Figure 2.6: Results from the oil refinery supply chain case study.

2.5 Conclusions and future research

Based on a wide range of cases studies, two of which were presented here, it is shown that models can successfully be developed using the "model factory" and that these models can be used to support decision makers in the area of socio-technical infrastructure systems. Through a shared ontology (which acts as an interface between model components, models and modelers), a library of building blocks and a set of procedures, models for Intelligent Infrastructures can be built re-using (parts of) existing models. Models built with this framework have already successfully been designed and implemented for various infrastructures, including transport, energy and industrial networks to support problems owners in those domains, as was demonstrated in Section 2.3.

In general, the agent paradigm is applicable for (conceptual) modeling of complex socio-technical systems if the following conditions are satisfied [22]:

- The problem has a distributed character;

- The subsystems (agents) operate in a highly dynamic environment;

- Subsystems (agents) have to interact in a flexible way, where flexibility means reactivity, pro-activeness, cooperativeness, and social ability.

The framework presented here supports modelers who are dealing with systems that match these criteria in an adequate and satisfactory way.

The models built with the "model factory" are inspired by real problems from the infrastructure domain. An equation-based model that was applied in the detailed benchmarking study to compare with the agent-based oil refinery supply chain model presented here, has been used in an actual oil refinery in Singapore. The freight hub case is a real policy case in the Queensland region in Australia. As such, the modeling approach presented here meets actual demand from problem owners.

The approach is designed, from the start, to deal with related problems in various infrastructures, and moreover to connect different infrastructures. It is an ongoing development through ongoing use. New people are using the approach for new projects and as such contribute to the shared framework. This is one of the key strengths of the approach: the more it is used, the more can be re-used.

Important directions for future research include analysis of the development of the framework over time to deduct lessons learnt from model development. Furthermore, a benchmarking study in a new application domain should be conducted to strengthen the conclusions on the advantages of the approach and add to verification and validation of the building blocks in the framework. The use of the models as decision support tools has to be expanded to include the optimization of different (conflicting) target functions by agents and to study how agents cooperate. Visualization of simulation results and of the interaction between the agents plays an important role in this, too. Finally, the inclusions of more detailed simulation of technical systems (to take the place of the input-output definition as is used in the OperationalConfigurations at the moment) should be studied. The ontology can serve as an interface between the agent-based model (describing the behavior of the actors) and the outcomes of specific simulations of the physical system.

Hereby the reader is invited to start thinking about problems in the infrastructure domain from a socio-technical and agent-based perspective and to map the system's elements onto the ontology presented here, so that the "model factory" can be used to effectively build better models.

References

1. Y. M. Bontekoning, C. Macharis, and J. J. Trip. Is a new applied transportation research field emerging? – A review of intermodal rail-truck freight transport literature. *Transportation Research Part A: Policy and Practice*, 38(1):1–34, January 2004.
2. E. J. L. Chappin, G. P. J. Dijkema, K. H. van Dam, and Z. Lukszo. Modeling strategic and operational decision-making – an agent-based model of electricity producers. In *Proceedings of the 21st annual European Simulation and Modelling Conference*, St. Julian's, Malta, October 2007.

3. F. Dignum, V. Dignum, and C. M. Jonker. Towards agents for policy making. In *Proceedings of the 9th International Workshop on Multi-Agent-Based Simulation*, Estoril, Portugal, May 2008.

4. J. H. Gennari, M. A. Musen, R. W. Fergerson, W. E. Grosso, M. Crubezy, H. Eriksson, N. F. Noy, and S. W. Tu. The evolution of Protege: An environment for knowledge-based systems development. *International Journal of Human-Computer Studies*, 58(1):89–123, January 2003.

5. T. R. Gruber. A translation approach to portable ontology specifications. *Knowledge Acquisition*, 5(2):199–220, June 1993.

6. R. J. Hansman, C. Magee, R. De Neufville, R. Robins, and D. Roos. Research agenda for an integrated approach to infrastructure planning, design, and management. *International Journal of Critical Infrastructures*, 2(2/3):146–159, March 2006.

7. B-M. Hodge, S. Aydogan-Cremaschi, G. Blau, J. Pekny, and G. Reklaitis. A prototype agent-based modeling approach for energy systems analysis. In B. Braunschweig and X. Joulia, editors, *Proceedings of the European Symposium on Computer Aided Process Engineering ESCAPE 18*, volume 25 of *Computer-Aided Chemical Engineering*, pages 1071–1076, Lyon, France, June 2008. Elsevier.

8. T. P. Hughes. The evolution of large technological systems. In W. E. Bijker, T. Hughes, and T. J. Pinch, editors, *The Social Construction of Technological Systems. New Directions in the Sociology and History of Technology*, pages 51–82. MIT Press, Cambridge, Massachusetts, March 1987.

9. N. R. Jennings. On agent based software engineering. *Artificial Intelligence*, 117(2):277–296, March 2000.

10. N. Julka, I. Karimi, and R. Srinivasan. Agent-based supply chain management – 2: A refinery application. *Computers and Chemical Engineering*, 26(12):1771–1781, December 2002.

11. J. Keirstead. SynCity – An integrated framework for modelling urban energy systems. Presentation at the International Symposium on Urban Energy and Carbon Management, Asian Institute of Technology, Pathumthani, Thailand, February 2008.

12. Z. Lukszo, K. H. van Dam, M. P. C. Weijnen, and G. P. J. Dijkema. Agent-based models for crisis management. In H. Bouwman, R. Bons, M. Hoogewegen, M. Janssen, and H. Pronk, editors, *Let a Thousand Flowers Bloom, Essays in commemoration of prof.dr. René Wagenaar*, pages 281–299. IOS Press, 2008.

13. Meyrick and Associates. National intermodal terminal study, 2006.

14. J. A. Nelder and R. Mead. A simplex method for function minimization. *Computer Journal*, 7(4):308–313, January 1965.

15. I. Nikolic, G. P. J. Dijkema, and K. H. van Dam. Understanding and shaping the evolution of sustainable large-scale socio-technical systems – towards a framework for action oriented industrial ecology. In M. Ruth and B. Davidsdottir, editors, *The Dynamics of Regions and Networks in Industrial Ecosystems*. Edward Elgar, July 2009.

16. M. A. Ortega-Vazquez and D. S. Kirschen. Assessment of generation expansion mechanisms using multi-agent systems. In *Proceedings of the IEEE Power and Energy Society General Meeting: Conversion and Delivery of Electrical Energy in the 21st Century*, pages 1–7, Baltimore, Maryland, April 2008.

17. M. Ottens, M. Franssen, P. Kroes, and I. van de Poel. Modelling infrastructures as socio-technical systems. *International Journal of Critical Infrastructures*, 2(2/3):133–145, March 2006.

18. S. S. Pitty, W. Li, A. Adhitya, R. Srinivasan, and I. A. Karimi. Decision support for integrated refinery supply chains. 1. Dynamic simulation. *Computers and Chemical Engineering*, 32(11):2767–2786, November 2008.

19. A. Sirikijpanichkul and L. Ferreira. Modelling intermodal freight hub location decisions. In *Proceedings of the 2006 IEEE International Conference on Systems, Man, and Cybernetics*, pages 890–895, Taipei, Taiwan, October 2006.

20. A. Sirikijpanichkul, K. H. van Dam, L. Ferreira, and Z. Lukszo. Optimizing the location of intermodal freight hubs: An overview of the agent based modelling approach. *Journal of Transportation Systems Engineering and Information Technology*, 7(4):71–81, August 2007.

21. K. H. van Dam, A. Adhitya, R. Srinivasan, and Z. Lukszo. Critical evaluation of paradigms for modelling integrated supply chains. *Journal of Computers and Chemical Engineering*, 2009.

22. K. H. van Dam and Z. Lukszo. Modelling energy and transport infrastructures as a multi-agent system using a generic ontology. In *Proceedings of the 2006 IEEE International Conference on Systems, Man, and Cybernetics*, pages 890–895, Taipei, Taiwan, October 2006.

23. K. H. van Dam, Z. Lukszo, and R. Srinivasan. Abnormal situation management in a refinery supply chain supported by an agent-based simulation model. In *Proceedings of the 10th International Symposium on Process Systems Engineering*, Bahia, Brazil, August 2009.

24. K. H. van Dam, J. A. Ottjes, G. Lodewijks, Z. Verwater-Lukszo, and R. Wagenaar. Intelligent Infrastructures: Distributed intelligence in transport system control – an illustrative example. In *Proceedings of the 2004 IEEE International Conference on Systems, Man and Cybernetics*, pages 4650–4654, The Hague, The Netherlands, October 2004.

25. M. P. C. Weijnen and I. Bouwmans. Innovation in networked infrastructures: Coping with complexity. *International Journal of Critical Infrastructures*, 2(2/3):121–132, March 2006.

Part II
Electricity Infrastructures

Chapter 3
Prevention of Emergency Voltage Collapses in Electric Power Networks using Hybrid Predictive Control

S. Leirens and R.R. Negenborn

Abstract The reliable operation of electricity transport and distribution networks plays a crucial role in modern societies. However, too often, when a fault occurs in electricity networks, such as a transmission line drop, loss of generation, or any other important failure, voltages start to decay, potentially leading to complete blackouts with dramatic consequences. Thus, techniques are required that improve the power grid operation in case of emergencies. In this chapter, to achieve this aim, an approach is presented that uses an adaptive predictive control scheme. Electric power transmission networks are hereby considered as large-scale interconnected dynamical systems. First, voltage instability issues are illustrated on a 9-bus benchmark system. Then, the details of the proposed approach are discussed: the power network modeling and the construction of a hybrid prediction model (i.e., including both continuous and discrete dynamics), and the formulation and the resolution of the adaptive predictive control problem. In simulation studies on the 9-bus benchmark system the performance of the proposed approach is illustrated in various emergency voltage control cases.

S. Leirens
Universidad de Los Andes, Departamento de Ingeniería Eléctrica y Electrónica, Bogotá, Colombia,
e-mail: sleirens@uniandes.edu.co

R.R. Negenborn
Delft University of Technology, Delft Center for Systems and Control, Delft, The Netherlands,
e-mail: r.r.negenborn@tudelft.nl

R.R. Negenborn et al. (eds.), *Intelligent Infrastructures*, Intelligent Systems, Control and
Automation: Science and Engineering 42, DOI 10.1007/978-90-481-3598-1_3,
© Springer Science+Business Media B.V. 2010

3.1 Introduction

The human and economic consequences of power outages have shown that the reliable operation of electricity transport and distribution networks plays a crucial role in modern societies, as illustrated by recent problems in the United States and Canada, Europe, and Latin America [52–54]. The safe operation of the electricity network has to be carried out both under regular operating conditions, and also when the system is operating close to its limits. A great part of current research efforts is devoted to explore new ways to improve the power grid operation in terms of efficiency, reliability, and robustness, while satisfying constraints on economy and environment.

This is necessary, since electric power networks are experiencing rapid and important changes, in particular in the way they are operated and managed:

- The environmental opposition against the expansion of the physical power transportation infrastructure is now stronger than before, and the consumption of electricity increases in areas that are already heavily loaded [16];
- The development of interconnections between countries (e.g., in Europe) leads to very complex large-scale dynamical systems [24];
- New economic regulations due to the growth of energy markets induce unpredicted power flows and demand for a relaxation of security margins [10].
- The number of actors in the network increases as the amount of distributed or embedded generation increases, e.g., as industrial suppliers and households start to feed electricity into the network [20].

Due to the increased complexity arising from these aspects, the consequences of failures, such as transmission line drops, losses of generation, or any other important failures in the system, become more significant. The conventional control schemes of the network operators have to be revised, renewed, or even replaced by control schemes that can manage the electric power network of the future.

3.1.1 Power systems issues

In general, a power system is a strongly nonlinear system that have to be controlled over a wide range of operating conditions, possibly far from equilibrium points, in particular in emergency situations such as voltage instability [55]. The behavior of power systems is characterized by so-called hybrid behavior [2], i.e., behavior resulting from the interaction between continuous and discrete dynamics. Continuous dynamics of power systems are mainly driven by components such as generators and loads, and are usually represented by systems of differential-algebraic equations (DAEs). The discrete behavior arises from the nature of connected elements, such as capacitor banks, line breakers, and limiters in voltage regulators, or the way in which such elements are controlled, e.g., via discrete shedding of load, *on* or *off* switching of generation units, and on-load tap changing of ratio control in transformers. Moreover, power networks typically span a wide range of time scales and large geographical areas. As a consequence, electric power networks are modeled as large-scale nonlinear hybrid systems.

Such complex systems are generally controlled in a hierarchical way and control takes place at different layers with a decomposition based on space and time divisions [5]. At the lowest layer, the controllers act directly on the actuators of the physical system with fast and localized control, e.g., single-input single-output controlled systems, such as automatic voltage regulators of synchronous machines. At the higher layers, supervisory controllers determine set-points for lower control layers in order to obtain coordination. Model-based approaches and global control for power systems become conceivable at these layers, since wide-area and phasor measurements are available [51], and utilities increasingly demand wide-area control, protection, and optimization systems [22].

Emergency voltage control deals with the problem of voltage instability leading to so-called *collapses* after disturbances [49, 50, 55]. The current protection schemes against voltage collapses are generally rule-based and consist of load shedding and reactive power compensation associated with strong operator training. However, the nonlinear behavior of power systems makes these rules strongly dependent on the operating conditions. After a disturbance such as breaking of a transmission line, the generation and transmission system may not have sufficient capacity to provide the loads with power. Voltage instability may then occur, in the worst case leading to total network blackouts. Voltage collapses are not only associated with weak systems such as power networks with low transfer capability, but are also a source of concern in highly developed networks that are heavily loaded. An illustration of this phenomenon is presented in Section 3.2.

The principal objective of the control system of an electricity network is to minimize the effects of the possible disturbances on the quality of the supplied energy [55], i.e.,

- voltages must remain within an acceptable range, e.g., within 5–10 % of the nominal value;

- requirements on physical limitations of interconnected elements must be satisfied,

while minimizing power losses, and achieving economic objectives, e.g., by minimizing the use of load shedding, since customers are then disconnected from the system, suffering great economical losses.

In the current power network operation, emergency voltage control is typically performed by human operators that are *in the loop*. The operators monitor the grid (typically power flows, voltage magnitudes, and angles) and take decisions following pre-established procedures. Decisions, such as on the *on* or *off* switching of equipment and the provision of set-points to lower control layers are based on offline studies, extensive experience, system conditions observed via telemetry, heuristics, knowledge bases, and state-estimator outputs. The control problem is then inherently complex and in general there is no possibility to rapidly change the operating conditions in an online and coordinated manner since the grid operation relies mainly on the operators (*dispatchers*). New online control systems become increasingly necessary to face the recent changes in electricity networks and to achieve improved performance.

3.1.2 Power network modeling and control

Power networks are large-scale interconnected dynamical systems and their physical modeling relies on analytic models of individual components and knowledge of the network structure. General models at component level can be expressed in terms of the nonlinear dynamics of the local states, and input and output variables. These input and output variables are subject to algebraic constraints defined by Kirchhoff's laws. Hence, the combination of the dynamic models of the individual components and these algebraic constraints takes the form of a system of nonlinear DAEs [19].

Electric power systems consist of two types of fundamental components: single-port equipment components and two-port transmission components. Single-port equipment components are components such as synchronous machines and loads, including their primary controllers (governors, exciters, switched-shunt capacitors, and reactors). Two-port transmission components are components such as high-voltage transmission lines that connect buses (nodes in the network) and their primary controllers (series capacitors, phase-shifting transformers, and on-load tap changers) [19]. Single-port components are connected to other single-port components via two-port components. Each of these components is hereby described by its constitutive relations, the complexity degree of which can be very high. E.g., obtaining dynamic models of synchronous machines is an extremely laborious process [19, 25]. The equations of the component models can be nonlinear, hybrid, differential, and non-smooth. Simplifications are generally made with respect to the phenomena of interest, e.g., in the case of quasi-static models for voltage stability studies, for which fast dynamics, such as electromechanical dynamics, are neglected and frequency is assumed to be constant [55].

Despite the fact that models consisting of systems of DAEs attract much interest, due to their importance as models for a large class of dynamical processes (e.g., in mechanics, robotics, and chemical engineering), such models present intrinsic numerical difficulties. Systems of DAEs are more difficult to solve than systems of ordinary differential equations because of issues related to their index [17] and the determination of consistent initial conditions [8, 45].

To deal with control and optimization of systems modeled by DAEs, several strategies have been proposed, such as simultaneous strategies [6, 9, 21], multiple shooting strategies [11, 12], and direct search methods [43]. The particular structure of power network models (see Section 3.3.4) can especially be used advantageously to set up tractable models for model-based control approaches, such as model predictive control (MPC) [31]. MPC is a control methodology that has been successfully applied to a wide range of control problems, including problems in industrial processes [40], steam networks [32], residential energy resources [42], greenhouse systems [47], drug delivery systems [7], and water systems [44].

In MPC, a control action is obtained at discrete control sample steps by solving an optimization problem that minimizes an objective function over a finite receding horizon subject to the equations of a prediction model and operational constraints. The main advantages of MPC are:

- the explicit way of integrating present and future soft as well as hard constraints, such as operational constraints on bus voltages, line currents, generator excitations, and bounded control inputs and states, such as transformer ratios and reactive compensation units;

- the ease of integrating forecasts to anticipate events in the infrastructure, in the case that an event is known in advance to occur, such as an overloaded line disconnection, loss of generation, consumption changes, and maintenance actions;

- the ability to update online the prediction model to manage fast changing conditions in an adaptive way;

- the straightforward design procedure (see Section 3.4).

Given a model of the system to be controlled, hard constraints can be incorporated directly as inequalities and soft constraints can also be accounted for in the objective function by using penalties for violations. Fundamental trade-offs between efficiency and priorities are handled through the minimization of a cost function to fulfill economic objectives and minimize power losses.

As stated above, power systems belong to the class of hybrid systems. Conventional methods, i.e., either purely continuous or purely discrete methods, cannot be used for control of systems with both continuous and discrete dynamics. The control approach that is presented in this chapter is called a hybrid approach in the sense that it deal with both continuous and discrete dynamics at the same time in an integrated way.

MPC has first been applied to a voltage control problem in [26], in which a coordinated system protection scheme against voltage collapse using search and predictive control is presented. To be tractable, this approach uses a single-step linearized prediction model, that is, a constant control input is assumed over the entire prediction interval. This approach cannot handle discrete behavior. It is a small signal approach, in which all control variables are discretized leading to a purely combinatorial optimization problem.

In [18], the mixed-logical dynamical framework [4] is used to model the hybrid behavior of a 4-bus power system. In this approach the continuous state-input space is divided into several polytopes; a different affine model may be used for each combination of the values of the discrete variables. The gridding of the continuous state-input space has to be tight enough for good accuracy and leads to a great number of affine models. A trade-off has to be made between combinatorial complexity and required accuracy of the model. Using mixed-logical dynamical prediction models for MPC results in solving at each control step a mixed-integer programming problem [27, 30, 37].

The approach that is presented below is based on an adaptive hybrid MPC scheme. The power network equations are symbolically linearized and the hybrid prediction model is updated online at the current operating point. Since the controller implements a model that is adapted to the operating point, the MPC-based hybrid strategy involves a reduced combinatorial complexity to the one induced by the discrete variables only, without partitioning the state-input space. A simple and suitable mixed optimization algorithm is proposed to solve the hybrid predictive control problem.

3.1.3 Outline

In Section 3.2 an illustration of the voltage instability phenomenon is presented using a 9-bus benchmark system, and current operating schemes are discussed. Section 3.3 presents the steps to build an accurate prediction model which captures the dynamics involved in voltage instability issues. The basic concepts of predictive control and the special features of the hybrid approach that is presented here as well as the solution of the hybrid MPC problem are presented in Section 3.4. In Section 3.5, the results of this approach applied to the 9-bus benchmark system are presented. Section 3.6 gives concluding remarks and directions for future research.

3.2 Power network operation

This section starts with illustrating voltage instability of power networks using a voltage collapse in a 9-bus benchmark system. This example shows that despite a set of corrective actions, the voltages collapse, mainly due to lack of coordination in the operation of the network. Then, operating schemes that could stabilize the network are presented.

3.2.1 Emergency scenario

A major source of power outages is voltage instability. This dynamic phenomenon arises when the individual controllers of the loads attempt to restore the consumed power beyond the capacity (physical limit) of the production-transport system [55]. This may occur following the outage of one or more components in the system, such that the load demand cannot be satisfied with a physically sustainable profile of the voltage plan, typically caused by voltage levels becoming very low. The reduced capacity of the network together with the requested load consumption requires coordinated corrective actions to avoid that involved dynamics drive the system into undesired or unacceptable states.

Consider the power system shown in Figure 3.1. This system represents a transmission network in which each generator is a simplified representation of an adjacent subnetwork. This benchmark system reproduces most of the phenomena of interest in power network dynamics, including in particular voltage instability and collapses. It is composed of the following elements and control devices:

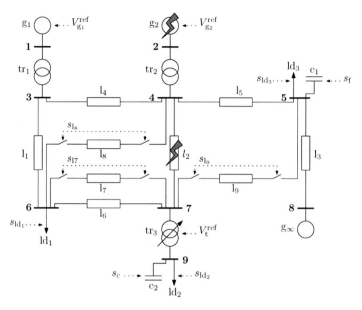

Figure 3.1: Overview of the 9-bus power network benchmark system.

- Generator g_∞ represents a large surrounding network and forces the voltage at bus 8 to stay constant.
- Generators g_1 and g_2 represent simplified models of adjacent subnetworks and are equipped with automatic voltage regulators and overexcitation limiters; they accept as control input the voltage references $V_{g_1}^{ref} \in [0.9, 1.1]$ and $V_{g_2}^{ref} \in [0.9, 1.1]$, respectively, for the automatic voltage regulators;
- Nine transmission lines, l_1–l_9, interconnect the components of the network. Three of them, l_7, l_8, and l_9, are equipped with controllable line breakers; these controllable lines have as control inputs s_{l_7}, s_{l_8}, and s_{l_9}, respectively, that can take on the values 1 (connected) or 0 (disconnected).
- A flexible AC transmission system c_1 is present for reactive power control; it is modeled as a continuously controlled capacitor with as control input the amount $s_f \in [0, 2]$ of reactive power injected into the network.
- A capacitor bank c_2 is present for additional reactive power compensation, using several separate units that can be connected or disconnected, with as control input the number $s_c \in \{0, 1, 2, 3\}$ of capacitor units connected to the network.
- The transformers tr_1 and tr_2 transform voltage magnitudes at fixed ratios.
- The transformer tr_3 is a transformer equipped with an on-load tap changer, which has as control input the voltage reference $V_t^{ref} \in [0.8, 1.2]$.
- The loads ld_1, ld_2, and ld_3 are dynamic loads representing groups of consumers with a voltage dependent behavior, i.e., active and reactive power recovery [23]. The loads can be shed, i.e., disconnected, using the control inputs s_{ld_1}, s_{ld_2}, and s_{ld_3}, respectively, that take on the value 0 (no power shedding) or 1 (5% power shedding).

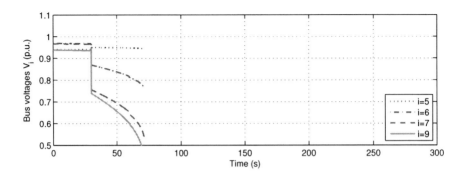

Figure 3.2: Voltage collapse after the loss of transmission line l_2.

The equations describing the elements of this network take on the form of a system of hybrid DAEs, in which the continuous dynamics arise from the loads (cf. Section 3.3.1.2), and discrete events occur when:

- a generator, g_1 or g_2, reaches the excitation limit (cf. Section 3.3.1);
- a line is connected or disconnected (l_7–l_9);
- a unit of capacitor banks c_2 is connected or disconnected;
- consumers are connected or disconnected from the grid (load shedding);
- the on-load tap changer of the transformer tr_3 changes the ratio tap by tap.

An event can be controllable or not. Line, capacitor and load switching events are controllable by inputs, whereas reaching the excitation limit or changing taps are uncontrollable events. A complete description of the parameters of the different models and numerical values can be found in [28].

Now consider a fault in line l_2 at $t = 30$ s. The breakers at both sides of the transmission line open and cause the transmission line to be disconnected. Figure 3.2 shows the evolution of the voltages when no control is employed. As can be seen, directly after the fault, voltages start to drop. Nothing is done to correct the evolution and the bus voltages collapse quickly.

3.2.2 Stabilizing operation

The purpose of emergency voltage control is to supply a set of corrective actions to apply to the system following an outage while fulfilling physical and operational constraints. Usually, the control objectives are specified as follows:

- to achieve a steady-state point of operation allowing the voltage plane to stay between 0.9 and 1.1 p.u, that is close to the nominal values, to fulfill requirements of safety and quality of energy;
- to optimize the use of control means to fulfill economic objectives, i.e., the use of reactive power compensation has to be preferred to load shedding, which is the ultimate control action since it disconnects consumers from the grid.

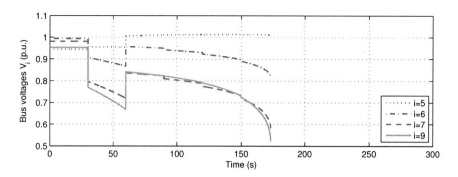

Figure 3.3: Voltage collapse under manual control after the loss of transmission line l_2.

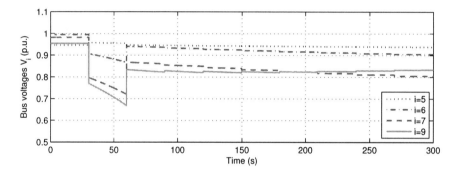

Figure 3.4: Stabilization of voltages after the loss of transmission line l_2.

The effects of corrective actions are illustrated with various cases. Consider the same fault as before, i.e., on line l_2. The breakers at both sides of the transmission line l_2 open at $t = 30$ s. At time $t = 60$ s, which is only 30 s after the disturbance, load shedding of 10% of all the consumption and maximum reactive power compensation (FACTS c_1 and capacitor bank c_2) are employed via *manual control*. As shown in Figure 3.3, this manual control is not sufficient to prevent a voltage collapse. The bus voltages start to collapse at the moment that the power consumption reaches the maximum transfer capability of the system (at $t \approx 100$ s).

Figures 3.4 and 3.5 illustrate the stabilization of the network dynamics in two different emergency situations. In Figure 3.4, 30 s after the loss of transmission line l_2, the topology of the network is modified with the connection of lines l_7 and l_8. The distribution of power flows is modified by the new configuration of the network and stability is recovered, but some buses have an unacceptably low voltage (V_7 is around 0.8 p.u.). Figure 3.5 shows the stabilization of the network dynamics after the loss of generator g_2. In this case, stability is recovered by using reactive power compensation (c_1 and c_2), modification of the reference voltage of the on-load tap changer of tr_3 and load shedding (ld_2 and ld_3).

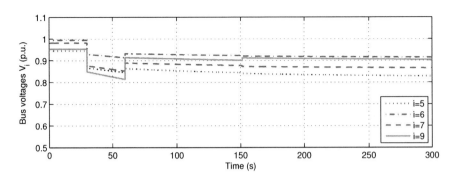

Figure 3.5: Stabilization of voltages after the loss of generator g_2.

In Section 3.3 and 3.4, an MPC-based method is proposed to determine the optimal corrective autonomously and automatically.

3.3 Hybrid dynamical models of power networks

The need for more accurate operation techniques as well as the available computation capabilities of modern computers call for new control methods and algorithms. The dynamics of the network hereby have to be taken into account as well as possible. This section presents the nonlinear and hybrid modeling aspects of electric power networks in the context of emergency voltage control.

3.3.1 Generation and consumption

The dynamics involved in voltage stability issues are said to be slow (the time constants are about 30–60 s) compared with electromechanical dynamics involved in frequency issues (transients lasting for a few seconds). Therefore, quasi-static or quasi-steady state models are commonly used in voltage stability studies [55]. In quasi-steady state models it is assumed that the frequency of the power network is constant, that is, fast dynamics are neglected and replaced with equilibrium equations. The in practice present three phases are assumed to be balanced reducing the models to equivalent one-line diagrams. Since the frequency is assumed to be constant, voltage, current, and state variables can be represented by so-called *phasors* [25], i.e., complex numbers that represent sinusoids.

In the remaining of this chapter, the following notations are used:

- Phasors are shown as capital letters with an overline, e.g., $\bar{V} = v_x + jv_y, \bar{I} = i_x + ji_y$.
- The magnitude of a phasor is shown by the capital letter of that phasor without the overline, e.g., $V = \sqrt{v_x^2 + v_y^2}$, $I = \sqrt{i_x^2 + i_y^2}$.
- Lowercase bold letters, e.g., $\mathbf{x}, \mathbf{y}$, correspond to column vectors. Superscript T denotes transpose. Therefore row vectors are denoted by $\mathbf{x}^\mathrm{T}, \mathbf{y}^\mathrm{T}$.

- A collection of phasors in a column vector is represented by a capital bold letter with an overline, e.g., $\bar{\mathbf{I}}$.

- Matrices are denoted by bold capital letters, e.g., $\mathbf{A}$.

- Time derivatives of variables are indicated with a dot, e.g., $\dot{x}$.

3.3.1.1 Generators

Generators are modeled using synchronous machine equations [25, 55]. Almost all variables and parameters are expressed in the per unit (p.u.) system [25], that is with respect to base quantities, i.e., the per unit system used for the stator is based on the three-phase nominal power and the voltage values of the machine.

In the following, indices d and q refer to direct and quadrature machine axis, respectively, and arrise from the Park transformation [25]. This transformation consists of replacing the three armature windings in a generator by three fictitious windings labelled d, q, and o, where the d and the q axis rotate together with the machine rotor. In balanced conditions, the o winding does not play any role and therefore it will not be considered here. Notice that also magnetic saturation is neglected.

Under the above mentioned assumptions, the generator is described by the differential equations [25, 55]:

$$\dot{\delta} = \omega - \omega_0 \tag{3.1}$$

$$\dot{\omega} = -\frac{D}{2H}\omega + \frac{\omega_0}{2H}\left(P_{\mathrm{m}} - P_{\mathrm{g}}\right) \tag{3.2}$$

$$\dot{E}_q' = \frac{-E_q' + E_{\mathrm{f}} - \left(X_d - X_d'\right)\left(i_{xg}\sin\delta - i_{yg}\cos\delta\right)}{T_{do}'}, \tag{3.3}$$

where for this generator, δ is the rotor angle (in rad), ω is the angular frequency (in rad/s), E_q' is the electromotive force (emf) behind the transient reactance (in p.u.), ω_0 is the nominal angular frequency (in rad/s), D is the damping coefficient (in p.u.), H is the inertia constant (in s), P_{m} is mechanical power (in p.u.) provided to the generator, and P_{g} is the active power produced by the generator (in p.u.), E_{f} is the exciter (or field) voltage (in p.u.), X_d is the direct-axis synchronous reactance (in p.u.), X_d' is the direct-axis transient reactance (in p.u.), $\bar{I}_{\mathrm{g}} = i_{xg} + ji_{yg}$ is the armature current phasor (in p.u.), and T_{do}' is the open-circuit transient time constant (in p.u.).

The active power produced by the generator is given by:

$$P_{\mathrm{g}} = v_{xg}i_{xg} + v_{yg}i_{yg}, \tag{3.4}$$

where $\bar{V}_{\mathrm{g}} = v_{xg} + jv_{yg}$ is the armature voltage phasor (in p.u.), and where the real and imaginary parts of the armature current are:

$$i_{xg} = \frac{\sin 2\delta}{2}\left(\frac{1}{X_q} - \frac{1}{X_d'}\right)\left(v_{xg} - E_q'\cos\delta\right)$$

$$-\left(\frac{\cos^2\delta}{X_q}+\frac{\sin^2\delta}{X_d'}\right)\left(v_{yg}-E_q'\sin\delta\right) \tag{3.5}$$

$$i_{yg}=\left(\frac{\sin^2\delta}{X_q}+\frac{\cos^2\delta}{X_d'}\right)\left(v_{xg}-E_q'\cos\delta\right)$$

$$+\frac{\sin 2\delta}{2}\left(\frac{1}{X_d'}-\frac{1}{X_q}\right)\left(v_{yg}-E_q'\sin\delta\right), \tag{3.6}$$

where X_q is the quadrature-axis synchronous reactance (in p.u.). The automatic voltage regulator that is considered is a proportional controller and the overexcitation limiter is modeled as a saturation element:

$$E_f=\min\left(G\left(V_g^{\text{ref}}-V_g\right),E_f^{\text{lim}}\right), \tag{3.7}$$

where G is the steady-state open-loop gain of the automatic voltage regulator (in p.u./p.u.), V_g^{ref} is the reference voltage of the automatic voltage regulator (in p.u.), and E_f^{lim} is the excitation limit of the overexcitation limiter (in p.u.).

The work presented in this chapter focuses on load dynamics since they are the *driving force* of voltage instability. Therefore a steady-state approximation of the generator equations (3.1)–(3.3) is used together with the algebraic equations (3.4)–(3.7).

Then the set of generator models of the network is grouped into the following nonlinear algebraic equation:

$$\mathbf{h}(\bar{\mathbf{V}}_g,\bar{\mathbf{I}}_g,\mathbf{V}_g^{\text{ref}})=\mathbf{0}, \tag{3.8}$$

where $\bar{\mathbf{V}}_g$ and $\bar{\mathbf{I}}_g$ are vectors of the voltage and the injected current phasors at the buses that connect the generators to the grid, respectively, and $\mathbf{V}_g^{\text{ref}}$ is a vector with reference voltages for the automatic voltage regulators.

3.3.1.2 Loads

Load is a common term for aggregates of many different devices that are mainly voltage dependent. Load dynamics are considered from the point of view of power recovery, i.e., after a voltage drop, the internal control systems of the load attempt to recover the consumed power at its nominal level [23]. Load dynamics are described by a smooth nonlinear differential equation:

$$T_P\dot{x}_P+x_P=P_s(V_{\text{ld}})+P_t(V_{\text{ld}}) \tag{3.9}$$

$$P_{\text{ld}}=(1-s_{\text{ld}}\sigma)(x_P+P_t(V_{\text{ld}})), \tag{3.10}$$

where x_P is a continuous state variable, V_{ld} is the load voltage magnitude, $P_s(V_{\text{ld}})=P_0V_{\text{ld}}^{\alpha_s}$ and $P_t(V_{\text{ld}})=P_0V_{\text{ld}}^{\alpha_t}$ are the steady-state and transient voltage dependencies, respectively, P_{ld} is the active power which is consumed by the load and T_P is the active power recovery time constant. A similar model is used for the reactive load power considering the following variables: x_Q, $Q_s(V_{ld}) = Q_0 V_{ld}^{\beta_s}$, $Q_t(V_{ld}) = Q_0 V_{ld}^{\beta_t}$, Q_{ld} and T_Q. Constant σ represents a constant load shedding step, and s_{ld} is a discrete control variable that takes its values in a bounded discrete set. The internal state of the load is defined as $\mathbf{x}_{ld} = \begin{bmatrix} x_P & x_Q \end{bmatrix}^T$. The set of load models of the network is grouped into the following nonlinear state-space equations:

$$\dot{\mathbf{x}} = \mathbf{f}(\mathbf{x}, \bar{\mathbf{V}}_{ld}) \tag{3.11}$$

$$\bar{\mathbf{I}}_{ld} = \mathbf{g}(\mathbf{x}, \bar{\mathbf{V}}_{ld}, \mathbf{s}_{ld}), \tag{3.12}$$

where $\mathbf{x}$ is the vector of internal states, $\bar{\mathbf{V}}_{ld}$ and $\bar{\mathbf{I}}_{ld}$ are the vectors of voltage and absorbed current phasors at the buses that connect the loads to the grid, respectively, and $\mathbf{s}_{ld}$ is the vector of load shedding inputs.

3.3.2 Transmission system

The transmission system interconnects the generators to the loads and is composed of transformers, compensation devices such as capacitor banks and FACTS, and transmission lines. The respective models are detailed below.

3.3.2.1 Transformers

Transformers are modeled using a complex impedance $\bar{Z}_t$ in series with an ideal transformer whose ratio is denoted by n_t. Variables $\bar{V}_{t1}$ and $\bar{V}_t$ refer to the primary and secondary voltage transformer phasor, respectively. Similar notations are used for the current phasors $\bar{I}_{t1}$ and $\bar{I}_t$. The transformer is described by linear equations with respect to voltage and current phasors:

$$\bar{V}_t = n_t \bar{V}_{t1} - n_t^2 \bar{Z}_t \bar{I}_t \tag{3.13}$$

$$\bar{I}_{t1} = n_t \bar{I}_t. \tag{3.14}$$

The presence of an on-load tap changer allows the ratio n_t to vary tap by tap within bounds. The sequential behavior of the on-load tap changer can be described by a discrete-time dynamic model:

$$n_t(k+1) = n_t(k) - n_{step} \, \xi(\Delta V(k)), \tag{3.15}$$

where $n_t(k)$ is the bounded transformer ratio at time step k, $\Delta V(k) = V_t(k) - V_t^{ref}(k)$ is the error between the secondary voltage magnitude V_t, i.e., at the output of the transformer, and the bounded reference voltage V_t^{ref} of the on-load tap changer, and n_{step} is the ratio step corresponding to one tap change. A function ξ (being a simplification of the function used in [48]) is used to determine when a tap change is made as follows:

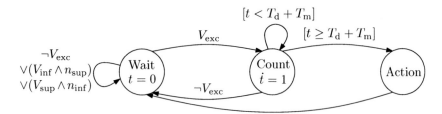

Figure 3.6: Finite-state machine for the on-load tap changer.

$$\xi(\Delta V) = \begin{cases} 1 & \text{if } \Delta V > \gamma \quad \text{for} \quad (T_d + T_m)\,\text{s} \\ -1 & \text{if } \Delta V < -\gamma \quad \text{for} \quad (T_d + T_m)\,\text{s} \\ 0 & \text{otherwise,} \end{cases} \tag{3.16}$$

where γ is the semi-deadband (or tolerance band) centered around V_t^{ref}, T_d is the time delay before a tap change is made, and T_m is the time necessary to perform the actual tap change. Depending on the type of on-load tap changer, T_d can be a constant or depend on ΔV. Since the size of a tap step is quite small (usually in the range of 0.5%–1.5% of the nominal ratio), to simplify the control design, n_t is considered as a bounded continuous variable.

The on-load tap changer used in the simulations presented in Section 3.5.2 is implemented using a finite state machine with three discrete states. Consider the logical variables (i.e., taking on values from the domain {true,false})

$$V_{\text{inf}} \equiv \left[V_t < V_t^{\text{ref}} - \gamma \right] \tag{3.17}$$

$$V_{\text{sup}} \equiv \left[V_t > V_t^{\text{ref}} + \gamma \right] \tag{3.18}$$

$$V_{\text{exc}} \equiv V_{\text{inf}} \lor V_{\text{sup}} \tag{3.19}$$

and

$$n_{\text{inf}} \equiv \left[n_t \leq n_t^{\text{min}} \right] \tag{3.20}$$

$$n_{\text{sup}} \equiv \left[n_t \geq n_t^{\text{max}} \right], \tag{3.21}$$

where the symbol $\lor$ denotes the logical operator OR. The finite-state machine representing the tap changer is shown in Figure 3.6, where t refers to the internal timer variable and symbols $\neg$ and $\land$ refer to logical operators NOT and AND respectively. As long as the output voltage of the transformer stays inside the limits defined by the deadband or as long as a tap change is not possible, the on-load tap changer stays in the state *Wait*. When the voltage exceeds the deadband limits, the on-load tap changer enters in the state *Count*. If the voltage returns inside of the deadband limits before the timer reaches $T_d + T_m$, the state returns to *Wait*. If the internal timer reaches $T_d + T_m$, the state becomes *Action*, a tap change is performed, and the state returns to *Wait*. More details about on-load tap changer dynamics can be found in [48] and [28]. The integration of the on-load tap changer dynamics in the prediction model is addressed in Section 3.3.4.

3.3.2.2 Reactive power compensation devices

Two kinds of reactive power compensation devices are considered: capacitor banks and FACTSs. A Capacitor bank is described by the following linear equation with respect to voltage and current phasors:

$$\bar{I}_c = j p_c s_c \bar{V}_c, \tag{3.22}$$

where a capacitor unit (susceptance) is represented by p_c and the number of connected units s_c can only take its values in a bounded discrete set.

A FACTS is considered here as a continuously varying capacitor and is described by the following linear equation with respect to voltage and current phasors:

$$\bar{I}_f = j p_f s_f \bar{V}_f, \tag{3.23}$$

where the total susceptance p_f is adjusted by a variable s_f that takes its value in a bounded continuous set.

3.3.2.3 Transmission lines

A transmission line is modeled using a complex impedance $\bar{Z}_l$:

$$\Delta \bar{V}_l = s_l \bar{Z}_l \bar{I}_l, \tag{3.24}$$

where $\Delta \bar{V}_l$ is the voltage across the line, i.e., the difference between two bus voltages, and $\bar{I}_l$ is the current through the line. If a line can be connected or disconnected in order to modify the network topology, a boolean control variable s_l is used to represent the state of the line: connected (1) or disconnected (0).

3.3.2.4 Complete transmission system

The transformer, compensation devices, and transmission line equations are linear with respect to voltage and current phasors. The complete transmission system model therefore consists of a system of linear algebraic equations:

$$\bar{\mathbf{w}}_{out} = \mathbf{M} \bar{\mathbf{w}}_{in}, \tag{3.25}$$

where $\bar{\mathbf{w}}_{out}$ and $\bar{\mathbf{w}}_{in}$ are generator and load voltages and currents phasors $\bar{\mathbf{V}}_{g,ld}$ and $\bar{\mathbf{I}}_{g,ld}$, respectively. Notice that the elements of matrix $\mathbf{M}$ depend on the variables $\mathbf{n}_t$, $\mathbf{s}_c$, $\mathbf{s}_f$ and $\mathbf{s}_l$.

3.3.3 Interconnected network

The hybrid nature of power systems is particularly characterized by the presence of two kinds of control inputs: discrete inputs arising from loads (s_{ld}), capacitor banks (s_c) and transmission lines (s_l), and continuous inputs arising from transformers (n_t) and FACTS devices (s_f). In the sequel, the discrete control vector is denoted by $\mathbf{u}_d = [s_{ld}^T \; s_c^T \; s_l^T]^T$ and the continuous control vector is denoted by $\mathbf{u}_c = [\mathbf{n}_t^T \; s_f^T]^T$.

Discrete disturbances, such as transmission line drop and generator loss, together with the state of the generators (maximum excitation limitation) and the discrete control inputs define the discrete operating mode i of the network. The general power network model, defined by (3.8), (3.11)–(3.12), and (3.25), takes the form of a system of nonlinear and hybrid DAEs in each mode i:

$$\dot{\mathbf{x}} = \varphi_i(\mathbf{x}, \mathbf{y}, \mathbf{u}_c) \tag{3.26}$$

$$0 = \psi_i(\mathbf{x}, \mathbf{y}, \mathbf{u}_c), \tag{3.27}$$

where $\mathbf{x}$ is the load state vector and the output vector $\mathbf{y}$ typically includes bus voltage magnitudes. This model is useful for performing simulations, as will be illustrated in Section 3.5. Nevertheless, simulating such a model requires dedicated algorithms, such as DASSL for solving the system of hybrid DAEs [8, 45], and particular special attention to the way in which the interactions between continuous and discrete dynamics are dealt with [13, 15].

3.3.4 Symbolic off-equilibrium linearization

Power systems are strongly nonlinear and to predict the system evolution, a feasible approach is to use locally a linear or, more generally, affine model. To obtain an accurate prediction model, (3.26)–(3.27) are symbolically linearized with respect to the continuous variables, i.e., the load state vector $\mathbf{x}$ and the continuous control input vector $\mathbf{u}_c$. Note that

- the operating point $(\mathbf{x}_0, \mathbf{u}_{c0})$ is not necessarily an equilibrium point. This kind of linearization is said to be *off-equilibrium*;
- the operating point $(\mathbf{x}_0, \mathbf{u}_{c0})$ and the discrete mode i are symbolic parameters of the linearized model.

The evolution of the network can be predicted with accuracy using the linearized model, since the model that is used to compute the prediction is adapted based on the current operating point. Moreover, this modeling framework allows to handle the hybrid aspects explicitly. In the sequel, the operating point is assumed to be available or estimated. In practice, the operating point data are not available directly. All the measurement data are computed by an estimator and research efforts are made to improve methods and algorithms to estimate the network state online [39].

For small-scale power system such symbolic linearization may work well. However, almost all power systems are large-scale systems. The symbolic computation

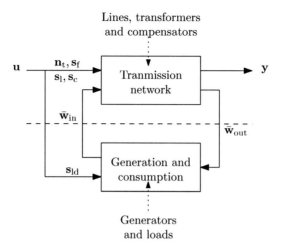

Figure 3.7: Decomposition of the power network into a linear and a nonlinear part.

required to directly linearize equations (3.26)–(3.27) becomes more and more complex with increasing network size. This complexity can be reduced by exploiting the particular structure of the equations. The generation and consumption part of the network (i.e., the generators and the loads) are modeled with a system of nonlinear equations, whereas the transmission part (i.e., the transmission lines, the compensation devices, and the transformers) are modeled with a system of linear equations. The model of the whole network can therefore be divided into two interconnected subsystems, as illustrated in Figure 3.7. The nonlinear subsystem is represented by (3.8) and (3.11)–(3.12). The linear subsystem is represented by (3.25). Note that by setting $\dot{\mathbf{x}} = 0$ in (3.11), a purely static problem that is similar to a load-flow calculation is obtained.

The linearized model is computed in two steps:

1. Equations (3.8) and (3.11)–(3.12) are symbolically linearized considering the symbolic parameters $\mathbf{x}_0$, $\mathbf{u}_{c0}$, and i. The computations are performed offline using the software package Mathematica [56] and lead to a set of Jacobian matrices parameterized by $\mathbf{x}_0$, $\mathbf{u}_{c0}$, and i.
2. The obtained symbolic model is updated online at each control step with the current values of the operating point $(\mathbf{x}_0, \mathbf{u}_{c0})$ and the discrete mode i. This update realizes the adaptation of the prediction model to the current operating conditions.

Time is discretized into discrete time steps $k = 0, 1, \ldots$, where discrete time step k corresponds to continuous time kT_s s, with T_s the sample period (s). Using the backward-Euler method to approximate the derivative, the following discrete-time affine model of the power network is obtained:

$$\mathbf{x}(k+1) = \mathbf{A}_i \mathbf{x}(k) + \mathbf{B}_i \mathbf{u}_c(k) + \mathbf{a}_i \qquad (3.28)$$

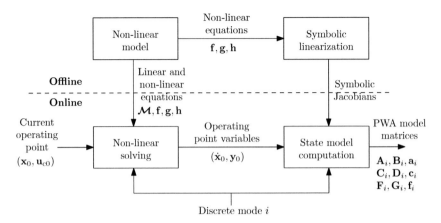

Figure 3.8: Overview of the computations of the predictions.

$$\mathbf{y}(k) = \mathbf{C}_i \mathbf{x}(k) + \mathbf{D}_i \mathbf{u}_c(k) + \mathbf{c}_i, \qquad (3.29)$$

where the matrices $\mathbf{A}_i$, $\mathbf{B}_i$, $\mathbf{C}_i$, $\mathbf{D}_i$, and the vectors $\mathbf{a}_i$ and $\mathbf{c}_i$ depend on the discrete mode $i(k)$ and operating point $(\mathbf{x}_0(k), \mathbf{u}_{c0}(k))$, e.g., $\mathbf{A}_i = \mathbf{A}(\mathbf{x}_0(k), \mathbf{u}_{c0}(k), i(k))$. Furthermore, $\mathbf{u}_c(k)$ is the vector of continuous valued control inputs applied to the system from instant k to instant $k+1$, assumed to be constant over the sample period.

Due to the presence of generator excitation limiters, it is relevant to take into account the state of the generators in the prediction computation by adding to the model (3.28)–(3.29) the constraints:

$$\mathbf{F}_i \mathbf{x}(k) + \mathbf{G}_i \mathbf{u}_c(k) \leq \mathbf{f}_i. \qquad (3.30)$$

These constraints in the continuous state-input space describe the state of the generators with respect to the discrete mode i. The resulting model has now a piecewise affine form and allows to predict the evolution of the state of the generators as well.

The linearization procedure is summarized in Figure 3.8. Note that the computational effort required by this approach is driven more by the number of connected generators and loads, than by the number of buses, since the computational effort depends strongly on the vector dimensions. The variables involved in these vectors are the voltage and current phasors that interconnect the generators and loads to the transmission system as defined in Section 3.3.2.

Special attention must be paid to the integration of the dynamics of on-load tap changers in the model. A good prediction requires that tap changes are synchronized with the control time steps, i.e., executed at the same time as control actions. A basic approach is to choose a sample time equal to the tap change delay and to introduce the following equation in the model (3.28)–(3.29):

$$n_t(k+1) = n_t(k) + \Delta n_t(k), \qquad (3.31)$$

where n_t is now a state variable and the tap change Δn_t is a discrete control input with $\Delta n_t \in \{-n_{\text{step}}, 0, n_{\text{step}}\}$. To control the on-load tap changer, a bang-bang control strategy is implemented locally:

$$V_t^{\text{ref}}(t) = \begin{cases} V_t^{\text{ref,min}} & \text{if} \quad \Delta n_t = -n_{\text{step}}, \\ V_t^{\text{ref,max}} & \text{if} \quad \Delta n_t = n_{\text{step}}, \\ V_t(t) & \text{if} \quad \Delta n_t = 0, \end{cases} \tag{3.32}$$

where V_t^{ref} is the on-load tap changer reference voltage, V_t is the secondary voltage of the transformer, $V_t^{\text{ref,min}}$ and $V_t^{\text{ref,max}}$ are the lower and upper bounds on V_t^{ref}, respectively, and n_{step} corresponds to one tap change of the transformer. The transformer ratio n_t between times k and $k+1$ is then consistent with the optimal value of Δn_t computed by the controller at time k. An extension to the case in which the sample time and tap change delay differ is proposed in [28].

A controller designed for the resulting piece-wise affine model is expected to control the behavior of the hybrid nonlinear system efficiently. At each time step, the model update requires a few computations and the controller only acts in a region close the operating point at which the system was linearized. In the next section a predictive control strategy for controlling power systems using the discussed modeling framework is presented.

3.4 Model predictive control

Model predictive control is based on solving online a finite time optimal control problem using a receding horizon approach, as summarized in the following steps [31]:

- At time step k and for the current state $\mathbf{x}(k)$, an open-loop optimal control problem over a future time interval is solved online, taking into account the current and future constraints on input, output, and state variables. This results in a sequence of actions over that future time interval that gives the best predicted performance.

- The first action in the optimal control sequence so obtained is applied to the system.

- The procedure is repeated at time $k+1$ using the newly obtained state $\mathbf{x}(k+1)$.

The solution is converted into a closed-loop strategy by using the measured or estimated value of $\mathbf{x}(k)$ as the current state. The stability of the resulting feedback system can be established by using the fact that the cost function can act as a Lyapunov function for the closed-loop system [36].

3.4.1 Formulation of the control problem

Let $\mathbf{U}_N$ be the sequence of control inputs over the prediction horizon with a length of N time steps:

$$\mathbf{U}_N = [\mathbf{u}^T(0) \; \mathbf{u}^T(1) \; \cdots \; \mathbf{u}^T(N-1)]^T, \tag{3.33}$$

where $\mathbf{u}(k) = [\mathbf{u}_c^T(k) \; \mathbf{u}_d^T(k)]^T$, and given the following cost function or performance index:

$$J_N(\mathbf{x}(0), \mathbf{U}_N) = F(\mathbf{x}(N)) + \sum_{k=0}^{N-1} L(\mathbf{x}(k), \mathbf{u}(k)). \tag{3.34}$$

In practice, the cost function to be optimized usually includes a term based on the state $\mathbf{x}(k)$ and a reference $\mathbf{x}_r(k)$, and a term based on the control inputs $\mathbf{u}(k)$:

$$L(\mathbf{x}(k), \mathbf{u}(k)) = \|\mathbf{x}(k+1) - \mathbf{x}_r(k)\|_{\mathbf{Q}_x} + \|\mathbf{u}(k)\|_{\mathbf{Q}_u}, \tag{3.35}$$

and a term based on the final state $\mathbf{x}(N)$:

$$F(\mathbf{x}(N)) = \|\mathbf{x}(N) - \mathbf{x}_r(N)\|_{\mathbf{Q}_f}, \tag{3.36}$$

where $\|\mathbf{w}\|_{\mathbf{Q}}$ denotes the 2-norm of a vector $\mathbf{w}$ with weighting matrix $\mathbf{Q}$. The weighting matrices are such that $\mathbf{Q}_x \geq 0$, $\mathbf{Q}_u > 0$ (by taking into account the constraints on $\mathbf{u}$, typically actuator constraints, it is sufficient for the matrix $\mathbf{Q}_u$ to be semi-positive definite) and $\mathbf{Q}_f \geq 0$.

At each time step, the following optimization problem $\mathcal{P}_N$ has to be solved, where the superscript o here refers to optimality:

$$\mathcal{P}_N(\mathbf{x}(0)) \; : \; J_N^o(\mathbf{x}(0)) = \min_{\mathbf{U}_N} J_N(\mathbf{x}(0), \mathbf{U}_N), \tag{3.37}$$

while satisfying the power network model constraints (3.28)–(3.30) over the prediction horizon[1] Additional constraints may allow to include some knowledge about the system that is not captured by the model, such as actuator limitations and physical limits on state variables. On the one hand, input constraints take into account actuator limits over the prediction horizon and thus are considered as hard constraints. On the other hand, output limits are generally not considered as hard constraints, since the optimization problem $\mathcal{P}_N$ could then become infeasible. The constraints on the outputs are therefore usually softened by adding slack variables $\mathbf{s}$ that represent the amount of constraint violation, and that are constrained as

$$\mathbf{y}_{\text{inf}} - \mathbf{s}(k) \leq \mathbf{y}(k) \leq \mathbf{y}_{\text{sup}} + \mathbf{s}(k), \tag{3.38}$$

---

[1] Note that the PWA model is time invariant over the prediction horizon, but not over several time steps: at each time step the matrices have to be updated to the current operating point.

and for which the penalty term $\|\mathbf{s}(k)\|_{Q_s}$ is added to (3.35). Thus, an optimal solution of the MPC optimization problem (3.37) can be found while minimizing the constraint violations. A final state constraint $\mathbf{x}(N) = \mathbf{x}_f$ can be added to guarantee the stability of the closed-loop system [36], but attention has to be paid to the feasibility of the optimization problem (3.37).

An important characteristic of the optimization problem (3.37) is its mixed nature. The presence of both the continuous inputs ($\mathbf{u}_c$) and the discrete inputs ($\mathbf{u}_d$) causes having to find the control sequence $\mathbf{U}_N$ as the sequence of both the continuous inputs $\mathbf{U}_{cN}$ and the discrete inputs $\mathbf{U}_{dN}$ over the prediction horizon with length N. Hence:

$$J_N^0(\mathbf{x}(0)) = \min_{\mathbf{U}_{dN}} \left(\min_{\mathbf{U}_{cN}} J_N(\mathbf{x}(0), (\mathbf{U}_{cN}, \mathbf{U}_{dN})) \right), \tag{3.39}$$

subject to the model constraints, for $k = 0, 1, \cdots, N-1$:

$$\mathbf{x}(k+1) = \mathbf{A}_i\mathbf{x}(k) + \mathbf{B}_i\mathbf{u}_c(k) + \mathbf{a}_i \tag{3.40}$$

$$\mathbf{y}(k) = \mathbf{C}_i\mathbf{x}(k) + \mathbf{D}_i\mathbf{u}_c(k) + \mathbf{c}_i \tag{3.41}$$

$$\mathbf{F}_i\mathbf{x}(k) + \mathbf{G}_i\mathbf{u}_c(k) \leq \mathbf{f}_i, \tag{3.42}$$

where $i(k)$ is a function of $\mathbf{x}(k)$, $\mathbf{u}_c(k)$ and $\mathbf{u}_d(k)$.

Let $\mathbf{I}_N = \begin{bmatrix} i(0) & i(1) & \cdots & i(N-1) \end{bmatrix}^T \in \mathbb{I}$ be a sequence of modes over the horizon N, where $\mathbb{I}$ is the set of admissible sequences. A sequence $\mathbf{I}_N$ defines $\mathbf{U}_{dN}$ and N sets of constraints (3.40)–(3.42) for $k = 0, 1, \cdots N-1$. The problem $\mathcal{P}_N$ can now be reformulated as:

$$J_N^0(\mathbf{x}(0)) = \min_{\mathbf{I}_N} \left(\min_{\mathbf{U}_{cN}} J_N(\mathbf{x}(0), (\mathbf{U}_{cN}, \mathbf{I}_N)) \right). \tag{3.43}$$

Then, for a given sequence of modes $\mathbf{I}_N$, the cost

$$J_N^*(\mathbf{x}(0), \mathbf{I}_N) = \min_{\mathbf{U}_{cN}} J_N(\mathbf{x}(0), (\mathbf{U}_{cN}, \mathbf{I}_N)) \tag{3.44}$$

is the optimal cost that is found by solving a continuous constrained optimization subproblem. This subproblem can easily be reformulated as a standard quadratic programming (QP) problem. The superscript $*$ here refers to optimality with respect to a given sequence of modes $\mathbf{I}_N$. However, the constrained optimization subproblem (3.44) is not necessarily feasible, i.e., for a given sequence $\mathbf{I}_N$ it could be the case that no solution satisfies the model constraints (3.40)–(3.42) over the prediction horizon.

Due to the presence of the discrete variables, $\mathcal{P}_N$ is categorized as an NP-hard problem. In addition, since $\mathcal{P}_N$ also involves continuous variables, it is a so-called mixed-integer problem. A branch-and-bound algorithm will be used to solve the problem $\mathcal{P}_N$ efficiently. This algorithm basically consists of a best-first descent strategy to reach suboptimal solutions and a branch-cutting strategy to profit from partial-horizon cost evaluation and infeasible subproblems. Below the details of this algorithm are discussed.

3.4.2 Mixed-integer optimization

The optimization problem associated with the predictive control of a piece-wise affine prediction model has been formulated above. This section is dedicated to the presentation of a simple, yet suitable mixed-integer optimization algorithm for hybrid MPC problems.

3.4.2.1 Exhaustive enumeration

The easiest way to find the optimal solution of the mixed-integer optimization problem consists of first enumerating all the possible sequences of modes over the prediction horizon and then solving the QP subproblems associated with each of these mode sequences. For a given sequence of modes, two situations with respect to a QP subproblem can occur: either the subproblem has no solution that satisfies the constraints (infeasibility), or the QP subproblem is feasible. For each feasible sequence $\mathbf{I}_N$ (defining a discrete control sequence $\mathbf{U}_{dN}$) an optimal continuous control sequence $\mathbf{U}_{cN}^*$ and a corresponding cost $J_N^*(\mathbf{x}(0), \mathbf{I}_N)$ can be obtained. The optimal solution is then given by the sequence of modes that minimizes (3.44):

$$J_N^o(\mathbf{x}(k)) = \min_{\mathbf{I}_N} \left(J_N^*(\mathbf{x}(k), \mathbf{I}_N) \right). \tag{3.45}$$

This method, which is referred to as exhaustive enumeration, quickly becomes useless when the number of modes, or the length of the prediction horizon increases, since the control problem to be solved is NP hard. The number of possible mode sequences to enumerate over grows exponentially with the number of modes and the length of the prediction horizon. Let p be the number of possible modes of the system. The exhaustive enumeration of all mode sequences requires p^N QP subproblems to be solved.

Figure 3.9 depicts as a tree the various possibilities of mode sequences. The depth of the tree of possibilities grows with the length of the horizon. For a given depth, the width of the tree is fixed by the number of possible modes of the system. Each leaf is a QP subproblem to be solved and one of them yields the optimum searched for. The dimension of the decision vector of all the QP subproblems is identical, i.e., $\dim(\mathbf{U}_c) = \dim(\mathbf{u}_c) \times N$. All leaves of the tree for which the associated QP subproblem is feasible are suboptimal solutions, except for that feasible subproblem that yields the solution with the lowest cost of all. This is the optimal solution.

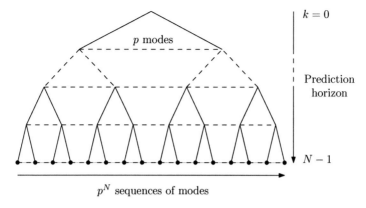

Figure 3.9: Illustration of exhaustive enumeration.

3.4.2.2 Partial enumeration

Exhaustive enumeration involves completely searching through the tree of possibilities. It is possible, however, to exploit the structure of the optimization problem associated with the hybrid predictive control scheme: the cost function of the optimization problem is additive, and has only positive terms. The key idea of the proposed partial enumeration algorithm is: given a suboptimal solution of (3.39)–(3.42), evaluate partial costs in order to prune the tree, i.e., cut branches that cannot lead to the optimum. This idea results in a branch-and-bound algorithm.

For a partial horizon P ($P < N$), i.e., at a depth P in the tree, the partial cost is defined as:

$$J_P(\mathbf{x}(0), \mathbf{U}_P) = \sum_{k=0}^{P-1} L(\mathbf{x}(k), \mathbf{u}(k)). \tag{3.46}$$

The proposed approach is a recursive algorithm that consists of a descent strategy and a branch-cutting criterion. The algorithm explores the tree according to the descent strategy, starting with a one step prediction horizon and increasing it step by step. The branch-cutting criterion allows to reduce the number of branches to consider and then to not explore the whole tree.

Descent strategy

Assume that the algorithm is at a depth P (with $P < N$) in the tree of possibilities, i.e., P time steps in the future. The proposed descent strategy is a *best-first* strategy:

- Compute the optimal costs J_{P+1} associated with the feasible subproblems for the possible choices of mode i.
- Choose the branch, i.e., the mode, that gives the minimal cost over the prediction horizon $P+1$ to continue the exploration.

Branch cutting

Assume that a first suboptimal solution is available, i.e., an upper bound on the optimal cost is available. Prune the tree by cutting the branches for which

- either the optimal cost on a partial horizon is greater than the cost of the suboptimum,
- or the subproblem is infeasible.

Cutting a particular branch means eliminating all branches originating from that particular branch. The best suboptimum so far is updated each time a leaf is evaluated and determined to have a cost that is lower than the cost of the previous best suboptimum.

3.4.2.3 Justification and illustration

Now some details are given to justify that the partial enumeration algorithm guarantees to find the optimum. Let a sequence of N modes $\mathbf{I}_N$ (the associated QP subproblem is assumed to be feasible) and a horizon P with $P < N$ be given. Below, the following notations are used:

- $\mathbf{I}_P^{(N)}$ is the sequence of the P first modes extracted from the sequence $\mathbf{I}_N$;
- $\mathbf{U}_{cP}^{(N)}$ is the continuous control sequence of length P extracted from the sequence $\mathbf{U}_{cN}$.

Recall that the superscript $*$ refers to optimality with respect to a given sequence of modes, i.e., the sequence $\mathbf{U}_{cN}^*$ is optimal with respect to a given sequence of modes $\mathbf{I}_N$. However, the extracted sequence $\mathbf{U}_{cP}^{*(N)}$ is not necessarily optimal over the partial horizon P.

Given a sequence of modes $\mathbf{I}_N$, for all $P < N$, the optimal cost that is obtained for the sequence $\mathbf{I}_N$ is greater than the optimal cost that is obtained for an extracted sequence $\mathbf{I}_P^{(N)}$:

$$\forall P < N, \quad J_N^*\big(\mathbf{x}(0), \mathbf{I}_N\big) \geq J_P^*\big(\mathbf{x}(0), \mathbf{I}_P^{(N)}\big). \tag{3.47}$$

For a detailed proof of this statement, see [29].

Assume the algorithm to be arrived at a depth $P < N$ and that the associated cost J_P is greater than a previously computed suboptimum. With respect to (3.47), the corresponding branch and all the following ones can be cut. An example of the execution of the proposed algorithm is illustrated in Figure 3.10.

The partial-enumeration algorithm is a branch-and-bound algorithm that leads to the optimal solution by taking advantage of the particular structure of the optimization problem associated with predictive control. General mixed-integer programming does not exploit this feature.

The proposed descent strategy is a heuristic that is likely to reach a first suboptimum close to the optimum. In fact, the suboptimal character comes from the choice of the sequence of modes (the best-first strategy at one prediction step), but the suboptimum is obtained by solving a QP subproblem over the full horizon with length N. In this partial enumeration approach, the dimension of the solution vector of the

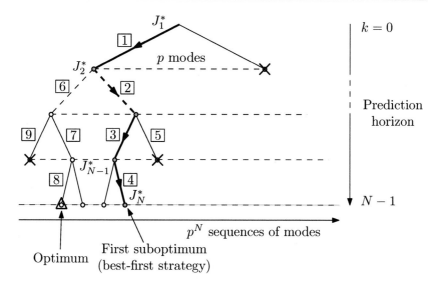

Figure 3.10: Illustration of execution of partial enumeration. *Numbers* indicate the route followed in the tree, i.e., the way in which the tree has been explored. *Bold lines* indicate the path to the first suboptimum (best-first strategy). The presence of a *cross* is the result of a branch cutting action: the cost were greater than the cost of the known suboptimum, or the subproblem was infeasible. The optimum search for is marked out by a *triangle*.

QP subproblems to be solved starts with $\dim(\mathbf{u}_c)$ at the top of the tree (one time step prediction) to grow with the horizon until reaching $\dim(\mathbf{u}_c) \times N$ at the bottom of the tree (using a horizon with length N).

3.5 Simulation studies

In this section, first the object-oriented concepts for modeling and simulation of the electricity network introduced in Section 3.2 are briefly presented. Then, simulation results of the proposed control approach are presented and discussed.

3.5.1 Simulation tools

To face the difficulty of developing complex power network models, object-oriented approaches for analysis and simulation of power systems have received increasing attention [33]. In object-oriented modeling, models are mapped as closely as possible to the corresponding physical subsystems that make up the overall system. Models are described in a declarative way, i.e., only local equations of the objects and the connections between the objects are defined. *Inheritance* and *composition* concepts enable proper structuring of models and generally lead to more flexible, modular, and reusable models. Extended models can be constructed by inheriting dynamics and properties of more basic models.

As stated above the dynamics of power networks involve continuous and discrete dynamics and are therefore hybrid. Each of the objects of a power network can therefore be modeled with a mixture of differential equations, algebraic equations, and discrete-event logic, e.g., in the form of if-then-else rules. The model of the overall system then consists of the models for the objects and in addition algebraic equations interconnecting the individual objects.

Several object-oriented approaches have been developed over the years, e.g., [3, 13, 35, 38, 46]. These approaches typically support both high-level modeling by composition and detailed component modeling using equations. Models of system components are typically organized in model libraries. A component model may be a composite model to support hierarchical modeling and specify the system topology in terms of components and connections between them. Using a graphical model editor, e.g., Dymola [13], a model can be defined by drawing a composition diagram by positioning icons that represent the models of the components, drawing connections, and giving parameter values in dialog boxes.

Some of the object-oriented simulation software packages, such as Simulink, assume that a system can be decomposed into sub-models with fixed causal interactions [1]. This means that the models can be expressed as the interconnection of sub-models with an explicit state-space form. Often a significant effort in terms of analysis and analytical transformations is required to obtain a model in this form [13]. In general, causality is not assigned in power networks. Setting the causality of an element of the power network, e.g., a transmission line, involves representing the model equations in an explicit input-output form. In a voltage-current formulation this means that currents are expressed as function of voltages, or vice versa. Non-causal modeling permits to relax the causality constraint and allows to focus on the elements and the way these elements are connected to each other, i.e., the system's topology. For an example of the use of a non-causal and object-oriented approach for power system modeling, see [41].

An environment that allows non-causal modeling, and that was used in this work, is Dymola [13], which implements the object-oriented modeling language Modelica [38]. Figure 3.11 and 3.12 illustrate the implementation of the 9-bus benchmark system in the Dymola environment. The Dymola-Simulink Interface [14] allows to easily use a simulation model (implemented in Dymola) in the Matlab-Simulink environment.

3.5.2 Simulation results

The hybrid predictive control algorithm presented in Section 3.4 has been implemented in Matlab and applied to the 9-bus power network introduced in Section 3.2 (cf. Figure 3.1).

This benchmark system has a strong combinatorial nature since almost all the control variables take on discrete values. The major difficulty in the mixed opti-

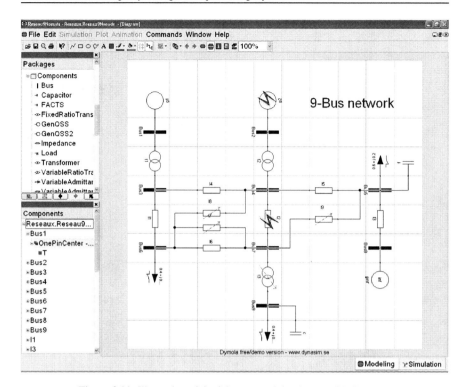

Figure 3.11: Illustration of the 9-bus network implemented in Dymola.

mization problem associated to the hybrid MPC formulation for this system is the number of discrete variables. Notice that an increase in the number of continuous variables does not significantly increase the complexity of the control problem.

To simulate the power system, a full nonlinear model has been implemented in Dymola [13] and simulations have been performed using the software package Matlab [34]. The sample period is $T_s = 30$ s and the prediction horizon is $N = 3$, such that the settling time of the load dynamics ($T_{P,Q} = 60$ s) is exceeded. The controller has been tuned using weighting matrices according to the formulation presented in Section 3.4. The use of reactive power compensation devices and transformer ratio changes is slightly penalized. On the contrary, line reconfiguration and load shedding is more penalized, since the economic cost of these actions is much higher. Voltage deviations of the buses to which consumers (loads) are connected are penalized most. A full description of the numeric values of the tuning parameters can be found in [28].

Below the results of transmission-line drop and generator-loss contingencies are discussed.

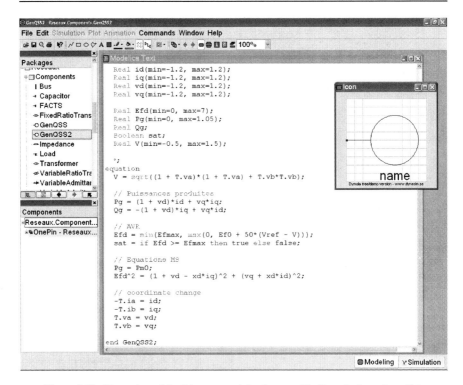

Figure 3.12: Illustration of the 9-bus network implemented in Dymola (equation side).

3.5.2.1 Scenario 1: Transmission line drop

In this scenario, transmission line l_2 is lost, i.e., unintentionally disconnected, at $t = 30$ s. Figures 3.13–3.14 show the results obtained when using the adaptive predictive control approach. In Figure 3.13, voltage stability is recovered 30 s after the fault has occurred, and all the voltages remain above 0.9 p.u. A new steady-state point of operation has been reached without using load shedding. Figure 3.14 shows the evolution of the control inputs over time. As can be observed, the solution determined by the controller mainly consists of injecting all the remaining reserve of reactive power (c_1 and c_2) and to connect transmission line l_9 at time $t = 60$ s. The controller acts in a coordinated manner on the other devices too, such as the generators and FACTS c_1 to keep the variables in acceptable ranges.

3.5.2.2 Scenario 2: Transmission line drop with parameter uncertainty

The same line drop scenario as above is considered, but now now the actual consumption of the loads at buses 5, 6 and 9 have been increased by 10%. The predictive controller is not aware of this, and, hence, uses values that not correct. In this way, the robustness of the controlled system to parametric uncertainty is illustrated.

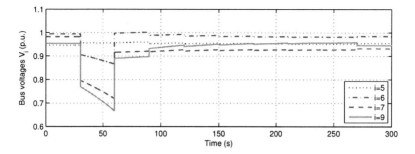

Figure 3.13: Bus voltage magnitudes (loss of transmission line l_2, nominal model).

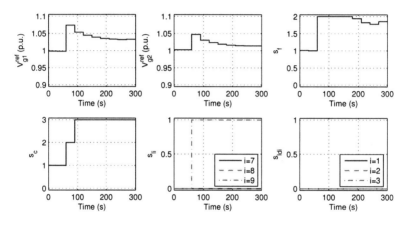

Figure 3.14: Control inputs (loss of transmission line l_2, nominal model).

Figures 3.15–3.16 indicate that for this scenario, the stability of the system is recovered, although the bus voltages cannot be maintained above 0.9 p.u. without using load shedding. If the actual load is more than 10% of the load assumed by the controller, the collapse is too fast for the controller to take adequate actions. In that case, the sample time T_s of the controller should be decreased if the systems has to stabilized.

3.5.2.3 Scenario 3: Generator loss

In this scenario, generator g_2 is lost, i.e., it is isolated from the grid. Figures 3.17–3.18 show the evolution of the network variables after the loss of the generator. To manage the loss of this generator, the controller uses load shedding of ld_2 and transmission line l_9 is connected at time $t = 60$ s. At the next time step, i.e., $t = 90$ s, the controller disconnects 5% of load ld_3. All the reactive compensation is used and the maximum excitation limit is reached for generator g_1. The network is stabilized at a new operating point, although some of the bus voltages are low, e.g., $V_5 < 0.9$ p.u.

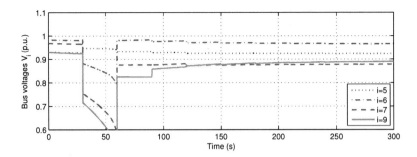

Figure 3.15: Bus voltage magnitudes (loss of transmission line l_2, 10 % consumption increase).

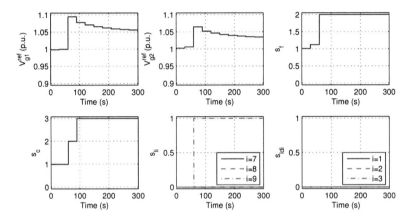

Figure 3.16: Control inputs (loss of transmission line l_2, 10 % consumption increase).

3.6 Conclusions and future research

This chapter has discussed the concepts of voltage stability of electric power networks and the possibly dramatic consequences of such instabilities, such as complete blackouts. Power networks are large-scale, nonlinear, and hybrid systems. An adaptive, model predictive control (MPC) approach for solving the emergency voltage control problem in such systems has been proposed. This approach uses a symbolically off-equilibrium linearized prediction model to deal efficiently with the hybrid and nonlinear characteristics of the dynamics involved. The symbolic linearization is based on a decomposition of the network into two interconnected subsystems and is performed offline to minimize the computational efforts required to update the hybrid prediction model at each time step online. Simulation studies on a 9-bus benchmark system have illustrated the performance of the proposed approach in line and generator loss case studies.

The proposed approach is not restricted to control of power networks alone. Also in other types of infrastructure systems, such as water networks and road traffic net-

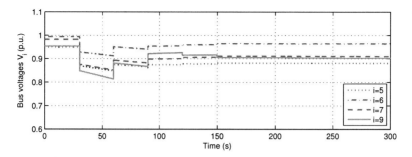

Figure 3.17: Bus voltage magnitudes (loss of generator g_2, nominal model).

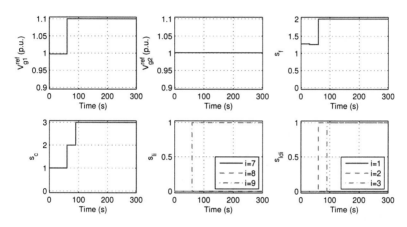

Figure 3.18: Control inputs (loss of generator g_2, nominal model).

works, the combination of discrete and continuous dynamics is found, and, hence, hybrid MPC problems have to be solved. The approach proposed here has the potential to do that.

Future research will focus on distributed predictive control, as the increase of the interconnections among countries leads to very large power networks with limited exchange of information. For industrial politics and security reasons, only partial knowledge of a subnetwork may be available to the others subnetworks, e.g., simplified subnetwork structures and dynamics. Control systems will have to deal efficiently with unpredictable disturbances, including unpredicted power flows and power outages in other subnetworks. Techniques from distributed optimization will be integrated in the approach proposed here to obtain an efficient distributed predictive control approach for hybrid systems.

Acknowledgements This research was partially carried out at Supélec–IETR, Hybrid Systems Control Group, Cesson-Sévigné, France. The authors would like to thank J. Buisson from Supélec-IETR, Rennes, France and J.L. Coullon from Areva Transport and Distribution, Massy, France for their support. This research is supported by the BSIK project "Next Generation Infrastructures (NGI)" and the Delft Research Center Next Generation Infrastructures.

References

1. E. Allen, N. LaWhite, Y. Yoon, J. Chapman, and M. Ilić. Interactive object-oriented simulation of interconnected power systems using Simulink. *IEEE Transactions on Education*, 44(1):87–95, February 2001.

2. P. J. Antsaklis. A brief introduction to the rheory and applications of hybrid systems. *IEEE Proceedings, Special Issue on Hybrid Systems: Theory and Applications*, 88(7):879–886, July 2000.

3. P. Barton and C. Pantelides. Modeling of combined discrete/continuous processes. *AIChE Journal*, 40(6):966–979, June 1994.

4. A. Bemporad and M. Morari. Control of systems integrating logic, dynamics, and constraints. *Automatica*, 35(3):407–427, March 1999.

5. J. Bernussou and A. Titli. *Interconnected Dynamical Systems: Stability, Decomposition and Decentralisation*. North-Holland Publishing Company, Amsterdam, The Netherlands, 1982.

6. L. T. Biegler, A. M. Cervantes, and A. Wachter. Advances in simultaneous strategies for dynamic process optimization. *Chemical Engineering Science*, 57(4):575–593, February 2002.

7. L. G. Bleris, P. D. Vouzis, J. G. Garcia, M. G. Arnold, and M. V. Kothare. Pathways for optimization-based drug delivery. *Control Engineering Practice*, 15(10):1280–1291, October 2007.

8. K. E. Brenan, S. L. Campbell, L. R., and Petzold. *Numerical Solution of Initial-Value Problems in Differential-Algebraic Equations*. SIAM, Philadelphia, Pennsylvania, 1996.

9. A. M. Cervantes and L. T. Biegler. Large-scale DAE optimization using a simultaneous NLP formulation. *AIChE Journal*, 44(5):1038–1050, May 1998.

10. I. K. Cho and H. Kim. Market power and network constraint in a deregulated electricity market. *The Energy Journal*, 28(2):1–34, April 2007.

11. M. Diehl, H. G. Bock, and J. P. Schlöder. A real-time iteration scheme for nonlinear optimization in optimal feedback control. *SIAM Journal on Control and Optimization*, 43(5):1714–1736, October 2005.

12. M. Diehl, H. G. Bock, J. P. Schlöder, R. Findeisen, Z. Nagy, and F. Allgöwer. Real-time optimization and nonlinear model predictive control of processes governed by differential-algebraic equations. *Journal of Process Control*, 12(4):577–585, June 2002.

13. Dynasim AB. *Dymola – Dynamic Modeling Laboratory, User's Manual*. Lund, Sweden, 2004.

14. Dynasim AB. Dymola-Simulink interface, 2004.

15. H. Elmqvist, S. E. Mattsson, and M. Otter. Modelica - a language for physical system modeling, visualization and interaction. In *Proceedings of the 1999 IEEE Symposium on Computer-Aided Control System Design*, pages 630–639, Kohala Coast, Hawaii, August 1999.

16. C. A. Falcone. Transmission in transition: bringing power to market. *IEEE Power Engineering Review*, 19(18):11–36, August 1999.

17. C. W. Gear. Differential-algebraic equation index transformation. *SIAM Journal on Scientific Computing*, 9(1):39–47, January 1988.

18. T. Geyer, M. Larsson, and M. Morari. Hybrid control of voltage collapse in power systems. In *Proceedings of the European Control Conference 2003*, volume 9, pages 157–166, Cambridge, UK, September 2003.

19. M. Ilić and J. Zaborsky. *Dynamics and Control of Large Electric Power Systems*. Wiley-Interscience, New York, New York, 2000.

20. N. Jenkins, R. Allan, P. Crossley, D. Kirschen, and G. Strbac. *Embedded Generation*. TJ International, Padstow, UK, 2000.

21. S. Kameswaran and L. T. Biegler. Simultaneous dynamic optimization strategies: Recent advances and challenges. *Computers and Chemical Engineering*, 30(10–12):1560–1575, September 2006.

22. D. Karlsson, M. Hemmingsson, and S. Lindahl. Wide area system monitoring and control - Terminology, phenomena, and solution implementation strategies. *IEEE Power and Energy Magazine*, 2(5):68–76, September–October 2004.

23. D. Karlsson and D.J. Hill. Modelling and identification of nonlinear dynamic loads in power systems. *IEEE Transactions on Power Systems*, 9(1):157–166, February 1994.

24. J. H. Kim, J. B. Park, J. K. Park, and S. K. Joo. A market-based analysis on the generation expansion planning strategies. In *Proceeding of the 13th International Conference on Intelligent Systems Application to Power Systems*, page 6, Washington, District of Columbia, November 2005.

25. P. Kundur. *Power System Stability and Control*. McGraw-Hill, New York, New York, 1994.

26. M. Larsson, D. J. Hill, and G. Olsson. Emergency voltage control using search and predictive control. *International Journal of Electric Power and Energy systems*, 24(2):121–130, February 2002.

27. R. Lazimy. Mixed-integer quadratic programming. *Mathematical Programming*, 22(1):332–349, December 1982.

28. S. Leirens. *Approche hybride pour la commade prédictive en tension d'un réseau d'énergie électrique*. PhD thesis, Supélec–Université de Rennes I, Rennes, France, December 2005.

29. S. Leirens and J. Buisson. An efficient algorithm for predictive control of piecewise affine systems with mixed inputs. In *Proceedings of the 2nd IFAC Conference on Analysis and Design of Hybrid Systems*, pages 309–314, Alghero, Italy, June 2006.

30. J. T. Linderoth and T. K. Ralphs. Noncommercial software for mixed-integer linear programming. Technical Report 04T-023, Department of Industrial and Systems Engineering, Lehigh University, Bethlehem, Pennsylvania, December 2004.

31. J. M. Maciejowski. *Predictive Control with Constraints*. Prentice-Hall, Harlow, UK, 2002.

32. Y. Majanne. Model predictive pressure control of steam networks. *Control Engineering Practice*, 13(12):1499–1505, December 2005.

33. A. Manzonic, A. S. de Silva, and I. C. Decker. Power systems, dynamics simulation using object-oriented programming. *IEEE Transactions on Power Systems*, 14(1):249–255, February 1999.

34. The MathWorks. MATLAB and Simulink for Technical Computing, 2009.

35. S. E. Mattsson, H. Elmqvist, and M. Otter. Physical system modeling with Modelica. *Control Engineering Practice*, 6(4):501–510, April 1998.

36. D. Q. Mayne, J. B. Rawlings, C. V. Rao, and P. O. M. Scocaert. Constrained model predictive control: Stability and optimality. *Automatica*, 36(6):789–814, June 2000.

37. D. Mignone. *Control and Estimation of Hybrid Systems with Mathematical Optimization*. PhD thesis, Swiss Federal Institute of Technology (ETH), Zürich, Switzerland, 2002.

38. Modelica Association. *Modelica - A Unified Object-Oriented Language for Physical Systems Modeling, Language Specification Version 3.0*, September 2007.

39. A. Monticelli. Electric power system state estimation. *Proceedings of the IEEE*, 88(2):262–282, February 2000.

40. M. Morari and J. H. Lee. Model predictive control: Past, present and future. *Computers and Chemical Engineering*, 23(4):667–682, 1999.

41. I. R. Navarro, M. Larsson, and G. Olsson. Object-oriented modeling and simulation of power systems using Modelica. In *Proceedings of the IEEE Power Engineering Society Winter Meeting*, pages 790–795, Singapore, January 2000.

42. R. R. Negenborn, M. Houwing, B. De Schutter, and J. Hellendoorn. Model predictive control for residential energy resources using a mixed-logical dynamic model. In *Proceedings of the 2009 IEEE International Conference on Networking, Sensing and Control*, pages 702–707, Okayama, Japan, March 2009.

43. R. R. Negenborn, S. Leirens, B. De Schutter, and J. Hellendoorn. Supervisory nonlinear MPC for emergency voltage control using pattern search. *Control Engineering Practice*, 17(7):841–848, July 2009.

44. R. R. Negenborn, P. J. van Overloop, T. Keviczky, and B. De Schutter. Distributed model predictive control for irrigation canals. *Networks and Heterogeneous Media*, 4(2):359–380, June 2009.

45. L. R. Petzold. A differential/algebraic system solver. In *Proceedings of the 10th World Congress on System Simulation and Scientific Computation*, pages 430–432, Montreal, Canada, August 1983.

46. P. C. Piela, T. G. Epperly, K. M. Westerberg, and A. W. Westerberg. ASCEND: an object-oriented computer environment for modeling and analysis: The modeling language. *Computers and Chemical Engineering*, 15(1):53–72, January 1991.

47. S. Piñón, E. F. Camacho, B. Kuchen, and M. Peña. Constrained predictive control of a greenhouse. *Computers and Electronics in Agriculture*, 49(3):317–329, December 2005.

48. D. H. Popović, D. J. Hill, and Q. Wu. Coordinated static and dynamic voltage control in large power systems. In *Proceedings of the Bulk Power System Dynamics and Control IV – Restructuring*, Santorini, Greece, August 1998.

49. Power System Stability Subcommittee. Voltage stability of power systems: Concepts, analytical tools and industry experience. Technical Report 90TH0358-2-PWR, IEEE/PES, 1990.

50. Power System Stability Subcommittee. Voltage stability assessment: Concepts, practices and tools. Technical Report SP101PSS, IEEE/PES, August 2002.

51. C. Rehtanz, M. Larsson, M. Zima, M. Kaba, and J. Bertsch. System for wide area protection, control and optimization based on phasor measurements. In *Proceedings of the International Conference on Power Systems and Communication Systems Infrastructures for the Future*, Beijing, China, September 2002.

52. C. A. Ruiz, N. J. Orrego, and J. F. Gutierrez. The Colombian 2007 blackout. In *Proceedings of the IEEE/PES Transmission and Distribution Conference and Exposition: Latin America*, pages 1–5, Bogotá, Colombia, aug 2008.

53. UCTE. Interim report of the investigation committee on the 28 September 2003 blackout in italy. Technical report, UCTE, 2004.

54. U.S.–Canada Power System Outage Task Force. Final report on the August 14, 2003 blackout in the USA and Canada: causes and recommendations. Technical report, U.S. Department of Energy, Natural Resources Canada, 2004.

55. T. Van Cutsem and C. Vournas. *Voltage Stability of Electric Power Systems*. Kluwer Academic Publishers, Dordrecht, The Netherlands, 1998.

56. Wolfram Research. Mathematica, Technical and Scientific Software, 2007.

Chapter 4
Module-Based Modeling and Stabilization of Electricity Infrastructure

L. Xie and M.D. Ilić

Abstract In this chapter we introduce a module-based approach to modeling and controlling electricity infrastructure to achieve reliability of services. Nonuniform generator or load components are defined as modules and are represented in terms of their internal state variables and the interaction state variables between the modules and the transmission network. Therefore, it is possible to specify the dynamical performance sub-objectives of each module for a given range of variations in interaction variables. It is also possible ensure that the local sub-objectives are met through a combination of local sensing, actuation and global communication. This approach, compared with the present off-line worst-case simulation approach, provides a systematic means of analyzing and stabilizing the infrastructure dynamics with increasing penetration of dispersed sustainable energy resources such as wind and solar. An interactive communication protocol between the distributed modules and control center could be implemented for operating the system with pre-specified stability performance. Sufficient conditions on network properties are derived under which this interactive protocol between the transmission networks and the modules converges to a system-wide stable operation. A five node example demonstrates the integration of wind power into the existing electricity infrastructure with pre-specified stability performance.

L. Xie
Carnegie Mellon University, Electrical & Computer Engineering Department, Pittsburgh, Pennsylvania, e-mail: lx@ece.cmu.edu

M.D. Ilić
Carnegie Mellon University, Electrical & Computer Engineering Department, Pittsburgh, Pennsylvania & Delft University of Technology, Faculty of Technology, Policy, and Management, Delft, The Netherlands, e-mail: milic@ece.cmu.edu

R.R. Negenborn et al. (eds.), *Intelligent Infrastructures*, Intelligent Systems, Control and Automation: Science and Engineering 42, DOI 10.1007/978-90-481-3598-1_4,
© Springer Science+Business Media B.V. 2010

4.1 Introduction

Electricity infrastructure has been undergoing profound changes throughout the world in recent years. Technically, the increasing presence of alternative energy resources, usually intermittent and geographically dispersed, poses challenges to conventionally centralized electric power system planning and operation. Institutionally, the deregulation of the electricity industry in past decades has unbundled the chain of electricity services. The generation, transmission, and distribution of electricity are no longer vertically integrated in regions with deregulated industry structure. The conventional centralized, top-down electric power industry is being transformed into a more distributed, bottom-up electric energy system, characterized by active participation of new energy resources [9]. Therefore, distributed decision making at the individual component level is required in future electricity infrastructure with presence of much more distributed energy-converting components.

The decentralization of the electricity infrastructure operation is facilitated by the development in information technology (IT) devices. The cost and performance of state-of-the-art sensing, actuation and computing devices (e.g., phasor measurement units, fast switching power electronics devices) provide opportunities for a more pro-active decision making framework for the changing electricity infrastructure, commonly referred to as "smart grids" [1, 19].

While efforts have been made to provide the component level with advanced sensing and actuation hardware and/or software, much remains to be done to understand the role of information necessary to achieve better performance of the system as a whole. Questions concerning type of information for improving specific performance measures of interest have not been studied systematically. This chapter is a step towards modeling and analyzing electricity infrastructure facilitated by the information processing devices. The major theoretical contribution of this chapter is a possible framework for the future distributed electric energy industry. Based on the proposed framework, problems such as stabilization of the electricity infrastructure are shown to be solvable using a distributed approach supported by specific information coordination.

In particular, the problem of interest in this chapter is how to construct a linearized electric power system dynamical model which lends itself to distributed criteria for system-wide stability. Conventional models of electric power system dynamics either assume constant power or constant impedance loads, which results a dense matrix characterizing the power system dynamics. With advanced sensing and data processing devices such as phasor measurement units (PMUs), we propose to model each load center dynamics via statistical model identification techniques. By taking the load center dynamics into consideration, the system matrix preserves the topology structure of the electricity network. Distributed stabilization criteria are derived based on the proposed structure-preserving dynamical model. Based on the distributed stabilization criteria, an interactive communication protocol is proposed for online stabilizing linearized system dynamics in the electricity infrastructure. All the energy-converting components in the system could cooperate towards a common goal of stabilizing the interconnected electricity infrastructure. The proposed protocol is a potential means to implementing the "plug-and-play" concept in future electricity infrastructure, which consists of much more dispersed energy resources [1].

This chapter is organized as follows. In Section 4.2, the existing literature related to both the theoretical and practical aspects of dispersed electricity infrastructure modeling and control is reviewed. In Section 4.3, our proposed framework for modeling and decision making in an electric energy system with distributed intermittent resources is introduced. A distributed criterion to check system bounded-input-bounded-state stability is derived. An online interactive communication protocol to ensure system-wide stability is proposed based on the criteria. Preliminary numerical results are illustrated on a five node electric power system in Section 4.4. In Section 4.5, the work in this chapter is summarized and future research directions are discussed.

4.2 Literature review

In the control theory community, decentralized modeling and control of large-scale linear infrastructure systems were extensively studied in the 1970s and 1980s [2, 16, 17]. The basic approach is to decompose a large-scale infrastructure system into subsystems and define stability criteria by limiting the strength of interconnections among subsystems. This approach requires weak coupling among the subsystems, which tends not to hold in many real-world systems.

To overcome these problems, a distributed cooperative control theory, which no longer assumes weak coupling between subsystems was recently proposed [6, 15]. In this new approach, subsystems typically communicate iteratively with neighboring subsystems in search of consensus of the global states, then distributed control action is taken at the individual subsystem level with the consensus of the global states. While it remains theoretically challenging to speed up the consensus convergence, a more practical framework for distributed decision making in large-scale physical systems, such as the electricity infrastructure, is yet to be developed.

In the domain application of electric power systems, active research is under way concerning the design and operation of more dispersed electricity infrastructures [14]. It has been observed from empirical studies [8] that the increased presence of distributed intermittent energy resources such as wind power could potentially destabilize the system dynamics. In parallel, efforts are under way to model the intermittent resources, such as wind power and photovoltaic panels [3, 4, 7, 14]. The challenge remains in the area of modeling and analysis of interactions with embedded dispersed generation. This chapter is intended to partially fill this gap.

4.3 Our approach

In this section, a module based dynamical model of electricity infrastructure is introduced. Each group of energy-converting components is defined as a module and is represented in terms of its own state variables and the interaction variables between the module and the rest of the system. Therefore, it is possible to specify the performance sub-objectives of each module for a given range of variations in interaction variables and to ensure that the local sub-objectives are met through local sensing and actuation. This approach, compared with today's operations and planning practices, which are based on the off-line worst-case specifications, could lead to a systematic means of analyzing the interactions between the electricity infrastructure and the dispersed energy resources.

In general, energy converting devices can be classified into two categories: power sources and power sinks. The devices are interconnected via the electric power grid. Furthermore, we classify power sources into two sub-categories: conventional generators whose output power is controllable and intermittent resources whose output is often not fully controllable. We consider these three major components of electric energy systems as modules, with well-defined internal state variables and associated interactions variables. Figures 4.1–4.3 depict a conventional governor-turbine-generator (G-T-G) , intermittent generator, and load, respectively. The operation of electricity infrastructure has two basic objectives: schedule available resources at minimum cost and keep the system dynamics stable within the technically acceptable limits. Scheduling is usually done every ten minutes or so. The stabilization is a much faster process. Therefore, stabilization and scheduling are usually modeled as separate processes. Stabilization is attempted for the assumed schedule; similarly, scheduling is done assuming stable fast dynamics. The stabilization problem is of concern in this chapter. In the next subsection, the proposed linearized module-based model of electricity infrastructure dynamics is described.

4.3.1 Linearized dynamics of electricity infrastructures

Linearized dynamics concern the response of electricity infrastructure systems to small perturbations around a given equilibrium. The complex system dynamics are represented by a set of nonlinear differential equations subject to algebraic constraints. Linearization around an equilibrium results in a set of linear ordinary differential equations. [1]

4.3.1.1 Linearized model of typical governor-turbine-generator modules

A typical Governor-Turbine-Generator (G-T-G) module can be modeled as a combination of two subsystems [11], which represent the electromechanical and the

---

[1] It should be noted that the linearized dynamical model is valid for small-signal stability studies. Module-based approach to modeling transient dynamics, which typically involve nonlinear equations, is subject to future research.

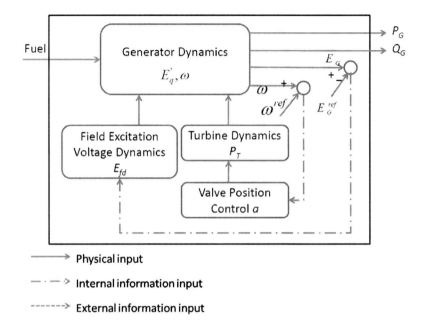

Figure 4.1: Dynamical model of typical governor-turbine-generator modules.

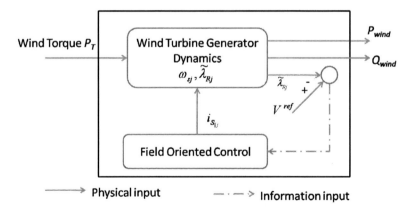

Figure 4.2: Wind turbine generator dynamics.

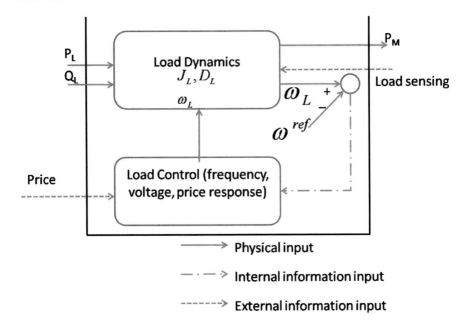

Figure 4.3: Load module dynamics.

electromagnetical dynamics of the machine, respectively. Governed by Newton's second law, the electromechanical dynamics of module i can be represented as:

$$J_i \dot{\omega}_{G_i} + D_i \omega_{G_i} = P_{T_i} + e_{T_i} a_i - P_{G_i} \tag{4.1}$$

$$T_{u_i} \dot{P}_{T_i} = -P_{T_i} + K_{t_i} a_i \tag{4.2}$$

$$T_{g_i} \dot{a}_i = -r_i a_i - \omega_{Gi} + \omega_{Gi}^{ref}, \tag{4.3}$$

where P_{T_i} and P_{G_i} refer to the mechanical and electrical real power deviation around an equilibrium of the turbine and the generator, respectively. J_i, D_i, T_{u_i}, and T_{g_i} refer to the moment of inertia of the generator, the damping coefficient of the generator, the time constant of the turbine and the generator, respectively. Two state variables ω_{Gi} and a_i are the generator frequency and the valve opening position deviation around an equilibrium point, respectively. The speed governor is the local primary controller intended to stabilize frequency around a desired set point ω_{Gi}^{ref}.

The electromagnetic subsystem is coupled with the electromechanical subsystem through the energy transfer across the magnetic field in the air gap of the synchronous generator. Reasonable assumptions can be made that the phase angle differences between the synchronous generator θ_i and the phase angle of the complex voltage behind the transient reactance δ_i are both small and do not change very much [11]. Therefore, the electromagnetic dynamics can be represented as:

$$\dot{V}_i = \frac{1}{T_{d_i}} \left[-V_i - \left(x_{d_i} - x'_{d_i} \right) i_{d_i} + E_{fd_i} \right], \qquad (4.4)$$

where V_i refers to the voltage magnitude at the output terminal of the machine. T_{d_i} refers to the transient time constant of the field rotor winding. x_{d_i} and x'_{d_i} represent the synchronous reactance and transient reactance of the machine, respectively. i_{d_i} corresponds to the direct projection of the machine armature current in the machine frame of reference. E_{fd_i} represents the field excitation control which is regulated by Automatic Voltage Regulator (AVR). In this chapter we assume that AVR functions well and no control limit is reached, therefore i_{d_i} can be used interchangeably with Q_{G_i}, which is the reactive power injection to the electricity grid by the generator i.

Therefore, the state variables of a closed-loop G-T-G module are

$$x_{G_i} = \begin{bmatrix} \omega_{Gi} & P_{T_i} & a_i & V_i \end{bmatrix}^T. \qquad (4.5)$$

The local local module control variables are

$$u_{G_i} = \begin{bmatrix} \omega_{Gi}^{ref} & E_{fd_i} \end{bmatrix}^T. \qquad (4.6)$$

The interaction variables between the G-T-G module and the rest of the system can be denoted as

$$x_{net_i} = \begin{bmatrix} P_{G_i} & Q_{G_i} \end{bmatrix}^T. \qquad (4.7)$$

By defining two notational coefficients $\alpha_{G_i} = -\frac{1}{J_{G_i}}$, and $\beta_{G_i} = -\frac{\left(x_{d_i} - x'_{d_i} \right)}{T_{d_i}}$, the linearized model of a typical G-T-G module dynamics can be represented as

$$\dot{x}_{G_i} = \begin{bmatrix} -\frac{D_i}{J_i} & \frac{1}{J_i} & \frac{e_{T_i}}{J_i} & 0 \\ 0 & -\frac{1}{T_{u_i}} & \frac{K_{t_i}}{T_{u_i}} & 0 \\ -\frac{1}{T_{g_i}} & 0 & -\frac{r_i}{T_{g_i}} & 0 \\ 0 & 0 & 0 & -\frac{1}{T_{d_i}} \end{bmatrix} x_{G_i} + \begin{bmatrix} 0 & 0 \\ 0 & 0 \\ \frac{1}{T_{g_i}} & 0 \\ 0 & \frac{1}{T_{d_i}} \end{bmatrix} u_{G_i} + \begin{bmatrix} \alpha_{G_i} & 0 \\ 0 & 0 \\ 0 & 0 \\ 0 & \beta_{G_i} \end{bmatrix} x_{net_i}$$

$$= A_i x_{G_i} + B_i u_{G_i} + N_i x_{net_i}. \qquad (4.8)$$

4.3.1.2 Linearized model of intermittent energy resource modules

Intermittent generation modules such as wind and photovoltaic power plants are not fully controllable compared with conventional G-T-G modules. Nevertheless, we can still represent intermittent energy resource module dynamics in terms of its internal state variables and interaction state variables with the rest of the system. In this chapter we consider an induction wind generator module as an example. For induction machines, much literature exists in field-oriented modeling and control techniques of induction machines, e.g., [13]. Field-oriented modeling of small

induction machines refers to changing the reference frame from the synchronous
stator direct and quadrature axis to the asynchronous rotor frame. The advantage of
field-oriented modeling is that the induction machine state-space model becomes a
linear time-invariant (LTI) system after the transformation, which lends itself well
to well established tools of LTI systems. The internal state variables for a wind
farm j are the rotor speed ω_{s_j} and the flux linkage projection onto the rotating frame
$\tilde{\lambda}_{R_j}$. Typically the flux linkage is not directly measurable, but with advanced sensing hardware and estimation software [5, 18], it is reasonable to choose $\tilde{\lambda}_{R_j}$ as a
state variable. Therefore, the state-space model of a wind-turbine generator module
j takes in the form

$$\dot{\omega}_{s_j} = \frac{1}{J_j}\left(P_{T_j} - P_{G_j}\right) \tag{4.9}$$

$$\dot{\tilde{\lambda}}_{R_j} = -\frac{R_{R_j}}{L_{R_j}}\tilde{\lambda}_{R_j} + \frac{R_{R_j}L_{M_j}}{L_{R_j}}i_{S_{\parallel j}}, \tag{4.10}$$

where P_{T_j} is the wind mechanical torque input to the generator. J_j is the moment of
inertia of the generator rotor and R_{R_j} is the rotor winding equivalent resistance. L_{R_j}
and L_{M_j} refer to the self inductance of the rotor winding and mutual inductance between the rotor and stator windings. $i_{S_{\parallel j}}$ represents the projection of stator current
onto the field reference. Under normal operating conditions, it is reasonable to assume that the machine terminal voltage magnitude is close to 1.0 per unit. Therefore
the reactive power produced by the wind generator Q_{G_j} can be used interchangeably
with $i_{S_{\parallel j}}$.

Consequently, the internal state variables of wind generator module j are

$$x_{WG_j} = \left[\omega_{s_j} \ \tilde{\lambda}_{R_j}\right]^T. \tag{4.11}$$

The local input variable is

$$u_{WG_j} = P_{T_j}. \tag{4.12}$$

The interaction variables with the rest of the system are

$$x_{net_j} = \left[P_{G_i} \ Q_{G_i}\right]^T. \tag{4.13}$$

Denoting $\alpha_{G_j} = -\frac{1}{J_j}$, and $\beta_{G_j} = \frac{R_{R_j}L_{M_j}}{L_{R_j}}$, a linearized dynamical model of a wind-
turbine generator module can be represented as

$$\dot{x}_{WG_j} = \begin{bmatrix} 0 & 0 \\ 0 & -\frac{R_{R_j}}{L_{R_j}} \end{bmatrix} x_{WG_i} + \begin{bmatrix} \frac{1}{J_j} \\ 0 \end{bmatrix} u_{WGj} + \begin{bmatrix} \alpha_{G_j} & 0 \\ 0 & \beta_{G_j} \end{bmatrix} x_{net_j} \tag{4.14}$$

$$= A_j x_{WG_i} + B_j u_{WGj} + N_j x_{net_j}. \tag{4.15}$$

4.3.1.3 Linearized model of load module dynamics

Viewed at the electric power grid level, load centers usually consist of a variety of heterogenous electric power consuming devices, e.g., light bulbs and electrical motors. Therefore, modeling the aggregate load dynamics from first principles is not realistic. An alternative approach to representing the power grid level load center dynamics is by statistical parametric modeling at each node [10]. Using this approach each load center module k is characterized by the real and reactive power P_{L_k} and Q_{L_k} delivered to the load from the grid, the load center frequency ω_{L_k}, the voltage magnitude at the load center V_{L_k}, and the actual power P_{M_k} consumed by the load.

We define J_{L_k} and D_{L_k} to be the equivalent moment of inertia and the damping coefficient of the postulated aggregate load dynamics. Let T_{d_k} be the transient time constant of voltage dynamics. Let x_{d_k} be the equivalent reactance of the load module. The postulated load module dynamics can then be represented as

$$J_{L_k}\dot{\omega}_{L_k} + D_{L_k}\omega_{L_k} = -P_{L_k} - P_{M_k} \tag{4.16}$$

$$\dot{P}_{M_k} = K_{M_k}P_{M_k} + \mu_{L_k} \tag{4.17}$$

$$\dot{V}_{L_k} = -\frac{1}{T_{d_k}}V_{L_k} - \frac{x_{d_k}}{T_{d_k}}Q_{Lk}, \tag{4.18}$$

where K_{M_k} is the coefficient derived from the auto-regression analysis of the observed data. The noise μ_{L_k} is represented by a Gaussian random variable, which could be parametrically identified [10]. Therefore the local state variables of load module k are

$$x_{L_k} = \begin{bmatrix} \omega_{L_k} & P_{M_k} & V_{L_k} \end{bmatrix}^T. \tag{4.19}$$

The interaction variables with the rest of the system are

$$x_{net_k} = \begin{bmatrix} P_{G_k} & Q_{G_k} \end{bmatrix}^T. \tag{4.20}$$

The linearized dynamics of the load module k can be represented as a third-order LTI model:

$$\dot{x}_{L_k} = A_{L_k}x_{L_k} + B_{L_k}u_{L_k} + N_k x_{net_k}, \tag{4.21}$$

where

$$A_{L_k} = \begin{bmatrix} -\dfrac{D_{L_k}}{J_{L_k}} & -\dfrac{1}{J_{L_k}} & 0 \\ 0 & K_{M_k} & 0 \\ 0 & 0 & -\dfrac{1}{T_{d_k}} \end{bmatrix}, \quad B_{L_k} = \begin{bmatrix} 0 \\ 1 \\ 0 \end{bmatrix} \tag{4.22}$$

$$N_k = \text{diag}\{\alpha_{L_k}, \beta_{L_k}\}, \quad \alpha_{L_k} = -\frac{1}{J_{L_k}}, \quad \beta_{L_k} = -\frac{x_{d_k}}{T_{d_k}}. \tag{4.23}$$

It should be noted that the proposed models of individual modules are conceptual. Further work is needed to verify these models and carry out model parameterizations.

4.3.1.4 Linearized model of an interconnected electricity infrastructure system

The modules are interconnected via the electric power grid subject to Kirchoff's Laws, which state the conservation of power at each node. By taking the first order Tayler series around an equilibrium point, the linearized power flow equations can be expressed as

$$\begin{bmatrix} \Delta P \\ \Delta Q \end{bmatrix} = H \begin{bmatrix} \Delta \theta \\ \Delta V \end{bmatrix}, \tag{4.24}$$

where $\Delta P = \begin{bmatrix} \Delta P_1 & \cdots & \Delta P_m \end{bmatrix}^T$ and $\Delta Q = \begin{bmatrix} \Delta Q_1 & \cdots & \Delta Q_m \end{bmatrix}^T$ refer to all the modules' real and reactive power injection deviations from an operating point around which the linearization is derived, respectively. $\Delta \theta = \begin{bmatrix} \Delta \theta_1 & \cdots & \Delta \theta_m \end{bmatrix}^T$ and $\Delta V = \begin{bmatrix} \Delta V_1 & \cdots & \Delta V_m \end{bmatrix}^T$ represent all the modules' voltage phase angle and voltage magnitude deviations from an operating point around which the linearization is derived, respectively. Matrix H refers to the power flow Jacobian matrix. By taking a limit operation of (4.24), we can get

$$\lim_{\Delta t \to 0} \begin{bmatrix} \frac{\Delta P}{\Delta t} \\ \frac{\Delta Q}{\Delta t} \end{bmatrix} = H \lim_{\Delta t \to 0} \begin{bmatrix} \frac{\Delta \theta}{\Delta t} \\ \frac{\Delta V}{\Delta t} \end{bmatrix}, \tag{4.25}$$

namely,

$$\begin{bmatrix} \dot{P} \\ \dot{Q} \end{bmatrix} = H \begin{bmatrix} \dot{\theta} \\ \dot{V} \end{bmatrix} \tag{4.26}$$

$$= A_{net} \begin{bmatrix} P \\ Q \end{bmatrix} + H \begin{bmatrix} \omega \\ \Lambda_V V \end{bmatrix}, \tag{4.27}$$

where

$$A_{net} = \begin{bmatrix} O & O \\ O & \Lambda_Q \end{bmatrix} \tag{4.28}$$

$$\Lambda_Q = \text{diag}\{\beta_1, \cdots, \beta_m\} \tag{4.29}$$

$$\Lambda_V = \mathrm{diag}\{A_1(n_1,n_1),\cdots,A_m(n_m,n_m)\} \tag{4.30}$$

$$A_i(n_i,n_i) = \text{entry of last row and column in } A_i. \tag{4.31}$$

By combining the dynamical equations of all the modules and the dynamical equations of interaction variables together, the dynamics of the entire electricity infrastructure is in the form of a set of ordinary differential equations (ODE): let $x_{mod} = \begin{bmatrix} x_1^T & x_2^T & \cdots & x_m^T \end{bmatrix}^T$ be the local state variables corresponding to the m existing modules in the system; let $u = \begin{bmatrix} u_1^T & u_2^T & \cdots & u_m^T \end{bmatrix}^T$ be the local control or input corresponding to the m modules; let $x_{net} = \begin{bmatrix} P^T & Q^T \end{bmatrix}^T$ be the network interaction variables. The dynamics of all the modules in the electricity infrastructure can be expressed as:

$$\begin{bmatrix} \dot{x}_{mod} \\ \dot{x}_{net} \end{bmatrix} = \begin{bmatrix} A & N \\ HE & A_{net} \end{bmatrix} \begin{bmatrix} x_{mod} \\ x_{net} \end{bmatrix} + \begin{bmatrix} B \\ O \end{bmatrix} u, \tag{4.32}$$

where

$$A = \mathrm{diag}\{A_1,\cdots,A_m\} \tag{4.33}$$

$$B = \mathrm{diag}\{B_1,\cdots,B_m\} \tag{4.34}$$

$$N = \mathrm{diag}\{N_1,\cdots,N_m\} \tag{4.35}$$

$$E = \mathrm{diag}\{E_{G_i}^P,\cdots,E_{WG_j}^P,\cdots,E_{L_k}^P,$$
$$E_{G_i}^Q,\cdots,E_{WG_j}^Q\cdots,E_{L_k}^Q\} \tag{4.36}$$

$$E_{G_i}^P = \mathrm{diag}\{1,0,0,0\} \tag{4.37}$$

$$E_{G_i}^Q = \mathrm{diag}\{0,0,0,-\frac{1}{T_{d_i}}\} \tag{4.38}$$

$$E_{WG_j}^P = \mathrm{diag}\{1,0\} \tag{4.39}$$

$$E_{WG_j}^Q = \mathrm{diag}\{0,-\frac{R_{R_j}}{L_{R_j}}\} \tag{4.40}$$

$$E_{L_k}^P = \mathrm{diag}\{1,0,0\} \tag{4.41}$$

$$E_{L_k}^Q = \mathrm{diag}\{0,0,-\frac{1}{T_{d_k}}\}. \tag{4.42}$$

The proposed dynamical model of the electricity infrastructure (4.32) preserves the structure of electric power network and lends itself to distributed sensing and control. This is the key starting model to developing distributed control and estimation for provable performances.

4.3.2 Distributed criteria for linearized system stability

Based on the proposed model from the previous subsection, sufficient conditions on system stability are derived under which an interactive protocol between the transmission networks and the modules could converge to a system-wide stable operation.

Theorem 4.1. *If $A \in \Re^{n \times n}, C \in \Re^{n \times m}, H \in \Re^{m \times n}, O \in \Re^{m \times m}, O_{ij} = 0, B \in \Re^{(n+m) \times l}$, the linearized system dynamics model*

$$\dot{x} = \begin{bmatrix} \dot{x}_{mod} \\ \dot{x}_{int} \end{bmatrix} = \begin{bmatrix} A & C \\ H & O \end{bmatrix} \begin{bmatrix} x_{mod} \\ x_{int} \end{bmatrix} + Bu \qquad (4.43)$$

is bounded-input-bounded-state (BIBS) stable if all the following three conditions are satisfied:
(1) all the eigenvalues of A have negative real parts, i.e. $\lambda_i^*(A) = Re(\lambda(A)) < 0, i = 1, \cdots, n$;
(2) matrix HC is symmetric, i.e., $(HC)^T = HC$;
(3) matrix HC's eigenvalues $\mu_j(HC)$ satisfy $0 > \mu_j > -\frac{1}{2}\lambda_1^{*2}, j = 1, 2, \cdots, m$, where $\lambda_1^* = maxRe(\lambda(A))$.

Proof. In the appendix of this chapter.

4.3.3 Interactive communication protocol to ensure system stability

The previous subsection proposed a theorem under which BIBS stability could be ensured in a distributed setting. An interactive communication protocol between modules and the grid is proposed in this subsection for operating the electricity infrastructure system with guaranteed linearized dynamical stability[2]. The objective of the proposed communication protocol is to achieve stability through distributed condition checking and cooperation among individual energy-converting components. The key features of the proposed protocol include: (1) iterative communication structure, in both centralized and distributed ways; and (2) quasi-static communication rate. Information (the range of real and reactive power injections at each node) needs only to be exchanged at the rate of minutes, similar to the rate of the present Supervisory Control And Data Acquisition (SCADA) system. This makes the proposed protocol practical to be deployed in the near future.

Define G_i to be the conventional G-T-G module i in the system. Define WG_j to be the proposed planned module j to be integrated into the system (e.g., a new wind power plant). Define L_k to be the load module k in the system. The interactive communication protocol for operating the electricity infrastructure with guaranteed dynamical stability performance is described below.

---

[2] Here we assume the real-reactive power decoupling, therefore the linearized system dynamical model takes the form as in (4.43). Future work should address stabilization in a coupled model.

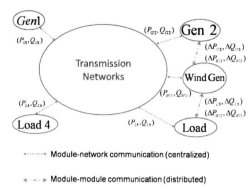

Figure 4.4: Communication protocol to ensure system-wide BIBS stability with wind integration.

Communication Protocol of Distributed System Operation with Guaranteed Linearized Dynamical Stability

- **Step 1**: At time step z, all the modules predict the range of real power generation (or demand) $\left[P_i^{\min}, P_i^{\max}\right]$ for the next time step interval. Each module should check its local control to guarantee the local module dynamics is stabilized, and send the predicted range of real power injection (or extraction) to the transmission system operator. Because the prediction is for the next time interval in the scale of minutes, it is reasonable to assume high confidence of prediction even for intermittent resources like wind generator modules.

- **Step 2**: The transmission system operator runs the steady-state load flow program and examines if condition (3) of Theorem 4.1 is satisfied for the range of power injections from all the nodes. If the condition is satisfied, go to step 4. Otherwise, the transmission operator will communicate back to the modules with the information of what needs to be adjusted in terms of of real power output range to guarantee that the load flow Jacobian matrix satisfies condition (3) of Theorem 4.1.

- **Step 3**: Once the modules receive the required adjustment of real power injection (or extraction) from the transmission system operator, distributed communication starts among the modules. They iteratively communicate and adjust their local controls to achieve an equilibrium under which no additional adjustment is needed to guarantee the requirement from the transmission system operator. For instance, the wind generator might want to ask its neighboring natural gas turbine generators to change output accordingly due to some unexpected wind speed changes.

- **Step 4**: It is guaranteed that the system remains BIBS stable for the next time interval. At the next time step $z + 1$, go back to step 1.

Figure 4.4 shows the information structure of the proposed protocol. This cyber protocol includes both centralized and distributed communication structure at a rate of minutes, which could guarantee the linearized system dynamical stability.

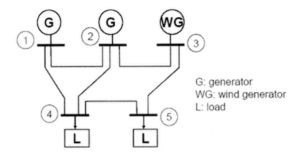

Figure 4.5: A five node electricity infrastructure system with wind power integration.

4.4 Numerical examples

In this section a five node electricity infrastructure system with one wind farm is used to illustrate the effectiveness of the proposed communication protocol. Figure 4.5 is a five-bus power system with one wind power plant. Wind generation consists of around 20 percent of the total real power generation. The deviation of wind torque input around the predicted value is considered a Gaussian random variable with mean value of 1.0 per unit and variance 0.8 per unit. The detailed parameters of this system are listed in Table 4.1.

Figure 4.6 and Figure 4.7 show the trajectory of frequency deviation around the equilibrium at the wind power plant location. As we can see from these figures, under random bounded wind torque input, the trajectories of frequency deviation stays bounded stable. The power flow Jacobian matrix satisfies the condition (3) in Theorem 4.1. The sufficient conditions stated in Theorem 4.1 guarantee the stability of the linearized interconnected electric power system dynamics. However, it should be noted that Theorem 4.1 provides sufficient but not necessary conditions for stability. Future work should investigate the conservativeness of the conditions in Theorem 4.1 [20].

Table 4.1: Parameters for the five node example.

	D	J	e_T	K_M	K_t	T_u	T_g	T_d	α	β
Gen 1	2.0	1.26	0.15	N/A	0.95	0.20	0.25	0.50	-1.20	-1.1
Gen 2	1.8	1.15	0.13	N/A	0.92	0.18	0.23	0.48	-1.0	-0.90
Load 4	4.5	2.3	N/A	-0.05	N/A	N/A	N/A	0.69	-0.43	-0.21
Load 5	5.5	2.5	N/A	-0.03	N/A	N/A	N/A	0.75	-0.40	-0.30

D_{WG}	J_{WG}	R_{WG}	$L_{J_{WG}}$	$L_{M_{WG}}$	X_{12}	X_{23}	X_{14}	X_{24}	X_{45}	X_{35}
0.28	0.34	0.10	0.43	0.10	0.10	0.20	0.10	0.10	0.30	0.20

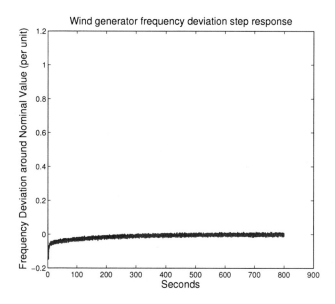

Figure 4.6: Frequency deviation at wind farm under bounded mechanical perturbation.

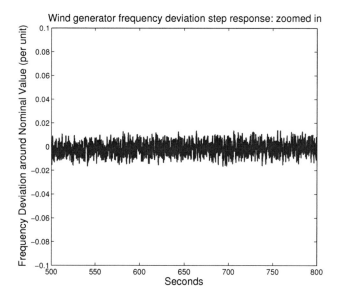

Figure 4.7: Frequency deviation at wind farm under bounded mechanical perturbation: zoomed in.

4.5 Conclusions and future research

In this chapter we recognize the fundamental needs of interactive systematic design of future electricity infrastructures with open-access to many dispersed renewable energy resources. We introduce a module-based dynamical model of interconnected electricity infrastructures. By modeling components as modules and specifying local state variables, local control, and network-related interaction variables, the proposed dynamical model preserves the network structure and lends itself to distributed monitoring and control. In particular, a distributed criterion to check system-wide BIBS stability is derived, based on which we propose an interactive communication protocol for online electric power system operation with quantifiable system stability. The proposed protocol requires only a minute-level communication rate, which is in line with the existing SCADA hardware and software systems in the electricity infrastructure. A five node example, which consists of a wind farm power plant, is simulated to demonstrate the effectiveness of the proposed communication protocol.

In this chapter, the distributed criterion of system stability is derived based on a linearized electric power system dynamical model, which is valid for small-signal stability studies. In light of the module-based approach, further work can be extended to study the transient dynamical studies. Also, the proposed criterion is a sufficient but not necessary condition to ensure system-wide dynamical stability. In particular, when sufficient interconnection level conditions are not met, future research is needed to establish more advanced communication protocols for supporting stable integration of distributed intermittent energy resources.

The proposed module-based dynamical model is motivated by the electricity infrastructure, nevertheless, this module-based approach is generalizable towards capturing the essential interactions between modules and networks in many other critical infrastructures. Future work includes more quantitative analysis on the tradeoffs among efficiency, stability and reliability of the interactive design of open-access infrastructure systems.

Appendix

Proof. Denote matrix $D = \begin{bmatrix} A_{n \times n} & C_{n \times m} \\ H_{m \times n} & O_{m \times m} \end{bmatrix}$. Let matrix $P_{n \times n}$ be the transformation matrix that diagonalizes matrix A, i.e., $P^{-1}AP = \Lambda$, where $\Lambda = diag\{\lambda_i(A)\}$. Define

$$M = \begin{bmatrix} P^{-1} & O_{n \times m} \\ O_{m \times n} & I_{m \times m} \end{bmatrix} D \begin{bmatrix} P & O_{n \times m} \\ O_{m \times n} & I_{m \times m} \end{bmatrix} = \begin{bmatrix} \Lambda & P^{-1}C \\ HP & O \end{bmatrix}.$$

Therefore, M is similar to D. Without loss of generality, assume that

$$0 > \lambda_1^* \geq \lambda_2^* \geq \cdots \geq \lambda_n^*.$$

M can be decomposed as the sum of M_1 and M_2, i.e.,

$$M = M_1 + M_2$$

$$
= \begin{bmatrix} \lambda_1^* - \frac{\lambda_1^*}{2} & & & & & & \\ & \lambda_2^* - \frac{\lambda_1^*}{2} & & & & & \\ & & \ddots & & & & \\ & & & \lambda_n^* - \frac{\lambda_1^*}{2} & & & \\ & & & & \frac{\lambda_1^*}{2} & & \\ & & & & & \ddots & \\ & & & & & & \frac{\lambda_1^*}{2} \end{bmatrix} + \begin{bmatrix} \frac{\lambda_1^*}{2} I_{n\times n} & P^{-1}C \\ HP & -\frac{\lambda_1^*}{2} I_{m\times m} \end{bmatrix}.
$$

The non-displayed off-diagonal entries of matrix M_1 are equal to zero. The eigenvalues of diagonal matrix M_1 are the diagonal elements $\{\lambda_1^* - \frac{\lambda_1^*}{2}, \lambda_2^* - \frac{\lambda_1^*}{2}, \cdots, \lambda_n^* - \frac{\lambda_1^*}{2}, \frac{\lambda_1^*}{2}, \cdots, \frac{\lambda_1^*}{2}\}$. We now derive the eigenvalues of matrix M_2. We can first augment matrix M_2 to a $2n \times 2n$ matrix by adding $(n-m)$ rows and columns of zeros:

$$
\hat{M}_2 = \begin{bmatrix} \frac{\lambda_1^*}{2} I_{n\times n} & P^{-1}C & O_{n\times(n-m)} \\ HP & -\frac{\lambda_1^*}{2} I_{m\times m} & \\ O_{(n-m)\times n+m} & & -\frac{\lambda_1^*}{2} I_{(n-m)\times(n-m)} \end{bmatrix}.
$$

We know that block diagonal matrix's spectrum is the union of spectrum of block matrices [12], therefore,

$$
eig(\hat{M}_2) = eig(M_2) \bigcup \{-\frac{\lambda_1^*}{2}, -\frac{\lambda_1^*}{2}, \cdots, -\frac{\lambda_1^*}{2}\}.
$$

Reorganize matrix $\hat{M}_2$ as four $n \times n$ block matrices, i.e.,

$$
\hat{M}_2 = \begin{bmatrix} Q & R \\ S & T \end{bmatrix},
$$

where

$$
Q = \begin{bmatrix} \frac{\lambda_1^*}{2} & & & \\ & \frac{\lambda_1^*}{2} & & \\ & & \ddots & \\ & & & \frac{\lambda_1^*}{2} \end{bmatrix}, \qquad R = \begin{bmatrix} P^{-1}C & O_{n\times(n-m)} \end{bmatrix}
$$

$$S = \begin{bmatrix} HP \\ O_{(n-m)\times n} \end{bmatrix}, \qquad T = \begin{bmatrix} -\frac{\lambda_1^*}{2} & & & \\ & -\frac{\lambda_1^*}{2} & & \\ & & \ddots & \\ & & & -\frac{\lambda_1^*}{2} \end{bmatrix}.$$

The eigenvalues of matrix $\hat{M}_2$ are computed by solving the roots for the following equation:

$$det(sI - \hat{M}_2) = det\left(\begin{bmatrix} sI - Q & -R \\ -S & sI - T \end{bmatrix} \right) = 0.$$

Since square matrices $sI - Q, -R, -S$, and $sI - T$ are of same size, this equation can be written as

$$det\{(sI - Q)(sI - T) - RS\} = 0.$$

After defining $z = s^2 - \frac{\lambda_1^{*2}}{4}$, this equation can be further simplified as [12]

$$det\left(zI - \begin{bmatrix} P^{-1} & \\ & I \end{bmatrix} \begin{bmatrix} HC & \\ & O \end{bmatrix} \begin{bmatrix} P & \\ & I \end{bmatrix} \right) = 0.$$

From this, one can see that z are the eigenvalues of the matrix $\begin{bmatrix} HC & \\ & O \end{bmatrix}$. Namely,

$$z = \{\mu_1, \mu_2, \cdots, \mu_m\} \bigcup \{0, 0, \cdots, 0\},$$

where the number of 0s is $(n-m)$. Recall that $z = s^2 - \frac{\lambda_1^{*2}}{4}$. So the eigenvalues of the matrix $\hat{M}_2$ are

$$s = \{\frac{\lambda_1^*}{2}, \frac{\lambda_1^*}{2}, \cdots, \frac{\lambda_1^*}{2}\} \bigcup \{-\frac{\lambda_1^*}{2}, -\frac{\lambda_1^*}{2}, \cdots, -\frac{\lambda_1^*}{2}\}$$

$$\bigcup \{\pm\sqrt{\frac{\lambda_1^{*2}}{4} + \mu_1}, \pm\sqrt{\frac{\lambda_1^{*2}}{4} + \mu_2}, \cdots, \pm\sqrt{\frac{\lambda_1^{*2}}{4} + \mu_m}\}$$

$$= \Phi_1 \bigcup \Phi_2 \bigcup \Phi_3,$$

where the cardinality of Φ_1, Φ_2, and Φ_3 is $n-m, n-m$, and $2m$, respectively. Recall that matrix $\hat{M}_2$ is a block diagonal matrix composed of M_2 and $-\frac{\lambda_1^*}{2} I_{(n-m)\times(n-m)}$. The eigenvalues of matrix $\hat{M}_2$ are simply the union of the spectrum of the two block matrices. Namely,

$$s = eig(M_2) \bigcup \{-\frac{\lambda_1^*}{2}, -\frac{\lambda_1^*}{2}, \cdots, -\frac{\lambda_1^*}{2}\} = eig(M_2) \bigcup \Phi_2.$$

Therefore, the eigenvalues of matrix M_2 are

$$eig(M_2) = \Phi_1 \bigcup \Phi_3.$$

Lemma 4.1. *Let $A, B \in C^{n \times n}$, with A, B diagonalizable. If A is diagonal with eigenvalues λ_1, λ_2, $\cdots$, λ_n and matrix B has eigenvalues μ_1, μ_2, $\cdots$, μ_n. Let μ be an eigenvalue of $A + B$, then μ lies in at least one of the disks*

$$\{z : |z - \lambda_j| \leq max_j |\mu_j|\}, j = 1, 2, \cdots, n.$$

Lemma 1 is from [12]. Because matrix $M = M_1 + M_2$, and M_1 is diagonal with eigenvalues $\{\lambda_1 - \frac{\lambda_1^*}{2}, \lambda_2 - \frac{\lambda_1^*}{2}, \cdots, \lambda_n - \frac{\lambda_1^*}{2}, \frac{\lambda_1^*}{2}, \cdots, \frac{\lambda_1^*}{2}\}$, matrix M_2 has eigenvalues $\mu_j = \Phi_1 \bigcup \Phi_3$. The eigenvalues of matrix M should lie in at least one of the disks:

$$\{z : |z - (\lambda_j - \frac{\lambda_1^*}{2})| \leq \frac{\lambda_1^*}{2}\}, j = 1, 2, \cdots, n.$$

From the given conditions in the theorem, all these disks lie in the left half plane. Therefore, all the eigenvalues of matrix M should lie in the left half plane. Hence, the energy system described by (4.43) is stable if all the three conditions are satisfied.

Acknowledgements This work was supported by the U.S. National Science Foundation ITR Project CNS0428404. The authors greatly appreciate this financial help. The first author also appreciates informative discussion with Professor Felix F. Wu from The University of Hong Kong and Dr. Haidong Yuan from Massachusetts Institute of Technology.

References

1. S. M. Amin and B. F. Wollenberg. Toward a smart grid. *IEEE Power & Energy Magazine*, 3(5):34–41, September/October 2005.
2. M. Aoki. On decentralized linear stochastic control problems with quadratic cost. *IEEE Transactions on Automatic Control*, 18:243–250, June 1973.
3. J. Apt. The spectrum of power from wind turbines. *Journal of Power Sources*, 169(2):369–374, June 2007.
4. F. Bouffard and F. D. Galiana. Stochastic security for operations planning with significant wind power generation. *IEEE Transactions on Power Systems*, 23(2):306–316, May 2008.
5. M. Cirrincione, M. Pucci, G. Cirrincione, and G.A. Capolino. A new adaptive integration methodology for estimating flux in induction machine drives. *IEEE Transactions on Power Electronics*, 19(1), January 2004.
6. W. B. Dunbar and R. M. Murray. Distributed receding horizon control for multi-vehicle formation stabilization. *Automatica*, 42(4):549–558, April 2006.
7. K. Furushima, Y. Nawata, and M. Sadatomi. Prediction of photovoltaic (pv) power output considering weather effects. In *Proceedings of the SOLAR*, July 2006.
8. R. T. Guttronmson. Modeling distributed energy resource dynamics on the transmission system. *IEEE Transactions on Power Systems*, 17(4):1148–1153, November 2002.

9. M. Ilić. From hierarchical to open access electric power systems. *Proceedings of the IEEE*, 95(5):1060–1084, May 2007.

10. M. D. Ilić, L. Xie, U. A. Khan, and J. M. F. Moura. Modeling future cyber-physical energy systems. In *Proceedings of IEEE Power and Energy Society General Meeting*, Pittsburgh, Pennsylvania, July 2008.

11. M. D. Ilić and J. Zaborszky. *Dynamics and Control of Large Electric Power Systems*. Wiley Interscience, New York, New York, 2000.

12. P. Lancaster and M. Tismenetsky. *The Theory of Matrices: Second Edition with Applications*. Academic Press, London, UK, 1984.

13. R. D. Lorenz. A simplified approach to continuous on-line tuning of field-oriented induction machine drives. *IEEE Transactions on Inducsty Applications*, 26(3), May 1990.

14. M. Milligan, K. Porter, B. Parsons, and J. Caldwell. Wind energy and power system operations: A survey of current research and regulatory actions. *The Electricity Journal*, 15(2):56–67, March 2002.

15. R. Olfati-Saber, J. A. Fax, , and R. M. Murray. Consensus and cooperation in networked multi-agent systems. *Proceedings of the IEEE*, 95(1):215–233, January 2007.

16. N. Sandell, P. Varaiya, M. thans, and M. Safonov. Survey of decentralized control methods for large scale systems. *IEEE Transactions on Automatic Control*, 23(2):108–128, April 1978.

17. D. D. Šiljak. *Large-scale Dynamic Systems*. North-Holland: New York, 1978.

18. G. Verghese and S. R. Sanders. Observers for flux estimation in induction machines. *IEEE Transactions on Inducstrial Electronics*, 35(1), February 1988.

19. F. F. Wu, K. Moslehi, and A. Bose. Power system control centers: Past, present, and future. *Proceedings of the IEEE*, 93(11):1890–1908, November 2005.

20. L. Xie and M. D. Ilić. Reachability analysis of stochastic hybrid systems by optimal control. In *NGInfra '08: Proceedings of the IEEE International Conference on Infrastructure Systems*, Rotterdam, The Netherlands, November 2008.

Chapter 5
Price-based Control of Electrical Power Systems

A. Jokić, M. Lazar, and P.P.J. van den Bosch

Abstract In this chapter we present the price-based control as a suitable approach to solve some of the challenging problems facing future, market-based power systems. On the example of economically optimal power balance and transmission network congestion control, we present how global objectives and constraints can in real-time be translated into time-varying prices which adequately reflect the current state of the physical system. Furthermore, we show how the price signals can be efficiently used for control purposes. Becoming the crucial control signals, the time-varying prices are employed to optimally shape, coordinate and synchronize local, profit-driven behaviors of producers/consumers to mutually reinforce and guarantee global objectives and constraints. The main focus in the chapter is on exploiting specific structural properties of the global system constraints in the synthesis of price-based controllers. The global constraints arise from sparse and highly structured power flow equations. Preserving this structure in the controller synthesis implies that the devised solutions can be implemented in a distributed fashion.

A. Jokić, M. Lazar, P.P.J. van den Bosch
Eindhoven University of Technology, Department of Electrical Engineering, Eindhoven, The Netherlands, e-mail: {a.jokic,m.lazar,p.p.j.v.d.bosch}@tue.nl

R.R. Negenborn et al. (eds.), *Intelligent Infrastructures*, Intelligent Systems, Control and Automation: Science and Engineering 42, DOI 10.1007/978-90-481-3598-1_5,
© Springer Science+Business Media B.V. 2010

5.1 Introduction

The aim of this chapter is to present, illustrate and discuss the role of prices in devising certain control solutions for electrical power systems. In particular we focus on the problem of *capacity management* in the sense of optimal utilization of scarce transmission network capacity. The term *price-based control*, as we use it in this chapter, can be considered as equivalent to *market-based* or *incentives-based* operation and control. In general terms, the main idea of the chapter is to present how the price signals can be used as the main control signals for *coordination* of many *local* behaviors (subsystems) to achieve some crucial *global* objectives.

As an introduction to the control problem considered in this chapter, we continue with pointing out to some of the changes that are taking place in the operation of today's power systems, and to some specific features of these systems which make their control an extremely challenging task.

5.1.1 Power systems restructuring

In spite of their immense complexity and inevitable lack of our full comprehension of all dynamic phenomena that are taking place in electrical power systems, to the present days these systems have shown an impressive level of performance and robustness. To a certain extent, this can be attributed to the long persistence of a traditional, regulated industry, which had a practice of rather conservative engineering, control and system operation. Another reason for their success is that traditional power systems are characterized by highly repetitive daily patterns of power flows, with a relatively small amount of suddenly occurring, uncertain fluctuations on the aggregated power demand side, and with well-controllable, large-scale power plants on the power production side. As a consequence, in traditional power systems, a large portion of power production could be efficiently scheduled in an open-loop manner, while the classical automatic generation control (AGC) scheme [17] sufficed for efficient real-time power balancing of uncertain demands.

Market-based operation

The most significant change that is taking place in power systems over the past decade is a liberalization and a policy shift towards competitive market mechanisms for their operation. From a monopolistic, one utility controlled operation, the system is being restructured to include many parties competing for energy production and consumption, and for provision of many of the *ancillary services* necessary for the system operation, e.g., provision of various classes of capacity reserves [30]. The main operational goal has shifted from centralized, utility cost minimization objective to decentralized profit maximization objectives of individual parties, e.g., of producers, consumers, retailers, energy brokers, etc. Fulfillment of crucial system constraints, such as global power balance and transmission network limits, has become a responsibility of market and system operators. The challenging task in designing the control and decision algorithms for these "global" operators is to ensure that the autonomous, profit driven behaviors of local subsystems will not act in a way that the system is driven to a highly unreliable and fragile state (or even instability), but will rather mutually reinforce on ensuring its integrity. The physical properties of electrical power systems play a prominent role in designing these markets and control architectures, and they are responsible for a very tight coupling in between economical and physical/technical layers of an electrical power system. They are also a reason why a straightforward transfer of knowledge and experience from deregulation, restructuring, operation and control of other sectors to the electric power system sector is often hampered or, even more often, is simply impossible.

Distributed generation and renewable energy sources

Another major change that is taking place in today's power systems is large-scale integration of distributed power generators (DG), many of which are based on intermittent renewable sources like wind and sun. Non-dependence on fossil fuels of many DG technologies, together with environmental issues, are the main driving forces for this change, and many countries have posed high targets concerning deployment of renewable sources over ten years horizons.

As a consequence, future power system will face a significant increase of uncertainties in any future system state prediction. Large and relatively fast fluctuations in production are likely to become normal operating conditions, standing in contrast to today's operating conditions characterized by highly repetitive, and therefore highly predictable, daily patterns. Note that the success of the present power systems heavily relies on this high predictability, while in the future, the need for fast acting, power balancing control loops will increase significantly.

5.1.2 Some specific features of power systems

Electrical power systems are one of the largest and most complex engineering systems ever created. They consist of thousands of generators and substations, and hundreds of millions of consumers all interconnected across circuits of continental scale. A distinguishing feature of electrical power systems, when compared for example to telecommunication networks, internet or road traffic networks, is that the subsystems in a electrical power system are all *physically* interconnected, i.e., dynamics of subsystems in the network are directly coupled.[1,2]

---

[1] Direct dynamical coupling is expressed through a set of equality constraints relating certain physical quantities among subsystems, e.g., coupling power flow equations in the electrical power system.

[2] Note that the networks in other infrastructures, e.g., telecommunication networks, internet or road traffic networks, could also be considered as *physically interconnected* in the sense that the subsystems in a network are related through certain physically realized links (which are possibly further characterized by some constraints), e.g., a highway in a road traffic network. However, the distinguishing feature of these systems, when compared to an electrical power system, is that

Large scale, physical interconnections, and the following specific features of electrical power systems makes mastering their complexity and devising an efficient operational and control solutions an extremely challenging task:

- *Heterogeneity and autonomy.* There is an enormous variety of physical devices interconnected to the network, with huge spectra of possible dynamical (physical) characteristics. All these *local*, almost exclusively nonlinear, dynamical characteristics of the subsystems are taking part in shaping the *global* dynamic behavior of the system, as they are all physically interconnected. In the economical layer of liberalized power systems, power producers and consumers (prosumers) are autonomous decision makers which are driven by their *local*, profit-driven objectives. As they are sharing the same power transmission network, which has a limited capacity, care has to be taken that the *local* and *autonomous* behaviors do not overload or destabilize the system. This generic *global* goal is not the natural aim of autonomously acting prosumers.
- *No free routing.* Unlike other transportation systems, which assume a free choice among alternative paths between source and destination, the flow of power in electrical energy transmission networks is governed by physical laws and is characterized by complex dependencies on nodal power injections (nodal productions and consumptions). Due to the complex relations, creating efficient congestion management schemes to cope with the transmission constraints is one of the toughest problems in design of market, operational and control architectures for power systems.
- *No buffering of commodities.* Electrical energy cannot be efficiently stored in large quantities, which implies that production has to meet rapidly changing demands immediately in *real-time*. This characteristic makes electricity a commodity with fast changing production costs, and is responsible for a tight coupling between economical and physical/techical layers of a power system.

5.1.3 Related work

The publications of Fred Schweppe and his co-workers can be considered as the first studies that systematically investigated the topic of price-based operation of electrical power systems. Many of the results from that period are summarized in [8, 26–28]. Ever since, there has been a tremendous amount of research devoted to a market-oriented approach for the electrical power system sector. For a detailed introduction and an overview, the interested reader is to referred to many books on the subject, e.g., [14, 18, 27, 29, 30]. In particular, for a detailed overview and some recent results concerning different approaches to price-based power balancing and congestion management of transmission systems we refer to [3, 6, 10, 13, 23, 31] and the references therein.

---

the dynamics of subsystems (e.g., cars on a highway) are not *directly coupled*, but possibly only indirectly, e.g., through some common performance objectives (common tasks) and/or inequality constraints (e.g., collision avoidance).

Probably the most closely related to some of the results presented in this chapter is the work of Alvarado and his co-workers [1–4, 11]. In [11], the authors have investigated how an independent system operator (ISO) could use electricity prices for congestion management without having an *a priori* knowledge about cost functions of the generators in the system. There, the authors illustrated how, in principle, a sequence of market observations could be used to estimate the parameters in the cost function of each generator. Based on these estimates, and by solving a suitably defined optimization problem, an ISO could issue the nodal prices causing congestion relieve. Although dealing with an intrinsically dynamical problem, the paper considered all the processes in a static framework. In [2, 3] the results of [11] have been extended by addressing possible issues of concern when price-based congestion management is treated as a dynamical process. The usage of price as a *dynamic feedback control* signal for power balance control has been investigated in [4]. There, the effects of interactions of price update dynamics and the dynamics of an underlying physical system (e.g., generators) on the stability of the overall system have been investigated. However, no congestion constraints have been considered and therefore only one, scalar valued, price signal was used to balance the power system.

In this chapter we present how nodal prices can be efficiently used for real-time power balance control and congestion management of a transmission network. These results, which treat the considered problem in a dynamical framework, present an extension of the above mentioned contributions. In particular, the emphasis in this chapter is on devising efficient control structures that exploit the specific structure of global power flow equations and constraints related to the transmission network. We show how preserving this structure in the proposed solutions results in a price-based control structure with an advantageous property that it can be readily implemented in a *distributed* fashion.

5.1.4 Nomenclature

The field of real numbers is denoted by $\mathbb{R}$, while $\mathbb{R}^{m \times n}$ denotes m by n matrices with elements in $\mathbb{R}$. For a matrix $A \in \mathbb{R}^{m \times n}$, $[A]_{ij}$ denotes the element in the i-th row and j-th column of A. For a vector $x \in \mathbb{R}^n$, $[x]_i$ denotes the i-th element of x. The transpose of a matrix A is denoted by $A^\top$. A vector $x \in \mathbb{R}^n$ is said to be nonnegative (nonpositive) if $[x]_i \geq 0$ ($[x]_i \leq 0$) for all $i \in \{1, \ldots n\}$, and in that case we write $x \geq 0$ ($x \leq 0$). For $u, v \in \mathbb{R}^k$ we write $u \perp v$ if $u^\top v = 0$. We use the compact notational form $0 \leq u \perp v \geq 0$ to denote the complementarity conditions $u \geq 0$, $v \geq 0$, $u \perp v$. $\mathrm{Ker} A$ and $\mathrm{Im} A$ denote the kernel and the image space of A, respectively. We use I_n and $\mathbf{1}_n$ to denote an identity matrix of dimension $n \times n$ and a column vector with n elements all being equal to 1, respectively. The operator $\mathrm{col}(\cdot, \ldots, \cdot)$ stacks its operands into a column vector, and $\mathrm{diag}(\cdot, \ldots, \cdot)$ denotes a square matrix with its operands on the main diagonal and zeros elsewhere. The matrix inequalities $A \succ B$ and $A \succeq B$ mean A and B are Hermitian and $A - B$ is positive definite and positive semi-definite, respectively. For a scalar-valued differentiable function $f : \mathbb{R}^n \to \mathbb{R}$,

$\nabla f(x)$ denotes its gradient at $x = \mathrm{col}(x_1,\ldots,x_n)$ and is defined as a *column vector.* *With a slight abuse of notation we will often use the same symbol to denote a signal, i.e., a function of time, as well as possible values that the signal may take at any time instant.*

5.2 Optimization decomposition: Price-based control

It is fair to say that the modern control systems theory is grounded on the following remarkable fact: virtually all control problems can be casted as optimization problems. It is insightful to realize that the same, far reaching statement, holds as well for the power systems: virtually all global operational goals of a power system can be formulated as constrained, time-varying optimization problems. Similarly as modern control theory accounts for efficiently solving these optimization problems (which is in most cases a far from trivial task), the same mathematical framework provides a systematic and rigorous scientific approach to shape operational and control architectures of power systems[3]. For illustration, in mathematical terms, a shift from monopolistic, one utility controlled system, to the market-based system is seen as a shift from explicitly solving a *primal problem* (e.g., economic dispatch at the control center) to solving its *dual problem* (e.g., operating real-time energy market). The former case can be called the cost-based operation, while the latter can be called the price-based operation. Before continuing with consideration of some specific problems in power systems, and their price-based solutions, we will first recall some basic notions from optimization theory. For the general introduction, closely related subjects and the state-of-the-art results on the distributed optimization, the interested reader is refereed to [5, 7, 9, 19, 22] and the references therein.

Consider the following structured, time varying[4], optimization problem

$$\min_{x_1,\ldots,x_N} \sum_{i=1}^{N} J_i(x_i), \tag{5.1a}$$

subject to

$$x_i \in \mathcal{X}_i, \qquad i = 1,\ldots,N, \tag{5.1b}$$

$$G(x_1,\ldots,x_N) \le 0, \tag{5.1c}$$

$$H(x_1,\ldots,x_N) = 0, \tag{5.1d}$$

where $x_i \in \mathbb{R}^{n_i}$, $i = 1,\ldots,N$ are the local decision variables, the functions $J_i : \mathbb{R}^{n_i} \to \mathbb{R}$, $i = 1,\ldots,N$, denote the local objective functions, while each set $\mathcal{X}_i \subseteq \mathbb{R}^{n_i}$ is defined through a set of *local* constraints on the corresponding local variable x_i as follows

---

[3] The interested reader is referred to the excellent paper [9] where the role of alternative ways for solving optimization problems is reflected in devising alternative operational structures for communication networks.

[4] For notational convenience, we have omitted the explicit reference to the time dependence.

$$\mathcal{X}_i = \{x_i \in \mathbb{R}^{n_i} \mid g_i(x_i) \leq 0,\ h_i(x_i) = 0\}, \qquad (5.2)$$

where $g_i(\cdot)$ and $h_i(\cdot)$ are suitably defined vector valued functions. The functions G and H, which respectively take values in $\mathbb{R}^k$ and $\mathbb{R}^l$, define *global* inequality and equality constraints.

Note that the optimization problem (5.1) is defined on the overall, global system level, where global objective function is sum of local objectives as indicated in (5.1a). Furthermore, note that if the global constraints (5.1c) and (5.1d) are omitted, the optimization problem (5.1) becomes separable in a sense that it is composed of N *independent* local problems which can be solved separately. For such a completely separable case, we say that the optimization problem can be solved in a *decentralized* way. For the future reference, we will call the problem (5.1) the *primal problem*.

Next, from (5.1) we formulate the *dual problem* as follows

$$\max_{\lambda,\mu} \quad \ell(\lambda,\mu) \qquad (5.3a)$$

$$\text{subject to} \quad \mu \geq 0, \qquad (5.3b)$$

where

$$\ell(\lambda,\mu) := \min_{x_1 \in \mathcal{X}_1,\dots,x_N \in \mathcal{X}_N} \left(\sum_{i=1}^{N} J_i(x_i) - \lambda^\top H(x_1,\dots,x_N) \right.$$

$$\left. + \mu^\top G(x_1,\dots,x_N) \right). \qquad (5.4)$$

In (5.3) and (5.4) $\lambda \in \mathbb{R}^k$ and $\mu \in \mathbb{R}^l$ are the dual variables (Lagrange multipliers) and have an interpretation of *prices* for satisfying the global constraints (5.1c) and (5.1d). If (5.1) is a convex optimization problem, it can be shown that the solutions of the primal and the dual problem coincide[5].

Remark 5.1. Decomposition and local optimization. If the functions H and G have an additive structure in local decision variables x_i, meaning that $H(x_1,\dots,x_N) = \sum_{i=1}^{N} \tilde{h}_i(x_i)$ and $G(x_1,\dots,x_N) = \sum_{i=1}^{N} \tilde{g}_i(x_i)$ with some given functions $\tilde{h}_i(\cdot)$, $\tilde{g}_i(\cdot)$, $i = 1,\dots,N$, then for a fixed λ and μ the optimization problem in (5.4) is separable and can be solved in a decentralized way. In that case the i-th local optimization problem is given by

$$\min_{x_i \in \mathcal{X}_i} \quad J_i(x_i) - \lambda^\top \tilde{h}_i(x_i) + \mu^\top \tilde{g}_i(x_i). \qquad (5.5)$$

□

---

[5] In fact, an additional mild condition, the so-called Slater's constraints qualification, is required for the solutions to coincide, see e.g., [7] for more details.

*Remark 5.2. **Coordination via global price determination.*** Updating the the dual variables (prices) λ, μ to solve the maximization problem in (5.3) can be achieved in a centralized way on a global level, e.g., at the central market operator which calculates the market clearing price. In some important cases, as it will be presented in the following section, the optimal prices (λ, μ) can also be efficiently calculated in a *distributed* way. This means that they can be calculated even if there is no one central unit that gathers information and communicates with all the subsystems (local optimization problems) in the network, but the optimal price calculations are based only on the locally available information and require only limited communication among neighboring systems.   □

Example 5.1. Loosely speaking, the market-based power system can be seen as solving the dual optimization problem (5.3). When the power limits in transmission network are not considered, this can be more precisely described as follows. Suppose that for each i, local decision variable x_i is a scalar and represents the power production ($x_i > 0$) of a power plant i, or a power consumption ($x_i < 0$) if the subsystem i is a consumer. Furthermore, let $J_i(x_i)$ denote power production costs when the i-th subsystem is a producer, and its negated benefit function when the i-th subsystem is a consumer. Since we do not consider the transmission network limits, the only global constraint is the power balance constraint $\sum_{i=1}^{N} x_i = 0$, i.e., in (5.1) and (5.3) we have that $H(x_1, \ldots, x_N) := \sum_{i=1}^{N} x_i$. Obviously, the *primal problem* (5.1) now corresponds to minimization of total production costs and maximization of total consumers benefit, while the power balance constraint is *explicitly* taken into account via (5.1d). Let us now consider the *price-based solution* through the corresponding *dual problem*. First note that minimization problem in (5.4) is in this case given by

$$\ell(\lambda) := \min_{x_1, \ldots, x_N} \sum_{i=1}^{n} \left(J_i(x_i) - \lambda x_i \right). \tag{5.6}$$

In (5.6) each term in the summation, i.e., $J_i(x_i) - \lambda x_i$, denotes the benefit of a subsystem i where λ denotes the price for electricity. Obviously, the dual problem (5.3) then amounts to maximizing the total benefit of the system. Note that in solving the dual problem, the power balance constraint is not explicitly taken into account. However, the corresponding maximum in (5.3) is attained precisely when the price λ is such that for the solution to the corresponding minimization problem in (5.6) it holds that $\sum_{i=1}^{N} x_i = 0$. In other words, the price λ which maximizes the total benefit of the system is precisely the price for which the system is in balance.

To summarize, while in primal solution the global constraints were *explicitly taken into account*, in the dual solution they are *enforced implicitly through the price λ*.   □

The observation from the above presented example can be generalized to the core idea of the price price-based control approach: In the price-based control, a price (Lagrange multiplier) is assigned to each crucial global constraint (i.e., each row in

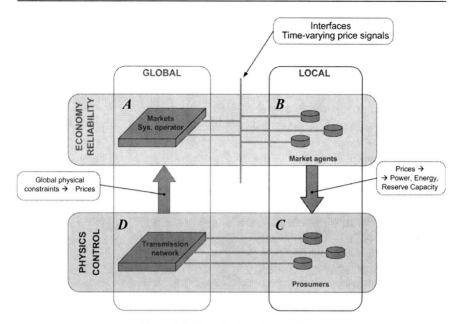

Figure 5.1: The price-based control loop.

(5.1c) and (5.1d), see (5.4)) and is used to implicitly enforce this constraint via local optimization problems (see e.g., (5.5)).

In a rather general sense, the price-based control loop can be illustrated as shown in Figure 5.1, and is encompassing the interplays between:
i.) *physical layer* of a power system (C and D in Figure 5.1), with time varying power flows as prominent signals; and *economical layer* (A and B in Figure 5.1) with time varying price signals as the prominent information carriers;
ii.) *local objectives* of producers/consumers (prosumers) (B and C in Figure 5.1, corresponding to (5.5)) and *global constraints*, e.g., power balance, transmission network limits and reliability constraints (A and D in Figure 5.1, corresponding to (5.1c) and (5.1d)).

Furthermore, prices are the signals used for *coordination* and *time synchronization* of actions from decentralized decision makers, so that the global system objectives are necessarily achieved in such that way that the total social welfare of the system is maximized, i.e., that they are achieved in the economically optimal way. It is also insightful to interpret the price-based solutions as *incentives-based* solutions, as prices λ are used to give incentives to the local subsystems so that their local objectives will make them behave in a way which serves global needs.

5.3 Preserving the structure: Distributed price-based control

5.3.1 Problem definition

Consider a connected undirected graph $\mathcal{G} = (\mathcal{V}, \mathcal{E}, A)$ as an abstraction of an electrical power network. $\mathcal{V} = \{\nu_1, \dots, \nu_n\}$ is the set of nodes, $\mathcal{E} \subseteq \mathcal{V} \times \mathcal{V}$ is the set of undirected edges, and A is a weighted adjacency matrix. Undirected edges are denoted as $\varepsilon_{ij} = (\nu_i, \nu_j)$, and the adjacency matrix $A \in \mathbb{R}^{n \times n}$ satisfies $[A]_{ij} \neq 0 \Leftrightarrow \varepsilon_{ij} \in \mathcal{E}$ and $[A]_{ij} = 0 \Leftrightarrow \varepsilon_{ij} \notin \mathcal{E}$. No self-connecting edges are allowed, i.e., $\varepsilon_{ii} \notin \mathcal{E}$. We associate the edges with the power lines of the electrical network and, for convenience, we set the weights in the adjacency matrix as follows: $[A]_{ij} = -b_{ij}$, where b_{ij} is the line susceptance. Note that the matrix A has zeros on its main diagonal and $A = A^\top$. The set of neighbors of a node ν_i is defined as $N_i \triangleq \{\nu_j \in \mathcal{V} \mid (\nu_i, \nu_j) \in \mathcal{E}\}$. Often we will use the index i to refer the node ν_i. We define $I(N_i)$ as the set of indices corresponding to the neighbors of node i, i.e., $I(N_i) \triangleq \{j \mid \nu_j \in N_i\}$. We associate the nodes with the buses in the electrical energy transmission network.

5.3.1.1 Primal: Optimal power flow problem

To define the optimal power flow problem as a primal optimization problem (5.1) on the global level, with each node ν_i we associate a set of local decision variables $\{p_i, \delta_i\}$, i.e., in (5.1) $x_i := \mathrm{col}(p_i, \delta_i)$, a singlet $\hat{p}_i$ and a triplet $(\underline{p}_i, \overline{p}_i, J_i)$. Here $p_i, \delta_i, \underline{p}_i, \overline{p}_i, \hat{p}_i \in \mathbb{R}$, $\underline{p}_i < \overline{p}_i$ and $J_i : \mathbb{R} \to \mathbb{R}$ is a strictly convex, continuously differentiable function. The values p_i and $\hat{p}_i$ denote the reference values for node power injections into the network, while δ_i denotes a voltage phase angle at the node ν_i. Positive values of p_i and $\hat{p}_i$ correspond to a flow of power into the network (production), while negative values denote power extracted from the network (consumption). Both p_i and $\hat{p}_i$ can take positive as well as negative values, and the only difference is that, in contrast to $\hat{p}_i$, the value p_i has an associated objective function J_i and a constraint $\underline{p}_i \leq p_i \leq \overline{p}_i$. In the case of a positive p_i, the function J_i represents the variable costs of production, while for negative values of p_i, it denotes the negated benefit function of a consumer. We will refer to p_i as the power from a price-elastic producer/consumer (or simply, power from a price-elastic unit), and to $\hat{p}_i$ as the power from a price-inelastic producer/consumer (price-inelastic unit).

Note that the assumption that one price-elastic unit and one price-inelastic unit are associated with each node is made to simplify the presentation and it does not result in any loss of generality of the presented results.

We use a "DC power flow" model[6] to determine the power flows in the network for given values of node power injections. The power flow in a line $\varepsilon_{ij} \in E$ is given by $p_{ij} = b_{ij}(\delta_i - \delta_j) = -p_{ji}$. If $p_{ij} > 0$, power in the line ε_{ij} flows from node ν_i to node ν_j. The power balance in a node yields $p_i + \hat{p}_i = \sum_{j \in I(N_i)} p_{ij}$. With the

---

[6] The DC power flow model is a linear approximation of a complex AC power flow model and is often used in practice. For a study comparing the AC and DC power flow models, and in particular the impact of the linear approximation on the nodal prices, the interested reader is referred to [20].

abbreviations $p = \text{col}(p_1,\ldots,p_n)$, $\hat{p} = \text{col}(\hat{p}_1,\ldots,\hat{p}_n)$, $\delta = \text{col}(\delta_1,\ldots,\delta_n)$ the overall network balance condition is $p + \hat{p} = B\delta$, where the matrix B is given by $B = A - \text{diag}(A\mathbf{1}_n)$.

Problem 5.1. *Optimal Power Flow (OPF) problem.*
For any constant value of $\hat{p}$,

$$\min_{p,\delta} J(p) \triangleq \min_{p,\delta} \sum_{i=1}^{n} J_i(p_i) \tag{5.7a}$$

subject to

$$p - B\delta + \hat{p} = 0, \tag{5.7b}$$

$$\underline{p} \le p \le \overline{p}, \tag{5.7c}$$

$$b_{ij}(\delta_i - \delta_j) \le \overline{p}_{ij}, \ \forall (i,j) \in I(N_i)), \tag{5.7d}$$

where $\underline{p} = \text{col}(\underline{p}_1,\ldots,\underline{p}_n)$, $\overline{p} = \text{col}(\overline{p}_1,\ldots,\overline{p}_n)$, and $\overline{p}_{ij} = \overline{p}_{ji}$ is the maximal allowed power flow in the line ε_{ij}. $\qquad\square$

We will refer to a vector p that solves the OPF problem as a *vector of optimal power injections*.

For an appropriately defined matrix L and a suitably defined vector of power line limits $\overline{p}_L$, the set of constraints in (5.11c) can be written in a more compact form as follows:

$$L\delta \le \overline{p}_L. \tag{5.8}$$

Note that the constraints (5.7b) and (5.7d) (or equivalently (5.8)) represent global equality and inequality constraints (5.1d) and (5.1c), respectively. Furthermore, for each node i, the corresponding constraint (row) in (5.7c) represent the local inequality constraint in (5.2).

Remark 5.3. In Problem 5.1 we have included δ explicitly as a decision variable, which will be crucial in the price-based control design. Another possibility, common in the literature, is to introduce a "slack bus" with zero voltage phase angle and to solve the equations for the line flows, completely eliminating δ from the problem formulation. However, in that case a specific structure, i.e., sparsity, of the power flow equations is lost. As we will see later in this chapter, preserving this sparsity will show to be beneficial for *distributed* controller implementation. $\qquad\square$

Remark 5.4. The matrix B is a singular matrix with rank deficiency one and with the kernel space spanned by the vector $\mathbf{1}_n$. Physically, this reflects the fact that only the relative voltage phase angles determine the power flow. $\qquad\square$

In traditional power system structures, where the production units are owned by one utility and there are little or no price elastic consumers, adjusting the production according to the solution of the OPF problem is one of the major operational goals of a utility. In such a system, the OPF problem is directly (explicitly) solved at a utility dispatch center, and the optimal reference values p are sent to the production units.

In contrast to this, in deregulated and liberalized power systems, the OPF problem is only indirectly (implicitly) solved by utilization of nodal prices. Next we define the optimal nodal prices problem as a central problem of liberalized, price-based operated systems.

5.3.1.2 Dual: Optimal nodal prices problem

According to price-based control approach we assign prices (Lagrange multipliers) to the global coupling constraints (5.7b) and (5.7d) (i.e., (5.8)) in Problem 5.1 to obtain the corresponding dual problem:

$$\max_{\mu \geq 0, \lambda} \quad \ell(\lambda, \mu) \tag{5.9}$$

where

$$\ell(\lambda, \mu) := \min_{\delta,\, p \in \{\tilde{p} | \underline{p} \leq \tilde{p} \leq \overline{p}\}} \quad \sum_{i=1}^{N} J_i(p_i) - \lambda^\top (p - B\delta + \hat{p}) + \mu^\top (L\delta - \overline{p}_L). \tag{5.10}$$

In (5.9) and (5.10) λ and μ are (vector) Lagrange multipliers.

Remark 5.5. The global coupling constraints (5.7b) and (5.8) do not have an additive structure in the decision variables p and δ, see Remark 5.1. Therefore for some fixed λ and μ the optimization problem in (5.10) is not separable (see Remark 5.1). However, when only the prices λ and the decision variables p are considered, the problem (5.10) becomes separable, i.e., it can readility be decomposed into n local problems, each assigned to one price-elastic unit.   □

With respect to Remark 5.5, it is insightful the reformulate the dual problem (5.9) to the following equivalent problem.

Problem 5.2. *Optimal nodal prices (ONP) problem.* For a constant value of $\hat{p}$,

$$\min_{\lambda, \delta} \sum_{i=1}^{n} J_i(\Upsilon_i(\lambda_i)) \tag{5.11a}$$

$$\text{subject to} \quad \Upsilon(\lambda) - B\delta + \hat{p} = 0, \tag{5.11b}$$

$$L\delta \leq \overline{p}_L, \tag{5.11c}$$

where $\lambda = \mathrm{col}(\lambda_1, \ldots, \lambda_n)$ is a vector of nodal prices,

$$\Upsilon(\lambda) \triangleq \mathrm{col}(\Upsilon_1(\lambda_1), \ldots, \Upsilon_n(\lambda_n))$$

and

$$\Upsilon_i(\lambda_i) \triangleq \arg\min_{\tilde{p}_i} \{ J_i(\tilde{p}_i) - \lambda_i \tilde{p}_i \mid \underline{p}_i \leq \tilde{p}_i \leq \overline{p}_i \}. \tag{5.12}$$

□

Although equivalent to (5.9), the problem formulation via (5.11) and (5.12) is insightful as it clearly indicates on one hand the role and *global* objectives of a system operator and on the other hand the *local* objectives of price elastic units. The letter is described by (5.12) and has the following interpretation: when a price elastic unit at node i receives a price λ_i for electricity at that particular node, it will adjust its production p_i to maximize its own benefit $J_i(p_i) - \lambda_i p_i$. The role of a system operator is to determine and issue a vector of nodal prices λ such that the overall system benefit is maximized (5.11a) while the system is in balance (5.11b) and while no line in the transmission system is congested (5.11c). A vector λ that solves the ONP problem is the *vector of optimal nodal prices*.

5.3.1.3 Price-based control problem

Consider a power network where each unit, i.e., producer/consumer, is a dynamical system, and assign to each such unit an appropriate model of its dynamics. Let G_i and $\hat{G}_i$ denote respectively a dynamical model of price-elastic and price-inelastic unit at node i as follows:

$$G_i : \dot{x}_i = f_i(x_i, p_i^A, p_i) = f_i(x_i, p_i^A, \Upsilon_i(\lambda_i)), \ \forall i, \tag{5.13a}$$

$$\hat{G}_i : \dot{z}_i = \hat{f}_i(z_i, \hat{p}_i^A, \hat{p}_i), \ \forall i, \tag{5.13b}$$

where x_i and z_i are the state vectors, p_i^A and $\hat{p}_i^A$ denote the *actual* node power injection from the system G_i and $\hat{G}_i$, respectively, into the network. As already mentioned, the input $p_i = \Upsilon_i(\lambda_i)$ denotes a price-dependent *reference* signal for power injection, i.e., $p_i = \Upsilon_i(\lambda_i)$ represents desired production/consumption of a price-elastic unit, while the input $\hat{p}_i$ denotes a *reference* value for the power injection of a price-inelastic unit. The desired production/consumption $\hat{p}_i$ of a price-inelastic unit does not depend on the electricity price λ_i, neither on any other signal from the power system.

Note that (5.11b) is always fulfilled when $\Upsilon(\lambda)$ and $\hat{p}$ are replaced respectively with $p^A = \text{col}(p_1^A, \ldots, p_n^A)$ and $\hat{p}^A = \text{col}(\hat{p}_1^A, \ldots, \hat{p}_n^A)$, since in this case (5.11b) represents the conservation law, i.e.,

$$p^A - B\delta + \hat{p}^A = 0. \tag{5.14}$$

To summarize, the complete dynamical model of a power system is described with the set of differential algebraic equations (5.13) and (5.14), with λ and $\hat{p}$ as inputs.

As opposed to the *actual* power injections, which are always in balance (5.14), keeping the balance in *reference* values (5.11b), i.e., balance in *desired* production and consumption, is a control problem. For future reference, we will always use the term *power balance* to refer to the power balance in sense of (5.11b), and not to the physical law (5.14).

To solve the power balance control problem, a measure of imbalance has to be available. The network frequency serves that purpose. Let $\Delta f \triangleq \text{col}(\Delta f_1, \ldots, \Delta f_n)$ denote the vector of nodal frequency deviations. In steady-state the network fre-

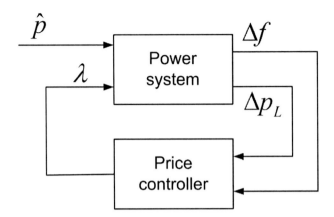

Figure 5.2: Price-based control scheme for real-time power balancing and congestion management.

quency is equal for all nodes in the system and the system is in balance if the network frequency is equal to its reference value, i.e., if $\Delta f = 0$. More precisely, if a system is in a steady-state with $\Delta f = 0$, then for each node (5.13a) implies $p_i = \Upsilon_i(\lambda_i) = p_i^A$, while (5.13b) implies $\hat{p}_i = \hat{p}_i^A$, and therefore (5.14) implies (5.11b).

In addition to controlling the power balance, nodal prices are used for congestion control, i.e., for fulfilment of the inequality constraints (5.8). For convenience we will define the vector of line overflows as $\Delta p_L \triangleq L\delta - \overline{p}_L$.

Finally, we are ready to define the control problem.

Problem 5.3. *Optimal price-based control problem.* For a power system (5.13) - (5.14), design a feedback controller that has the network frequency deviation vector Δf and the vector of line overflows Δp_L as inputs, and the nodal prices λ as output (see Figure 5.2), such that the following objective is met: for any constant value of $\hat{p}$ such that the ONP problem is feasible, the state of the closed-loop system converges to an equilibrium point where the nodal prices are the *optimal nodal prices* as defined in Problem 5.2.   □

5.3.2 Distributed price-based controller

In this subsection we first present an algebraic characterization (the Karush-Kuhn-Tucker optimality conditions) of optimal nodal prices in Problem 5.2 and study the structure of the matrices B and L which define the global coupling constraints (5.7b) and (5.7d) (i.e., (5.8)). As the main point, we show that this structure is preserved in the algebraic characterization of optimal nodal prices. Secondly, we show how an appropriate dynamic extension of these algebraic optimality conditions can be used as a solution to Problem 5.3.

5.3.2.1 Algebraic characterization of optimal nodal prices: the KKT conditions

The optimal power flow problem (5.7) is a convex problem which satisfies Slater's constraint qualification [7]. Therefore, for this problem the strong duality holds and the first-order Karush-Kuhn-Tucker (KKT) conditions [7] are *necessary and sufficient* conditions for optimality, and present us with the following characterization of optimal nodal prices.

Consider some constant value $\hat{p}$ such that the ONP problem (and therefore the optimal power flow problem (5.7)) is feasible. The KKT conditions for the optimal power flow problem (5.7) are given by:

$$p - B\delta + \hat{p} = 0, \tag{5.15a}$$

$$B\lambda + L^{\top}\mu = 0, \tag{5.15b}$$

$$\nabla J(p) - \lambda + \nu^{+} - \nu^{-} = 0, \tag{5.15c}$$

$$0 \leq (-L\delta + \overline{p}_L) \perp \mu \geq 0, \tag{5.15d}$$

$$0 \leq (-p + \overline{p}) \perp \gamma^{+} \geq 0, \tag{5.15e}$$

$$0 \leq (p + \underline{p}) \perp \gamma^{-} \geq 0, \tag{5.15f}$$

where λ and μ are Lagrange multipliers associated to global constraints (5.7b) and (5.7d) (i.e., (5.8)) as before in (5.9),(5.10), while γ^{+} and γ^{-} the Lagrange multipliers associated with the local inequality constraints $p \leq \overline{p}$ and $p \geq \underline{p}$, respectively. Recall that the for $a, b \in R^n$, the expression $0 \leq a \perp b \geq 0$ means $a \geq 0$, $b \geq 0$ and $a^{\top}b = 0$.

Notice that if no line is congested in the system, then the Lagrange multiplier μ in (5.15) is equal to zero and (5.15b) yields $B\lambda = 0$. This implies $\lambda \in \operatorname{Ker} B$ or $\lambda = \mathbf{1}_n \lambda^\star$, $\lambda^\star \in \mathbb{R}$ (see Remark 5.4), i.e., at the optimum, there is one price in the network for all nodes. In the case that at least one line in the system is congested, it follows that the optimal nodal prices will in general be different for each node in the system.

Remark 5.6. The only "direct" coupling of the elements in of optimal Lagrange multipliers λ and μ is present in equality (5.15b) and is completely determined by matrices B and L, while the "indirect" coupling beween elements of λ and μ is via p and δ and through (5.15a), (5.15c) and (5.15d).   □

Example 5.2. Consider a simple network depicted in Figure 5.3 and let $\overline{p}_{12}$ and $\overline{p}_{13}$ denote the line flow limits in the lines ε_{12} and ε_{13}, respectively. With μ_{12} and μ_{13} denoting the corresponding Lagrange multipliers from (5.15d), the optimality condition (5.15b) relates the optimal nodal prices with the following equality:

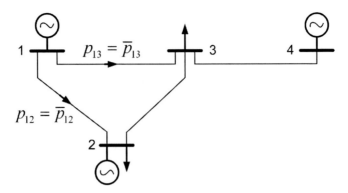

Figure 5.3: An example of a simple congested network.

$$\begin{bmatrix} b_{12,13} & -b_{12} & -b_{13} & 0 & b_{12} & b_{13} \\ -b_{12} & b_{12,23} & -b_{23} & 0 & -b_{12} & 0 \\ -b_{13} & -b_{23} & b_{13,23,34} & -b_{34} & 0 & -b_{13} \\ 0 & 0 & -b_{34} & b_{34} & 0 & 0 \end{bmatrix} \begin{bmatrix} \lambda_1 \\ \lambda_2 \\ \lambda_3 \\ \lambda_4 \\ \overline{\mu}_{12} \\ \mu_{13} \end{bmatrix} = 0, \qquad (5.16)$$

where $b_{12,13} = b_{12} + b_{13}$ and so on. Each row in (5.16) represents an equality related to the corresponding node in the network, i.e., the first row is related to the first node etc. Note that the i-th row directly relates the nodal price λ_i only with the nodal prices of its neighboring nodes, i.e., with λ_j, $j \in I(N_i)$, and that only the nodal prices in the nodes corresponding to the congested line ε_{ij} are directly related to the corresponding Lagrange multiplier μ_{ij}.   □

5.3.2.2 Price-based controller

Next, we present the price-based controller that solves Problem 5.3. Let K_λ, K_f, K_p and K_o be positive definite diagonal matrices, such that $K_f = \alpha K_\lambda$, $\alpha \in \mathbb{R}$ and $\alpha > 0$. Consider the following dynamic linear complementarity[7] controller:

$$\begin{bmatrix} \dot{x}_\lambda \\ \dot{x}_\mu \end{bmatrix} = \begin{bmatrix} -K_\lambda B & -K_\lambda L^\top \\ 0 & 0 \end{bmatrix} \begin{bmatrix} x_\lambda \\ x_\mu \end{bmatrix} + \begin{bmatrix} -K_f & 0 \\ 0 & K_p \end{bmatrix} \begin{bmatrix} \Delta f \\ \Delta p_L + w \end{bmatrix}, \qquad (5.17a)$$

$$0 \leq w \perp K_o x_\mu + \Delta p_L + w \geq 0, \qquad (5.17b)$$

$$\lambda = \begin{bmatrix} I_n & 0 \end{bmatrix} \begin{bmatrix} x_\lambda \\ x_\mu \end{bmatrix}, \qquad (5.17c)$$

---

[7] For an introduction to complementarity systems, the interested reader is referred to e.g., [12, 25] and the references therein.

where x_λ and x_μ denote the controller states, $\mathrm{col}(\Delta f, \Delta p_L)$ and w denote inputs to the controller, while λ denotes the output. The matrices K_λ, K_f, K_p and K_o represent the controller gains. The input $\mathrm{col}(\Delta f, \Delta p_L)$, which collects the nodal frequency and line overflow vectors, is an exogenous input to the controller, while the input w is required to be a solution to the finite dimensional complementarity problem (5.17b). The output λ is a vector of nodal prices.

Assumption 5.1. The closed-loop system resulting from the interconnection of the controller (5.17) with the power system (5.13) - (5.14) is globally asymptotically stable for any constant value of $\hat{p}$ (i.e., with respect to the corresponding steady-state) such that the ONP problem is feasible.   □

Theorem 5.1. *Suppose that* Assumption 5.1 *holds. Then the dynamic controller* (5.17) *solves the optimal power balance and congestion control problem, as defined in Problem 5.3.*

The proof of Theorem 5.1 follows from straightforward algebraic manipulations on the steady-state relations of the closed-loop system, i.e., of the power system (5.13),(5.14) in the closed-loop with the controller (5.17), where it can be shown that these steady-state relations necessarily include the KKT optimality conditions (5.15). The complete proof is omitted here for brevity, and for all the details, as well as for an approach how to verify Assumption 5.1, the interested reader is referred to [15] and [16].

Note that the controller (5.17) is in fact nothing else than a suitable dynamic extension of the optimality condition (5.15b), which is further appropriately updated by input signals $\mathrm{col}(\Delta f, \Delta p_L)$. With a reference to Remark 5.6, we have the following insightful interpretation of (5.17): the controller (5.17) explicitly includes the "direct" coupling among the elements in λ and μ, while the "indirect" coupling is obtained by adjustment of λ and μ to the inputs Δf and Δp_L which respectively carry the information if the constraints (optimality conditions) (5.15a) and (5.15d) are satisfied or not. The remaining optimality conditions (5.15c), (5.15e) and (5.15f), from (5.15) are satisfied on the local level through profit maximization behavior of price-elastic units as defined by (5.12).

Remark 5.7. The *only* system parameters that are explicitly included in the controller (5.17) are the transmission network parameters, i.e., the network topology and line impedances, which define the matrices B and L. To provide the correct nodal prices, the controller requires no knowledge of cost/benefit functions J_i and of power injection limits $(\underline{p}_i, \overline{p}_i)$ of producers/consumers in the system (neither is it based on their estimates). Furthermore, note that in practice often only a relatively small subset of all lines is critical concerning congestion, and for the controller (5.17) it suffices to include only these critical lines by appropriately choosing Δp_L and L.   □

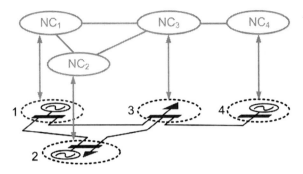

Figure 5.4: Distributed control scheme for power balance and congestion control.

Example 5.3. Consider again a simple network depicted in Figure 5.3 and described in Example 5.2. The highly structured relations from the optimality condition (5.15b) are as well present in the proposed controller (5.17), allowing for its *distributed* implementation. This means that the control law (5.17) can be implemented through a set of *nodal controllers*, where a nodal controller (*NC*) is assigned to each node in the network, and each *NC* communicates only with the *NC*'s of the neighboring nodes. From (5.17) and (5.16) it is easy to derive that the *NC* corresponding to node 1 in the network depicted in Figure 5.3 is given by:

$$
\begin{bmatrix} \dot{x}_{\lambda_1} \\ \dot{x}_{\mu_{12}} \\ \dot{x}_{\mu_{13}} \end{bmatrix} = \begin{bmatrix} -k_{\lambda_1}b_{12,13} & k_{\lambda_1}b_{12} & k_{\lambda_1}b_{13} \\ 0 & 0 & 0 \\ 0 & 0 & 0 \end{bmatrix} \begin{bmatrix} x_{\lambda_1} \\ x_{\mu_{12}} \\ x_{\mu_{13}} \end{bmatrix}
$$
$$
+ \begin{bmatrix} k_{\lambda_1}b_{12} & k_{\lambda_1}b_{13} & -k_{f_1} & 0 & 0 \\ 0 & 0 & 0 & k_{p_{12}} & 0 \\ 0 & 0 & 0 & 0 & k_{p_{13}} \end{bmatrix} \begin{bmatrix} x_{\lambda_2} \\ x_{\lambda_3} \\ \Delta f_1 \\ \Delta p_{12}+w_{12} \\ \Delta p_{13}+w_{13} \end{bmatrix}, \qquad (5.18a)
$$

$$
0 \le \begin{bmatrix} w_{12} \\ w_{13} \end{bmatrix} \perp \begin{bmatrix} k_{o_{12}}x_{\mu_{12}} \\ k_{o_{13}}x_{\mu_{13}} \end{bmatrix} + \begin{bmatrix} \Delta p_{12} \\ \Delta p_{13} \end{bmatrix} + \begin{bmatrix} w_{12} \\ w_{13} \end{bmatrix} \ge 0, \qquad (5.18b)
$$

$$
\lambda_1 = \begin{bmatrix} 1 & 0 & 0 \end{bmatrix} \begin{bmatrix} x_{\lambda_1} \\ x_{\mu_{12}} \\ x_{\mu_{13}} \end{bmatrix}, \qquad (5.18c)
$$

where $k_{\lambda_1} = [K_\lambda]_{11}$, $k_{f_1} = [K_f]_{11}$, and $k_{p_{12}}$, $k_{p_{13}}$, $k_{o_{12}}$, $k_{o_{13}}$ are the corresponding elements from the gain matrix K_p and K_o in (5.17c). Note that the state $x_{\mu_{ij}}$ is present only in one of the adjacent nodal controllers, i.e., in node i or in node j, and is communicated to the *NC* in the other node. □

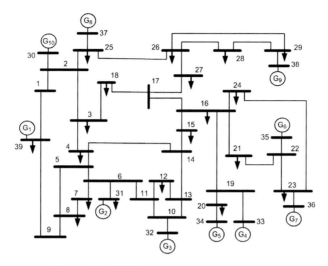

Figure 5.5: IEEE 39-bus New England test system.

The distributed implementation of the developed controller is graphically illustrated in Figure 5.4. The communication network graph among NC's is the same as the graph of the underlying physical network. Any change in the network topology requires only simple adjustments in NC's that are in close proximity to the location of the change. A distributed control structure is specially advantageous taking into account the large-scale of electrical power systems. Since in practice B is usually sparse, the number of neighbors for most of the nodes is small, e.g., two to four.

5.3.3 Illustrative example

To illustrate the potential of the developed, distributed price-based control method-ology, we consider the widely used IEEE 39-bus New England test network. The network topology, generators and loads are depicted in Figure 5.5. The complete network data, including reactance of each line and load values can be found in [21]. All generators in the system are modeled using the standard third order model used in automatic generation control studies [17]. The parameter values, in per units, are taken to be in the $\pm 20\%$ interval from the values given in [24], pp. 545. Each generator is taken to be equipped with a proportional feedback controller for frequency control with the gain in the interval [18, 24].

We have used quadratic functions to represent the variable production costs, i.e., $J_i(p_i) = \frac{1}{2}c_{g,i}p_i^2 + b_{g,i}p_i$, where the values of parameters $c_{g,i}$, $b_{g,i}$, with $i = 30, 31, \ldots, 39$ denoting the indices of generator busses, are taken from [4] and are listed in Table 5.1. For simplicity, no saturation limits $\underline{p}$, $\overline{p}$ have been considered. All loads are taken to be price-inelastic, with the values from [21].

The proposed distributed controller (5.17) was implemented with the following values of the gain matrices: $K_\lambda = 3I_{39}$, $K_f = 8I_{39}$. For simplicity of exposition, the

Table 5.1: Production cost parameters for generator buses.

Bus number i	$c_{g,i}$	$b_{g,i}$
30	0.8	30.00
31	0.7	35.99
32	0.7	35.45
33	0.8	34.94
34	0.8	35.94
35	0.8	34.80
36	1.0	34.40
37	0.8	35.68
38	0.8	33.36
39	0.6	34.00

line power flow limit was assigned only for the line connecting nodes 25 and 26, and both corresponding gains K_p and K_o in the controller were set equal to 1. The simulation results are presented in Figure 5.6 and Figure 5.7. In the beginning of the simulation, the line flow limit $\overline{p}_{25,26}$ was set to infinity, and the corresponding steady-state operating point is characterized by the unique price of 39.16 for all nodes. At time instant 5s, the line limit constraint $\overline{p}_{25,26} = 1.5$ was imposed. The solid line in Figure 5.6 represents the simulated trajectory of the line power flow $p_{25,26}$. In the same figure, the dotted line indicates the limits on the power flow $\overline{p}_{25,26}$. The solid lines in Figure 5.7 are simulated trajectories of nodal prices for the generator buses, i.e., for buses 30 to 39, which is where the generators are connected. In the same figure, dotted lines indicate the off-line calculated values of the corresponding steady-state optimal nodal prices. For clarity, the trajectories of the remaining 29 nodal prices were not plotted. In the simulation, all these trajectories converge to the corresponding optimal values of nodal prices as well. The optimal nodal prices for all buses are presented in Figure 5.8. In this figure, the nodal prices corresponding to generator buses $30-39$ are emphasized with the gray shaded bars. The obtained simulation results clearly illustrate the efficiency of the proposed distributed control scheme.

5.4 Conclusions and future research

In this chapter we have presented and illustrated on examples the price-based control paradigm as a suitable approach to solve some of the challenging problems facing future, market-based power systems. It was illustrated how global objectives and constraints, updated from the on-line measurements of the physical power system state, can be optimally translated into time-varying prices. The real-time varying price signals are guaranteed to adequately reflect the state of the physical system, present the signals that optimally shape, coordinate and in real or near real-time syn-

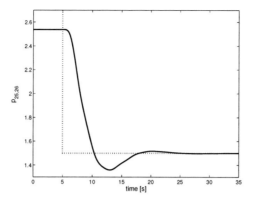

Figure 5.6: Power flow in the line connecting buses 25 and 26.

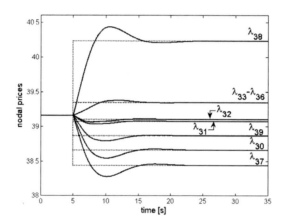

Figure 5.7: Trajectories of nodal prices for generator buses, i.e., for busses 30–39 where the generators are connected.

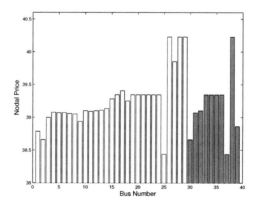

Figure 5.8: Optimal nodal prices in the case of congestion. The nodal prices corresponding to generator busses 30–39 are emphasized with the gray shaded bars.

chronize local, profit driven behaviors of producers/consumers to mutually reinforce and guarantee global objectives and constraints.

Future research will be devoted to modification of the devised price-based control scheme so that the prices are updated on the time scale of 5–15 minutes, rather than on the scale of seconds. Instead of using rapidly changing network frequency deviations as an indication of power imbalance in the system, one possibility is to use deviation of power production reference values to the generators which originate from (slightly modified) automatic generation control loops over the sampling period (i.e., over 5–15 minutes). These deviations can be used as a measure of imbalance in the system.

As a final remark, we would like to point out that in its core idea the price-based control approach presented in this chapter, which is based on a suitable dynamic extension of the Karush-Kuhn-Tucker (KKT) optimality conditions, is suitable for application in some other types of infrastructures as well. More precisely, when the system's objectives are characterized in terms of steady-state related constrained optimization problems, the time-varying price signals can be efficiently used for control purposes. In particular, the proposed approach is suitable for solving problems of economically optimal load sharing among various production units in a network. Examples of such systems include smart power grids in energy-aware buildings, industrial plants, large ships, islands, space stations or isolated geographical areas; water pumps, furnaces or boilers in parallel operation, etc. A distinguishing and advantageous feature of the presented approach is that the dynamic extension of the KKT optimality conditions preserves the structure of the underlying optimization problem, which implies that the corresponding price-based control structure can be implemented in a distributed fashion.

References

1. F. Alvarado. The stability of power system markets. *IEEE Transactions on Power Systems*, 14(2):505–511, 1999.
2. F. L. Alvarado. Is system control entirely by price feasible? In *Proceedings of the 36th Annual Hawaii International Conference on System Sciences*, Big Island, Hawaii, 2003.
3. F. L. Alvarado. Controlling power systems with price signals. *Decision Support Systems*, 40(3):495–504, 2005.
4. F. L. Alvarado, J. Meng, C. L. DeMarco, and W. S. Mota. Stability analysis of interconnected power systems coupled with market dynamics. *IEEE Transactions on Power Systems*, 16(4):695–701, 2001.
5. D. P. Bertsekas. *Nonlinear Programming*. Athena Scientific, 1999.
6. E. Bompard, P. Correia, and G. Gross. Congestion-management schemes: A comparative analysis under a unified framework. *IEEE Transactions on Power Systems*, 18(1):346–352, 2003.
7. S. Boyd and L. Vandenberghe. *Convex optimization*. Cambridge University Press, 2004.
8. M. C. Caramanis, R. E. Bohn, and F. C. Schweppe. Optimal spot pricing, theory and practice. *IEEE Transactions on Power Apparatus and Systems*, 109(9):3234–3245, 1982.

9. M. Chiang, S. H. Low, A. R. Calderbank, and J. C. Doyle. Layering as optimization decomposition: A mathematical theory of network architectures. *Proceedings of the IEEE*, 95(1):255–213, 2007.

10. R. D. Christie, B. F. Wollenberg, and I. Wangensteen. Transmission management in the deregulated enviroment. *Proceedings of the IEEE*, 88(2):170–195, 2000.

11. H. Glavitsch and F. Alvarado. Management of multiple congested conditions in unbundled operation of a power system. *IEEE Transactions on Power Systems*, 13(3):1013–1019, 1998.

12. W. P. M. H. Heemels, J. M. Schumacher, and S. Weiland. Linear complementarity systems. *SIAM Journal on Applied Mathematics*, 60(4):1234–1269, 2000.

13. W. W. Hogan. Contract networks for electric power transmission. *Journal of Regulatory Economics*, 4:211–242, 1992.

14. M. Ilic, F. Geliana, and L. H. Fink. *Power Systems Restructuring: Engineering and Economics*. Kluwer Academic Publishers, 1998.

15. A. Jokic. *Price-based optimal control of electrical power systems*. PhD thesis, Eindhoven University of Technology, Eindhoven, The Netherlands, 2007.

16. A. Jokic, M. Lazar, and P. P. J. van den Bosch. Real-time control of power systems using nodal prices. In *Proceedings of the 16th Power Systems Computation Conference*, Glasgow, UK, 2008.

17. P. Kundur. *Power System Stability and Control*. McGraw-Hill, 1994.

18. L. L. Lai. *Power System Restructuring and Deregulation*. John Wiley and Sons, 2001.

19. A. Nedic and A. Ozdaglar. Distributed subgradient methods for multi-agent optimization. *IEEE Transactions on Automatic Control*, 54(1):48–61, 2009.

20. T. J. Overbye, X. Cheng, and Y. Y. Sun. A comparison of the AC and DC power flow models for LMP calculations. In *Proceedings of the 37th Hawaii International Conference on System Sciences*, Big Island, Hawaii, 2004.

21. M. A. Pai. *Energy Function Analysis for Power System Stability*. Kluwer Academic Publishers, 1989.

22. D. P. Palomar and M. Chiang. Alternative distributed algorithms for network utility maximization: Framework and applications. *IEEE Transactions on Automatic Control*, 52(12):2254–2269, 2007.

23. F. Rubio-Oderiz and I. Perez-Arriaga. Marginal pricing of transmission services: A comparative analysis of network cost allocation methods. *IEEE Transactions on Power Systems*, 15:448–454, 2000.

24. H. Saadat. *Power System Analysis*. McGraw-Hill, 1999.

25. J. M. Schumacher. Complementarity systems in optimization. *Mathematical programming B*, 101:263–296, 2004.

26. F. C. Schweppe, M. C. Caramanis, and R. D. Tabors. Evaluation of spot price based electricity rates. *IEEE Transactions on Power Apparatus and Systems*, 104(7):1644–1655, 1985.

27. F. C. Schweppe, M. C. Caraminis, R. D. Tabors, and R. E. Bohn. *Spot Pricing of Electricity*. Kluwer Academic Publishers, 1988.

28. F. C. Schweppe, R. D. Tabors, J. L. Kirtley, H. R. Outhred, F. H. Pickel, and A. J. Cox. Homeostatic utility control. *IEEE Transactions on Power Apparatus and Systems*, 99(3):1151–1163, 1980.

29. Y.-H. Song and X.-F. Wang. *Operation of market oriented power systems*. Springer, 2003.

30. S. Stoft. *Power System Economics: Designing Markets for Electricity*. Kluwer Academic Publishers, 2002.

31. F. F. Wu and P. Varaiya. Coordinated multilateral trades for electric power networks: Theory and implementation. *International Journal of Electrical Power Energy Systems*, 21(2):75–102, 1999.

Chapter 6
Survivability and Reciprocal Altruism: Two Strategies for Intelligent Infrastructure with Applications to Power Grids

P. Hines

Abstract While electric power grids are generally robust to small failures and thus provide a fairly high level of reliability, they are notably vulnerable to large, spectacular cascading failures. Single component failures rarely impede the ability of a power grid to serve its customers. But larger sets of concurrent outages can produce blackouts with sizes that are highly improbable from the perspective of Gaussian statistics. Because of the number of components in a power grid it is impossible to plan for and mitigate all sets of failures. Maintaining a high level of reliability in the midst of this risk is challenging. As market forces, variable sources (e.g., wind and solar power) and new loads (e.g., electric cars) increase stress on electricity infrastructure, the challenge of managing grid reliability and costs will certainly increase. Therefore we need strategies that enable the most important services that depend on electricity infrastructure to continue in the midst of risks. This chapter discusses two strategies for enabling the most important services that depend on electricity continue in the midst of significant systemic vulnerability. The first, as proposed by Talukdar et al. [26] is survivability, in which backup electricity sources provide a very high level of reliability for services that are economically and socially vital. The second, as proposed by Hines et al. [16], is Reciprocal Altruism, under which agents that manage the infrastructure are encouraged to align personal goals with those of the system as a whole.

P. Hines
University of Vermont, School of Engineering, Burlington, Vermont,
e-mail: paul.hines@uvm.edu

R.R. Negenborn et al. (eds.), *Intelligent Infrastructures*, Intelligent Systems, Control and
Automation: Science and Engineering 42, DOI 10.1007/978-90-481-3598-1_6,
© Springer Science+Business Media B.V. 2010

6.1 Introduction

About 25% of primary energy is consumed in the production of electricity. After conversion losses, about half of this is delivered to consumers over the global electricity infrastructure [5, 11]. This percentage is substantially higher in many developed countries, indicating that electricity consumption will increase as less developed countries industrialize (see Figure 6.1). With growing interest in electric-drive vehicles, power grids may be asked to carry some of the energy burden that has traditionally been carried by liquid fuels. And with growing concern about the climate impact of fossil fuel combustion, the electricity infrastructure is increasingly being asked to transport energy from non-traditional sources, such as wind and solar. In order to facilitate continued economic growth, particularly in less developed countries, we need a reliable electricity infrastructure. Since the electricity infrastructure is the primary transport network for reduced carbon energy sources (i.e., energy sources such as solar, wind, geothermal, hydro, nuclear, coal with carbon capture and sequestration), a reliable electricity infrastructure is also vital to environmental goals.

Electricity infrastructures, or power grids, can be divided into three sets of physical components: generation, transmission, and distribution. Generators convert primary energy into electric energy and inject that energy, through transformers, into the high-voltage transmission system. The transmission system moves the energy, as an alternating current (AC) wave, over high voltage (50 kV–1000 kV) transmission lines that can move large amounts of energy with small (about 3%) losses. The distribution system delivers the energy from the high-voltage system to customer locations over medium voltage (5 kV–50 kV) lines. At customer locations, smaller transformers bring the medium voltage electricity down to safer low voltages at which electric energy is consumed. Electricity infrastructure is thus the combination of the physical infrastructure of generation, transmission, and distribution, along with the the human organizations and information technology (IT) that work to manage these assets.

6.1.1 Properties of power networks

Electric energy networks differ from other transport networks in a number of important ways. For one, electricity networks cannot store any appreciable quantity of product at a given point in time. For example, a 100 km, 1 GW transmission line carries approximately 300 MJ at any point of time, whereas a 100 km (0.5 m) petroleum pipeline can hold approximately 900 TJ. Without significant storage in the network, generation and consumption rates must match at every instant in time. In order to keep this balance between supply and demand, operators constantly adjust mechanical power inputs through a variety of market and technical mechanisms. When the balance between mechanical power input (P_m) and electrical power output (P_g) is not maintained, generator speeds (ω) deviate from nominal according to the differential equation:

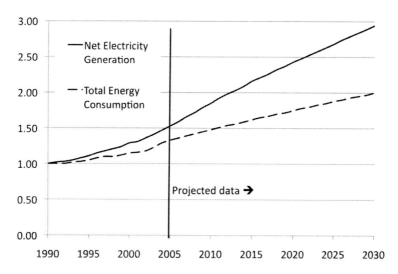

Figure 6.1: Projected growth of electricity consumption relative to total global energy consumption, relative to 1990 levels (1990 consumption = 1). Source: US Dept. of Energy [11].

$$P_m = P_g + K_D \omega + K_H \frac{d\omega}{dt} \ , \tag{6.1}$$

where K_D and K_H are damping and inertia constants for each generator [19]. If this balance is not maintained generator speeds (ω) deviate from nominal, which can result in relays disconnecting (tripping) generators from the grid. Generator failures can in turn further exacerbate power imbalances and initiate a cascading sequence of component failures. Cascading failures generally end with the failure of service to customers: a blackout. Figure 1 shows the dramatic frequency fluctuation, resulting from power imbalances, during the European blackout of 28 September 2003. Within competitive electricity markets the balance between supply and demand is maintained through frequent auctions for energy and ancillary services. Because there is very little storage in the system, prices for electric energy fluctuate more rapidly than for other energy forms (see Figure 6.3).

A second peculiar property of electricity networks is the inability to direct the flow of the product (electric energy) through the network. In traffic, airline, and even some pipeline networks operators control a product's path of travel between its source and the destination. In electricity networks, energy flows are governed by a set of non-linear network equations, namely Kirchhoff's laws and Ohm's law. These equations can be written using a complex vector of nodal power injections ($\mathbf{S} = \mathbf{P} + j\mathbf{Q}^1$), which is a function of power supply ($\mathbf{S}_G = \mathbf{P}_G + j\mathbf{Q}_G$) and demand ($\mathbf{S}_D = \mathbf{P}_D + j\mathbf{Q}_D$), a complex vector of node (bus) voltages ($\mathbf{V}$), and a network structure matrix ($\mathbf{Y}$) that comes from the locations and properties of network connections

---

[1] In this chapter the electrical engineering convention of using j to represent the complex number $j = \sqrt{-1}$ is followed. i is used either as an index x_i or to represent current: I.

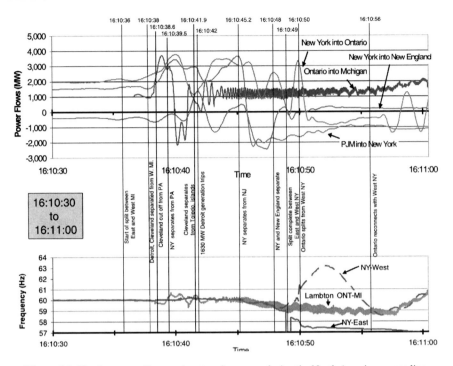

Figure 6.2: Varying power flows and system frequency during the North American cascading failure of 14 August 2003. Source: US-Canada Power System Outage Task Force [29].

as:[2]

$$S(S_G, S_D) = V \odot (YV)^* . \tag{6.2}$$

Equations (6.1) and (6.2) link together through the real portion of the power injection at each generator bus, $P_g = \Re(S_g)$, the electromagnetic connection between the rotor angle of each generator ($\delta_i = \int \omega_i dt$), and the voltage phase angle (θ_i) at the terminals of a generator. The phase angle is the angle of the complex node voltages V:

$$V_i = |V_i|(\cos \theta_i + j \sin \theta_i) = |V_i|e^{j\theta_i} . \tag{6.3}$$

Using voltage in its polar form (6.3), Equation 6.2 becomes:

$$P_i = \Re(S_i) = |V_i| \sum_{j=1}^{n} \left(g_{ij}|V_j|\cos(\theta_i - \theta_j) + b_{ij}|V_j|\sin(\theta_i - \theta_j) \right) \tag{6.4}$$

$$Q_i = \Im(S_i) = |V_i| \sum_{j=1}^{n} \left(g_{ij}|V_j|\sin(\theta_i - \theta_j) - b_{ij}|V_j|\cos(\theta_i - \theta_j) \right) . \tag{6.5}$$

---

[2] In this chapter, $x \odot y$ represents the element-wise product of vectors x and y and x^* represents the complex conjugate of complex number x.

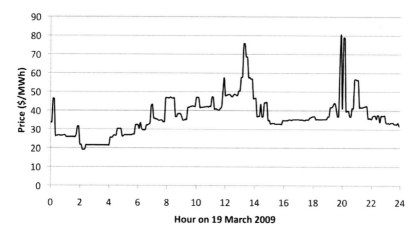

Figure 6.3: The spot price of electricity at an arbitrary node in the Northeastern US. While most energy prices are volatile, electricity prices frequently change by an order of magnitude or more within a single day. Source: ISO-New England.

A third important property of electricity networks is the discrete dynamics created by relays that protect the components of a power grid. Hundreds of thousands of relays monitor for measures of stress, such as high current, low voltage, off-nominal frequency or temperature, within transmission lines, generators, and transformers. Relays are essentially selfish devices. When they see excessive stress on their assigned component, they disconnect the device from the power grid. As relays were in existence long before communications technology they typically do not communicate with one another, or with operators, before making decisions. Because of their limited communications, traditional relays do not, and cannot, consider the impact of a given decision on the system as a whole. Rather they seek to protect their assigned component independent of the effect their actions will have on the global infrastructure.

Finally, because of the complex physical dynamics of power networks, and the importance of electricity to daily life, thousands of humans and human organizations constantly interact with the physical grid and with each other in electricity-related transactions. Power plant operators manage generators, transmission system operators manage transformers and transmission lines, energy traders facilitate financial transactions and regulators seek to enforce rules. As evidence of these social dynamics, Box 6.1 shows a conversation between grid operators during the August 13, 2003 blackout in North America. Each of these human agents must have an understanding of both the physical network over which energy is transported and the social networks that link the agents in cooperation and competition. These social dynamics form a vital, but little understood, component of the electricity infrastructure.

The result is an infrastructure system with complex continuous, discrete, and social dynamics. When they function well power grids reliably deliver useful energy to consumers at low costs and with minimal environmental impact. When not performing well, electricity causes massive pollution, extreme price swings that can result in dramatic political change, and blackouts that can trigger massive social upheaval. The high electricity prices in Western North America in 2000 and 2001 resulted in emergency elections, which brought California Governor Arnold Schwarzenegger into power in 2003. The New York city blackout of 1977 triggered massive riots and looting, which led to police action and about 3000 arrests.

Box 6.1: A telephone conversation between two Midwestern United States power grid operators at 17:49 on 14 August 2003, as the blackout was unfolding. The social interactions among grid operators are an important component of the electricity infrastructure. Source: US House of Representatives [31].

MISO/Tim Johns: Midwest ISO, this is Tim.

Hoosier Energy/Bob: Yes, this is Bob at Hoosier.

MISO/Tim Johns: Hey Bob.

Hoosier Energy/Bob: What do you know, buddy?

MISO/Tim Johns: Same old stuff, man.

Hoosier Energy/Bob: Having just a quiet night, kicking back, watching TV. Is that what's going on up there?

MISO/Tim Johns: What, watching TV? Sure. Yes.

Hoosier Energy/Bob: I understand. Busy, man. This is kind of a strange thing, man.

MISO/Tim Johns: Pretty much.

Hoosier Energy/Bob: Yes, it is. No, Tim, I just came in a little bit, you know. Just ciphering things up.

MISO/Tim Johns: We're still ciphering up here.

Hoosier Energy/Bob: Do you have any kind of a — kind of a mock diagram of that region that's affected showing transformers and power flow and stuff like that that you could like fax, just to look at or something?

MISO/Tim Johns: I don't have one, no.

Hoosier Energy/Bob: Oh, okay.

MISO/Tim Johns: I mean we've got big huge ones, but I don't have one that I could fax anywhere.

Hoosier Energy/Bob: Oh, okay.

6.2 Stress, blackouts, and complexity

Understanding the physical and social dynamics of a given infrastructure is critical to the development of intelligent methods for improving the performance of that infrastructure. As previously noted, large blackouts are an important outcome of these dynamics in electricity systems. Blackouts come from a wide variety of causes, which can be divided into two categories: exogenous causes and cascading failures. All blackouts begin with a set of exogenous events that bring stress to the grid, such as human errors or attacks, ice storms, heat waves or hurricanes. In some cases these initial causes remove enough of the grid from service to result in an immediate loss of service (a blackout). When high winds damage distribution lines, which are typically radial in structure, customers will immediately loose electricity services. The transmission system, on the other hand, is designed and operated such that single component failures do not result in service interruptions (see, for example, the North American Electric Reliability Corp., NERC, reliability standard TOP-001-1 [22]). This "$n-1$" security rule does not prevent multiple simultaneous failures from causing a blackout. Exogenous failures can cause immediate blackouts in both the transmission and distribution systems.

However, some exogenous failures do not immediately cause blackouts, but rather create stress in the power grid in the form of high currents, low voltages and off-nominal frequency. This initial stress can expose hidden failures in system components. When the "selfish" relays in a power grid act to protect their assigned local components this stress is shifted, via Kirchhoff's laws (6.2), to other areas of the grid. This relocated stress can trigger more relay actions, which again relocate stress. This cycle, known as a cascading failure, can substantially exacerbate the affects of the initial exogenous failures. For example, the 10 August 1996 blackout in Western North America was triggered by a fault between a transmission line and a tree. This exposed a hidden relay failure, which disconnected a parallel line. This change in transmission capacity led to the failure of a large generator, which triggered a long sequence of transmission and generation failures, and the loss of service to 7.5 million customers [35]. Table 6.1 describes several of the largest cascading failures in Europe and North America.

In a number of infrastructure systems technology and policy changes have resulted in notable improvements in the frequency of large failures. As a result of improved air traffic control systems in the US the number of fatal commercial airline accidents has decreased from 27 fatalities per 100 million kilometers in 1960 to 0.7 in 2000 [2]. However, at least for the period for which data are available, the electricity infrastructure has not seen this same improvement. The median number of large blackouts (those larger than 300 MW, after adjusting for demand growth) has increased from 11 per year for 1984-1995 to 17 per year for 1996-2006 [15]. Despite the improvements in technology over this time period, the frequency of large blackouts is not decreasing.

If we look at the empirical cumulative distribution function (cdf) of blackout size data, we find a particularly important trend. The largest North American blackouts (those 1000 MW or larger, after adjusting for demand growth) follow a power-law, or Pareto probability distribution (see Figure 6.4). If one ignores this fact and performs decision analysis assuming exponential statistics or even using empirical means to calculate expected values, the value of technology and policy that reduce the size of large blackouts will be grossly underestimated. To make good decisions about investments in intelligent electricity infrastructure, this statistical trend needs to be included in the calculation of benefits and costs (see [15] for example calculations).

Power-law distributions appear in a number of contexts, including in the degree (# of connections) of devices in the physical Internet, the fatality of wars in the 20th

Table 6.1: Some notable cascading failures in North America and Europe. The far right column shows remedial actions that would likely have substantially reduced the social costs associated with the event.

Date	Location	Size	Causes and outcomes	Remedial actions that could have reduced costs
9-Nov-1965	Northeast US	30,000,000 pers. 20,000 MW	A relay on lines from Niagara Falls to Toronto was set too low and tripped, triggering a cascade throughout the region. The event led to the creation of NYPP and NERC.	Immediate reduction of load in Toronto and generation at Niagara would likely have reduced the consequences [13, 30].
13-July-1977	Northeast US	9,000,000 pers.	Three transmission lines opened due to lightning, resulting in the loss of generation in NY. Con Edison system separated from grid within 30 min. Led to widespread use of UFLS relays.	Shed load and/or increase gen. in the NYPP area after the initial generation loss [12].
29-Feb-1984	Western US	3,160,000 cust.	Transmission line fault in OR initiated cascading failure. Controlled separation scheme did not operate as intended.	Automatic load shedding reduced the consequences of the event after separation. Quickly reducing N-S flows may have prevented the separation [10].
13-Mar-1989	Hydro Quebec, Northeast US	19,400 MW	Solar flare caused 5 735 kV transmission lines to trip, initiating a cascading failure.	Quickly reduce HQ area demand by 9000 MW [10].
10-Aug-1996	Western US	7,500,000 cust.	500 kV line sagged into a tree. The line and one parallel to it tripped, initiating a cascade.	Reduce flows on the CA-OR intertie within 5 min. of initial events. Increase reactive power in North along Columbia R. [18, 35].
14-Aug-2003	Northeast US, Canada	50,000,000 pers. 57,669 MW	Transmission cables in OH contacted trees, initiating a cascading failure.	Load reduction, and/or reactive power voltage support near Cleveland would have dramatically reduced the size of the event [29].
28-Sept-2003	Southern Europe	57,000,000 pers.	Transmission line contacted tree, initiating a cascade, resulting in lost service to most Italian customers, and some of Italy, Switzerland.	After initial event, operators reduced imports by 300 MW. Imports should have been reduced by much more [9, 28].
4-Nov-2006	Germany, France	15,000,000 cust.	Operators disconnected a double circuit line over the Ems River to allow a ship to pass, triggering a cascading failure, splitting the European grid into 3 regions.	30 min. after initial event E.ON operators implemented incorrect remedial switching actions. Reducing flow on the Landesbergen-Wehrendorf line would have reduced blackout size [32].

century, and the severity of terrorist attacks [8]. Albert et al. [1] found a causal relationship between the structure of networks with power-law connectivity (scale free networks) and the sizes of failures within those networks. One could thus conjecture that the observed power-law distribution in failure sizes results from a power-law in grid topology. However, in power systems, this relationship does not exist. The distribution of node connectivity (degree distribution) in power grids is clearly exponential (see Figure 6.4), whereas the distribution of failure sizes clearly follows a power law. Therefore, we need other explanations for the surprisingly high probability of large failures. Carreras et al. [7] built a long-term model of cascading failures and concluded that this pattern is the result of self-organized criticality [4]. They found that the processes of optimal dispatch and investments in response to large failures leads to a system that self-organizes to a state of near collapse, and thus gives a power-law distribution in blackout sizes. Carreras et al. noted that, "apparently sensible attempts to mitigate failures in complex systems can have adverse effects and therefore must be approached with care."

The observation that power grids frequently operate at or near the point of collapse is particularly important as the power grid is used to transport more energy from variable sources such as wind and solar. When the output of power sources varies raipdly with time the system will likely approach operating limits more frequently. Intelligent systems that can mitigate the risks associated with fluctuations will have tremendous value in such systems. Using the standard definition of risk as probability times cost, we need to find methods to reduce the probability and/or cost of large blackouts. It is nearly impossible to reduce the probability of exogenous failures, as the forces of nature and human error or malice will continue to affect infrastructure damage on occasion. However, there are strategies that can reduce the cost of the blackout that results from a given set of exogenous failures. This chapter focuses on two strategies for reducing blackout costs: survivability and reciprocal altruism.

6.3 Survivability

According to Lipson and Fisher [20], survivability is the "ability of a system to fulfill its mission, in a timely manner, in the presence of attacks, failures, or accidents." Electricity systems serve a wide variety of life and economically critical services. Hospitals, police stations, emergency response call centers and even traffic lights can be life critical in an urban area, particular in the case of an urban emergency such as a blackout. Noting this application, Talukdar et al. [26] suggest a seven step process applying the survivability concept to services that rely on electric energy. This process was tested and refined in an analysis of infrastructure in the US city of Pittsburgh [3, 21]. These seven steps are as follows:

1. *Identify life critical and economically important missions (services) that require electric energy.* In the Pittsburgh study [21] we identified the following services: hospitals, police, emergency call centers (911), water, sewage,

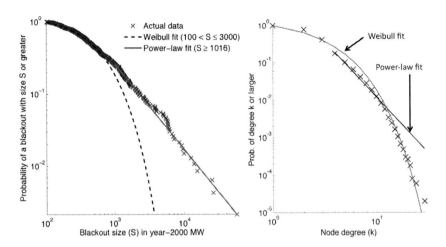

Figure 6.4: Large blackouts are more frequent than one would expect from exponential statistics (left), but power grids show a clearly exponential degree distribution (node connectivity). The figure at left shows the cumulative distribution of blackout sizes in North America (Source: Hines, Apt and Talukdar [15], confirming results reported in [7]). The figure at right shows the cumulative distribution of node degree, or the number of connections per bus, for the Eastern North American power grid.

telecommunications, petroleum (gasoline, natural gas, diesel) distribution, solid waste collection, transportation (air, roads, tunnels and public transport), grocery stores, banking, elevators and building lighting.

2. *Identify a set of reference blackout scenarios.* In [21] three reference events were identified: a single neighborhood blackout that lasted for hours, a city-wide event that lasted for 2.5 days, and a regional event that lasted for two weeks.

3. *Prioritize critical services for each blackout scenario.* Priorities may differ, depending on the scenario. For example traffic lights are particularly vital during the first few hours of a blackout to ensure continued flow of automotive traffic in the immediate aftermath of a power grid failure. After this point, other services, such as hospitals, police services and heating and cooling become particularly important.

4. *Determine the extent to which services have adequate survivability.* Many hospitals, for example, have well-maintained backup generators. Many, but not all, police stations and EMS call centers have adequate backup energy plans. Many transportation systems however do not have adequate survivability.

5. *Identify systems that require additional investment or procedural changes to ensure an adequate level of service after a blackout.* The most common technologies for ensuring services include battery systems for relatively short-term storage and petroleum-based backup generators for services that need to survive over a longer time period.

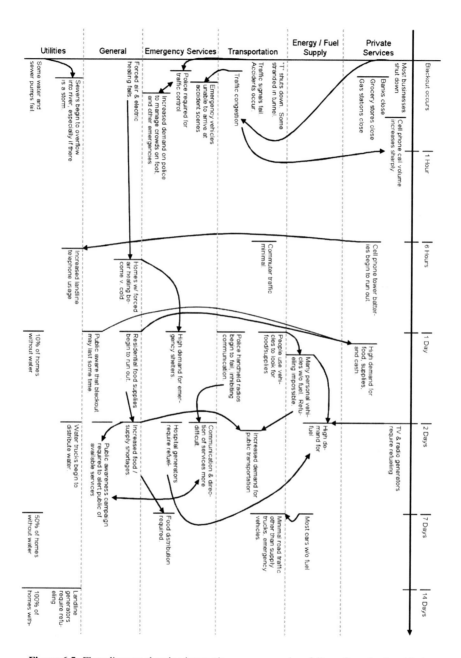

Figure 6.5: Flow diagram showing interactions among services failures for a daytime black-out in the Pittsburgh region. These interactions were identified from extensive interviews with city officials. Source: Meisterling et al. [21].

6. *Design and implement infrastructure and policy changes.* Since resources are finite, it is not possible to install backup generation or batteries for all services, so the weighing of costs and benefits becomes particularly important. The list of priorities described in step three becomes particularly important at this stage. In some cases emergency energy sources, such as a diesel generator or a gas combustion turbine, can have secondary benefits. In the US city of Portland, Oregon, the local electricity supplier (PGE) has developed an information technology network that allows PGE to remotely activate backup generators when the price of electricity exceeds the cost of operating these backup generators [23]. As broadband communications become increasingly available to appliances connected to the power grid, the possibility for multiple-use energy technology becomes increasingly feasible. Distributed combined heat and power generators can become particularly beneficial as an asset that provides thermal energy, a hedge against high electricity prices, and survivability. In some cases policy changes are needed to facilitate adoption of such technology, as is the case with distributed generation in many US locations.

7. *Design systems for allocating scarce resources during a blackout.* For example, methods to distribute food, water, blankets, and diesel fuel become particularly important if the traditional commercial outlets for these services fail.

The Pittsburgh study identified a number of new of areas where the expected benefits of additional survivability investments exceeded costs. One of the most important is in the area of traffic signaling. As cities replace incandescent traffic lights with low-power LED ones, it is possible to keep traffic lights operating for hours after a blackout with a relatively small battery in each intersection. The installation of batteries for traffic lights along vital urban exit corridors and along routes that are near medical and emergency services are good candidates for additional investment. Also, we found that most gasoline service stations did not have backup generators, which could lead to potentially vital energy being unavailable in subsurface storage tanks.

Finally, in the Pittsburgh study we found that the interactions among service failures can lead to problems that would not result from any single service failure. The loss of electricity in an urban area can lead to the failure of other systems that do not immediately rely on electricity. Figure 6.5 describes potential interactions among service failures for an extended blackout in the Pittsburgh region.

6.4 Adaptive reciprocal altruism

As previously noted, the relays that protect the components of a power grid are, in the vast majority of cases, selfish devices. They make decisions with local information with the narrow goal of minimizing damage to the physical infrastructure. The reasons for this are many. Historically, there has been very little technology in power grids enabling devices to share information. IT enabled Supervisory Control And Data Acquisition (SCADA) and Energy Management Systems (EMS) have enabled increased information sharing but much of this sharing is centralized; information is collected periodically at a central location, from which control decisions are made. Because SCADA systems tend to collect data on rather slow time scales (seconds rather than microseconds), time-critical decisions must be made by local agents. Relays (many of which are still electro-mechanical, as opposed to solid-state electronic, agents) needed to make time-critical decisions and thus needed to make decisions using only locally available information. Advances in communications technology have enabled some limited cooperative behaviour among agents in power grids. Pilot relays for example use dedicated communications channels between opposite ends of a transmission line to exchange information that allows them to more quickly differentiate between high currents resulting from faults and high currents resulting from unusual operating conditions. In the case of pilot relaying, a small amount of cooperation can have a significant, socially valuable, effect in terms of reducing the chance of a very large blackout.

Special Protection Schemes (SPS) and Remedial Action Schemes (RAS) are a broad class of technology aimed at improving the resistance of a power system (as a whole) to cascading failures. Zima et al. [37] provide a review of state-of-the-art SPS technology. Most of the schemes that have been implemented in industry are pre-programed schemes operated using SCADA data from a central location. They are not adaptive in that they do not respond well to the constantly changing state of the power grid. Some researchers have proposed SPS designs that are more adaptive. Most of these are optimization-based methods, which choose control actions based on measured state information and a set of goals deemed appropriate to the operation of power grids under stress [17, 24, 25, 36].

However, even if one can improve the adaptivity of power systems with goal-based methods and feedback, centralized schemes have significant limitations in terms of scalability, actuation speed, security and robustness to failures. A well-designed decentralized method would be less vulnerable to attacks or random failures at the central location, would be able to react more quickly to local conditions, and would have the security benefits associated with a diversity of agents. Finally, and most importantly, a well designed decentralized scheme would directly address the problem of selfish local agents by augmenting local goals to more carefully align with global system goals.

The following sections describe the global goals for a power grid under stress, and describe how reciprocally altruistic agents can be used to solve this problem.

6.4.1 Global goals for operating power grids under stress

In order to mitigate stress before cascading failures result, it is important to define a set of global goals for the optimal operation of electricity infrastructures under stress. The methods described here are based on the following goals (objectives and constraints) for operating a power grid under stress:

1. Minimize the risk associated with cascading failure.
2. Minimize the costs associated with mitigating risks.

3. Operate the physical infrastructure within its physical limits.

If we consider a discrete time horizon $t_0...t_K$, where $\mathbf{u}_k$ and $\mathbf{x}_k$ are control and state vectors respectively, these goals can be formulated as a Model Predictive Control [6] problem as follows:

$$\min_{\mathbf{u}_1...\mathbf{u}_K} \quad R(\mathbf{x}_0...\mathbf{x}_K) + C(\mathbf{u}_0...\mathbf{u}_K) \tag{6.6}$$

$$\text{subject to} \quad \mathbf{x}_k = f(\mathbf{x}_{k-1}, \mathbf{u}_{k-1}, \mathbf{u}_k) \; \forall k \tag{6.7}$$

$$\underline{\mathbf{x}} \leq \mathbf{x}_k \leq \overline{\mathbf{x}} \; \forall k \tag{6.8}$$

$$\underline{\mathbf{u}} \leq \mathbf{u}_k \leq \overline{\mathbf{u}} \; \forall k \tag{6.9}$$

$$\underline{\Delta \mathbf{u}} \leq \mathbf{u}_k - \mathbf{u}_{k-1} \leq \overline{\Delta \mathbf{u}} \; \forall k . \tag{6.10}$$

This problem formulation assumes that the risk (R) and cost (C) functions can be formulated such that the sum represents the aggregate expected cost in some way.

In Hines and Talukdar [14, 16] this formulation is specifically adapted to the problem of optimally planning for and choosing emergency risk mitigation actions in a power grid. In the example results shown below the measures of risk (stress variables) are low voltages and high currents and the control variables include generator power output, generator voltage set points (or reactive power generation), and manual load reduction (load shedding). With this general problem formulation it is not difficult to add additional control or state variables to the problem, or even to adapt the formulation to that of mitigating stress in other infrastructure systems.

6.4.2 Reciprocally altruistic agents

According to Trivers [27] altruism is, "behavior that benefits another organism, not closely related, while being apparently detrimental to the organism performing the behavior." In other words an agent acts altruistically when it considers the goals of other agents when making local decisions, even if doing so could be detrimental to its local goals. Reciprocal altruism occurs when agents choose to consider the goals of other agents (though not necessarily all other agents), while assuming that some (but not all) other agents will act reciprocally. Trivers [27] provides examples in a number of biological systems including pseudo-parasitic fish species and humans. Wilkinson [33, 34] found that Vampire bats had evolved a particularly interesting form of Reciprocal Altruism in which individuals would share food with agents who had not been successful in finding food, even when there was not a direct genetic connection between the agents. In other words agents spontaneously choose to consider the goals of neighbors while making local decisions.

In the power grid case, we have a system in which selfish agents can effectively manage local goals (relays and other autonomous control agents), but sometimes exacerbate global problems, which sometimes leads to cascading failures. It would be useful to get these agents to act a bit more altruistically.

In [14, 16] the author describes a method for taking global goals, such as those described in (6.6)-(6.10), assigning them to control agents in a power grid and de-

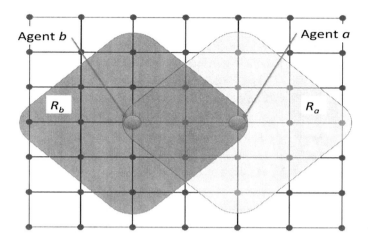

Figure 6.6: Agents a and b with their reciprocal sets (R_a, R_b) chosen based on a graph distance of $r = 2$.

signing the agents to share goals and measurements such that with a reasonable level of inter-agent information exchange the agents can arrive at solutions that approach what one would get from an omniscient optimizing agent. Figure 6.6 describes this process pictorially. Essentially we place one agent at each node in a power grid. Each agent chooses a set of neighboring agents to share goals and information with. When risky circumstances are observed in local measurements or in data from neighbors the agents use a linearized version of (6.6)-(6.10) to calculate mitigating control actions. The result of each calculation is an estimate of what needs to be done now and in the future, and what actions should be taken locally and by remote agents. After calculating mitigating actions each agent engages in a simple negotiation protocol with its neighbors to refine its estimate of what needs to be done, implements the actions that need to be taken locally, and then waits to measure the effects of these actions before repeating the process. The agents act reciprocally altruistically in that they share goals with their neighbors, and they work adaptively by using feedback from local and remote sensors to update models that are used to predict the effects of mitigating control actions.

6.4.3 Example results

The performance of these reciprocally altruistic control agents is measured by simulating agents interacting with a simulation of the IEEE 300 bus power network. Ten permutations of the IEEE 300 bus power network provide an initial set of random, unstressed conditions. To each of these, we apply 10 random sets of 4 and 15 branch outages to produce 100 cascading failure scenarios. Without any mitigating control actions the simulated cascading failures result in an average demand loss of 7,593 MW, or 32.6% of the initial demand. The simulated blackout costs, which are

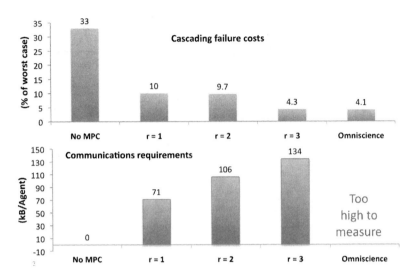

Figure 6.7: Average blackout costs and communications requirements for 100 cascading failure simulations. Each case is simulated with no mitigating control actions (left), with reciprocally altruistic agents (center) and with one omniscient agent (right). As the amount of inter-agent communication increases the quality of the control results increases to approach what one obtains from a global controller with infinite computational and communications abilities.

based on randomly assigned demand-loss costs, range from \$39 to \$147,000,000, with a mean of \$5,600,000.

After calculating blackout sizes without mitigating control actions we simulate the 100 cases with adaptive reciprocally altruistic control agents with varying amounts of inter-agent communications. The agent-to-agent communication is varied by increasing the size of the agents' reciprocal sets from nearest neighbor ($r = 1$) to third neighbor ($r = 3$). Figure 6.7 shows the outcome of these experiments, in terms of blackout costs and communications requirements.

6.5 Conclusions and future research

Survivability and reciprocally altruistic agents can be powerful guiding principles in the design of intelligent infrastructure systems. Both have the effect of reducing the costs associated with large infrastructure failures and thus increasing the robustness of the infrastructure to failures. However as power grids and other infrastructures are engineered to become more robust to failures this additional robustness can be redirected to increase the economic and/or environmental performance of the infrastructure. As the grid is more able to deal with stress with minimal costs it becomes increasingly feasible to push the infrastructure closer to its physical capacity limits. Certainly this process has already occurred to some extent, particularly in North America and Europe. With competition among electric energy suppliers, long distance transactions are increasingly common, pushing the physical infrastructure closer to its limits, whereas very little new transmission capacity has been built. There are risks to the process of increasing robustness and then using this robustness to increase efficiency. Carrerras et al. [7] conclude from a simple multi-decade power system model that this process is precisely what causes power-grids to see large failures more frequently than we would expect from exponential statistics. Thus caution is necessary when using gains from increased robustness to increase other dimensions of performance.

While it is difficult to precisely measure the nature of the long term benefits of survivability and reciprocally altruistic agents it is fairly clear that substantial benefits can come from these strategies. Survivability enables critical services to continue in the midst of electricity infrastructure failure. Reciprocally altruistic agents would give the infrastructure good reflexes so that the components of the grid react to stress in ways that better align with global goals. These two strategies together can be powerful tools in the efforts to improve infrastructure systems.

Future research will focus on the development of theoretically tractable models of reciprocal altruism, and the identification of other applications for the proposed approaches.

References

1. R. Albert, H. Jeong, and A.-L. Barabasi. Error and attack tolerance of complex networks. *Nature*, 406:378–382, 2000.
2. J. Apt, L. B. Lave, S. Talukdar, M. G. Morgan, and M. Ilić. Electrical blackouts: A systemic problem. *Issues in Science and Technology*, 20(4):55–61, 2004.
3. J. Apt, M. G. Morgan, P. Hines, D. King, N. McCullar, K. Meisterling, S. Vajjhala, H. Zerriffi, P. Fischbeck, M. Ilić, L. Lave, D. Perekhodstev, and S. Talukdar. Critical electric power issues in Pennsylvania: Transmission, distributed generation and continuing services when the grid fails. Technical report, Carnegie Mellon University Electricity Industry Center, 2004.
4. P. Bak, C. Tang, and K. Wiesenfeld. Self-organized criticality. *Physical Review A*, 38:364–374, 1988.
5. BP. BP statistical review of world energy.
6. E. F. Camacho and C. Bordons. *Model Predictive Control*. Springer, London, UK, 2004.
7. B. A. Carreras, D. E. Newman, I. Dobson, and A. B. Poole. Evidence for self-organized criticality in a time series of electric power system blackouts. *IEEE Transactions on Circuits and Systems – I: Regular Papers*, 51:1733–1740, 2004.
8. A. Clauset, C.R. Shalizi, and M.E.J. Newman. Power-law distributions in empirical data. *SIAM Review*, (to appear), 2009.
9. Commission de Regulation de L'Energie and Autorita per l'energia electrica e il gas. Report on the events of September 28th, 2003 culminating in the separation of the Italian power system from the other UCTE networks, 2004.
10. Disturbance Analysis Working Group. DAWG database: Disturbances, load reductions, and unusual occurrences. Technical report, North American Electric Reliability Council, 2006.
11. EIA. International energy outlook. Technical report, US DOE Energy Information Agency, 2008.
12. FERC. The con Edison power failure of July 13 and 14, 1977. Technical report, US Department of Energy Federal Energy Regulatory Commission, 1978.

13. G. D. Friedlander. What went wrong VIII: The great blackout of '65. *IEEE Spectrum*, 1976.
14. P. Hines. *A Decentralized Approach to Reducing the Social Costs of Cascading Failures*. PhD thesis, Carnegie Mellon University, 2007.
15. P. Hines, J. Apt, and S. Talukdar. Large blackouts in North America: Historical trends and policy implications. CEIC Working paper, January 2009.
16. P. Hines and S. Talukdar. Reciprocally altruistic agents for the mitigation of cascading failures in electrical power networks. In *Proceedings of the International Conference on Infrastructure Systems*, Rotterdam, The Netherlands, 2008.
17. I. A. Hiskens and B. Gong. MPC-based load shedding for voltage stability enhancement. In *Proceedings of the 44th IEEE Conference on Decision and Control*, Seville, Spain, 2005.
18. D. N. Kosterev, C. W. Taylor, and W. A. Mittelstadt. Model validation for the August 10, 1996 WSCC system outage. *IEEE Transactions on Power Systems*, 14:967–979, 1999.
19. P. Kundur. *Power System Stability and Control*. Electric Power Research Institute/McGraw-Hill, 1993.
20. H. F. Lipson and D. A. Fisher. Survivability: A new technical and business perspective on security. In *Proceedings of the 1999 New Security Pardigms Workshop*, Ontario, Canada, 1999.
21. K. Meisterling and P. Hines. Sustaining Pittsburgh's vital services when the power goes out. Technical report, Carnegie Mellon University, 2004.
22. NERC. *Reliability standards for the bulk electric systems of North America*. North American Electric Reliability Corporation, May 2009.
23. M. Osborn. Virtual peaking network: Customer owned – PGE managed. In *Proceedings of the Grid-Interop Forum*, Albuquerque, New Mexico, 2007.
24. W. Shao and V. Vittal. Corrective switching algorithm for relieving overloads and voltage violations. *IEEE Transactions on Power Systems*, 20:1877–1885, 2005.
25. W. Shao and V. Vittal. LP-based OPF for corrective FACTS cotrol to relieve overloads and voltage violations. *IEEE Transactions on Power Systems*, 21:1832–1839, 2006.
26. S. N. Talukdar, J. Apt, M. Ilić, L. B. Lave, and M. G. Morgan. Cascading failures: survival versus prevention. *The Electricity Journal*, 16(9):25–31, November 2003.
27. R. L. Trivers. The evolution of reciprocal altruism. *Quarterly Review of Biology*, 46:35–57, 1971.
28. Union for the Co-ordination of Transmission of Electricity. Final report of the investigation committee on the 28 September 2003 blackout in Italy, 2004.
29. US-Canada Power System Outage Task Force. Final rport on the August 14, 2003 blackout in the United States and Canada, 2004.
30. US Federal Power Commission. Northeast power failure: November 9 and 10, 1965. Technical report, US Federal Power Commission, 1965.
31. USHEC. Blackout 2003: How did it happen and why?: House energy committee meeting transcript. In *US House of Representatives, Energy Committee*, September 2003.
32. UTCE. Final Report System Disturbance on 4 November 2006. Technical report, Union for the Co-ordination of Transmission of Electricity, 2007.
33. G. S. Wilkinson. Reciprocal food sharing in the vampire bat. *Nature*, 308:181–84, 1984.
34. G. S. Wilkinson. Food sharing in vampire bats. *Scientific American*, 262:76–82, 1990.
35. WSCC Operations Committee. Western Systems Coordinating Council Disturbance Report For the Power System Outages that Occurred on the Western Interconnection on August 10, 1996. Technical report, Western Systems Coordinating Council, 1996.
36. H. You, V. Vittal, and X. Wang. Slow coherency-based islanding. *IEEE Transactions on Power Systems*, 19:483–491, 2004.
37. M. Zima, M. Larsson, P. Korba, C. Rehtanz, and G. Andersson. Design aspects for wide-area monitoring and control systems. *Proceedings of the IEEE*, 93:980–996, 2005.

Chapter 7
Multi-agent Coordination for Market Environments
Towards a Next Generation Electricity Infrastructure based on Microgrids

R. Duan and G. Deconinck

Abstract The financial crisis has deflated oil prices, prolonging the attractiveness of fossil fuel combustion as a method of energy generation. However, mankind faces a future of a hot, flat, and crowded world [2], making a critical transformation away from the use of fossil fuels imperative. After years of research and experimentation, a number of Renewable Energy Sources (RES) have become technically available as alternatives. Yet a pivotal task which still needs to be carried out is that of adapting the existing electricity infrastructure – still a very efficient energy delivery facility – to allow it to incorporate emerging RES openly and equally. To stimulate the widespread adoption of RES, which would result in the evolution to next generation infrastructures for electricity, incentives should include economic and political measures rather than only technology. In this chapter we summarize the properties of different RES and introduce the 'microgrid', a grid architecture allowing high RES penetration. We also analyze the prevailing electricity market structure and describe existing economic incentives for RES accommodation. Most importantly, we elaborate on the multi-agent model of electricity infrastructures based on the microgrid and its coordination mechanism within the market environment.

R. Duan, G. Deconinck
Katholieke Universiteit, Department of Electrical Engineering, ELECTA, Leuven, Belgium,
e-mail: {Rui.Duan, Geert.Deconinck}@esat.kuleuven.ac.be

R.R. Negenborn et al. (eds.), *Intelligent Infrastructures*, Intelligent Systems, Control and Automation: Science and Engineering 42, DOI 10.1007/978-90-481-3598-1_7,
© Springer Science+Business Media B.V. 2010

7.1 Introduction

7.1.1 Driving forces of power system evolution

Several forces are driving the change in worldwide energy supply. One is that the proportion of electricity in the total energy supply is expected to rise substantially. Another is the growing market penetration of Distributed Generation (DG) sources, which can be defined as sources of electric power connected to the distribution network or to a customer site (behind the meter). This approach is fundamentally distinct from the traditional central plant model for electricity generation and delivery. Driving forces behind the growing penetration of DG are:

- **Environmental concerns**: RES are to a large extent distributed generators. Photovoltaic solar cells and wind turbines are examples of this type of generator. Apart from large-scale wind farms, these generators are connected to low voltage distribution networks or at customer sites. The aims of governments to increase the proportion of sustainable energy in their national energy mixes have been translated into incentives and tax policies to promote the uptake of RES in European countries as well as in many other parts of the world.

- **Liberalization of the electricity market**: As a result of liberalization, the long-term prospects for large-scale investments in electricity generation are unclear at this moment. As a result, a shift in investor interest from large-scale power plants to medium and small-sized generation is happening. Investments in DG are lower and typically have shorter payback periods than those of traditional central power plants. Capital exposure and risk are reduced and unnecessary capital expenditure is avoided by matching capacity increase with local demand growth.

- **Diversification of energy sources**: Diversification of energy sources is a way to reduce the economic vulnerability to external factors. In particular, a higher proportion of sustainable energy in the energy mix would reduce the dependency on fossil fuels from politically instable regions. Currently, European energy demand is largely met through imports from outside the EU. As energy demand continues to grow, this external dependence could grow from 50 to 70% in 25 years or less.

- **Energy autonomy**: A sufficient amount of production capacity in a local electricity network opens the possibility of intentional islanding. Intentional islanding is the transition to stand-alone operation during abnormal conditions on the externally connected network, such as outages or instabilities, e.g., during a technical emergency. In this manner, autonomy can be achieved on different scales, from single buildings to wide-area subsystems.

- **Energy efficiency (i)**: In general, distributed generation reduces energy transmission losses. Estimates of power lost in the long-range transmission and distribution systems of Western economies are in the order of 7%. By producing electricity in the vicinity of where it is consumed, transport losses are avoided. In cases where the local production outgrows the local consumption, the transmission losses again begin to rise. However, in the greater part of the European distribution network we are far from reaching that point.

- **Energy efficiency (ii)**: Heat production from natural gas can reach higher efficiency by using Combined Heat-Power generation (CHP) instead of traditional burners. CHP is a growing category of DG, especially in regions where natural gas is used for heating. In Northern Europe, for instance, CHP is already commonly used in the heating of large buildings, greenhouses and residential areas. The use of micro-CHP for domestic heating in single dwellings is expected to rise steeply in the next 10 to 20 years.

7.1.2 Emerging generation technologies

Over the past decades, several kinds of environmentally friendly electricity generation have been developed. Apart from those with unpredictable energy output, such as wind farm, photovoltaic cell and combined heat-power, which may diminish the balancing ability of a power system, two sorts of distributed generation with controllable production – namely the microturbine and fuel cell – are introduced here. In addition to their capabilities and advantages, the resolvent to the special problems with regard to the interconnection of DG to the distribution grid is also presented.

7.1.2.1 Generation Methods

Microturbines are an important emerging technology. They are mechanically simple, single-shaft devices with air bearings and no lubricants. They are designed to combine the reliability of commercial aircraft auxiliary power units with the low cost of automotive turbochargers. The generator is usually a permanent magnet machine operating at variable speeds (50,000–100,000 rounds per minute). This variable speed operation requires power electronics to interface to the electrical system. Examples include Capstone's 30 kW and 60 kW systems and products from Bowman and Turbec. Sophisticated combustion systems, low turbine temperatures and lean fuel-to-air ratios results in NO_x emissions of less than 10 ppmv and inherently low carbon monoxide emissions. The traditional larger gas turbines, reciprocating engines and reformers all involve higher temperatures that result in much higher NO_x production. Microturbines can operate using a number of different fuels including natural gas and such liquid fuels like gasoline with efficiencies in the 28–30% range.

Fuel cells, which produce electricity from hydrogen and oxygen, emit only water vapor. NO_x and CO_2 emissions are associated with the reforming of natural gas or other fuels to produce the fuel cell's hydrogen supply. Fuel cells offer higher efficiency than microturbines with low emissions but are currently expensive. Phosphoric acid cells are commercially available in the 200 kW range, and high temperature solid-oxide and molten-carbonate cells have been demonstrated and are particularly promising for distributed applications. A major development effort by automotive companies has focused on the possibility of using on-board reforming of gasoline or other common fuels to hydrogen, to be used in low temperature proton exchange membrane fuel cells. Automotive fuel cells will have a major impact on stationary power if the automotive cost goal of $100/kW is achieved.

DG includes more than small generators and fuel cells. Storage technologies such as batteries, ultra-capacitors and flywheels play an important role. Combining storage with microsources provides peak power and ride-through capabilities during disturbances. Storage systems are far more efficient than they used to be. Flywheel systems can deliver 700 kW for 5 seconds, while 28 cell ultra-capacitors can provide up to 12.5 kW for a few seconds.

7.1.2.2 Microsource issues

These technologies have very different issues than those normally found in traditional power sources. For example, they are applied at close proximity to the customers, load at a voltage of 480 volts or less, and require power electronics and different methods of control and dispatch. All of theses energy technologies provide DC output, which requires power electronics in order to interface with the power network and its loads. In most cases the conversion is performed using a voltage source converter with a possibility of pulse width modulation to provide fast control of voltage magnitude.

Power electronic interfaces introduce new control issues and new possibilities. A system with clusters of microsources and storage could be designed to operate both in island mode and connected to the power grid. One large class of problems is related to the fact that fuel cells have slow response times and microturbines are inertia-less. It must be remembered that the current power systems have storage capacity in generators' inertia. When a new load comes online the initial energy balance is satisfied by the system's inertia. This results in a slight reduction in system frequency. A system with clusters of microsources could be designed to operate in island mode if it included some form of storage that could provide the initial energy balance.

The control of inverters used to supply power to an AC system in a distributed environment should be based on information available locally at the inverter. In a system with many microsources, communication of information between systems is impractical. Communication of information may be used to enhance system performance but must not be critical for system operation. Essentially, this implies that the inverter control should be based on terminal quantities.

In many cases there are also major barriers to connecting small generators to the electrical network. These barriers range from technical issues to business practices and regulatory issues. In the National Renewable Energy Laboratory report NRELBR-200-28053, it was found that interconnection cost ranged from zero to over $1000/kW. The high end of this range would more than double the cost of most microturbines.

7.2 Emerging model of the next generation electricity infrastructure

7.2.1 Microgrid concept

To overcome the issues discussed above, a microgrid concept is presented as a way to bring value to both the utility and the customer. A microgrid is a cluster of microsources, storage systems and loads which present themselves to the grid as a single entity that can respond to central control. The heart of this concept is the notion of a flexible yet controllable interface between the microgrid and the wider power system. This interface essentially isolates the two sides electrically and yet connects them economically. The conditions and quality of service are determined by the microgrid, while flows across the dividing line are motivated by the prevailing need of the transmission system.

From the customer side of the interface, the microgrid should appear as an autonomous power system functioning optimally to meet the requirements of the customer. Such issues as local voltage, reliability, losses and quality of power should be those that support the customers' objectives. From the power system side, however, the microgrid should appear as a good citizen to the power grid.

The microgrid structure stratifies the current, strictly hierarchical, centralized control of the power system into at least two layers. The upper layer is the one with which current power engineers are familiar; that is, the high voltage meshed power grid. A centralized control center dispatches a limited set of assets to fulfill contracts established between buyers and sellers of electricity and ancillary services while maintaining the energy balance and power quality, thus protecting the system and ensuring reliability.

The customer views the microgrid as a low voltage neighborhood of the power system that obeys the upper layer's central command only to the extent that its behavior at the node complies with the rigorous requirements of the grid. Within the microgrid, standards of operation and methods of control could diverge significantly from the norms traditional power systems use to meet the needs of the user.

Microgrids have specific internal protection requirements. It is evident that the protection requirements are different from that of a traditional utility setup, both at the grid interface and internally. While traditional protection schemes may be adequate at the interface, a different set of considerations may govern the internal protection of microgrids, one which is highly dependent upon the short circuit currents provided by the microsources and power grid. For example, the microsource can be rapidly isolated from the microgrid, followed by a long restoration process. With storage on the DC bus, the inverter can control the short circuit current without shutting down the microgrid while the fault is cleared. Equally important is how we can design the inverter's control along with the protection to minimize the contribution of the microgrid to fault current sensed in the transmission grid.

The control of the microgrid also has specific requirements. The basic assumption is that of plug and play for the fast response needs and a control/communication system to provide the 'acceptable characteristics' to the grid for slower responses.

The success of microgrids depends on how well each microsource integrates into a cluster of similar sources. In fact, the preferred model is one of "plug-and-play" or "peer-to-peer". In such a model, techniques for the control of DG which depend on SCADA (supervisory control and data acquisition), such as centralized coordination, will not be effective. Control of the microsource should be based on information available locally. In a system with many microsources, fast communication among sources is impractical. Communication may be used to enhance system performance but must not be critical for system operation.

At the same time, the microgrid should appear as an autonomous power system which meets the requirements of the customer. Voltage, reliability performance and quality of power should support the customers' objectives. From a control prospective, techniques need to be developed which significantly lower the system complexity encountered with the addition of extra microsources to a microgrid. The presence of inverter interfaces in fuel cells, photovoltaic, microturbines and storage technologies creates a situation different from that of more conventional synchronous generators in power sources and standby emergency power systems. Taking advantage of the properties of the power electronic interface to provide additional functionality to the microgrid and localized inverter control technology, along with a minimal amount of short energy storage at the DC bus, forms the basis for the approach. Various features of the inverter control could include: plug-and-play features, seamless connection and isolation from the electric grid, independent control of reactive and active power, ability to correct voltage sags and system imbalances – all of which are critical for the creation of a microgrid.

7.2.2 Microgrid architecture

To realize the emerging potential of DG, one must take a system approach that views generation and associated loads as a microgrid. During disturbances, the generation and corresponding loads can separate from the distribution system to isolate the microgrid's load from the disturbance (thereby maintaining service) without harming the transmission grid's integrity.

The difficult task is to achieve this functionality without extensive custom engineering and still have high system reliability and generation placement flexibility. To achieve this, a peer-to-peer and plug-and-play model for each component of the microgrid is promoted [8]. The peer-to-peer concept ensures that there are no components, such as a master controller or central storage unit, that are in themselves critical to the operation of the microgrid. This implies that the microgrid can continue operating with the loss of any component or generator. With one additional source, complete functionality can be secured with the loss of any source. Plug-and-play implies that a unit can be placed at any point on the electrical system without the need for re-engineering the controls.

Plug-and-play functionality is much akin to the flexibility one has when using a home appliance; that is, it can be attached to the electrical system at the location where it is needed. The traditional model is to cluster generation at a single point

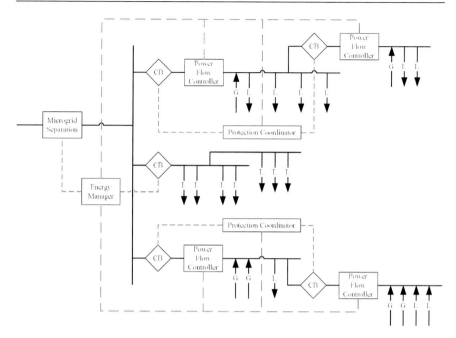

Figure 7.1: Microgrid architecture.

that makes the electrical application simpler. The plug-and-play model facilitates placing generators near the heat loads, thereby allowing more effective use of waste heat without complex heat distribution systems such as steam and chilled water pipes.

This ability to island together generation and loads has the potential to provide a higher local reliability than that provided by the power system as a whole. Smaller units with power ratings in thousands of watts can provide even higher reliability and fuel efficiency. These units can create microgrid services at customer sites such as office buildings, industrial parks and homes. Since the smaller units are modular, site management could decide to have more units than required by the electrical/heat load, providing local, online backup if one or more of the operating units were to fail. It is also much easier to place small generators near the heat loads, thereby allowing more effective use of waste heat.

Figure 7.1 illustrated the basic microgrid architecture. In this example the electrical system is assumed to be radial with three feeders and a collection of loads. The radial system is connected to the distribution system through a separation device, usually a static switch. Each feeder has circuit breakers and power flow controllers that regulate feeder power flow at a level prescribed by the Energy Manger. As downstream loads change, the local microsources increase or decrease their power output to keep the power flow constant. When there are power quality problems on the distribution system, the microgrid can island by using the separation device shown in the figure.

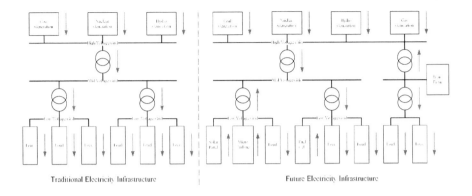

Figure 7.2: Traditional and future electricity infrastructure.

7.2.3 Future power grid based on microgrids

The growing share of DG in the electricity system may evolve in three distinct stages [4]:

1. **Accommodation**: Distributed generation is accommodated in the current market. Distributed units run free, while centralized control of the network remains in place.

2. **Decentralization**: The share of DG increases. Virtual utilities optimize the services of decentralized providers through the use of common ICT systems. Central monitoring and control is still needed.

3. **Dispersal**: Distributed power takes over the electricity market. Local low-voltage network segments provide their own supply with limited exchange of energy with the rest of the network. The central network operator operates more as a coordinating agent between separate systems than as a controller of the system.

In some places there are already signs of decentralization. In the future, a substantial proportion of electricity will be fed into the power grid at medium and low voltage levels, as illustrated in Figure 7.2. During the second and third stages, the lower parts of the power grid are expected to evolve from a hierarchical top-down structure into a network of microgrids, in which a vast number of components influence one another. A combination of technologies such as DG, electricity storage, demand response, real-time pricing, and intelligent control opens the possibility of optimization regarding economics, dependability and sustainability. In this case, the traditional paradigm of centralized control used in current electricity infrastructure will no longer be sufficient. The number of system components actively involved in the coordination will be huge. Centralized control of such a complex system will reach the limits of scalability and complexity.

In addition to the described technical evolution of the electricity infrastructure, there is also an ongoing evolution in the market structure. The electricity supply will no longer be in the hands of a small group of big players but will be spread out over a vast number of market players, big ones as well as small ones. This will give rise to new business models in electricity production and consumption. In regions with a highly deregulated energy system, like the Scandinavian countries, the United Kingdom and the USA, coordination mechanisms based on economic principles have been introduced at a central level. Market mechanisms are used in the planning of large-scale production via day-ahead power exchange trading, as well as for real-time balancing via spinning-reserve auctions held by the Transport System Operators (TSOs). The coordination mechanisms for the low end of the grid hierarchy, which become necessary during the second and third stage of DG growth, need to comply with the constraints imposed by the changing market structure. Consequently, these control mechanisms must also be based on economic principles.

7.3 Electricity market structure and mechanism

7.3.1 Nature of the electricity market

Electricity is by nature difficult to store and must be available on demand. Consequently, unlike other products, it is not possible under normal operating conditions to keep it in stock, ration it or have customers queue for it. Furthermore, demand and supply vary continuously. Therefore, there is a physical requirement for a control agent, the indepent or transmission system operator (TSO), to coordinate the dispatch of generating units to meet the expected demand of the system across the transmission grid. If there is a mismatch between supply and demand, the generators will speed up or slow down, causing the system frequency (either 50 or 60 Hz) to increase or decrease. If the frequency falls outside a predetermined range, the system operator will act to add or remove either generation or load. In addition, the laws of physics determine how electricity flows through the electricity network. Hence, the extent of electricity lost in transmission and the level of congestion on any particular branch of the network will influence the economic dispatch of the generation units.

In economic terms, electricity is a commodity capable of being bought and sold. An electricity market is a system for effecting the purchase and sale of electricity, quoting supply and demand to set the price, see Figure 7.3 Markets for certain related commodities required by (and paid for by) various grid operators to ensure reliability, such as spinning reserve, operating reserves, and installed capacity, are also typically managed by the grid operator. In addition, for most major grids there are markets for electricity derivatives, such as electricity futures and options, which are actively traded. These markets developed as a result of the deregulation of electric power systems around the world. This process has often gone on in parallel with the deregulation of natural gas markets. In different deregulation processes the in-

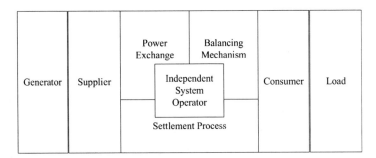

Figure 7.3: Electricity market structure.

stitutions and market designs were often very different, but many of the underlying concepts were the same. These are: separating the contestable functions of generation and retail from the natural monopoly functions of transmission and distribution; and establishing a wholesale electricity market and a retail electricity market.

A wholesale electricity market exists when competing generators offer their electricity output to retailers. The role of the wholesale market is to allow trading between generators, retailers, and other financial intermediaries both for short-term delivery of electricity (spot price) and for future delivery periods (forward price). Wholesale transactions are typically cleared and settled by the grid operator or a special-purpose independent agent charged exclusively with that function. Generally, electricity retail reform follows electricity wholesale reform. However, it is possible to have a single electricity generation company and still have retail competition. If a wholesale price can be established at a node on the transmission grid and the electricity quantities at that node can be reconciled, competition for retail customers within the distribution system beyond the node is possible. A retail electricity market exists when end-use customers can choose their suppliers from competing electricity retailers. Demand response based on pricing mechanisms and security constraints can be used to reduce peak demand.

7.3.2 Model of a prevailing electricity market

In the following sections, the pivotal components and mechanisms of a prevailing electricity market in England and Wales are delineated. Such a market arrangement is based on bilateral trading between generators, suppliers, traders, and customers, which takes place in the forward markets before gate closure (3.5 hours before real time, see Figure 7.4). Power Exchanges (PX) among these four parties are set up to facilitate this. The Balancing Mechanism (BM) works as a market in which an Independent System Operator (ISO) buys and sells *Increments* (*incs*) or *Decrements* (*decs*) of electricity in order to balance the system as a whole in cases of individual generators and suppliers being out of balance.

During a Settlement Process (SP), the ISO will compare the Contract Positions (quantities contracted) with the Actual Position (quantities generated or consumed)

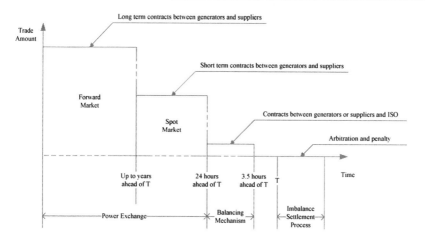

Figure 7.4: Electricity market mechanism.

for each of the consumers and generators to calculate the *imbalances*. The imbalance may be a *spillage* (if a plant generates more than it contracted or if a load consumes less than it contracted) or a *top-up* (if a plant generates less than it contracted or if a load consumes more than it contracted). For both types of imbalance, there is a price: if an entity is spilling, it will receive payment for the marginal generation at the System Selling Price (SSP); if an entity is topping up, it will pay at the System Buying Price (SBP). The spread between the two prices is intended to provide a penalty for being out of balance, for which SSP is expected to be considerably lower and SBP higher than the prices in the forward markets. The day-ahead and within-day balancing schedules are as follows [10].

ISO Day-Ahead Balancing Process

1. By 09:00, publishes the day-ahead demand forecast;
2. By 11:00, receives the Initial Physical Notifications (IPN);
3. Calculates the available national plant margin or shortfall;
4. Verifies system security with demand predictions, IPN, and planned transmission outage;
5. By 12:00, issues the total system plant margin data to the market for the day ahead;
6. Forecasts constraint costs based on the estimated Final Physical Notifications (FPN), bid (offer) prices and volumes;
7. If necessary, calls the most economic balancing service contracts to ensure system security;
8. During the following 11 hours, receives updates of Physical Notifications (PN);
9. By 16:00, publishes the revised national plant margin and zonal margin.

ISO Within-Day Balancing Process

1. Publishes averaged demand forecasts every half hour for a defined period until gate closure;
2. As participants become aware of changes to their physical position, they will advise the ISO;
3. At defined times, the zonal and national margins are reassessed and provided to the market;
4. Undertake security analysis and reassesses the requirements of balancing services contracts;
5. At gate closure, PN becomes FPN and ISO has received bids (offers) with prices and volumes from the participants in BM;
6. During BM, ISO balances the system with regard to technical constraints, dynamic operating characteristics of supply and demand balancing services, and uncertainty in demand.

7.3.3 Impact of electricity market arrangements

Such an electricity market arrangement is a costly system, partly because it is a reflection of the inherent complexity of its rules and partly because of the costs and risks of doing business in a market overshadowed by intentionally penal imbalance prices [5]. This is a trading environment in which the big players enjoy special advantages: they have the deep pockets to pay the entry and on-going costs of participating in it, and their diverse portfolios make them less susceptible to the vagaries of the imbalance settlement process. The obvious losers from all this are CHP and RES, which are inevitably relatively small and, in the case of some CHP plants and wind generation, especially exposed to the penalties of the imbalance settlement process. Aggregation in the supply business prompted by high costs has reduced the number of potential customers for embedded generation and added to these difficulties. The government has responded to the plight of RES by introducing financial incentives in the form of Renewable Obligation Certificates (ROCs), which place a premium on renewable generation [9], negating the detrimental effects of the very market arrangement.

This brings us to the critical issue of how to facilitate RES penetration into the distribution system with market forces. It can be easily seen that due to the unpredictable nature of RES, microgrids consisting of clustered microsources and loads that behave as fluctuating Virtual Power Plants (VPP) or virtual loads are not applicable in the bulk energy trade of PX with bilateral contracts. Yet the quick response of microgrids will allow their competence in BM with short-term supplementary services. Moreover, the considerably high SBP will guarantee an obvious profitability of the winning VPP in balancing markets during SP. From the inside of microgrids, SBP and SSP will be regarded as bids offered by outside consumers and suppliers. The uniformity of overall future markets and regional spot markets can be realized by transferring SBP and/or SSP to P_{OC} (outside consumer buying price) and P_{OS} (outside supplier selling price). This, in addition to the benefit from

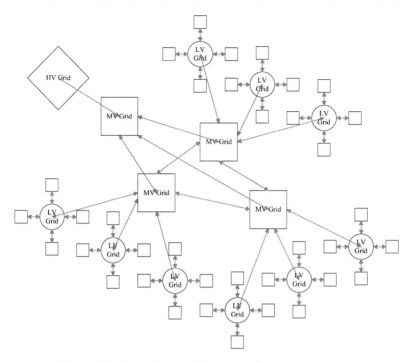

Figure 7.5: Future electricity infrastructure based on microgrids.

ROCs, will enhance the economic incentives for facilities to equip using RES. One of the indispensable conditions for such a market arrangement to be deployed is the establishment of the electronic market and the delegation of agents as well as their coordination mechanisms, which can enable the active involvement of various RES and flexible loads into the competition derived from the current prevailing electricity market. A multi-agent system carrying out the required functionality is elaborated in the following section.

7.4 A promising multi-agent system framework for coordination

In this section we construct a coordinating system based on multi-agent technology with microeconomics principles for the future power grid mentioned above. This system matches the supply and demand of electric power both within the micro-grids and among them, fulfilling the primary control requirement for the electricity infrastructure to be able to operate steadily.

The future electricity infrastructure can be generally viewed to have a hierarchical structure of three levels (see Figure 7.5): the low-voltage level, being the radial networks of a microgrid which distributes power supply from mid-voltage

Figure 7.6: Future electricity infrastructure modeled as multi-agent system.

grid and coordinates power demand from local devices within this microgrid; the mid-voltage level, a meshed network of microgrids performing as individual nodes that seek partners with matching supply or demand patterns; and the high-voltage level, which is the traditional centralized generation with a high-voltage transmission system that acts as the backup supplier for the failure of certain nodes in the free power market of the mid-voltage distribution network.

Two fundamental issues have to be addressed in this chapter. The first is the need to design resource agents that can bid on behalf of each generating, consuming, and storage device within a microgrid (see Figure 7.6). These resource agents often have the same general structure of a simple reactive agent but are equipped with different bidding strategies according to the various properties of the devices they serve. With this method, the behaviors of devices in the microgrid can be economically optimized according to their flexibility in terms of energy supply or demand over time.

The second is the task of designing market agents that supervise the microgrid, detecting whether there are resource agents being added to or removed from the underneath microgrid, initiating a market period for all the resource agents and confirming deals between bidding rounds. In the case of a microgrid that is rich in electricity consumers, a node of a virtual load will appear on the level of market agents in the meshed mid-voltage grid. Correspondingly, a node of a virtual generator will appear if the underlying microgrid is rich in suppliers. The market agent should also generalize and issue a demand function for a virtual load (or supply function for a virtual generator in the form of a negative demand) to outside suppliers of the microgrid that it covers, which can be either feeders from high-voltage transmission systems or peer microgrids behaving as a VPP. In addition, the market agent does not need to differentiate bids from underlying market agents or resource agents, allowing more flexibility for the multi-agent system to be deployed on real-life complex grid architectures.

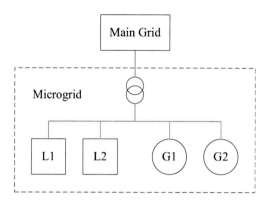

Figure 7.7: Simplified circuit model of a microgrid.

7.4.1 Microgrid modeled as multi-agent system

Let us first consider a simplified model of a microgrid (see Figure 7.7, which is composed of two distributed generators, G_1 and G_2, and two loads L_1 and L_2. Here we do not take storage devices into account, because they behave either as a load or as a generator at any certain moment. Each of the above actors participating in the market competition of electric energy within the microgrid has its own capacity Q (either purchasing price or selling price).

In a generation rare microgrid where $Q_{L1} + Q_{L2} > Q_{G1} + Q_{G2}$, loads will actively bid against each other to compete for local low cost supply. The extra demand that cannot be fulfilled by local generators will be satisfied by outside suppliers, such as peer microgrids with superfluous generation in nearby segments of the distribution system or centralized power plants on the other end of the transmission system. It is natural that the prices from local DG are lower than those from outside suppliers, due to the lower fuel cost and transmission tariff. On the other hand, in a generation rich microgrid where $Q_{L1} + Q_{L2} \leq Q_{G1} + Q_{G2}$, generators will actively bid against each other to compete for local high price demand. The superfluous supply that cannot be consumed by local loads will be delivered to outside consumers, when the status of the distribution system in the vicinity allows it to do so. Otherwise, the local DG will only have to reduce its production. It is also natural that the priorities of local loads are higher than those of outside consumers, due to the reliability and security concerns of the microgrid.

As such a simple model, the environment within a microgrid can be partially accessible [14] (see Figure 7.8, which means load agents can acquire all the information from generator agents, e.g., P_{G1}, P_{G2}, P_{OS} (price of outside supplier), and Q_{G1}, Q_{G2}, Q_{OS} (quantity of outside supplier), while generator agents can acquire all the information from load agents, e.g., P_{L1}, P_{L2}, P_{OC} (price of outside consumer), and Q_{L1}, Q_{L2}, Q_{OC} (quantity of outside consumer). However, the same kind of resource agents (either loads or generators) should be imperceivable to each other, due to their private ownership and competing relationship for resources (either consumers or producers) within the microgrid. Moreover, although all the above variables alter

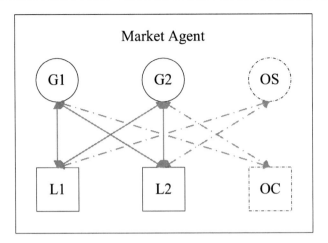

Figure 7.8: Market model of a microgrid.

during each market period, they will not change during each single bidding round, which results in a static environment [14]. These facts constitute the fundamental basis on which the coordination system based on MAS is built.

7.4.2 Market operation and principles

In order to deploy agents for the coordination system, the market model of a microgrid is used. For the purpose of adopting reasonable operation mechanisms for the market model, three properties of a microgrid should be realized:

1. The commodity of the auction market is energy (kWh), rather than power (kW). This happens because it is difficult for small DG units to keep their output steady for a long period. Energy is the duration of power, which is easier to produce in an exact amount within a certain period (15 minutes in our case).
2. Whenever the microgrid is connect to a transmission system, there is no limit on the energy that can be sold to or bought from outside consumers or suppliers (as long as there are no technical constraints), since the capacity of the main grid is almost infinite as compared to any individual microgrid.
3. A major issue in microgrid operation is the estimation of power demand or supply from each device for the next period. In our application, it is determined in two ways depending on the device type. For controllable units, like diesel generators or water heaters, it is considered to be the nominal capacity. For uncontrollable units, like photovoltaic panels or wind generators, it is assumed that the average for the next 15 minutes will be the same as the current one.

Based on the above preconditions, the FIPA [1] English Auction Protocol can be adapted the overall procedure of the market mechanism of microgrid operation, which is as follows (see Figure 7.9:

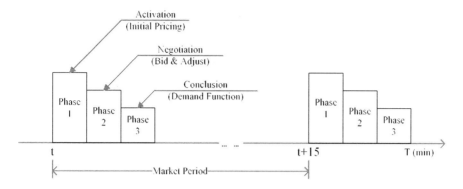

Figure 7.9: Market procedure of a microgrid.

1. When the first phase of a market period starts, the market agent of the micro-grid broadcasts price information from the main grid to all the resource agents within the microgrid, which means every load will be aware of P_{OS}, and every generator will be aware of P_{OC}. Normally, $P_{OS} > P_{OC}$. According to P_{OS} and P_{OC}, each resource agent determines initial P_L and Q_L or P_G and Q_G for the next 15 minutes, with respect to its own physical characteristics. There should be $P_L > P_{OC}$ and $P_L < P_{OS}$, as well as $P_G < P_{OS}$ and $P_G > P_{OC}$, which will guar-antee than an intra-grid transaction will have higher priority than an inter-grid one. Moreover, the market agent will create a virtual resource agent with the price of P_{OS} or P_{OC} and unlimited capacity, depending on whether extra supply or demand is needed to fulfill the requirement of each real resource agent. This will insure that the total supply is equal to the total demand in the microgrid, which is its primary operation prerequisite.

2. During the second phase, resource agents bid with one another to compete for the resources (either supply or demand) they need. At the end of each bidding round, those resource agents that make successful deals with their matched part-ners will refresh their resources. If a resource agent no longer has any of its re-source, it will quit the market; if it still has some remaining resource, it will stay in the market without adjusting the price. Resource agents that fail to obtain any resource in the auction market will adjust their price as long as $P_L < P_{L\max}$ or $P_G > P_{G\min}$ and will join the next bidding round. ($P_{L\max}$ and $P_{G\min}$ are the ex-treme prices that are acceptable to relevant resource agents. Normally there will be $P_{L\max} \leq P_{OS}$ and $P_{G\min} \geq P_{OC}$.) The adjustment will be made in such a pat-tern as $P' = P + Adj \times \frac{Q'}{Q}$, where Adj is the fundamental adjustment determined by the nature of the resource agents, which is positive for loads and negative for generators, while Q is the initial quantity and Q' is the quantity still needed to be traded.

3. At the end of the second phase, each resource agent will fulfill its desire for resources with peer resource agents within the microgrid or virtual resource agents on behalf of outside traders, except those that have strong flexibility on

energy consumption or production over time but weak affordability or high cost in the auction market. These withdraw from competition temporarily and will rejoin in the next period. Hereby, in the last phase, all resource agents, including the virtual one, will have fixed their energy schedule for the next 15 minutes, which means that the market agent can issue the demand function Q_{OS} or Q_{OC} of the microgrid to an upper-level market.

A critical point that must be clarified here is the behavior patterns of resource agents inside the microgrid, which depend on the two sorts of market operation. In a collaborative market, all resource agents cooperate for a common goal. This case exists only if the agents have the same owner or a very strong operational target. In a competitive market, each resource agent has its own interest, while it does not necessarily mean that other agents are opponents. Our research focuses on the latter case.

The core of market competition is the decision mechanism of market players – in our case, the answer to the question of how each resource agent selects other resource agents to provide the resource it needs. For the purpose of studying in detail the competition during each bidding round, we must return to the simplified market model introduced in the previous section. In such a situation, competition takes place between L1 and L2, as well as G1 and G2. The concrete question then evolves into why L1 and G1 choose each other.

The goal of a load in market competition is to satisfy Q_L with minimum cost, while for generators the goal is to obtain maximum profit with certain Q_G. However, it is very probably that some resource agents need to join more than one bidding round, due to the failure of obtaining enough resources in a single round. Moreover, between every two rounds, each resource agent that has to join the next round will adjust its bid according to its distinct intrinsic properties. Hence, if the complexity of the model increases, it will be an enormous computation to estimate total cost or profit of all possible series of transactions of those resource agents. Therefore it is very practical to introduce an index that can reflect the contribution of a single transaction to the total gain.

Let us take load L1 as an example. It can be aware of

$$Cst(L1, G1) = \begin{cases} Q_{L1} \cdot P(L1, G1) & \text{for } Q_{L1} \leq Q_{G1} \\ Q_{G1} \cdot P(L1, G1) & \text{for } Q_{L1} > Q_{G1}, \end{cases} \qquad (7.1)$$

which is the cost for L1 if it makes a deal with G1 in the current bidding round. G1 can acquire

$$Pft(G1, L1) = \begin{cases} Q_{L1} \cdot P(G1, L1) & \text{for } Q_{G1} \geq Q_{L1} \\ Q_{G1} \cdot P(G1, L1) & \text{for } Q_{G1} < Q_{L1}, \end{cases} \qquad (7.2)$$

which is the profit of G1 if it makes a deal with L1 in the current bidding round. $P(L1, G1)$ and $P(G1, L1)$ are the actual trading prices resulting from L1 and G1's acceptance of each other's bids. Next, we need to define the following two functions:

$$Pfm(L1,G1) = \begin{cases} 1 & \text{for } Q_{L1} \leq Q_{G1} \\ Q_{G1}/Q_{L1} & \text{for } Q_{L1} > Q_{G1} \end{cases} \tag{7.3}$$

$$Pfm(G1,L1) = \begin{cases} Q_{L1}/Q_{G1} & \text{for } Q_{G1} \geq Q_{L1} \\ 1 & \text{for } Q_{G1} < Q_{L1}. \end{cases} \tag{7.4}$$

Both of these reflect the contribution of a single transaction to the total desire for the resources of relevant resource agents. Then the real benefit of the very transaction can be determined. For L1, it will be:

$$Bft(L1,G1) = \frac{Pfm(L1,G1)}{Cst(L1,G1)} \tag{7.5}$$

$$= \begin{cases} \frac{1}{Q_{L1} \cdot P(L1,G1)} & \text{for } Q_{L1} \leq Q_{G1} \\ \frac{1}{Q_{L1} \cdot P(L1,G1)} & \text{for } Q_{L1} > Q_{G1} \end{cases} \tag{7.6}$$

$$= \frac{1}{Q_{L1} \cdot P(L1,G1)}. \tag{7.7}$$

For G1, it will be:

$$Bft(G1,L1) = \frac{Pft(G1,L1)}{Pfm(G1,L1)} \tag{7.8}$$

$$= \begin{cases} Q_{G1} \cdot P(G1,L1) & \text{for } Q_{G1} \geq Q_{L1} \\ Q_{G1} \cdot P(G1,L1) & \text{for } Q_{G1} < Q_{L1} \end{cases} \tag{7.9}$$

$$= Q_{G1} \cdot P(G1,L1). \tag{7.10}$$

It can be easily seen that since the quantity of the resource desired by any resource agent is fixed during each bidding round, the only decisive factor for the actual trading price is the choice of the load for a generator and vice versa. Loads are inclined to make deals with cheap generators, while generators prefer expensive loads. However, in such a free market it is impossible for all competitors to obtain optimal solutions, e.g., the most modest seller and the most generous buyer. In order to avoid conflicts and endless negotiations, extra guidance rules must be set up according to the status and emphasis of the market. In our case of electricity infrastructures based on microgrids, it would be more reliable if each microgrid's independence were to be enhanced and made into a self-contained unit without dependence on the outside transmission system. In a generation-rich microgrid, such a guidance rule should encourage consumption, meaning it should allow the load with the biggest demand to have priority over those with the same buying price when competing for low-price supply. A load-rich microgrid, on the other hand, should encourage production, allowing the generator with the biggest supply to have priority over others with the same selling price when competing for high-price demand.

7.4.3 Coordination strategy and auction types

In such a hierarchically distributed environment, for each resource (load or generator), resource agents should be introduced that are responsible for managing their functions respectively. On the one hand, a Generator Agent (GA) controls the output of its associated DER resources to ensure that the demand of Load Agents (LA) is satisfied. On the other hand, an LA determines the maximum unit price that the supervised load resource is willing to pay for a quantity of power required during a specific period of time, based on internal prediction algorithms that incorporate real-time and historical data. At the same time, the GA interacts with its affiliated DER to determine unit production cost, preferred markup and the quantity of power available for supply. Production costs are calculated based on operation, fuel and maintenance overheads. During a given time period, GA and LA cooperate with each other to achieve their objectives. Primarily, both LA and GA wish to maximize their utility. LA want to minimize their cost by purchasing energy below their unit price, whereas GA wants to maximize their profits by selling energy above the unit cost and maximizing markup. For the problem of energy allocation, an auction-based mechanism has been shown to work well [11], which implements bilateral contracts between GA and LA.

Since market prices can fluctuate significantly, undesirable stabilities in trading may occur. Consequently, a single microgrid could be best served by bilateral contracts, while the distribution system could also use a pool-based mechanism. In essence, each bilateral contract guarantees the energy allocation of one bidder within a group, facilitated by an auctioneer. The auctioneer specifies the quantity of goods and a starting unit price. Bidders make bids after analyzing the feasibility of the auctioneer's parameters. In practice, there are two possible auction scenarios in microgrids: the GA acts as auctioneer, auctioning energy to participating LA; or an LA acts as auctioneer, auctioning the right of providing energy to participating GA. Concentrating on four types of auctions [12], First-price Sealed-bid (FPSB), Vickrey, English, and Dutch, the first one is the most suitable mechanism in our application on microgrids, as explained later:

1. **First-price Sealed-bid**: FPSB is the most popular action type for virtual environments. The process is a single shot, sealed-bid process, where the good is awarded to the highest bidder. Since bids are sealed, participants make decisions without knowledge of other bids. Game theory shows that a bidder's best strategy is to bid less than its utility.

2. **Vickrey**: Vickrey auctions, proposed by William Vickrey in 1961, are a variation on FPSB, with the winner being the highest bidder but paying the second highest price. As a result, the participant's best strategy, as proven by game theory, is to bid its utility for the goods.

3. **English**: English Auctions are the most popular type of auctions in the physical world. They use an iterative open-cry process whereby the price is awarded to the highest bidder. As a result, the price increases with each bid. The bid-

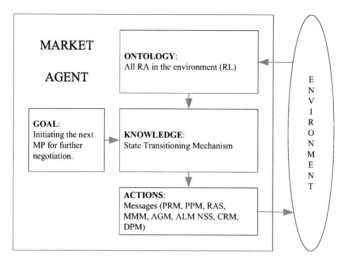

Figure 7.10: Market agent architecture.

der's best strategy is to increase its bid until it reaches its utility. Unfortunately, English Auctions can suffer from the Winner's Curse, because participants over-estimate the value of the goods and bid more than their utility.

4. **Dutch**: Dutch auctions are similar to English auctions because they also use an iterative open-cry process. In this case, however, the auctioneer reduces the price until a participant makes the first (highest) bid. Game theory shows that Dutch auctions are strategically equal to First Price Sealed Bid auctions, which means the bidder should wait until the price of the goods is slightly less than its utility before bidding. Dutch auctions can be more efficient than English auctions, since the auctioneer can rapidly decrease the price, but they can still suffer from the Winner's Curse.

7.4.4 Detailed agent behaviors

In order to design the specific software architecture for each agent in our coordination system, agent behaviors that take place in each phase of a Market Period (MP) should be considered in detail. A critical point that must be clarified is that any Resource Agent (RA), when newly added into the microgrid, must send its ID to the Market Agent (MA) to which it subordinates as a Resource Added Message (RAM) before being activated in its first MP. Moreover, any RA, when it is about to be removed from the microgrid, must also send a Resource Removed Message (RRM) to its MA. These operations will help MA to keep a complete updated Resource List (RL), composed of the ID and status of all RA with in the microgrid. Such a list will allow agents to remain fully aware of their common environment, which can greatly facilitate the negotiation among them. Figures 7.10 and 7.11 give an overview of the architectures of a market agent and a resource agent.

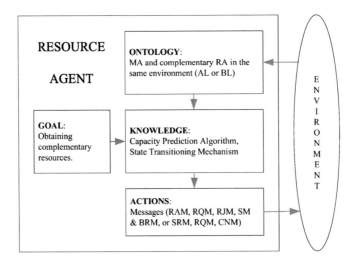

Figure 7.11: Resource agent architecture.

Phase 1: Activation

At the beginning of the Activation Phase (AP), an MA will send out a Price Request Message (PRM) to fetch P_{OS} and P_{OC} from its upper adjacent distribution system. Then it will broadcast them to every RA via Price Publication Messages (PPM), combined with Resource Activation Signals (RAS).

After receiving both PPM and RAS, RA with different extents of flexibility in energy production or consumption over time will decide whether to actively participate in the current MP or remain inactive. According to the different types of activated RA, Q_{Gi} ($i \in [1,M]$, M is the total number of active generator resources) and Q_{Lj} ($j \in [1,N]$, N is the total number of active load resources) will be calculated with their internal predictive algorithms. Then each of them will return Q_{Gi} or Q_{Lj} and its ID to the MA via a Resource Joining Message (RJM).

Upon receiving all RJM, the MA will refresh RL by changing the status of the responding RA from Inactive to Active, and compare the value of $\sum_{i=1}^{M} Q_{Gi}$ with $\sum_{j=1}^{N} Q_{Lj}$. If $\sum_{i=1}^{M} Q_{Gi} > \sum_{j=1}^{N} Q_{Lj}$, the energy market of the microgrid will be in Generation Rich (GR) mode during the next MP; if $\sum_{i=1}^{M} Q_{Gi} \leq \sum_{j=1}^{N} Q_{Lj}$, it will be in Load Rich (LR) mode. The result will be sent to all active RA via Market Mode Messages (MMM).

Moreover, the MA will broadcast the ID of all active generators to all active loads via Active Generator Messages (AGM), and it will also broadcast the ID of all active loads to all active generators via Active Load Messages (ALM). Finally, a Negotiation Start Signal (NSS) is sent to all active RA, this being the last agent action during AP.

Based on PPM and MMM, each active RA calculates $P_{Gi\min}$ (the lowest price with which generator G_i will sell its production) or $P_{Lj\max}$ (the highest price with which load L_j will buy its consumable). In principle, there should be $P_{Gi} \in$

$[P_{Gi\min}, P_{Gi\max}] \subseteq [P_{OC}, P_{OS}]$ and $P_{Lj} \in [P_{Lj\min}, P_{Lj\max}] \subseteq [P_{OC}, P_{OS}]$. If so, the Negotiation Phase (NP) starts. If the energy market is in LR mode, active generators will set their RA to be Auctioneers, and active loads will set theirs to be Bidders. If it is in GR mode, active generators will be bidders, while active loads will be auctioneers. According to AGM or ALM that have been received by each active RA respectively, bidders will create an Auctioneer List (AL) and auctioneers will create a Bidder List (BL).

Phase 2: Negotiation

The first step of our agents' negotiation in an auction market for energy allocation in the microgrid is that each of the auctioneers will send out Summon Messages (SM) to all of the RA on its BL. The message will include the ID of the auctioneer, its capacity (Q_{Gi} or Q_{Lj}) and the Entry Price (EP) for its resource ($P_{Gi\min}$ or $P_{Lj\max}$).

Once a bidder has received SM from all of the RA on its AL, it will compare its affordability ($P_{Lj\max}$ or $P_{Gi\min}$) with the EP of each auctioneer. For those auctioneers that it can afford (generator RA which have $P_{Gi\min} \leq P_{Lj\max}$ or load RA which have $P_{Lj\max} \geq P_{Gi\min}$), the bidder will rearrange its AL. Acceptable auctioneers are ranked according to their capacities in the upper section of AL. The rule is that the one with larger capacities ranks higher. Taking into account the EP of the most preferred auctioneer, the bidder will set its status in AL to be active and will calculate a Bidding Price (BP, P_{Gi1} or P_{Lj1}). Then the bidder will send the BP together with its ID and capacity (Q_{Gi} or Q_{Lj}) to that very auctioneer as a Bidding Message (BM). Regarding the other acceptable auctioneers in AL which are not on the top position, the bidder will set their status in AL to be 'pending' and send Pending Messages (PM) to them. As for those unacceptable auctioneers in the lower section of AL, the bidder will set their status in AL to 'rejected' and will send Summon Rejection Messages (SRM) to them.

When an auctioneer received responses to its SM from all the bidders, it will rearrange its BL. If the energy market is in GR mode, the bidder with the lowest price has priority over the other ones; if the energy market is in LR mode, the one with the highest price ranks first.

The auctioneer compares its capacity with that of the top bidder. If $Q_{auci1} > Q_{bidj1}$, the auctioneer will refresh its capacity as $Q_{auci2} = Q_{auci1} - Q_{bidj1}$ and send an Intention Message (IM) to revise the bidder's capacity to be zero. It will also refresh its BL and compare its capacity with that of the updated top bidder again. If $Q_{auci1} < Q_{bidj1}$, the auctioneer will send an IM to refresh the bidder's capacity as $Q_{bidj2} = Q_{bidj1} - Q_{auci1}$ and revise the capacity of itself to be zero. If $Q_{auci1} = Q_{bidj1}$, the auctioneer will revise its capacity to be zero and also send an IM to revise the bidder's capacity to be zero.

After receiving the IM, the bidder will immediately refresh its AL according. If its updated capacity is not zero, the bidder will send a BM, PM, and SRM again according to the new AL without adjusting price.

Whenever the capacity of any active RA reaches zero, it will broadcast Resource Quitting Messages (RQM) to all the corresponding RA on its AL or BL, which will clear the RA with no capacity from the BL or AL of its partners. Such an RQM will also be sent to the MA, refreshing the status on the RL of the very RA to be inactive. Finally, it will deactivate itself, discarding any messages from other RA.

As to the bidders that are not ranked first in BL, the auctioneer will send its updated capacity to them in Bid Rejection Messages (BRM). Upon receiving BRM, the bidder will immediately refresh its AL according to it. The bidder will also adjust its BP and send a BM, PM, and SRM to corresponding auctioneers again.

If an auctioneer did not receive any bids, but only PM and SRM, there will be no effective bidder on its BL, but only pending or rejecting ones. The auctioneer will do nothing except wait for further response. Those that only received an SRM will have all rejecting bidders on their BL. Such auctioneers will adjust their EP and send an SM again according to their current BL.

Phase 3: Conclusion

The NP continues until the MA perceives that there is only one active RA on the RL. At this moment, all other RA, which have once taken part in the negotiation, have already made deals with their peers by auctions and successfully quit the energy market. This begins the Conclusion Phase (CP) of the current MP. The MA will send a Capacity Request Message (CRM) to the last RA. When the RA receives the CRM, it will return its rest resource with a Capacity Notifying Message (CNM), followed by an RQM. When there is no active RA on the RL, the MA will issue a Demand Publication Message (DPM) according to CNM and send it together with PRM, initiating the next MP.

The behaviors of the market agent and a resource agent are summarized in Figures 7.12 and 7.13.

7.5 Preliminary results

A software simulator which emulates the coordination among agents within a microgrid as introduced above is programmed in JAVA, in which the parameters are preset as follows:

1. 8 generators and 8 loads within 1 microgrid;
2. P_{OS} and P_{OC} are randomized from [0,1000], according to $P_{OS} > P_{OC}$;
3. 4 GA will join when $P_{OC} \geq 400$, 4 LA will join when $P_{OS} \leq 600$, flexibility of other RA are randomized;
4. Q_{Gi} and Q_{Lj} are randomized from [0,100];
5. Initial prices of RA are randomized according to $P_{Gi} \in [P_{Gi\min}, P_{Gi\max}] \subseteq [P_{OC}, P_{OS}]$ and $P_{Lj} \in [P_{Lj\min}, P_{Lj\max}] \subseteq [P_{OC}, P_{OS}]$.

It can be easily seen from the simulation results with different market scenarios that for FPSB or Vickrey auction, among every 100 successfully finished MP, the average amount of messages an RA sends and receives is around 180; as to English or Dutch auction, among every 100 successfully finished MP, the average amount of

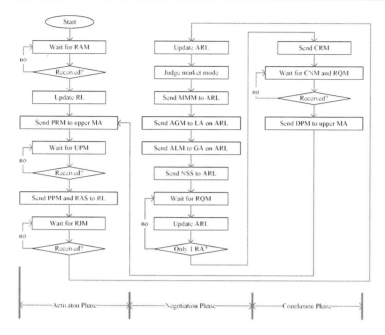

Figure 7.12: Market agent behaviors.

messages an RA sends and receives is around 150. According to the similar amount of data exchange among the MAS, price sealed bids can be more appealing to our application on coordination of microgrids for the sake of the privacy and security of communication between different market players.

7.6 Related work

Our work is related to that of two main groups of researchers who apply MAS to power engineering. Energy research Center of the Netherlands (ECN) devised the PowerMatcher [7], a market-based control concept for matching supply and demand in electricity networks that is able to reduce imbalance costs in commercial portfolios [6]. National Technical University of Athens (NTUA) studies the operation of a multi-agent system for the control of a microgrid. This system optimizes energy exchange between the production units of the microgrid and the local loads as well as the main grid [3]. While the former one places emphasis on the structure of MAS for power system and the software architecture of different agents, the latter attaches great importance to the market operation of the microgrid and the bidding mechanism of resource agents [13]. Our work presented in this chapter aims at incorporating the interests of both previous studies, as well as constructing a distributed open software platform for the future electricity infrastructure, enabling the involvement of other delegate MAS with operation goals of power quality and protection.

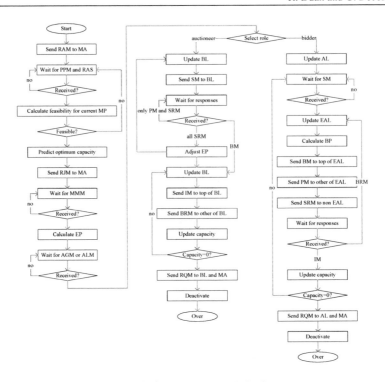

Figure 7.13: Resource agent behaviors.

7.7 Conclusions and future research

Aided with advanced computer science and communication technology, the future electricity infrastructure based on microgrids with MAS coordination will work in a way distinctly different from that of the traditional power system. Due to the independence of each microgrid from the main grid, derived from its capability of intentional islanding in cases of disruptions either inside or outside the microgrid, the future power system will be much more reliable than its predecessor. Since the supply and demand matching of devices within a microgrid or between peer microgrids is based on microeconomic principles, a global optimization of energy consumption will appear beyond the ownership of each market player, which can obviously dampen peak demand, hence reducing the construction cost of a transmission system. Moreover, with the automated negotiation and coordination enabled by applying MAS, a much shorter market period (such as 15 minutes) can be realized on the future electricity infrastructure, which will allow more RES with less predictable power output to be deployed.

Future research focuses on further developing and assessing the approach proposed in this chapter. A survey into the detailed profiles of different types of generators and loads will be carried out first. In this way, the practicality of our multi-agent model of the future electricity infrastructure will be consolidated with more numerical evaluation. According to the results of further simulation, the optimal bidding strategy of each type of supplier and consumer under the competitive electricity market arrangements will be generalized based on various typical scenarios that are investigated. Moreover, the physical realization of agent communication protocols for coordination will also be addressed.

References

1. Foundation of Intelligent Physical Agents. URL: http://www.fipa.org/.
2. T. L. Friedman. *Hot, Flat, and Crowded.* Farrar, Straus & Giroux, 2008.
3. N. D. Hatziargyriou, A. Dimeas, A. G. Tsikalakis, J. A. Pecas Lopes, G. Kariniotakis, and J. Oyarzabal. Management of microgrids in market environment. In *Proceedings of the International Conference on Future Power Systems*, Amsterdam, The Netherlands, November 2005.
4. A. Henney, J. Bower, and D. Newberry. An independent review of NETA, November 2002.
5. International Energy Agency IEA. Distributed generation in liberalized electricity markets, 2002.
6. I. G. Kamphuis, M. Hommelberg, C. J. Warmer, F. J. Kuijper, and J. K. Kok. Software agents for matching of power supply and demand: A field test with a real-time, automated imbalance reduction system. In *Proceedings of the International Conference on Future Power Systems*, Amsterdam, The Netherlands, November 2005.
7. J. K. Kok, C. J. Warmer, and I. G. Kamphuis. PowerMatcher: Multiagent control in the electricity infrastructure. In *Proceedings of the 4th International Joint Conference on Autonomous Agents and Multiagent Systems*, Utrecht, The Netherlands, 2005.
8. R. H. Lasseter and P. Paigi. Microgrid: A conceptual solution. In *Proceedings of the 35th Annual IEEE Power Electronics Specialists Conference*, Aachen, Germany, 2004.
9. Office of Gas and Electricity Markets. URL: http://www.ofgem.gov.uk/.
10. Ofgem. The review of the first year of neta, July 2002.
11. Reticular Systems Inc. Using intelligent agents to implement an electronic auction for buying and selling electric power. URL: http://www.aesc-inc.com/download/epri.pdf, August 1999.
12. T. W. Sandholm. Distributed rational decision making. In *Multiagent Systems: A Modern Approach to Distributed Artificial Intelligence*. The MIT Press, Cambridge, Massachusetts, 2001.
13. A. Tsikalakis and N. Hatziargyriou. Economic scheduling functions of a microgrid participating in energy markets. In *Proceedings of the DG CIGRE Symposium*, Athens, Greece, April 2005.
14. M. J. Wooldridge. *An Introduction to Multiagent Systems*. Wiley, Reading, Massachusetts, 2002.

Chapter 8
Intelligence in Electricity Networks for Embedding Renewables and Distributed Generation

J.K. Kok, M.J.J. Scheepers, and I.G. Kamphuis

Abstract Over the course of the 20th century, electrical power systems have become one of the most complex systems created by mankind. Electricity has made a transition from a novelty, to a convenience, to an advantage, and finally to an absolute necessity. The electricity infrastructure consists of two highly-interrelated subsystems for commodity trade and physical delivery. To ensure the infrastructure is up and running in the first place, the increasing electricity demand poses a serious threat. Additionally, two other trends force a change in infrastructure management. Firstly, there is a shift toward intermittent sources, which gives rise to a higher influence of weather patterns on generation. At the same time, introducing more combined heat and power generation (CHP) couples electricity production to heat demand patterns. Secondly, the location of electricity generation relative to the load centers is changing. Large-scale generation from wind is migrating towards and into the seas and oceans, and, with the increase of *distributed generators* (DG), the generation capacity embedded in the (medium and low voltage) distribution networks is rising. Due to these developments, intelligent distributed coordination will be essential to ensure the efficient operation of this critical infrastructure in the future. As compared to traditional grids, operated in a top-down manner, these novel grids will require bottom-up control. As field test results have shown, intelligent distributed coordination can be beneficial to both energy trade and active network management. In future power grids, these functions need to be combined in a dual-objective coordination mechanism. To exert this type of control, alignment of power systems with communication network technology as well as computer hardware and software in shared information architectures will be necessary.

J.K. Kok, M.J.J. Scheepers, I.G. Kamphuis
Energy research Centre of the Netherlands (ECN), Power Systems and Information Technology, Petten, The Netherlands, e-mail: {j.kok, scheepers, kamphuis}@ecn.nl.

R.R. Negenborn et al. (eds.), *Intelligent Infrastructures*, Intelligent Systems, Control and
Automation: Science and Engineering 42, DOI 10.1007/978-90-481-3598-1_8,
© Springer Science+Business Media B.V. 2010

8.1 Introduction

In the year 1888, Nikola Tesla presented his "New System of Alternate Current Motors and Transformers", laying the foundation for today's electricity infrastructure. Tesla's 'new system' made it possible to transmit electrical power over long distances and to use one single infrastructure for all power delivery. Previously, generators needed to be located near their loads due to highly-inefficient transmission. Furthermore, multiple electric lines were needed for each application class (lighting, mechanical loads, etc) requiring different voltage levels. Over the course of the 20th century, the electrical power systems of industrialized economies have become one of the most complex systems created by mankind. In the same period, electricity made a transition from a novelty, to a convenience, to an advantage, and finally to an absolute necessity. World-wide electricity use has been ever-growing. Especially, three major trends are accelerating its growth [2]:

- The rapid expansion of world population – the growth in the number of people needing electricity.
- The "electrification of everything" – the growth in the number of devices that require electricity.
- "Expectation inflation" – the growth in the sense of entitlement that turns electrical conveniences into essentials demanded by all.

The impact of these factors can be seen in Table 8.1 showing some related growth trends.

The worldwide electric power generation is expected to grow 2.4% a year at least until 2030. In spite of this relatively small annual increase, world electricity generation nearly doubles over the 2004 to 2030 period – from 16,424 billion kiloWatt hours (kWh) in 2004 to 30,364 billion kWh by 2030 [5].

In this chapter we will look into the electricity infrastructure, its peculiarities, its highly interrelated subsystems for commodity trade and physical delivery, and the intelligence that is essential to ensure this critical infrastructure runs efficiently in the future. To ensure the infrastructure is up and running in the first place, the

Table 8.1: Examples of Electricity Growth Trends [2].

Category	1950	2000	2050 (est.)
World Population	2.6B	6.2B	8.3B
Electricity as % of total energy	10.4%	25.3%	33%
Televisions	0.6B	1.4B	2B
Personal Computers	0	500M to 1B	6B to 8B
Cell Phone Connections* (USA)	0	0.8B	5B
Electric hybrid vehicles	0	55,800	3M

B = billion; M = million.
*Including machine to machine connections, e.g.: telemetering and telecontrol.

ever-increasing electricity demand poses a serious threat. Additionally, there are a number of other ongoing changes in the electricity system that are forcing a change in infrastructure management. Firstly, there is a shift to intermittent sources: a larger share of renewables in the energy mix means a higher influence of weather patterns on generation. At the same time, introducing more combined heat and power generation (CHP) couples electricity production to heat demand patterns. Secondly, the location of electricity generation relative to the load centers is changing. Large-scale generation from wind is migrating towards and into the seas and oceans, away from the locations of high electricity demand. On the other hand, with the increase of *distributed generators* (DG) the generation capacity embedded in the (medium and low voltage) distribution networks is rising. This form of generation is relatively small in individual capacities, but (very) high in numbers. These are, for example, medium and small-sized wind turbines, domestic photo-voltaic solar panels, and CHPs in homes and utility buildings. A large part of this chapter focusses on how to keep the balance between demand and supply in these future power systems.

In Section 8.2 we describe some of the peculiarities of electricity and its infrastructure. In Section 8.3 the cohesion between the physical infrastructure and the commodity markets is described, while Section 8.4 treats the changing nature of electricity generation and the implications for infrastructure management. Section 8.5 describes coordination intelligence for electrical power systems, its requirements, the ICT technologies available to meet these and a specific dedicated implementation: the PowerMatcher. The chapter ends with a description of two field experiments (Section 8.6) and an outlook into an architecture for intelligent multi-goal coordination in electrical power systems (Section 8.7).

8.2 On the special nature of electricity and its infrastructure

Electricity is an example of a *flow commodity*: physical flows that are continuous, i.e., the commodity is (virtually) infinitely divisible. Other examples of flow commodities are physical streams in the gas or liquid phase such as natural gas, industrially-applied steam and water for heat transport or drinking. The continuous nature of flow commodities makes their infrastructure behavior fundamentally different from infrastructures transporting discrete objects such as cars or data packets. On top of which, electricity and its infrastructure has some peculiarities not found in any other infrastructure. In the following we will dive deeper in the electricity system's special features.

8.2.1 Power loop flows

The (greater part of) current electricity networks are passive networks, there is no way of directing the flow to follow a particular path. Instead, the commodity follows

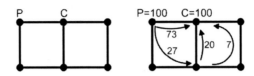

Figure 8.1: Power loop flows in an electrical network. *Left*: simple example electricity network with six nodes and seven lines. At node P electricity is produced, at node C it is consumed, while the other nodes are passive. *Right*: the physical flows resulting from producing 100 units at node P while consuming the same amount at node C. The (resistance) characteristics of all seven lines have been chosen equal, while, for the sake of the example, transport losses have been neglected.

the path of least resistance, possibly using a number of parallel trajectories. This may cause unexpected *power loop flows* as is shown in Figure 8.1.

This behavior has great implications for network planning and day-to-day operations of the grid, for instance. To illustrate this, consider the lines in Figure 8.1 and suppose each of the seven lines has a maximum capacity of 73. Then, although the total sum of line capacity between nodes P and C would be 146, the transport capacity from P to C would be limited to 100.

In December 2004 and January 2005, such a loop flow nearly caused a black-out in Northwestern Europe. An unexpected surplus of wind energy in the North of Germany flew to the South of that country via the neighboring grids of The Netherlands and Belgium. The Dutch operator of the transmission network needed to take special measures to ensure the stability of the network.

8.2.2 Contract paths and transport paths

In *directed* networks, such as those of road transportation and packet-switched information flows, the transport path follows the contract path quite closely. In the field of logistics, for instance, a specific post packet shipped to a particular destination will, if all goes well, arrive at that location. In contrast, in electricity, the contract path does not dictate the actual flow of the commodity, as is shown in Figure 8.2.

This has special implications for the way management is done in the electricity system. There are two separate sub-systems with limited interaction: the physical infrastructure and the electricity markets. We will delve deeper into this separation in Section 8.3.

8.2.3 Absence of infrastructure-inherent storage

Unlike many other infrastructures, in electricity, there is virtually no storage or buffering of the commodity in the network itself. In an electricity network, the supply/demand balance has to be maintained at all times to prevent instabilities, which would eventually result in a black-out. On a timescale of seconds, some

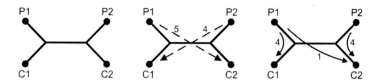

Figure 8.2: Contract paths and transport paths in electricity. *Left*: simple example electricity network with two producers (P1, P2) and two consumers (C1, C2). *Middle*: two possible contract paths: P1 sells an amount of 5 to C2, while P2 sells 4 to C1. *Right*: the resulting physical flows. Only 1 out of the 9 produced units flows physically to the contracted customer. In spite of this, all actors meet their contractual obligations. Note that the flows in the right subfigure are the result of the *superposition* of the contracted flows. For the sake of the example, transport losses have been neglected here.

infrastructure-inherent storage is present due to the rotating mass in power plant turbines. The inertia of this mass allows for small deviations in the momentary supply demand balance. However, these small deviations need to be compensated for on a seconds to minutes timescale to prevent the system sliding towards a black-out.

Consequently, electricity needs to be produced at exact the same time it is consumed. This is a feature unique to electricity. In Section 8.3.4 we will discuss the implications of this characteristic for the interaction between network management and electricity markets.

8.3 Electricity networks and electricity markets

The special nature of electricity has consequences for the way the highly-complex electricity systems of developed economies are organized. One important consequence is the separation between commodity trades and network operations, both performed in two separate sub-systems with limited, yet crucial, interaction.

8.3.1 The electricity system

The term *electricity system* is used to denote the collection of all systems and actors involved in electricity production, transport, delivery and trading. The electricity system consists of two subsystems: the physical subsystem, centered around the production, transmission, and distribution of electricity, and the commodity subsystem, in which the energy product is traded.

Figures 8.3 and 8.4 present a model[1] of the electricity system. In this text, the financial flows that result from the electricity trade are referred to as the "commodity transaction", to distinguish it from transactions related to the physical electricity flows. In the figures, the physical and commodity subsystems have been separated.

---

[1] This model and its description is adapted from [21], Section 3.1.

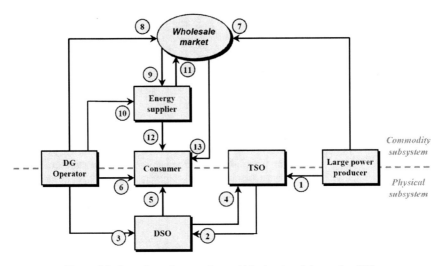

Figure 8.3: Overview of transactions within the electricity market [21].

Note that the two subsystems are related but they are not linked one to one. A generator with a constant output may have fluctuating revenues as a result of variations in market price. Both subsystems need to operate within certain technical and regulatory constrains, such as safety limits, construction permits, operating licenses and emission permits for the physical sub-system, and competition law and market rules for the commodity subsystem. It is important to note that in the figures, for simplicity, different actors of the same type (such as different DSOs) are aggregated into one presented actor.

In the liberalized electricity market, several relevant parties can be distinguished (parties and their definitions are based on European regulations [14]):

- The *producer* is responsible for generating electricity (large power producers, as well as DG-operators who produce electricity with small-scale distributed generation).

- The *transmission system operator* (TSO) is responsible for operating the transmission system in a given area and, where applicable, its interconnections with other systems. It ensures the long term ability of the system to meet reasonable demands for the transmission of electricity. To carry out these responsibilities, the TSO ensures the maintenance and, when necessary, the development of the transmission system. In this context transmission stands for the transport of electricity on the high-voltage interconnected system, the transmission grid.

  The TSO is also responsible for providing system services in his control area. System services consist of balancing services (i.e., compensating the difference in the demand and supply, see also Section 8.3.4), reserve capacity (i.e., compensating shortfall in power generating capacity), power quality (e.g., frequency control), reactive power supply, and black start capability.

- The *distribution system operator* (DSO) is responsible for operating the distribution system in a given area and the connections to the transmission grid. It ensures the long term ability of the system to meet reasonable demands for the distribution of electricity. To carry out these responsibilities, the DSO ensures the maintenance and, where necessary, the development of the distribution grid. In this context distribution means the transport of electricity on high-voltage, medium voltage and low voltage distribution systems with a view to its delivery to customers, but not including supply. The DSO is also responsible for system services, e.g., power quality.

- The *supplier* is responsible for the sale of electricity to customers (retail). Producer and supplier can be the same entity but this is not always the case. A supplier can also be a wholesale customer or independent trader who purchases electricity with the purpose to resell it within the system.

- The *final customer* purchases electricity for their own use and, in a liberalized market, is free to purchase electricity from the supplier of their choice. For different functions (lighting, heating, cooling, cleaning, entertainment, etc.) the final customer uses different electrical appliances.

8.3.2 The physical subsystem

The physical subsystem consists of all hardware that physically produces and transports electricity to customers, as well as all equipment that uses the electricity. The structure of the physical subsystem is determined by the nature of the components that make up the electricity supply system: generators (large power producers and DG operators), transmission network (TSO), distribution networks (DSOs) and loads (consumers) [4]. The physical subsystem is depicted in the lower part of Figure 8.3. The large power producers generate electricity that is fed into the transmission grid. Relation 1 represents the (regulated) agreement between the large power producer and the TSO. The power producer pays a *connection charge* (and sometimes also a *use of system charge*) for the transport of the produced electricity to the DSOs (2), who in turn, distribute it to the final consumer. Relation 5 represents the payment of the connection and use of system charges by the consumer to the DSO for the delivery of the electricity and system services. The figure shows that electricity generated by DG operators is directly fed into the distribution network based on a (regulated) agreement between the DSO and the DG operators (3). The DG operator pays a connection charge and sometimes also a use of system charge to the DSO for electricity transport and for system services. Most of this electricity is then distributed to the consumer by the DSOs (5), but due to the growing amount of DG capacity, a local situation can occur in which supply exceeds demand. In this case the surplus of electricity is fed upwards into the transmission grid (4), after which the TSO transports it to other distribution networks (2). The last relevant physical stream concerns the auto-production of DG electricity (6). This is the direct consumption of electricity produced on-site by a consumer, omitting the commodity purchase and sales process through the energy supplier.

8.3.3 The commodity subsystem

In contrast with the physical power streams, the economic transactions related to the commodity flow are merely administrative and depicted in the upper part of Figure 8.3. Its goal is an efficient allocation of costs and benefits, within the constraints imposed by the physical system. The commodity subsystem is defined as the actors involved in the production, trade or consumption of electricity, in supporting activities or their regulation and mutual relations [4]. The commodity subsystem controls the physical subsystem, but is constrained by it at as well. Large power producers (7) and some large DG operators (8) offer the commodity on the wholesale market, where the commodity is traded between different actors. Large electricity consumers (e.g., industrial customers) can buy the commodity directly on the wholesale market (13). Next to those consumers, energy suppliers buy commodity in the wholesale market (9) on the basis of wholesale contracts to serve smaller consumers. The trade on the wholesale market provides a payment for the produced electricity. Additional to the wholesale market trade, the energy supplier extracts the commodity directly via (small) DG operators (10). The energy supplier subsequently delivers the commodity from the wholesale market and the DG operators to the consumers (12) who pay for it. As energy suppliers are often 'long' (i.e., they have contracted more commodity than they plan to offer to consumers), there is a commodity stream backwards to the wholesale market (11). Therefore, the energy supplier is a third party trader that offers commodity to the wholesale market.

In the situation that the energy supplier has accurately forecasted the actual amount of electricity which his consumers use, the received payment for the commodity (12) perfectly corresponds to the amount of delivered electricity (5). However, deviations from forecasted use or planned generation often occur, and, due to the failing of the mechanism to balance supply and demand on the short-term, they create the need for an additional short-term balancing mechanism.

8.3.4 The balancing market

System operators and contractors have to estimate demand in order to make sure that sufficient supply is available on short (seconds and minutes), medium (hours), and long (days, months, years) timescales. As the electricity system is liberalized, the market itself is responsible for matching supply and demand on the long and medium terms. As stated before, the electricity supply (output from all generators including import) must be controlled very closely to the demand. This has to be maintained on the timescale of seconds. Maintaining the short and medium term balance is the responsibility of the system operator, which for this purpose uses forecasts of electricity production and demand which are submitted by market players (called energy programs or physical notifications). Deviations between elec-

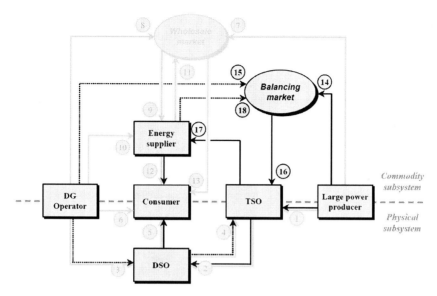

Figure 8.4: Overview of transactions within the electricity market, including the balancing market [21].

tricity demand and production during the actual moment of execution of the energy programs become visible to the TSO as an exchange of electrical power with neighboring control areas, different from the agreed international exchange programs (involuntary or unintentional exchange). In this way the TSO has insight in the actual balance of the total system. In the case of shortage, the system is balanced by producing additional electricity (upward adjustment of production units) or making use of demand response (i.e., lower consumption). In the case of a surplus, production units are adjusted downwards.

In many European control areas the liberalization of the energy market has led to the establishment of a separate balancing market, apart from the wholesale and retail market. This market is controlled by the TSO, being the sole buyer. Access to the supply side of the balancing market is mainly limited to the large power producers, but DG operators (in particular large CHP-units) and energy suppliers also have access. Figure 8.4 shows the impact of the balancing market. The transactions that are less common in existing electricity markets are shown with dotted lines. As soon as a situation of shortage arises, the TSO corrects this by buying the lowest priced commodity offer in the balancing market (16). Most offers come from the large power producers (14), but sometimes DG operators offer electricity as well (15, CHP units), just as energy suppliers (18). The TSO charges the energy supplier(s) which caused the imbalance (17) on basis of the (relatively high) price that it has paid on the balancing market. In the case of a surplus of produced electricity, the TSO accepts and receives the highest bid in the balancing market for adjusting generating units downwards. Also, in this case, the energy supplier(s) pay the TSO so-called imbalance charges. Handling these imbalance charges is arranged in the energy contracts between all market players, but mostly energy suppliers are responsible for the demand of their contracted consumers and contracted DG-operators. Therefore, the energy supplier must pay the balancing costs in case there is a deviation of the forecasted use of its consumers or forecasted generation of its contracted DG operators. If a large power producer does not comply with its contracts, e.g., there is a malfunctioning of a generating facility, they must pay the balancing costs themselves, as large power producers are responsible for their own energy program. To stimulate market players to make their forecasts of electricity production and demand as accurate as possible and to act in accordance with these energy programs, the price for balancing power (imbalance charges) must be above the market price for electricity. Since balancing power is typically provided by units with high marginal costs, this is, in practice, always automatically the case.

8.3.5 System support services

Another relevant issue in the electricity system is the delivery of *system support services* or *ancillary services*, i.e., all services necessary for the operation of a transmission or distribution system. It comprises compensation for energy losses, frequency control (automated, local fast control and coordinated slow control), voltage and flow control (reactive power, active power, and regulation devices), and restoration of supply (black start, temporary island operation). These services are required to provide system reliability and power quality. They are provided by generators and system operators.

8.4 Changing nature of electricity generation

In electricity generation two inter-related movements can be seen, both of paramount importance for the way the electricity system will be managed in the future:

1. The increase of electricity generated from **sustainable energy sources**.

2. **Decentralization** of electricity generation: electricity generating units are growing in numbers and moving closer to the load centers.

In the next two subsections we will describe these changes in more detail, and in the third subsection we will describe the impact on management of the electricity system.

8.4.1 Sustainable electricity sources

Worldwide, two thirds of the electricity is still produced from fossil fuels (natural gas, oil and coal) while approximately 15% originates from nuclear sources [6]. Of the sustainable options for electricity generation, hydro energy is currently most significant in the world wide power production (17%). Other sustainable energy sources (wind, solar, biomass, and geothermal) contribute for only about 2% to the world wide electricity generation[2].

However, there are important drivers to reduce the fossil fuel dependency and to substitute fossil fuels for sustainable energy sources. Two important drivers behind this are:

- **Environmental concerns**: pollution and climate change. Most fossil fuels are used as input for a combustion process which emit pollutants such as aerosols (e.g., soot), sulfur oxides and nitrogen oxides. Further, fossil fuel usage is one of the greatest contributors to global warming due to greenhouse gas emissions.

- **Diversification of energy sources**: the energy need of most western economies is largely imported from outside those economies. As energy demand continues to grow, this external dependence could grow steeply in the next decades. Moreover, a substantial portion of fossil fuels are imported from politically unstable regions. A higher portion of sustainable energy in the energy mix reduces this dependency.

Hydro energy is the only sustainable energy source with a substantial share in today's electricity supply. Worldwide, approximately 17% of electricity is generated by hydro power generators. However, the growth potential for hydro power is limited. Instead of large hydro power plants, in many countries there is a capacity increase because of new small hydro power facilities. These generators are connected to the medium voltage distribution grid.

With an annual growth of 25 to 30%, wind energy is becoming the second largest sustainable energy source for power generation. In 2008 the worldwide installed capacity was 121 GW [16] (3.2% of total power generation capacity world wide). With an annual growth of 25%, the wind generation capacity in 2020 will be 1750 GW, i.e., a share of at least 25% of the world wide power generation capacity. In 2008 Germany had 24 MW wind generation capacity installed with a production share of 7.5%. Among the countries with the largest wind generation capacity in 2008 are the USA (25 GW), Spain (17 GW) and China (17 GW). Initially wind turbines with a capacity up to 1000 kW (solitaire or in a wind park) were connected to the distribution grid. Today, very large wind turbines with a generation up to 5 MW each are installed offshore in large wind parks. Since the total generation capacity of these wind parks is often more than 100 MW, they are connected to the transmission grid. At the same time there is a trend towards smaller wind turbines, i.e., turbines with a capacity of less than 50 kW. These turbines are situated near dwellings and connected to the low voltage distribution grid.

The most abounded sustainable energy source world wide is solar energy. Solar energy can be converted to electricity through a thermal route using a steam cycle, as in conventional power plants, and through photo voltaic (PV) cells. The thermal technique is used on large plants (some hundreds of MW), so called concentrated solar power. Panels with PV cells are used in urban areas, mounted to the roofs of

---

[2] Sustainable Electricity Sources are also refered to as *Renewable Energy Sources* (RES). In the remainder of this text we will use these terms interchangeably.

buildings and dwellings, and connected to the low voltage distribution grid. The total installed capacity of PV world wide in 2007 was 9100 MWpeak [24] (of which 40% in Germany). If the average annual growth factor of about 30% continues, the installed total world wide generation capacity in 2020 may become 275 GWpeak. Although this will be only a few percent of the total installed generation capacity world wide, locally the share of electricity production from PV may be much larger.

Biomass (wood, organic waste, etc.) has been used for power generation on a limited scale for decades. There is a large growth potential for this sustainable energy source. Different kinds of biomass can be co-fired in coal fired power plants (10 to 30%). Biomass can also be converted into electricity in dedicated biomass plants. The size of these plants is smaller than conventional power plants, i.e., up to a few hundred MW. Another form of bioenergy is biogas. Biogas, from waste water treatment or anaerobic digestion of manure, can be used as a fuel for gas engines producing electrical power. These units have a capacity of some MWs and are connected to the medium voltage distribution grid.

Other sustainable energy sources are geothermal energy and wave and tidal energy. These energy sources are only available in specific regions, but there they may be of significant importance, as is the geothermal electricity generation in Iceland.

8.4.2 Distributed generation

Another ongoing change in electricity generation is the growing generation capacity located in the distribution part of the physical infrastructure. This trend breaks with the traditional central plant model for electricity generation and delivery. For this type of generation the term *distributed generation* (DG) is used:

Definition 8.1. Distributed Generation (DG) is the production of electricity by units connected to the distribution network or to a customer site.

Thus, DG units supply their generated power to the distribution network either directly or indirectly via a customer's private network (i.e., the network on the end-customer's premises, behind the electricity meter). Consequently, the generation capacities of individual DG units are small as compared to central generation units which are directly connected to the transmission network. On the other hand, their numbers are much higher than central generation and their growth is expected to continue.

Sustainable or renewable energy sources (RES) connected to the distribution grid fall under the definition of DG. However not all RES are DG as large-scale renewables, e.g., off-shore wind electricity generation, is connected to the transmission network. The same holds for Combined Heat and Power production (CHP – or Co-generation). A CHP unit is an installation for generating both electricity and useable heat simultaneously. Dependent of their size CHP units are either connected to the distribution grid (and, thus, fall under the definition of DG) or to the transmission grid. Table 8.2 categorizes different forms of CHP and RES into either large-scale generation or distributed generation.

Table 8.2: Characterization of Distributed Generation (adapted from [20]).

	Combined Heat and Power	Renewable energy sources
Large-scale Generation	- Large district heating* - Large industrial CHP	- Large hydro** - Off-shore wind - Co-firing biomass in coal power plants - Geothermal energy - Concentrated solar power
Distributed Generation	- Medium district heating - Medium industrial CHP - Utility building CHP - Micro CHP	- Medium and small hydro - On-shore wind - Tidal energy - Biomass and waste incineration - Biomass and waste gasification - PV solar energy

*Typically > 50MW_e; ** Typically > 10MW_e.

There are a number of drivers behind the growing penetration of DG [7]. Here, two important drivers, namely "Environmental concerns" and "Diversification of energy sources", are shared with the drivers for RES increase (see Section 8.4.1 for a description). Additional to these, important drivers for DG are:

- **Deregulation of the electricity market**: As a result of the deregulation, the long-term prospects for large-scale investments in power generation have become less apparent. Therefore, a shift of interest of investors from large-scale power generation plants to medium and small-sized generation can be seen. Investments in DG are lower and typically have shorter payback periods than those of the more traditional central power plants. Capital exposure and risk is reduced and unnecessary capital expenditure can be avoided by matching capacity increase with local demand growth.

- **Energy autonomy**: A sufficient amount of producing capacity situated in a local electricity network opens the possibility of intentional islanding. Intentional islanding is the transition of a sub-network to stand-alone operation during abnormal conditions on the externally connected network, such as outages or instabilities, e.g., during a technical emergency. In this manner, autonomy can be achieved on different scales, from single buildings to wide-area subsystems.

- **Energy Efficiency (i)**: In general, distributed generation reduces energy transmission losses. Estimates of power lost in long-range transmission and distribution systems of western economies are of the order of 7%. By producing electricity in the vicinity of consumption area, transport losses are avoided. There is, however, a concern that in cases where the local production outgrows the local consumption the transmission losses start rising again. But in the greater part of the world's distribution network we are far from reaching that point.

- **Energy Efficiency (ii)**: Heat production out of natural gas can reach higher efficiency rates by using combined heat-power generation (CHP) instead of traditional furnace burners. CHP is a growing category of distributed generation, especially in regions where natural gas is used for heating. In Northern Europe, for instance, CHP is already commonly used in heating of large buildings, green houses and residential areas. The use of micro-CHP for domestic heating in single dwellings is also expected to breakthrough in the coming few years.

The growing share of DG in the electricity system may evolve in three distinct stages (adapted and extended from [8]):

1. **Accommodation**: Distributed generation is accommodated in the existing electricity system, i.e., network and markets. Distributed units are running free, beyond the control of the transmission grid operator or market-party to which the generated energy is delivered. The centralized control of the networks remains in place. Electricity supply companies treat DG as being negative demand: it is non-controllable and to a certain extend forecastable.

2. **Decentralization**: The share of DG increases. Clustered operation of DG devices gives an added value. Supply companies optimize the services of decentralized providers through the use of common ICT–systems (Virtual Power Plant concept). Distribution grid operators use the services of decentralized providers for grid operational goals, like congestion management. Central monitoring and control is still needed.

3. **Dispersal**: Distributed power takes over the electricity market. Local low–voltage network segments provide their own supply with limited exchange of energy with the rest of the network. The central network operator functions more like a coordinating agent between separate systems rather than controller of the system.

8.4.3 Implications for infrastructure management

Both the rising share of renewable energy sources and the decentralization are changing the characteristics of power generation in three aspects:

- **Intermittancy**: The energy sources for conventional power generation are continuously available and can be adjusted according to the electricity demand. Electricity from sustainable energy sources, such as wind and solar energy, can only be produced if the primary energy source is available. Additionally, electricity from those CHP units which are operated to follow heat demand are intermittent in nature as well. With the growing share of these intermittent energy sources it becomes more difficult to follow the changing electricity demand.

- **Cardinality**: The number of electricity generating units is growing rapidly while individual capacities are decreasing.

- **Location**: The location of power generation relative to the load centers is changing. Due to decentralization, the distance between generation units in the grid relative to the location of electricity consumption is becoming smaller. However, central generation from wind is moving further away from the load centers as large-scale wind farms are being built off-shore.

These changes are expected to have a number of implications for infrastructure management such as: power quality problems, e.g., local voltage increase due to in-feeding by DG units or voltage distortion due to injection of higher harmonics by inverter-based DG units; Changing current directions in distribution grids disturbing, for instance, systems for short-circuit localization; Stability problems due to a decreasing amount of rotating mass in the system, etc.

In this chapter text, the main focus is on the important topic of maintaining the balance between supply and demand in the electricity system. In the status quo, this balance is maintained by a relative small number of big central power plants following inflexible and partially unpredictable load patterns. As the supply side becomes more inflexible, a need emerges to utilize the flexibility potential of the demand side. With that, the nature of coordination within the electricity system is changing from centrally controlling a few central power plants into coordinating among high numbers of generators and responsive loads, with varying levels of flexibility and having a great variety in (production and consumption) capacity.

As distributed generation gradually levels with central generation as a main electricity source, distributed coordination will be needed alongside central coordination. The standard paradigm of centralized control, which is used in the current electricity infrastructure, will no longer be sufficient. The number of system components actively involved in the coordination task will be huge. Centralized control of such a complex system will reach the limits of scalability, computational complexity, and communication overhead.

As demand response plays such a crucial role in such a system, we will expand on that topic in the next subsection.

8.4.4 Demand response

Crucial for an efficient and stable operation of future electricity networks is *demand response* (DR).

Definition 8.2. Demand response is the ability of electricity consuming installations and appliances to alter their operations in response to (price) signals from the energy markets or electricity network operators in (near-)real time.

Demand response can be achieved through avoidance of electricity use and/or by shifting load to another time period.

At present, *price elasticity* of electricity demand is very low in the electricity markets. This means that the quantity in demand stays cinstant with a changing price. Higher elasticity in electricity demand would lead to lower market power of

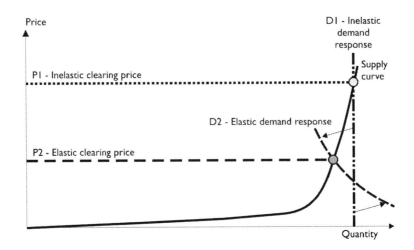

Figure 8.5: Impacts of Demand Elasticity on Wholesale Price [9].

producers and to a lower electricity price (see Figure 8.5). During the California
energy crisis, a demand reduction of 5% during the periods of the highest price
peaks would have reduced these prices by 50% [9].

Typical large flexible loads include different types of industrial processes, e.g.,
groundwood plants and mechanical pulping plants, electrolysis, arc furnaces, rolling
mill, grinding plants, extruders, gas compressors, etc. In the commercial and residential sectors flexible loads can include space heating, water heating, cooling,
ventilation, washing and drying, lighting, etc.

From the viewpoint of controllability, DG and DR are equivalent: increasing
production has the same effect on the supply and demand balance as decreasing
consumption, and vice versa. Due to this, demand response is sometimes treated
as being a resource. As a result of the common nature of DG and DR (and distribution network connected electricity storage), the overarching term *Distributed
Energy Resources* (DER) is used to refer to this threesome: DG, DR and storage.

8.5 Intelligent distributed coordination in electricity

As a result of the electricity evolution, as described above, the electricity infrastructure will get more and more inter-linked with ICT-infrastructures. The architecture and algorithmics of this ICT-infrastructure must be adapted to the technical
structure of the (future) electricity net and the connected producing and consuming installations, but also to the structure of the liberalized energy market. This
ICT-architecture and associated algorithms must be designed using a strong system-
wide viewpoint, but must also consider stakes of local actors in the system. In other
words, there is a need for a multi-actor coordination system, which optimizes global
system objectives (such as stability, power quality, and security of supply), in co-

herence with the interests of local actors in the form of installations for electricity production, consumption, and storage as well as their owners. These local actors vary greatly in characteristics defined by process type, purpose and size, and so do their specific constraints and objectives. This leads to a major shift in the requirements of the coordination system.

8.5.1 High-level requirements of the needed coordination system

Specific information system requirements of the ICT-infrastructure needed for the expected electricity evolution include [11, 12]:

- **Scalability**: A huge number of systems spread-out over a vast area will have to be involved in the coordination task. Especially on the level of the distribution grids, huge growth in the number of components actively involved in the coordination is expected. The ICT system must be able to accommodate this growth.
  This poses an important scalability requirement for the information systems architecture performing this task. Following the sector's paradigm of centralized control, the system may reach the limits of communication overhead rapidly.

- **Openness**: The information system architecture must be open: individual DER units can connect and disconnect at will and future types of DER – with own and specific operational characteristics – need to be able to connect without changing the implementation of the system as a whole. Therefore, communication between system parts must be uniform and stripped from all information specific to the local situation.

- **Multi-level Stakes**: The information system must facilitate a multi-actor interaction and balance the stakes on the global level (i.e., the aggregated behavior: reaction to energy market situation and/or network operator needs) and on the local level (i.e., DER operational goals).

- **Autonomy and Privacy**: In most cases, different system parts are owned or operated by different legal persons, so the coordination mechanism must be suitable to work over boundaries of ownership. Accordingly, the power to make decisions on local issues must stay with each individual local actor.

8.5.2 Multi-agent systems

The advanced technology of multi-agent systems (MAS) provides a well-researched way of implementing complex distributed, scalable, and open ICT systems. A multi-agent system is a system of multiple interacting software agents. A software agent is a self-contained software program that acts as a representative of something or someone (e.g., a device or a user). A software agent is goal-oriented: it carries out a task, and embodies knowledge for this purpose. For this task, it uses information from and performs actions in its local environment or context. Further, it is able to communicate with other entities (agents, systems, humans) for its tasks.

In multi-agent systems, a large number of actors are able to interact. Local agents focus on the interests of local sub-systems and influence the whole system via negotiations with other software agents. While the complexity of individual agents remains low, the intelligence level of the global system is high. In this way, multi-agent systems implement distributed decision-making systems in an open, flexible, and extensible way. Communication between actors can be minimized to a generic and uniform information exchange.

8.5.2.1 Electronic markets

The interactions of individual agents in multi-agent systems can be made more efficient by using *electronic markets*, which provide a framework for distributed decision making based on microeconomics. Microeconomics is a branch of economics that studies how economic agents (i.e., individuals, households, and firms) make decisions to allocate limited resources, typically in markets where goods or services are being bought and sold. One of the goals of microeconomics is to analyze market mechanisms that establish relative prices amongst goods and services and allocation of limited resources amongst many alternative uses [13]. Whereas, economists use microeconomic theory to model phenomena observed in the real world, computer scientists use the same theory to let distributed software systems behave in a desired way. Market-based computing is becoming a central paradigm in the design of distributed systems that need to act in complex environments. Market mechanisms provide a way to incentivize parties (in this case software agents), that are not under direct control of a central authority, to behave in a certain way [3, 19]. A microeconomic theory commonly used in MAS is that of general equilibrium. In general equilibrium markets, or exchange markets, all agents respond to the same price, that is determined by searching for the price that balances all demand and supply in the system. From a computational point of view, electronic equilibrium markets are distributed search algorithms aimed at finding the best trade-offs in a multidimensional search space defined by the preferences of all agents participating in the market [23, 26]. The market outcome is *Pareto* optimal, a social optimal outcome for which no other outcome exists that makes one agent better-off while making other agents worse-off.

8.5.2.2 Market-based control

In *Market-based Control*, agents in a MAS are competing for resources on an equilibrium market whilst performing a local control task (e.g., classical feedback control of a physical process) that needs the resource as an input. For this type of MAS, it has been shown by formal proof that the market-based solution is identical to that of a centralized omniscient optimizer [1]. From the viewpoint of scalability and openness of the information architecture, this is an important feature. In the centralized optimization all relevant information (i.e., local state histories, local

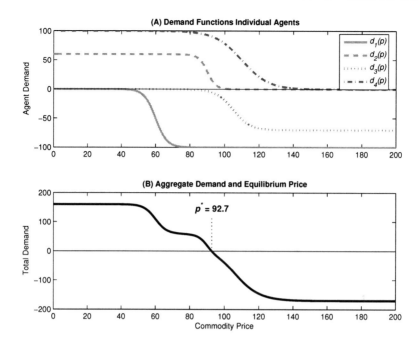

Figure 8.6: Example general equilibrium market outcome. (A) Demand functions of the four agents participating in the market. (B) Aggregate demand function and general equilibrium price p^*.

control characteristics, and objectives) needs to be known at the central level in order to optimize over all local and global control goals. While in the market-based optimization the same optimal solution is found by communicating uniform market information (i.e., market bids stating volume-price relations), running an electronic equilibrium market and communicating the resulting market price back to the local control agents. In this way, price is used as the control signal. It is important to note that, whether – in a specific application – the price has a monetary value or is virtual and solely used as a control signal depends on the particular implementation and on the business case behind the application.

In a typical application of market-based coordination, there are several entities producing and/or consuming a certain commodity or good[3]. Each of these entities is represented by a local control agent that communicates with a market agent (auctioneer). Each market round, the control agents create their market bids, dependent on their state history, and send these to the market agent. These bids are ordinary, or *Walrasian*, demand functions $d(p)$, stating the amount of the commodity the agent wishes to consume (or produce) at a price of p. The demand function is negative in the case of production. After collecting all bids, the market agent searches for the equilibrium price p^*, i.e., the price that clears the market :

---

[3] Or a series of commodities. Here we treat the single-commodity case for simplicity.

$$\sum_{a=1}^{N} d_a(p^*) = 0, \qquad\qquad (8.1)$$

where N is the number of participating agents and $d_a(p)$, the demand function of agent a. The price is broadcast to all agents. Individual agents can determine their allocated production or consumption from this price and their own bid.

Figure 8.6 shows a typical small-scale example of price forming in a (single-commodity) general equilibrium market with four agents. The demand functions of the individual agents are depicted in graph (A). There are two consuming agents whose demand decreases gradually to zero above a certain market price. Further, there are two producers whose supply, above a certain price, increases gradually to an individual maximum. Note that supply is treated as negative demand. The solid line in (B) shows the aggregate demand function. The equilibrium price p^* is determined by searching for the root of this function, i.e., the point where total demand equals total supply.

8.5.3 A decentralized control systems design

This section describes a novel control concept for automatic matching of demand and supply in electricity networks with a high share of distributed generation. In this concept, DG, demand response, and electricity storage are integrated using the advanced ICT technology of market-based distributed control.

This concept has been coined *PowerMatcher*. Since its incarnation in 2004, the PowerMatcher has been implemented in three major software versions. In a spiral approach, each software version was implemented from scratch with the first two versions being tested in simulations and field experiments [10, 11, 18]. The third version is currently under development and is planned to be deployed in a number of field experiments [17] and real-life demonstrations with a positive business case.

The PowerMatcher is a general purpose coordination mechanism for balancing demand and supply in clusters of *Distributed Energy Resources* (DER, distributed generation, demand response, and distribution grid-coupled electricity storage). These 'clusters' might be electricity networks with a high share of distributed generation or commercial trading portfolios with high levels of renewable electricity sources, to name a few.

The PowerMatcher implements *supply and demand matching* (SDM) using a multi-agent systems and market-based control approach. SDM is concerned with optimally using the possibilities of electricity producing and consuming devices to alter their operation in order to increase the over-all match between electricity production and consumption.

8.5.3.1 Logical structure and agent roles

Within a PowerMatcher cluster the agents are organized into a logical tree. The leafs of this tree are a number of *local device agents* and, optionally, a unique *objective*

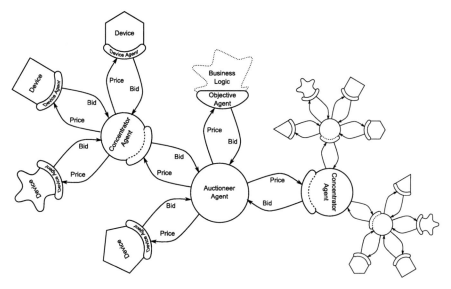

Figure 8.7: Example PowerMatcher agent cluster. See the text for a detailed description.

agent. The root of the tree is formed by the *auctioneer agent*, a unique agent that handles the price forming, i.e., the search for the equilibrium price. In order to obtain scalability, *concentrator agents* can be added to the structure as tree nodes. A more detailed description of the agent roles is as follows:

- **Local device agent**: Representative of a DER device. A control agent which tries to operate the process associated with the device in an economical opti-mal way. This agent coordinates its actions with all other agents in the cluster by buying or selling the electricity consumed or produced by the device on an electronic market. In order to do so, the agent communicates its latest bid (i.e., a demand function, see Section 8.5.2.2) to the auctioneer and receives price up-dates from the auctioneer. Its own latest bid, together with the current price, determines the amount of power the agent is obliged to produce or consume.

- **Auctioneer agent**: Agent that performs the price-forming process. The auc-tioneer concentrates the bids of all agents directly connected to it into one single bid, searches for the equilibrium price and communicates a price update back whenever there is a significant price change.

- **Concentrator agent**: Representative of a sub-cluster of local device agents. It concentrates the market bids of the agents it represents into one bid and commu-nicates this to the auctioneer. In the opposite direction, it passes price updates to the agents in its sub-cluster. This agent uses 'role playing'. On the auctioneer's side it mimics a device agent: sending bid updates to the auctioneer whenever necessary and receiving price updates from the auctioneer. Towards the sub-cluster agents directly connected to it, it mimics the auctioneer: receiving bid updates and providing price updates.

- **Objective agent**: The objective agent gives a cluster its purpose. When the objective agent is absent, the goal of the cluster is to balance itself, i.e., it strives for an equal supply and demand within the cluster itself. Depending on the specific application, the goal of the cluster might be different. If the cluster has to operate as a *virtual power plant*, for example, it needs to follow a certain externally provided setpoint schedule. Such an externally imposed objective can be realized by implementing an objective agent. The objective agent interfaces to the *business logic* behind the specific application.

This logical structure follows the CoTree algorithm [25]. By aggregating the demand functions of the individual agents in a binary tree, the computational complexity of the market algorithm becomes $O(\lg a)$, where a is the number of device agents. In other words, when the number of device agents doubles it takes only one extra concentrator processing step to find the equilibrium price. Furthermore, this structure opens the possibility for running the optimization distributed over a series of computers in a network in a complimentary fashion to power systems architectures.

8.5.3.2 Device agent types and strategies

From the viewpoint of supply and demand matching, DER devices can be categorized by their type of controllability into the following classes:

- **Stochastic operation devices**: devices such as solar and wind energy systems of which the power exchanged with the grid behaves stochastically. In general, the output power of these devices cannot be controlled, the device agent must accept any market price.

- **Shiftable operation devices**: batch-type devices whose operation is shiftable within certain limits, for example (domestic or industrial) washing and drying processes. Processes that need to run for a certain amount of time regardless of the exact moment, such as assimilation lights in greenhouses and ventilation systems in utility buildings. The total demand or supply is fixed over time.

- **External resource buffering devices**: devices that produce a resource, other than electricity, that are subject to some kind of buffering. Examples of these devices are heating or cooling processes, whose operation objective is to keep a certain temperature within two limits. By changing the standard on/off-type control into price-driven control allows for shifting operation to economically attractive moments, while operating limits can still be obeyed (see Figure 8.8). Devices in this category can both be electricity consumers (electrical heating, heat pump devices) and producers (combined generation of heat and power).

- **Electricity storage devices**: conventional batteries or advances technologies such as flywheels and super-capacitors coupled to the grid via a bi-directional connection. Grid-coupled electricity storage is widely regarded as a future en-

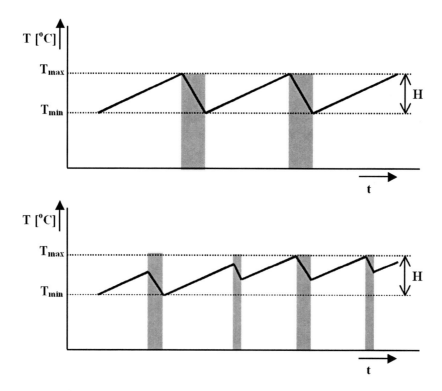

Figure 8.8: Operation shifting in a cooling process whilst obeying process state limits.

abling technology allowing the penetration of distributed generation technologies to increase at reasonable economic and environmental cost. Grid-coupled storage devices can only be economically viable if their operation is reactive to a time-variable electricity tariff, as is present in the PowerMatcher concept. The agent bidding strategy is buying energy at low prices and selling it later at high prices.

- **Freely-controllable devices**: devices that are controllable within certain limits (e.g., a diesel generator). The agent bidding strategy is closely related to the marginal costs of the electricity production.

- **User-action devices**: devices whose operation is a direct result of a user action. Domestic examples are: audio, video, lighting, and computers. These devices are comparable to the stochastic operation devices: their operation is to a great extent unpredictable and has no inherent flexibility. Thus, the agent must accept any market price to let them operate.

In all described device categories, agent bidding strategies are aimed at carrying out the specific process of the device in an economically optimal way, but within the constraints given by the specific process.

8.5.3.3 Cluster-level behavior

The self-interested behavior of local agents causes electricity consumption to shift towards moments of low electricity prices and production towards moments of high prices. As a result of this, the emergence of supply and demand matching can be seen on the global system level.

The PowerMatcher technology can be the basis of a *virtual power plant* (VPP). A VPP is a flexible representation of a portfolio of Distributed Energy Resources (DER), i.e.: distributed generation, demand response, and electricity storage [15]. One of the key activities of a VPP is the delivery of (near-) real-time balancing services, e.g., delivering reserve regulating power to the TSO, delivering active network management services to the DSO, or minimizing the imbalance costs of a commercial party. The aggregated, or concentrated, bid of all local control agents in the cluster – as is held by the auctioneer agent – can be regarded as a dynamic merit-order list of all DER participating in the VPP. On the basis of this list, the VPP is able to operate such the (near-)real-time coordination activity optimally.

8.6 Field test results

8.6.1 Commercial portfolio balancing

This subsection describes the first field experiment performed using the Power-Matcher. A more comprehensive and detailed description of the test, including an analysis of the business model, can be found in [10].

8.6.1.1 Value driver: Balancing responsibility

One of the tasks of a Transmission System Operator is maintaining the instantaneous demand and supply balance in the network. As described in Section 8.3.4, the system of balancing responsibility gives wholesale trading parties incentives to maintain their own portfolio balance. This system provides means to charge the costs made by the TSO when maintaining the real-time system balance to those wholesale market parties responsible of the unbalance. Central to this mechanism is the notion of *Balancing responsibility*, i.e., the obligation of wholesalers to plan their production and consumption and to make this plan available to the TSO. Parties having this responsibility are referred to as *balancing responsible parties* (BRPs).

As may be clear, the system of balancing responsibility imposes imbalance risks to the market parties. Among BRPs, this risk will vary with the predictability and controllability of the total portfolio of the BRP. BRPs with low portfolio predictability are faced with higher imbalance risks. Typically, wind power suffers from low predictability. This gives higher imbalance costs resulting in a lower market value for electricity produced by wind turbines. Using specialized forecasting techniques as post-processors to high-resolution meteorological models, the day-ahead pre-

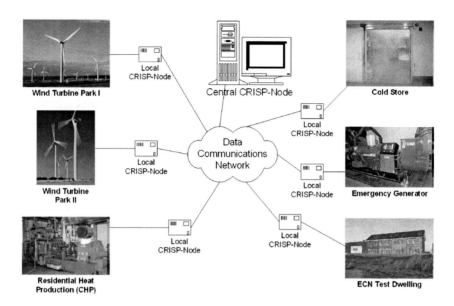

Figure 8.9: Configuration of the imbalance reduction field test.

dictability of wind energy production has been improved substantially in the last few years. However, a substantial error margin remains.

8.6.1.2 Field test set-up

For the purpose of the field test, five different installations were brought together in the portfolio of a virtual BRP. In reality, the installations represent a small part of the portfolios of two different BRPs, but for the sake of the experiment they were assumed to represent the full portfolio of one single BRP. Figure 8.9 gives the configuration of the field test. To all DER sites, hardware was added to run the local control agents on. These agents interacted with the existing local measurement and control system. Further, the local agents communicated with the auctioneer using a virtual private network running over a standard ADSL internet connection or (in one case) a UMTS wireless data connection.

Table 8.3 gives an overview of the capacities of the individual installations included in the test. In order to give the smaller sized installations a good influential balance compared to the larger ones, two of the sites were scaled up via an on-line simulation.

8.6.1.3 Imbalance reduction results

The field test ran for a number of months in the first half year of 2006. In the real-life DER portfolio, with a wind power dominated imbalance characteristic, the imbalance reductions varied between 40 and 43%. As seen from an electricity market perspective, these benefits are substantial. This makes the approach a good addition to the current options for handling wind power unpredictability, like wind/diesel combinations, balancing by conventional power plants and large-scale electricity storage. Topics that need further research include, the factors that influence the flexibility level of the aggregate and the system behavior when the number of attached DERs is increased substantially. The operation of a similar cluster as a commercial VPP, by adding an appropriate objective agent to the agent set, is part of the current research.

8.6.2 Congestion management

This subsection describes the second field experiment performed using the Power-Matcher. A more comprehensive and detailed description of the test can be found in [18] or [22].

8.6.2.1 Value driver: Deferral of grid reinforcements

In the Northwestern region of Europe, decentralized generation of heat and power by micro-CHP units in households is expected to penetrate the market at high speed in the coming years. When the number of micro-CHP units in a region exceeds a certain limit, added value can be gained by clustered coordination via common ICT systems. In a field test a cluster of five Stirling based micro-CHP units of 1 kW electric each has been operated as a virtual power plant[4]. The main goal of the field test was to demonstrate the ability of such a VPP to reduce the local peak load on the single low-voltage grid segment the micro-CHP units were connected to. In this way the VPP supports the local distribution system operator (DSO) to defer reinforcements in the grid infrastructure (substations and cables) when local demand is rising. Although not all micro-CHP units included in the field test were connected to the same low-voltage cable, during the trial a connection to a common substation (i.e., low-voltage to mid-voltage transformer) was assumed.

Table 8.3: Production (P) and Consumption (C) Capacities of the Field Test Installations.

Site	P/C	Capacity	Simulated
Wind Turbine	P	2.5 MW	-
CHP	P	6 MW	-
Cold Store	C	15 kW	1.5 MW
Emergency Generator	P	200 kW	-
Heat Pump	C	0.8 kW	80 kW

P = Production; C = Consumption.

---

[4] In total 10 micro-CHPs were equipped to be part of the VPP. The results presented are realized with 5 of these 10 participating.

8.6.2.2 Field test set-up

The field test focused on the network utilization factor of the local distribution grid in three different settings:

- **Baseline**: domestic load profile of 5 households.

- **Fit-and-Forget**: load profile of 5 households plus micro-CHPs controlled in standard heat-demand driven manner (thermostat).

- **VPP operation**: CHP operation coordinated by PowerMatcher intelligent control to reduce peak-load, without any intrusion on comfort for consumers.

In the third setting, the micro-CHPs were controlled by local PowerMatcher control agents. These agents were clustered together with an objective agent monitoring the load on the shared transformer and demanding CHP electricity production when it exceeded a safety level.

The households participating in the field test were equipped with a Whispergen micro-CHP for heating of living space and tap water. For the latter, these systems were equipped with a tap water buffer of 120 liter. For the field test, the systems were extended with a virtual power plant node or VPP-node. The local agents ran on these VPP-nodes, communicating with the local infrastructure (micro-CHP, thermostat, and electricity meter) through power line communications and with the auctioneer agent through a TCP/IP connection. The end users communicated with the system by means of the thermostat.

The local agents aimed at producing CHP electricity in high-priced periods with a hard constraint of not infringing the user's thermal comfort. When the transformer load exceeded the safety level, the objective agent issued a demand bid aiming at steering the load back to the safety level. This increase in demand caused a price rise on the electronic market, which, in turn, triggered those agents most fit to respond (i.e., the ones having the highest heat demand at that moment) to run their CHP. The micro-CHP units were only operated in case of local heat demand, either for space heating or for tap water heating. No heat was dumped. An additional simulation study was done to verify the findings in the field test and to investigate circumstances not engaged in the field experiment, such as winter conditions.

8.6.2.3 Congestion management results

The field test was conducted in May 2007, which was an exceptionally warm month for The Netherlands. Therefore there was no space heating demand in the households, only demand for tap water heating. Figure 8.10 shows a typical day pattern during the field test when five micro-CHPs were participating in the VPP. The PowerMatcher shifts the micro-CHP production so that electricity is produced when there is a high demand for electricity. This lowers the peak load on the substation.

The main findings of the field experiment and additional simulation studies were:

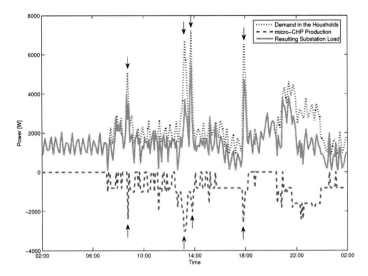

Figure 8.10: Typical measured day patterns for 5 micro-CHPs with PowerMatcher coordination: synchronisation of CHP output (dashed line) with domestic peak-demand (dotted) leading to peak load reduction at the transformer (solid line).

- The Fit-and-Forget policy did not provide benefits to the DSO in comparison to the baseline case. The load-duration curve was lowered on average by adding the micro-CHPs. However, the peak load remained virtually unchanged.
- Adding VPP operation, based on PowerMatcher intelligent control, led to a load-peak reduction of 30% in summer (field test result) and 50% in winter (simulation outcome).

8.7 Conclusions and future research

In the previous sections, we argued about the necessity of introducing distributed control in the electricity infrastructure in order to cope with the interrelated trends of increasing sustainable electricity sources and distributed generation. We have shown how a specific implementation of distributed control can be used for commercial portfolio balancing as well as for DSO congestion management. An important remaining question is: how to combine the two?

Such a dual-objective coordination mechanism needs to be designed for a future electricity system characterized by:

- Distributed Generation and Demand Response are a substantial factor in the electricity markets.
- A substantial portion of central generation is off-shore wind.
- Market parties and network operators optimize their stakes using the DER in their portfolio, or in their network area, respectively. Dependent on the situa-

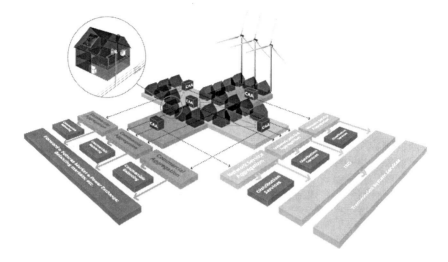

Figure 8.11: Orthogonal dual market-based architecture for commercial and technical VPPs in future electricity systems.

tion, these stakes may be conflicting at one time and non-conflicting at another time.

- Incentives to market parties (generators, suppliers, and end users alike) reflect the true costs of both generation and infrastructure. On the one hand, this will increase efficient usage of the infrastructure (network load factor optimization) and on the other hand it gives the right market signals for investment decisions (Generation against Demand Response against Infrastructural investments).

Figure 8.11 shows an architecture that supports the market situation described above. It is a setting with multiple Market Parties (Balancing Responsible Parties, BRPs), each running a commercial virtual power plant (CVPP), and multiple Distribution System Operators (DSOs), each running a technical virtual Power Plant (TVPP). In the figure the CVPPs are represented by the blocks labeled "Commercial Aggregation" and the TVPPs by those labeled "Network Service Aggregation".

A BRP has special interests:

- Desire to aggregate a high number of DER units, as this smoothens-out the stochastic behavior of the individual DER.
- Aspiration to spread its DER portfolio over a big (national) area to increase spatial smoothing of weather influences on DG and on responsive loads.
- Has no locational aspects attached to the desired portfolio behavior for most of its operational parameters.
- Avoids balancing costs when their portfolio as a whole is in balance.

As a result the commercial portfolio of a BRP is most likely located in the grid area of more than one DSO.

A DSO has special interests as well:

- Preference to address only the DER units in its grid area, sometimes even dependent on individual grid cells or segments.
- Desire to incentivise DER to deliver system management services.
- Has a locational aspect in the desired behavior or the DER in its network.
- Avoids investments in infrastructural components by active management of the DER in their network.

In the orthogonal dual architecture, one CVPP has to deal with several TVPPs and one TVPP with several CVPPs. The individual DER units at the premises of one customer, in the figure represented by a house, communicate with CVPP components only. The Commercial Aggregating Agent (CAA) aggregates all DERs in the portfolio of the corresponding BRP located in a common grid area. Each CAA provides commercial services directly to its CVPP, but it also provides local grid services to the DSO. Thus, each CAA responds to incentives of both the CVPP it is part of and the TVPP that covers its grid area. The stakes of BRP and DSO come together at this point. When these stakes are non-counteracting, the CAA can deliver the services requested by the DSO for a lower price compared to the situation in which the stakes do counteract. Accordingly, those CAAs without an internal conflict will respond to both the CVPP and TVPP request first. In this way, flexibility services from DER will be used based on merit order and the stakes of the different parties will be balanced automatically against each other.

Acknowledgements The authors would like to thank everyone who contributed in some way to the development of the PowerMatcher technology and/or the field test results presented in this paper. Further, we would like to thank Pamela Macdougall for proof-reading this text. The writing of this chapter text was partly funded by the EU as part of the SmartHouse/SmartGrid project (Grant no.: FP7-ICT-2007-224628).

References

1. H. Akkermans, J. Schreinemakers, and K. Kok. Microeconomic distributed control: Theory and application of multi-agent electronic markets. In *Proceedings of the 2nd International Conference on Critical Infrastructures*, Grenoble, France, 2004.
2. J. Berst, P. Bane, M. Burkhalter, and A. Zheng. The electricity economy. White paper, Global Environment Fund, August 2008.
3. R. K. Dash, D. C. Parkes, and N. R. Jennings. Computational mechanism design: A call to arms. *IEEE Intelligent Systems*, 18(6):40–47, November/December 2003.
4. L. J. de Vries. *Securing the public interest in electricity generation markets, The myths of the invisible hand and the copper plate*. PhD thesis, Delft University of Technology, Delft, The Netherlands, 2004.
5. Energy Information Administration. *International Energy Outlook 2007*. EIA, Paris, France, 2007.
6. Energy Information Administration. http://www.eia.doe.gov/emeu/international/RecentElectricityGenerationByType.xls, December 2008.
7. ENIRDGnet. Concepts and opportunities of distributed generation: The driving European forces and trends. Project Deliverable D3, ENIRDGnet, 2003.

8. International Energy Agency. *Distributed Generation in Liberalised Electricity Markets.* International Energy Agency, Paris, France, 2002.

9. International Energy Agency. *The Power to Choose – Demand Response in Liberalized Electricity Markets.* International Energy Agency, Paris, France, 2003.

10. K. Kok, Z. Derzsi, J. Gordijn, M. Hommelberg, C. Warmer, R. Kamphuis, and H. Akkermans. Agent-based electricity balancing with distributed energy resources, A multiperspective case study. In R. H. Sprague, editor, *Proceedings of the 41st Annual Hawaii International Conference on System Sciences*, page 173, Los Alamitos, California, 2008.

11. K. Kok, C. Warmer, and R. Kamphuis. PowerMatcher: Multiagent control in the electricity infrastructure. In *Proceedings of the 4th International Joint Conference on Autonomous Agents and Multiagent Systems*, pages 75–82, Utrecht, The Netherlands, 2005.

12. K. Kok, C. Warmer, and R. Kamphuis. The PowerMatcher: Multiagent control of electricity demand and supply. *IEEE Intelligent Systems*, 21(2):89–90, March/April 2006.

13. A. Mas-Colell, M. Whinston, and J. R. Green. *Microeconomic Theory.* Oxford University Press, 1995.

14. European Parliament and Council. Common rules for the internal market in electricity. EU Directive 2003/54/EC, June 2003.

15. D. Pudjianto, C. Ramsay, and G. Strbac. Virtual power plant and system integration of distributed energy resources. *Renewable Power Generation*, 1(1):10–16, 2007.

16. A. Pullen, L. Qiao, and S. Sawyer. Global wind 2008 report. Market report, Global Wind Energy Council, March 2009.

17. B. Roossien. Field-test upscaling of multi-agent coordination in the electricity grid. In *Proceedings of the 20th International Conference on Electricity Distribution CIRED.* IET-CIRED, 2009.

18. B. Roossien, M. Hommelberg, C. Warmer, K. Kok, and J. W. Turkstra. Virtual power plant field experiment using 10 micro-CHP units at consumer premises. In *SmartGrids for Distribution, CIRED Seminar*, number 86, 2008.

19. T. W. Sandholm. Distributed rational decision making. In G. Weiss, editor, *Multiagent Systems: A Modern Approach to Distributed Artificial Intelligence*, pages 201–258. The MIT Press, Cambridge, Massachusetts, 1999.

20. M. ten Donkelaar and M. J. J. Scheepers. A socio-economic analysis of technical solutions and practices for the integration of distributed generation. Technical Report ECN-C–04-011, ECN, 2004.

21. M. J. N. van Werven and M. J .J. Scheepers. The changing role of energy suppliers and distribution system operators in the deployment of distributed generation in liberalised electricity markets. Technical Report ECN-C–05-048, ECN, June 2005.

22. C. Warmer, M. Hommelberg, B. Roossien, K. Kok, and J. W. Turkstra. A field test using agents for coordination of residential micro-chp. In *Proceedings of the 14th International Conference on Intelligent System Applications to Power Systems*, 2007.

23. M. P. Wellman. A market-oriented programming environment and its application to distributed multicommodity flow problems. *Journal of Artificial Intelligence Research*, 1:1–23, 1993.

24. C. Wolfsegger, M. Latour, and M. Annett. Global market outlook for photovoltaics until 2012 – Facing a sunny future. Market report, European Photovoltaic Industry Association, February 2008.

25. F. Ygge. *Market-Oriented Programming and its Application to Power Load Management.* PhD thesis, Department of Computer Science, Lund University, Lund, Sweden, 1998.

26. F. Ygge and H. Akkermans. Resource-oriented multicommodity market algorithms. *Autonomous Agents and Multi-Agent Systems*, 3(1):53–71, 2000.

Chapter 9
Social and Cyber Factors Interacting over the Infrastructures: A MAS Framework for Security Analysis

E. Bompard, R. Napoli, and F. Xue

Abstract Critical infrastructures are usually characterized by a network structure in which many technical devices interact on a physical layer, being monitored and controlled throughout a cyber network in charge of conveying commands/measurements from/to the decision making centers run by the human decision makers or automatic controllers at the top of this complex system.

This chapter presents a general framework for the analysis of the security of critical infrastructures in terms of three different interacting layers: the *physical* layer, the *cyber* layer, and the *decision-making* layer. In this framework a multi-agent system is introduced to model the interaction of the various players, and the analysis of the security control against natural failures or malicious attacks is conceptually discussed with reference to the interconnected power systems. With each layer is associated a set of metrics able to characterize the layer with respect to its operation and security and that can account for the interactions among the various layers. In this chapter we will show a possible application for the quantitative assessment of the impact of information in system security, by comparing different information scenarios and then identifying and ranking the most critical information.

The aim of the framework is to provide a comprehensive viewpoint of the system robustness or security which takes into account not only physical operation but also the cyber and social (organizational and human) factors to support better security analysis for critical infrastructures.

E. Bompard (corresponding author), R. Napoli, F. Xue
Politecnico di Torino, Dipartimento di Ingegneria Elettrica, Torino, Italy,
e-mail: ettore.bompard@polito.it

R.R. Negenborn et al. (eds.), *Intelligent Infrastructures*, Intelligent Systems, Control and
Automation: Science and Engineering 42, DOI 10.1007/978-90-481-3598-1_9,
© Springer Science+Business Media B.V. 2010

9.1 Introduction

Critical infrastructures are of the utmost importance for modern societies; accidental failures and intentional attacks that may threaten public infrastructures can cause disastrous social and economic consequences. Among public facilities, the infrastructural systems for electric power supply have particular importance since they are widely distributed and key to most social activities. Power system outages may have severe impacts on a country in many respects [4].

Researchers working on security of critical infrastructures or risk analysis of complex technical systems have experienced a long-term development and adaptation in their perspective. Initially, the analysis and corresponding safety design methods were mostly hardware-driven, i.e., completely focused on the technical issues of physical systems. Gradually, it was recognized that human errors were often involved in accidents or failures; therefore, Human Reliability Analysis (HRA) become an important issue in security analysis. Furthermore, investigations of major accidents often cite management and organizational factors as major causes of human errors in operating technical systems. This led to the integration of Human and Organizational Factors (HOF) into risk analysis. HOF combined with the complex technical systems came to be called socio-technical systems (STS) [19].

The term STS recognizes the interaction between people and technology, e.g., interactions between infrastructures and human operations. The approaches proposed by HRA belong to two different generations [19]. In the first generation, human errors were identified as among the essential causes of many accidents and failures and were proposed to be considered in security design and regulations; here error was described as a deviation from the normative and rational performance. It was also recognized that the human performance in operation during accidents and emergency situations may be related to the long term process of management and organizational behaviors that took place long before. Human behavior and psychology have been shown to be strongly related to daily work contexts and organizational environment [31]. Therefore, HRA must be put in an organizational context as HOF to better perform in problem detection. Furthermore, "error" may not be easily identified and may not be the only factor in HOF that can cause accidents; sometimes, in complex infrastructural systems operated by various operators with conflicting interests, the superposition of many "normative correct" or "rational" decisions, from the point of view of each human operator, will lead to major failures if information, communication or coordination is lacking. For example, with the deregulation of the electrical power industry, operators may have more freedom in their decisions and be constrained only by the coordination rules issued by supervisory organizations, such as UCTE (Union for the Co-ordination of Transmission of Electricity) in Europe; in this situation, inappropriate, although correct and rational behaviors of human operators in case of emergency may cause a blackout of the entire network of interconnected systems due to insufficient or untimely coordination and communication. This has prompted the second generation of approaches, which change the perspective from normative performance to actual human behavior, from prescriptive models to descriptive models and from static analysis to dynamic response processes. In the field of socio-technical systems, much outstanding work has been done to address HOF in risk analysis [7, 9, 13, 20, 25, 30].

In STS the "social" aspects have traditionally been considered in terms of HOF relative to the technical factors. In traditional engineering approaches, infrastructures are considered from a hardware perspective (in power systems, for example, as generators, transmission lines and transformers). However, the study of the intersection of social and technical issues needs to be considered; modern infrastructure systems are mostly under both human and automatic control which are not completely independent. Automatic control is the self-running procedure that needs to be modeled with specific algorithms from human knowledge and experience representing human intent [28]; it may involve both human and technical factors. Furthermore, in research about security problems, it was recognized that information and cyber systems, as the carrier of information, played a crucial role [3] and that the performance of the overall system relies on a proper coordination of the automatic actions of the distributed local controllers with the centralized automatic and human decision making [22]. The impacts of those aspects, especially with reference to cyber and communication problems, on human and organizational behaviors have not been specifically addressed in a socio-technical framework. Therefore, besides distinguishing only between social and technical factors, when analyzing infrastructures the additional cyber dimension should be considered independently. HOF can be included, from a more general perspective, in the decision-making dimension. This will be introduced in detail in the next section.

In this chapter, we propose a general framework for the security analysis of critical infrastructures to capture common features and issues not related to any specific technology. Building on the socio-technical conceptual frame, this framework focuses on three different layers, i.e., *physical*, *cyber*, and *decision-making*. Furthermore, a multi-agent system (MAS) model is introduced to simulate the actual behaviors of the system operators (SO) in interconnected power systems in case of failures or attacks. The comprehensive platform can provide indices useful in quantitatively assessing the role of information, in terms of the impacts of its unavailability, on the system performance.

The chapter is organized as follows: Section 9.2 discusses the general framework for power system security analysis with the three different layers. Section 9.3 proposes a MAS model for the on-line coordination of distributed system operators in case of system failures or attacks. In Section 9.4, further methods for analyzing information impact are discussed. Conclusions are drawn in Section 9.5.

9.2 A general framework for security analysis

9.2.1 The general framework

As we mentioned, the security of critical infrastructures depends on various aspects related to the physical structure, to the communication of information and to the au-

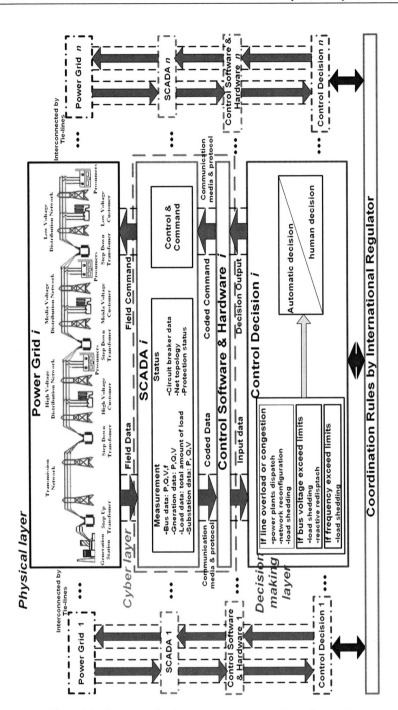

Figure 9.1: Framework for the power systems operation and control.

tomatic and human decision making. To capture the complex nature of the problem we considered a comprehensive framework composed of three layers:

- **Physical layer**: System structure & operative condition;

- **Cyber layer**: Delivery of information & commands;

- **Decision-making layer**: Human & organizational factors.

The application of this framework for power system operation and control is described in Figure 9.1.

9.2.1.1 Physical layer

The security issues related to the physical parts of the infrastructure have a longer history of research than those related to the cyber security or socio-technical aspects. For power systems, the physical layer refers to the network in terms of buses and lines, reserve margin and availability of ancillary services for security management. Moreover, given the network structure, the operational condition of the systems, both in terms of (un)availability of components (lines, power plants, etc.) due to maintenance or outages and the level of load and its localization on the system, is also a related physical issue for system security.

Power systems are subject to very strict physical and operational constraints, such as *just-in-time* production (an instantaneous balance between power generated and load plus losses needs to be assured), *variability of the demand* (power demand is continuously changing between peak and off-peak values), *no storage* (electricity cannot be stored in significant amounts), and *dynamic constraints* (limits connected to the system stability).

The key components of the physical layer of power systems can be categorized, according to their functions and structures, as *transmission network* (220 kV or over transmission lines), *high voltage* (HV) distribution network (132kV high voltage distribution lines), *substations* (nodes of connection and power interchange in which various lines and power plants converge), and *power plants* (units with synchronous generators injecting power into the grid).

Power systems are complex, dynamic networks whose security can be threatened by various types of failure such as *cascading failure of lines due to overload*, *loss of synchronism due to phase instability*, *large oscillations of electrical quantities*, *increase or decrease of the frequency beyond allowed limits*, and *voltage collapse*.

The proper operation of power systems at the physical layer relies on a bunch of different hardware devices such as generators, transmission lines, and transformers. The settings, the local and remote control and the automatic and human-driven operation are key factors in assuring system security which closely interface with the other layers.

The state of the physical layer, as a large-scale, non-linear system, is identified by some crucial parameters such as *real* and *reactive power*, *voltage magnitude* and *angle*, and *frequency*.

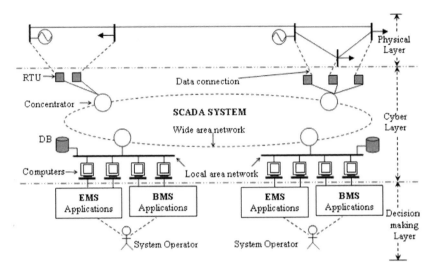

Figure 9.2: Conventional control systems.

9.2.1.2 Cyber layer

Most critical infrastructures, and power systems as well, are called "information intensive" critical infrastructures [3] in the sense that the recognition of either normal states or anomalies and failures strongly depends on the availability of a mass of information. Therefore, the supervision and control of power systems have evolved along with the evolution of the information and communication technologies (ICT), which are providing increasingly better technologies and tools to improve system operators' ability to monitor and control the physical systems. Figure 9.2 shows a typical structure of a conventional power system control architecture; every control center supervises a part of the electrical transmission grid for its proper functioning or failures. Remote Terminal Units (RTU) are devoted to collecting various data; a single RTU controls and acquires data from a substation, and more RTUs are connected to a shared remote concentrator. These real-time data are transmitted through dedicated transmission channels to control centers. The SCADA (Supervisory Control And Data Acquisition) system can thus be viewed as a cyber bridge between the physical system and the control centers.

Control centers equipped with Advanced Applications Software for state estimation, network analysis and generation control are called Energy Management Systems (EMS) [34]. The part of the control center functions that is responsible for business applications is called the Business Management System (BMS) [34].

In our framework, the cyber layer includes the information sets and the control commands to be delivered and executed, as well as their carrier and executor, i.e., the cyber infrastructures (computers, networks and databases), ancillary operation and communication software and protocols. The computers or workstations in the control centers are considered to be a part of the cyber layer; however, the applications for power system control in EMS or BMS installed in these computers also belong to the decision-making layer, as indicated in Figure 9.2.

9.2.1.3 Decision-making layer

The decision-making layer represents the automatic or human factors and the regulatory constraints that are involved in making decisions, according to the information available and the definition of control actions to be transferred and executed to the physical fields through the cyber layer.

The essential function of the decision-making layer is to assess information and produce control decisions. The appropriate performance of this function depends on the ability nested in the "intelligence" based on algorithms, knowledge and human experience. The decision making can be classified as *automatic* and *human*. The *automatic decision making*, in the control centers, is based on software applications which are designed to undertake automatic control actions. From the viewpoint of the system's user, a control center fulfills certain functions in the operation of a power system [34]. The first group of functions is for power system operation and largely inherits from the traditional EMS. Functions can further be grouped into data acquisition, generation control, and network (security) analysis and control. The second group of functions, the BMS, is for business applications, including market clearing price determination, congestion management, financial management and information management. The *human decision making* refers to the "intelligence" of human beings who may be involved directly in providing control decisions. Generally, persons involved in decision making would have different roles and rights depending on the organizational structures of the power system as a whole. Human decision making also participates in both energy management and business management according to the decision makers' responsibilities. Irrespective of their various roles, these operators should have certain common features to be taken into consideration as a part of the control process in studying the security issues. First, considering differences in personalities, a basic assumption about their performance should be their *rationality*. Second, to construct reasonable utility functions, the basic feature of *self-interest* in human decisions, e.g., maximizing benefits for one's own system or organization, should be taken into account. Third, the way in which *information* is used in the process of analysis and decisions by human intelligence should be considered. Lastly, possible error or *inappropriate human performance* due to emotion, fatigue or carelessness cannot be neglected.

The *organizational environment* is of the utmost importance in decision making; it is the set of rules and mechanisms that define how to organize and integrate decision-making entities and factors throughout the decision-making process, and that determines the division of roles and rights involved in the control process.

The decision-making process for power systems has undergone fundamental changes in the past few decades; the restructuring of the power industry in the second half of the 1990s transformed the power system's structure and operation, moving it from centralized to coordinated, decentralized decision-making. Within the new decentralized environment, the decision making of a multitude of self-interested individuals competing within a national and international market, and the coordination of various SOs became crucial issues. With competition, the self-interested behavior of participants is encouraged since it may provide higher overall social welfare; however, it may also cause more difficulty and uncertainty in coordinating their response to accidental failure and attacks.

9.2.2 Security analysis of power systems: Natural and malicious menaces

Power systems operation can be classified into several different states [10]: *normal, alert, emergency, extreme, restorative*. To keep a power system secure means allowing it to function in the normal state and maintaining the ability to restore it to this state after an external disturbance has changed its status. The external disturbances can be categorized as *natural, accidental,* or *malicious*. Natural threats are represented by disturbances related to natural events (lighting, floods, thunderstorms, etc.), accidental threats are related to the failure of devices (component breakdown, improper operation of relays, etc.) and malicious threats are connected to deliberate attacks with the aim of causing harm (e.g., terrorist attacks). Between the first two and the third there is an enormous and fundamental difference. In malicious threats, strategic interaction among players determines the probability and the real occurrence of an attack in time and space, while naturally based threats to a power system occur randomly (nature has no specific intent to cause harm; nature is a "random" player) [5]. Specific, recent malicious threats that deserved much greater attention should be discussed separately.

The most obvious difference in scenarios that can be identified here stems from the difference between threats and attacks. First it is necessary to make a distinction between a *threat* and an *attack*. A *threat* is the potential implementation of an attack, due to the intention of malicious entities to cause harm, or the possibility of harm from natural disasters or accidental errors. The *attack* is the implementation of a credible threat by malicious entities. Preventing a threat is basically *"off-line"* (detection of structural vulnerability, operational deficiencies or possible targets; enforcing protection, adding components); dealing with an ongoing attack is *"on-line"* and relies on the (coordinated) control actions undertaken by the various system operators. Therefore, we can generally classify the security analysis as off-line or on-line problems.

9.2.2.1 Off-line analysis

An intentional threat is the intention of hostile entities (e.g., terrorists, criminal organizations) to cause harm to society. Governmental organizations (e.g., police, the military, civil authorities) can provide countermeasures to those plans. The targets that are most likely to result in huge damages and losses would be more likely to be attacked; likewise, the probability of a target's being attacked would be reduced if it were protected.

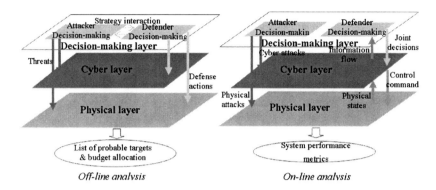

Figure 9.3: Framework of the security analysis.

The response to a threat, apart from general counter-terrorism measures, is largely off-line. It consists essentially of a structural vulnerability analysis and intervention (system reinforcement) on the network for which we also need an analysis of strategic interactions, allowing for a budget allocation to the most sensitive targets.

The framework of the off-line analysis is depicted in Figure 9.3. The threats from attackers may focus on both the cyber and physical systems to produce catastrophic consequences. The decision-making layer is the most important factor, which can be divided into decision making of attackers and defenders. The strategic interaction between them as shown by the two opposite arrows has drawn a great deal of research attention [4, 5, 15].

9.2.2.2 On-line analysis

The most credible threats, based on the previous off-line analysis, can be considered as possible attacks that can occur under specific operative conditions. In this case the response needs to be on-line and relies on the (coordinated) control actions undertaken by the various system operators whose interactions on the physical system in response to a malicious attack, given the information known, determine the performance of each system and of the interconnected power system as a whole. This performance can be expressed by a set of meaningful metrics (risk of loss of load, generation cost, etc.) [2, 21, 26]. These measures enable us to quantify the system vulnerability for a given physical structure and operational configuration, in the case that a credible set of attacks (determined on the basis of threat analysis), both physical and cyber, were undertaken. The emphases of this process are in how much information can be available to the decision makers, how they make final decisions based on internal strategic interactions, and how the coordination rules work in this joint decision process. The framework of on-line analysis is depicted in Figure 9.3.

For an ongoing attack, the three layers would interact more tightly. The existing work about the remedial actions of defenders in response to attacks is mainly from a purely physical perspective to theoretically determine the best responses.

However, there are still two main issues that cannot be completely captured from this pure physical perspective. First, there is still the necessity for systematical and mathematical methodologies for taking into account the interaction between the timely communication of information and the physical behaviors of power systems. Second, in order to identify viable corrective actions in practice under attack to minimize the impacts on system operators, the decision-making chain and the coordination among autonomous system operators should be taken into account.

9.3 Multi-agent model for the analysis of coordination and control

As discussed, the second generation approaches for socio-technical problems related to HOF changed the perspective from prescriptive to descriptive. The objective of the descriptive perspective is not to assure, with a normative approach, proper system performance, but to predict all possible actual behaviors of humans under emergency conditions. In this context, the organizational environment of an infrastructure system can be modeled by taking into account its different actors, roles and mutual influences. For example, a Bayesian Belief Network model has been applied to maritime transportation for this purpose [30]. Multi-agent system (MAS) models are another promising technique.

A MAS is a system composed of a population of agents simultaneously pursuing individual objectives. Hereby, an agent is a rational entity who interacts with an environment, choosing actions and getting rewards; from this process the agent learns to make choices based on past experience. Agent-based modeling is based on a population of idealized individuals; a set of parameters is assigned to each of them, to describe their behavior in a computer-simulated environment which includes the initial organizational patterns representing the interrelationships among all the agents.

A system composed of multiple agents who have the inherent ability of adaptation by evolving and learning from experience through mutual interactions can also be classified as a Complex Adaptive System (CAS) [33]. Adaptation can be defined as the capacity for modification of goal-oriented individuals, or collective behavior in response to changes in the environment. This capacity is mainly achieved through active learning by the agent. Agent learning is not a new topic and has been addressed by many different methods, such as various reinforcement learning techniques. However, the learning of a single agent is quite different from the learning of multiple agents in parallel. Multi-agent learning (MAL) has become an interesting but also difficult problem. Generally, we can divide MAL into five distinct areas [29]: (1) *Computational*; (2) *descriptive*; (3) *normative*; (4) *prescriptive-cooperative*; (5) *prescriptive-non-cooperative*.

From an engineering point of view, one of the main benefits of multi-agent learning is its potential applicability as a design methodology for distributed control, a branch of control theory that deals with the design and analysis of multiple controllers that operate together to satisfy certain design requirements [18], such as different MAS control systems in power systems [24]. Obviously, this area can be considered an aspect of artificial CAS, which is designed to take advantage of mutual complex interactions of distributed simple components performing and adapting in normative patterns to achieve expected objectives.

In engineering, especially for socio-technical systems, the descriptive perspective is also very important. If the various actors in STS can be described as agents with different roles, goals and functions, and the technical systems and organizational structures can be considered the environment, a MAS model can be reconfigured to explore the actual and dynamic behaviors of HOF and their impacts on the technical systems, pointed out by the second generation socio-technical problem. Moreover, such decisions are related to certain physical systems that have their own operational and physical constraints. The possibility of modeling the distributed decision-making process and its final outcome under a different set of rules and contingencies becomes of the utmost importance.

The continent-wide size of power systems in which different Transmission System Operators (TSOs) run a portion of an interconnected continental grid, possessing full authority over it and complying with a set of rules issued by UCTE, is an example of a distributed decision-making process. The outcomes of this process, especially when contingencies – natural or malicious – occur, are a key issue for the smooth functioning of the system, crucial to modern societies. The possibility of testing ex-ante different sets of coordination rules for the TSOs and various strategies in reaction to some credible contingency would be a major advantage.

According to the general framework we have proposed, the design of such a MAS model for the analysis of on-line security control of interconnected power systems needs to address particular issues such as: *How to define the utility functions of decision makers, how to simulate the decision-making process, how to consider the influence of information in decision making, how to assess the impacts of information quantitatively,* and *how to devise solutions for strategic interactions among independent decision makers.*

In the rest of this section, we will discuss some possible methods to build a MAS model that can address the issues previously mentioned and provide a viable way to model the complex scenario of coordination of power transmission systems.

9.3.1 A MAS model for on-line security analysis of interconnected power systems control

9.3.1.1 Physical layer model

The model adopted for representing the physical layers of power systems can be chosen with different levels of detail and accuracy; the choice of the model influences the information set transferred through the cyber layers to the decision-making layers in terms of the number, type and refreshing of parameters representing the information to be communicated to the decision-making layer through the cyber layer.

The *steady-state model* of the system, for example, can be represented by the simplified DC (Direct Current) or by the more exact AC (Alternating Current) Power Flow model. The DC power flow model may be the simplest way to simulate the steady-state behaviors of a power transmission system; it is a linear model based on a number of assumptions [1]:

- The **DC power flow model** neglects the influences of reactive power, line losses and voltage; the only state indication considered is the real power flow distributed across various transmission lines. Therefore, if we use a DC power flow model to describe the behavior of the physical layer, the only information to be exploited in decision making would be *real power flow*, and the security state of the physical layer would be described by real power congestions in different transmission lines. Although the DC power flow model makes too many simplifications, it has the advantage of calculation speed and is helpful in providing some insights into power systems behavior. For this reason, it will be adopted in our modeling.

- The **AC power flow model** is more detailed, taking into account reactive power and voltage levels in the system [12]. In this case, the set of parameters that represent the system status and that would be communicated and considered in decision making includes *real and reactive power flows and injections* and *voltage phases and magnitudes*. Correspondingly, the secure state of the physical layer would be described in terms of limitations of real power and reactive power of generators, power congestions of transmission lines, and limitations on voltage magnitudes. The advantages in terms of a more detailed model of the physical layer correspond to a higher calculation burden.

The dynamics of power systems operation and security are a key issue. The loss of dynamic and transient stability [16] is a common reason for blackouts. *Dynamic models* provide two additional parameters: the frequency and rotor angle of the synchronous machine [12]. The dynamic models are difficult to apply due to their large computation burden.

9.3.1.2 Cyber (information) layer model

Cyber security has been identified as an important issue in critical infrastructures, and much excellent work has been done on this topic in the field of information technologies [3, 4, 17]. We will focus on the cyber layer as an information bridge between the physical systems and the decision making, disregarding the specific structures, transmission carriers and protocols. We define an *information scenario* as a set of values of the parameters describing the physical layer, that are used for the decision making. The information is related to the physical layer and can be communicated to various system operators. The information scenario is represented by the matrix:

$$
F = \begin{bmatrix} F_{11} & \cdots & F_{1n} \\ \vdots & \ddots & \vdots \\ F_{n1} & \cdots & F_{nn} \end{bmatrix}, \tag{9.1}
$$

where a non diagonal element F_{ij} represents the aggregated information communicated from SO i to SO j. For example, in a DC model, the information related to the real power flows F_{ij} can be expressed as:

$$
F_{ij} = (p_{x,y}, p_{r,s}, \dots) \qquad \text{for } i = 1, \dots, n, \ j = 1, \dots, n, \ i \ne j, \tag{9.2}
$$

where n is the number of SO; each component $p_{x,y}$ of F_{ij} denotes the real power flow between bus x and bus y; $p_{x,y} = 0$ if there is no transmission line between these two buses; x and y, or r and s, denote any bus in the network. A diagonal element F_{ii} in the matrix denotes the information of SO i related to his own system. F^{ϕ} denotes no information exchanged between all SOs. F_{ij}^{ϕ} denotes that there is no information exchanged between operator i and operator j. F^{f} denotes a full information scenario (each operator has full knowledge of the whole interconnected system). F_{ij}^{f} denotes that all information from SO i to SO j has been communicated. F_{ii}^{f} denotes that all information of system i is available to SO i.

In the case of a partial information scenario in which real power flow in some transmission lines is not available, the SOs will consider those flows to be zero in the decision making.

9.3.1.3 Decision-making layer model

Structure and states

The SO of each power system, interconnected to at least one of the other n subsystems through the tie-line, is an agent that:

- gets information about its local system (e.g. the power flow of all transmission lines);
- gets information about the other systems;
- determines which action to perform on its local system, based on the maximization of an objective that represents its utility.

The interconnected power system is simulated as the physical layer with the relevant parameters (cfr. Section 9.3.1.1), depending on the specific models adopted.

We use a simple, two-state model to classify the power system states as *Secure* (no security constraint is violated) or *Emergency* (at least one constraint is violated) (Figure 9.4); for a given structure and a given secure operating condition, an attack (or failure) is considered successful (or damaging) if the system falls into the emergency state. Similarly, a control action is considered successful if the system is brought back to a secure state. As an example, if the model adopted is DC power flow and the parameter considered is the line flow, *Secure* can be defined as the state

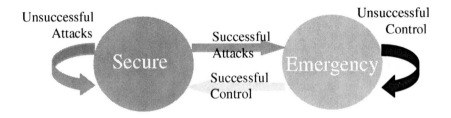

Unsuccessful
Attacks

Successful
Attacks

Successful
Control

Unsuccessful
Control

Secure

Emergency

Figure 9.4: Transition of states [6].

in which no lines are overloaded, and *Emergency* as the state in which at least one line is overloaded.

Control strategies

In real systems, a system operator may have many different methods and types of equipment to control the power system in order to maintain the system in a secure state. In our simulation model, a system operator's strategies depend on the degree of detail simulated in the physical layer and what kinds of security indices are considered. As an example, if the physical layer is simulated by a DC model, the secure state is defined according to real power congestion in transmission lines, while voltage and frequency, although very important, cannot be reflected in such a physical model. Therefore, the control strategies related to voltage and frequency will not be included in the set of control strategies. In fact, after a successful attack or a failure, if congestion arises, each SO must choose an individual *control action*. Let us suppose that the i-th SO has $\mathcal{X}_i = \{\alpha_i^1, \alpha_i^2, \ldots, \alpha_i^k, \ldots\}$ alternatives, each one corresponding to the shedding of some loads in its system. We model each action as $\alpha_i^k = [e_1^k, e_2^k, \ldots, e_j^k, \ldots]$ with $e_j^k = 0$ or 1, where 0 means that no load is shed at bus j and 1 means that all the load at bus j is shed. In the simulation, for simplicity, only none, one, or two of the coefficient e_j^k were allowed to be 1. Hence for action α_i^k the total load shed $L(\alpha_i^k)$ is easily found. Given an operative configuration and a successful attack, let $\mathcal{G}$ be the set of all possible joint actions (a joint action is composed by a set of individual actions, one from each agent) of all the n SOs; each joint action g_k may be represented as $g_k = \{\alpha_1^k, \alpha_2^k, \ldots, \alpha_i^k, \ldots, \alpha_n^k\}$, where α_i^k is the action of the i-th SO. The SOs have estimations about the influence on the line flows of the loads at each bus, based on specific distribution factors of different buses (such as PTDFs [1]) which reflect the sensitivities of the real power flows with respect to the variation of loads at different buses. The SO may, then, select the most sensitive load buses for the action sets and consequently, the joint actions set composed by the combination of all individual actions will be defined. More sophisticated sets of joint actions may be considered if additional parameters for modeling the physical layer are introduced according to the detailed models of the power systems.

Utility functions

To make decisions, each rational agent must have a clear utility function as an objective to maximize. Scientists have studied extensively the role and the definition of utility issues. For example, Patterson and Apostolakis developed a performance index (PI) for decision makers via multi-attribute utility theory (MAUT) [23]. PI is the sum of weighted disutility of different performance measures (PMs). PMs represent the differing concerns of stakeholders and are structured together to indicate overall objectives. Therefore, the selection of PMs and their structures may depend on the specific targeted infrastructural system and its related stakeholders. In the contexts of our framework, we mainly consider two performance aspects, i.e., the security index in a physical system and the costs of control strategies to maintain security states. However, these two aspects may not be given the same importance in different situations. Generally speaking, when the systems are in the emergency states, the SO will first try to restore the system to a secure state (or at least mitigate the extent of emergency) without overly considering the costs of available strategies; when the system is driven back to secure states, the SO would choose the most cost-effective strategies to control the system. As we have mentioned, the security indices and control strategies depend on which model is applied in simulating the physical layer; for example, if the physical layer is simulated by a DC model, the security index is related to the overload of transmission lines, and successful control actions are those that remove the overloads. To quantify globally the post control congestion for action g_k under specific state s for SO i, let us first define the congestion V^l for the generic line l (for simplicity we will drop in the following the action index k and state s):

$$V^l = \begin{cases} 0 & \text{if } P^l \leq P^l_{\max} \\ (P^l_{\max} - P^l)/P^l_{\max} & \text{if } P^l > P^l_{\max}, \end{cases} \tag{9.3}$$

where P^l is the power flow in line l after the control action g_k, and $P^l_{\max}$ is the maximum power limit in the line. By taking into account the set of all internal lines $\mathcal{L}_i$, (both terminal buses in the local system) and the set of tie-lines $\mathcal{T}_i$ (one terminal bus in the local system and the other in another system), the security index as defined by an individual congestion rate V_i^I for SO i is:

$$V_i^I = \sum_{l \in \mathcal{L}_i} V^l + \sum_{l \in \mathcal{T}_i} 0.5 V^l, \tag{9.4}$$

where coefficient 0.5 means that only half of the congestion of a tie-line would be considered by SO i and the other half would be considered by the other SO connected by the tie-line. In the operation of real systems, how an operator considers the security indices of his own system and those of the other systems is debatable. Intuitively, each SO is only responsible for the operation of his own system, therefore the estimated overall congestion rate V_i^E for the whole interconnected system may only consider its own individual congestion rate:

$$V_i^E = V_i^I. \tag{9.5}$$

From another point of view, since all the subsystems are interconnected, the emergency happening in another system could potentially propagate to the local system; therefore, the local system operator would also be willing to consider the security states of the other systems in his decision-making process if they are known. In this case we can estimate the overall congestion rate V_i^E as:

$$V_i^E = k_1 V_i^I + k_2 \left(\sum_{j=1, j \neq i}^{n} \varphi_j^i V_j^I \right). \tag{9.6}$$

Each SO i estimates V_i^E after the g_k action by considering the individual congestion rate of his own system and, with a proper weighting, of the systems of its neighbors, according to the socially rational agents approach [14]. The balance between "individual congestion" (of each SO) and "social congestions" (individual congestions of other SOs) is calculated by introducing the two weighting parameters k_1 and k_2, with $k_1 + k_2 = 1$; φ_j^i is a coupling factor whose value depends on how tight the interaction between each pair of SO i and j is. We use the total power exchange $P_{i,j}$ between the two systems as the parameter to evaluate their corresponding interaction. If the interaction between SO i and another SO j is stronger, then SO i would be likely to consider the congestion rate of SO j as more important, and φ_j^i can be expressed as:

$$\varphi_j^i = P_{i,j} / \sum k = 1, k \neq i^n P_{i,k}. \tag{9.7}$$

There are no universal rules of coordination about how an operator should deal with the relation between the emergency in his own system and that in his neighbors'. We assume the utility U_i of action g_k under state s for SO i can be defined as:

$$U_i = \begin{cases} V_i^E & \text{if } V_i^E < 0 \\ [M_i - L(\alpha_i^k)]/M_i & \text{if } V_i^E = 0, \end{cases} \tag{9.8}$$

where

$$M_i = \max_k L(\alpha_i^k). \tag{9.9}$$

The best action for SO i then is the one that provides the greatest utility. As we can see, the cost of control will be considered in terms of loads not served only if the congestion, in terms of line overload, is completely relieved.

Objective attainment by learning

When only partial information about the physical layer is available, SOs may not possess enough information about neighboring systems to perform an analysis of the entire interconnected system. Moreover, they may not know the current policies of their neighbor SOs, so they might not be able to determine directly the utilities of their joint control actions. Reinforcement learning provides a suitable method for

modeling such a context. We can apply the classic Q-learning method (a reinforcement learning technique that works by learning an action-value function that gives
the expected utility of taking a given action in a given state and following a fixed
policy thereafter) to the learning process of all the agents, which can be considered
as a repeated one-stage game [8]. For each SO i, when he gets the reward U_i by
executing g_k at state s_t, the updating formula for all Q-values of time step t is:

$$Q^i_{t+1}(s, g_k) = \begin{cases} Q^i_t(s, g_k) + \beta[U_i - Q^i_t(s, g_k)] & \text{if } s = s_t \text{ and } g_k = g_t \\ Q^i_t(s, g_k) & \text{otherwise,} \end{cases} \tag{9.10}$$

where $i = 1, 2, \ldots, n$.

The learning rate β decays in the learning process to guarantee the convergence
[32]. The criterion for the convergence of the Q value is that the difference between
two consecutive iterations is less than a specified small, arbitrary value. The initial
utility values are assigned to the different action schemes according to the quantity
of loads to be shed; the learning results would be different when the SO has different
sets of available information, since they will have different estimations about U_i.

Equilibrium search by fictitious play

After obtaining the estimations about the system performance following the possible control actions according to the available information, each SO makes decisions
based on his rational self-interest and on the strategic interaction with other operators. To check whether the learning process has reached a solution we resort to
fictitious play to find a Nash equilibrium, which, if it exists, represents the final
joint decision. A *fictitious play* is a process where each player believes that each
opponent is using a stationary, mixed strategy based on empirical distribution of
past actions until the strategies reach equilibrium [27]. This approach is appropriate
for the problems where the players are not aware of the utilities of other opponents
because of the lack of information and can only make decisions based on their own
experiences. Fictitious play is integrated with the model of socially rational agents:
each agent i keeps a count $C^i(\alpha_j)$ for each individual action α_j of the number of
times agent j has used it in the past. Agent i treats the relative frequencies of each
agent j's actions as indication of the current strategy of j. Agent i assumes that j
plays action α_j with probability:

$$R^i_{\alpha_j} = C^i(\alpha_j) / [\sum_{b_j \in \mathcal{X}_j} C^i(b_j)]. \tag{9.11}$$

For each player i, let his current action choice α_i be selected according to some pure
strategy which maximizes the weighted evaluation of utility:

$$U^W(\alpha_i) = \sum_{\alpha_{-i} \in \mathcal{X}_{-i}} Q(\alpha_{-i} \cup \alpha_i) \prod_{j \neq i} \{R^i_{\alpha_{-i}[j]}\}, \tag{9.12}$$

where α_{-i} denotes all the actions implemented by other agents except agent i. The process of fictitious play is integrated with the reinforcement learning, which keeps updating the Q values of the joint actions. When all the maximum frequencies of individual actions (or estimated probabilities) of each agent are higher than a criterion, the process can be considered as converged to a pure strategy equilibrium formed by the individual actions of each agent with maximum probabilities. Similarly, it is also possible to extend this approach to obtain a mixed-strategy equilibrium.

9.4 Analysis of information impact

9.4.1 Construction of scenarios

To analyze the MAS model, different scenarios for the three different layers are constructed. For the physical layer, in the physical model, specific attacks or failures are considered; for example, we can consider physical attacks that cause the failure of tie-lines or internal transmission lines. Since most transmission systems obey the N-1 criterion in operation (the disruption of one line does not cause an insecure state), the lines being attacked may be multiple. A discussion of possible attacks of other components such as power plants and substations can be found in [26]. Our model applies, as well, to those attacking scenarios in which the failure of a substation can be modeled as the failures of all lines connected with it.

For the cyber layer, we can consider cyber attacks as: *attacks on the availability or on the integrity of the information or attacks on the SCADA system* (to impair the automatic control ability or to alter the control). According to what some existing models have shown [11], cyber attacks may be more difficult to conduct successfully than physical attacks. However, cyber attacks can be accessorial in enlarging the damage of physical attacks. Unrelated to any ICT technical issue, a specific cyber attack in our framework can be converted and described by an information scenario defined in the last section. A multiple attack can combine both physical and cyber attacks. For the decision-making layer, any specific coordination rules can be embedded in the decision-making process in terms of the utility function, learning algorithm or construction of strategies.

9.4.2 Quantitative assessment of information impacts

To assess the impacts of information on the outputs of the coordinated decision making we need a metric for comparing various information sets available to the SOs. We propose the idea of *information distance* that characterizes the quantitative differences among different information scenarios under the same physical conditions and attacking pattern. This gives us insight and the means to compare different information scenarios on a common basis. In the decision-making process, the operators utilize the information available in the current scenario to choose their action and estimate the resulting system performance as measured by their utilities. Different

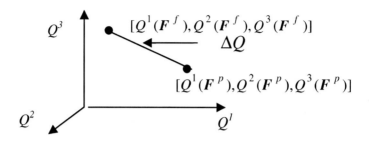

Figure 9.5: Transition of states [6].

information sets correspond to different levels of accuracy in the system knowledge and arrive at different estimations of the expected utility of the same actions. The variation of the expected utility under the same physical system conditions provides a way to compare quantitatively different information scenarios regardless of the human and organizational factors involved in the decision making.

There would be n different utility values of one joint action g for n different agents, from the reinforcement learning in a specific information scenario F. From (9.10), we can extend Q^i as a function of information scenario F: $Q^i(s,g,F)$. We consider the estimated utility values $[Q^1(F^f), Q^2(F^f), \ldots, Q^n(F^f)]$ for joint action g based on full information scenario F^f where information about the whole interconnected system could be common knowledge of all operators as the co-ordinates of a point in an n-dimensional space, and the estimated utility values, $[Q^1(F^p), Q^2(F^p), \ldots, Q^n(F^p)]$ for the same joint action g, based on another partial information scenario F^p as another point in the same space; as an example, Figure 9.5 indicates the situation for three agents in a 3-dimensional space.

The variation of the estimated utilities of the joint action g caused by the reduction of the available information can be considered as the Euclidean distance between these two points. For a general evaluation of an information scenario in terms of accuracy in estimating the system performances, we define the error of utility estimation in scenario F^p, H_p:

$$H_p = \sum_{g \in \mathcal{G}} \sqrt{\sum_{i=1}^{n} [Q^i(p) - Q^i(f)]^2}. \qquad (9.13)$$

Then the information distance between scenarios F^f and F^p could be defined as:

$$D_p = H_p / H_\phi. \qquad (9.14)$$

H_ϕ corresponds to information scenario F^ϕ, where there is no information exchange between SOs and is taken as a reference.

9.4.3 Information sensitivity

To reflect how sensitive the SO decisions on the physical systems are to the available information, we propose an index to assess the vulnerability of the system with reference to the availability of information. The vulnerability of a system needs to refer to a physical configuration and to a defined attacking pattern, assessing the information impacts in such a defined scenario.

The physical configuration is characterized both in terms of structure and operation. The power system structure is identified in terms of buses and lines, reserve margin, and availability of ancillary services for security management, while the power system operation refers both to the availability of components (such as power plants) due to maintenance or outages to the load level and its distribution.

Given a physical configuration and an attacking pattern, the role played by information is assessed in terms of how the availability of information among the SOs impacts their utilities. We assume that an information corruption (failure in a communication channel or inaccurate measurement) provided faked values of real parameters over the system, leading to a variation $\Delta Q(F)$ in the learning result for each utility gotten by each SO with respect to the one gotten when all the information is available and correct.

We consider a corruption mechanism that would make the value of each communicated parameter $\hat{P}$ (line flow as an example) lower than the real value P by d times (from 0 to 1; 1 represents that the communication is completely disabled):

$$\hat{P} = (1-d)P. \tag{9.15}$$

Then, we can define $(Q_g^1, Q_g^2, \ldots, Q_g^n)$ as the learning result of the utility for each SO for the joint action g under full information F^f, and $(\hat{Q}_g^1, \hat{Q}_g^2, \ldots, \hat{Q}_g^n)$ as the learning result for the utility when information is corrupted. We use the Euclidean distance ΔE to quantify the information impact on the evaluation of utilities:

$$\Delta E = \sum_{g \in \mathcal{G}} \sqrt{\sum_{i=1}^{n} (Q_g^i - \hat{Q}_g^i)^2}. \tag{9.16}$$

Then, under the full information scenario F^f with the same fictitious corruption, the *average information sensitivity* can be defined as:

$$E = \frac{\Delta E}{dN}, \tag{9.17}$$

where N is the total number of joint actions. This definition of average information sensitivity is an average and general measurement of information impact on the evaluation of utilities. If we consider two different cases of physical configurations a and b, we have the *relative information sensitivity* as:

$$E_{ab} = \frac{E_a}{E_b} = \frac{\Delta E_a}{\Delta E_b} \frac{N_b}{N_a}. \tag{9.18}$$

The fictitious corruption d has been removed in the relative information sensitivity, which is an inherent relationship between the two cases of physical configurations. It is the comparison of the two cases of physical configuration that gives a quantitative metric showing how much the SOs depend on the exchanged information in coordinating the control action after a failure or attack.

9.4.4 Identification and rank of critical information

We can show an example of how to identify and rank the critical information for power systems with a DC power flow representation of the physical layer in which the information communicated is the real power flows over the transmission lines. The procedure can be articulated in the following steps:

- **Step 1**: *Partition the information set*. We can partition the information set Y of the whole system into two sets, the *critical information* set S and the *non-critical information* set U ($S \cup U = Y$). The partition should fulfill the following policies:

  1. In the specific information scenario, where all information in S is available and all information in U is not available, fictitious play can be successful to find an equilibrium to relieve all the congestions;
  2. In the specific information scenario, where all information in S is not available and all information in U is available, the equilibrium found by fictitious play fails to relieve all the congestions;
  3. If information about any line in S is transferred to the set U, the output equilibrium in the scenario of policy 2 will be successful in relieving all the congestions.

- **Step 2**: *Rank critical information*. After getting the set S, for each element of information in S (the power flow of line i), F^i is the information scenario where only this information is available for neighbors and all other information is not available. The critical information would be ranked according to the output equilibrium of each of the aforementioned scenarios. The ranking should comply with the following policies:

  1. Information about line i is more important than information about line j, if the output equilibrium of F^i succeeds in relieving all congestions and that of F^j fails;
  2. Information about line i is more important than information about line j, if the output equilibria of F^i and F^j both succeed in relieving all congestions and more loads can be kept in service in F^i.
  3. Information about line i is more important than information about line j, if the output equilibria of F^i and F^j both failed in relieving all congestions and the overload rate of F^i is lower than that of F^j.

Following the policies above, the identified critical information can be specially protected according to their different importance in ranking.

9.5 Conclusions and future research

The vulnerability of critical infrastructure generally, and of power systems particularly, is a crucial concern for modern societies. Power systems need to be operated under very strict physical and operational constraints in a competitive market environment, and their smooth functioning must be assured even after accidental, natural or malicious events that can threaten the system's feasibility. The framework of the socio-technical system provides a valuable perspective for considering critical infrastructure systems. However, this perspective may not clearly distinguish possible intersections related to the automatic control of power systems. Therefore, it does not capture the special effects of information and communication as well as the related cyber systems which are behind the physical systems.

The ability to keep the power systems operating after a failure due to an accidental or malicious event is related to a number of different aspects: human, technological, physical and ICT, all of which need to be properly captured. In this respect, the three-layer model seems best able to provide a comprehensive framework.

In this chapter we depicted a possible comprehensive framework and some MAS tools that can be usefully applied to gain greater insight into the vulnerability of power systems as a result of the joint and somehow coordinated action of the various system operators over a bulk, interconnected power grid. Physical security indices, information distance and output equilibrium can be considered as respective metrics in the three layers and jointly constitute a complete description of the situation as a whole.

It should be stressed that the method needs further improvement. With a more detailed model in the physical layer, the calculation burden would be heavier. It is still difficult to completely capture the features and patterns of human decision making in an autonomous agent model and convert any coordination rules into an MAS mechanism. The proposed framework and corresponding MAS model can also be extended to other infrastructure systems with adaptation for specific physical models, communication systems and organizational environments.

References

1. S. Arianos, E. Bompard, A. Carbone, and F. Xue. Power grid vulnerability: A complex network approach. *CHAOS*, 19(1), February 2009.
2. J. M. Arroyo and F. D. Galiana. On the solution of the bilevel programming formulation of the terrorist threat problem. *IEEE Transactions on Power Systems*, 20(2):789–797, May 2005.
3. C. Balducelli, S. Bologna, L. Lavalle, and G. Vicoli. Safeguarding information intensive critical infrastructures against novel types of emerging failures. *Reliability Engineering & System Safety*, 92(9):1218–1229, September 2007.
4. E. Bompard, C. Gao, M. Masera, R. Napoli, A. Russo, A. Stefanini, and F. Xue. Approaches to the security analysis of power systems: Defence strategies against malicious threats. Technical report, Office for Official Publications of the European Communities, Luxembourg, 2007.
5. E. Bompard, C. Gao, R. Napoli, A. Russo, M. Masera, and A. Stefanini. Risk assessment of malicious attacks against power systems. *IEEE Transactions on Systems, Man, and Cybernetics – Part A: Systems and Humans*, 2009. In press.

6. E. Bompard, R. Napoli, and F. Xue. Assessment of information impacts in power system security against malicious attacks in a general framework. *Reliability Engineering & System Safety*, 94(6):1087–1094, June 2009.

7. P. C. Cacciabue. Human error risk management for engineering systems: a methodology for design, safety assessment, accident investigation and training. *Reliability Engineering & System Safety*, 83(2):229–240, February 2004.

8. C. Claus and C. Boutilier. The dynamics of reinforcement learning in cooperative multiagent systems. In *Proceedings of the 15th National Conference on Artificial Intelligence*, Madison, Wisconsin, July 1998.

9. S. Colombo and M. Demichela. The systematic integration of human factors into safety analyses: An integrated engineering approach. *Reliability Engineering & System Safety*, 93(12):1911–1921, December 2008.

10. O. I. Elerd. *Electric Energy Systems Theory*. McGraw-Hill, 1983.

11. B. J. Garrick, J. E. Hallb, M. Kilgerc, J. C. McDonaldd, T. O'Toolee, P. S. Probstf, E. Rindskopf Parkerg, R. Rosenthalh, A. W. Trivelpiecei, L. A. Van Arsdalej, and E. L. Zebroskik. Confronting the risks of terrorism: Making the right decisions. *Reliability Engineering & System Safety*, 86(2):129–176, November 2004.

12. J. D. Glover and M. Sarma. *Power System Analysis and Design*. PWD, Boston, Massachusetts, 2007.

13. D. D. Hee, B. D. Pickrell, R. G. Bea, K. H. Roberts, and R. B. Williamson. Safety management assessment system (smas): A process for identifying and evaluating human and organization factors in marine system operations with field test results. *Reliability Engineering & System Safety*, 65(2):125–140, August 1999.

14. L. Hogg and N. R. Jennings. Socially rational agents. In *Proceedings of the AAAI Fall Symposium on Socially Intelligent Agents*, pages 61–63, Boston, Massachusetts, November 1997.

15. Å. J. Holmgren, E. Jenelius, and J. Westin. Evaluating strategies for defending electric power networks against antagonistic attacks. *IEEE Transactions on Power Systems*, 22(1):76–84, February 2007.

16. P. Kundur. *Power System Stability and Control*. McGraw-Hill Professional, 1994.

17. S. Landau and M. R. Stytz. Overview of cyber security: A crisis of prioritization. *IEEE Security & Privacy*, 3(3):9–11, May/June 2005.

18. S. Mannor and J. S. Shamma. Multi-agent learning for engineers. *Artificial Intelligence*, 171(7):417–422, May 2007.

19. Z. Mohaghegh, R. Kazemi, and A. Mosleh. Incorporating organizational factors into Probabilistic Risk Assessment (PRA) of complex socio-technical systems: A hybrid technique formalization. *Reliability Engineering & System Safety*, 94(5):1000–1018, May 2009.

20. A. Mosleh and Y. H. Chang. Model-based human reliability analysis: Prospects and requirements. *Reliability Engineering & System Safety*, 83(2):241–253, February 2004.

21. A. L. Motto, J. M. Arroyo, and F. D. Galiana. A mixed-integer LP procedure for the analysis of electric grid security under disruptive threat. *IEEE Transactions on Power Systems*, 20(3):1357–1365, August 2005.

22. R. R. Negenborn, P. J. van Overloop, T. Keviczky, and B. De Schutter. Distributed model predictive control for irrigation canals. *Networks and Heterogeneous Media*, 4(2):359–380, June 2009.

23. S. A. Patterson and G. E. Apostolakis. Identification of critical locations across multiple infrastructures for terrorist actions. *Reliability Engineering & System Safety*, 92(9):1183–1203, September 2007.

24. C. Rehtanz. *Autonomous Systems and Intelligent Agents in Power System Control and Operation*. Springer-Verlag, Berlin, Germany, 2003.

25. L. Rognin and J.-P. Blanquart. Human communication, mutual awareness and system dependability. Lessons learnt from air-traffic control field studies. *Reliability Engineering & System Safety*, 71(3):327–336, March 2001.

26. J. Salmeron, K. Wood, and R. Baldick. Analysis of electric grid security under terrorist threat. *IEEE Transactions on Power Systems*, 19(2):905–912, May 2004.
27. J. S. Shamma and G. Arslan. Dynamic fictitious play, dynamic gradient play, and distributed convergence to Nash equilibria. *IEEE Transactions on Automatic Control*, 50(3):312–327, March 2005.
28. T. B. Sheridan. Human and automation: System design and research issues. *Reliability Engineering & System Safety*, 81(1):111–114, July 2003.
29. Y. Shoham, R. Powers, and T. Grenager. If multi-agent learning is the answer, what is the question? *Artificial Intelligence*, 171(7):365–377, May 2007.
30. P. Trucco, E. Cagno, F. Ruggeri, and O. Grande. A Bayesian belief network modeling of organizational factors in risk analysis: A case study in maritime transportation. *Reliability Engineering & System Safety*, 93(6):845–856, June 2008.
31. K. van Dam, S. Oreg, and B. Schyns. Daily work contexts and resistance to organisational change: The role of leader-member exchange, development climate, and change process characteristics. *Applied Psychology: An International Review*, 57(2):313–334, 2008.
32. C. J. C. H. Watkins and P. Dayan. Technical note Q-learning. *Machine Learning*, 8(3–4):279–292, May 1992.
33. A. M. Wildberger. Complex adaptive systems: Concepts and power industry applications. *IEEE Control Systems*, 17(6):77–88, December 1997.
34. F. F. Wu, K. Moslehi, and A. Bose. Power system control centers: Past, present and future. *Proceedings of the IEEE*, 93(11):1890–1908, November 2005.

Chapter 10
Distributed Predictive Control for Energy Hub Coordination in Coupled Electricity and Gas Networks

M. Arnold, R.R. Negenborn, G. Andersson, and B. De Schutter

Abstract In this chapter, the operation and optimization of integrated electricity and natural gas systems is investigated. The couplings between these different infrastructures are modeled by the use of energy hubs. These serve as interface between the energy consumers on the one hand and the energy sources and transmission lines on the other hand. In previous work, we have applied a distributed control scheme to a static three-hub benchmark system, which did not involve any dynamics. In this chapter, we propose a scheme for distributed control of energy hubs that do include dynamics. The considered dynamics are caused by storage devices present in the multi-carrier system. For optimally incorporating these storage devices in the operation of the infrastructure, their capacity constraints and dynamics have to be taken into account explicitly. Therefore, we propose a distributed Model Predictive Control (MPC) scheme for improving the operation of the multi-carrier system by taking into account predicted behavior and operational constraints. Simulations in which the proposed scheme is applied to the three-hub benchmark system illustrate the potential of the approach.

M. Arnold, G. Andersson
ETH Zürich, Power Systems Laboratory, Zürich, Switzerland,
e-mail: {arnold, andersson}@eeh.ee.ethz.ch

R.R. Negenborn
Delft University of Technology, Delft Center for Systems and Control, Delft, The Netherlands,
e-mail: r.r.negenborn@tudelft.nl

B. De Schutter
Delft University of Technology, Delft Center for Systems and Control & Marine and Transport
Technology, Delft, The Netherlands, e-mail: b@deschutter.info

R.R. Negenborn et al. (eds.), *Intelligent Infrastructures*, Intelligent Systems, Control and
Automation: Science and Engineering 42, DOI 10.1007/978-90-481-3598-1_10,
© Springer Science+Business Media B.V. 2010

10.1 Introduction

10.1.1 Multi-carrier systems

Most of today's energy infrastructures evolved during the second part of the last century and it is questionable whether these infrastructures will meet tomorrow's requirements on flexibility and reliability if their operation is not made more intelligent. The on-going liberalization of the energy markets involves extended cross-border electricity trading and exchange activities, which implicate that electricity networks have to operate closer and closer to their capacity limits. In addition, issues such as the continuously growing energy demand, the dependency on limited fossil energy resources, the restructuring of power industries, and the increasing societal desire to utilize more sustainable and environmentally friendly energy sources represent future challenges for both energy system planning and operation.

Nowadays, different types of infrastructures, such as electricity, natural gas, and local district heating infrastructures, are mostly planned and operated independently of each another. However, the integration of distributed generation plants, such as so-called co-generation and tri-generation plants [7, 13] links these different types of infrastructures. E.g., small-scale combined heat and power plants (μCHP) consume natural gas to produce electricity and heat simultaneously. In this way, such systems affect infrastructures for electricity and gas networks, as well as infrastructures for district heating. As the number of such generation units increases, the different infrastructures become more and more coupled.

Several conceptual approaches have been examined for describing systems including various forms of energy. Besides "energy-services supply systems" [11], "basic units" [5], and "micro grids" [18], so-called "hybrid energy hubs" [10] are proposed to address these kind of systems. The latter formulation has been established within the project "Vision of Future Energy Networks", which has been initiated at ETH Zürich. In this project, a general modeling and optimization framework is developed for multi-carrier energy systems, so-called "hybrid energy systems", where the the term "hybrid" indicates the usage of multiple energy carriers. The couplings between the different energy carriers are taken into account by the energy hub concept, with which storage of different forms of energy and conversion between them is described. Principally, energy hubs serve as interface between the consumers and the transmission infrastructures of the different types of energy systems.

Because of the increasing number of distributed generation facilities with mostly intermittent energy infeed (generation profiles), the issue of storing energy becomes more important. Electric energy storage devices are expensive and their operation causes energy losses. A more effective option is the operation of a μCHP device in combination with a heat storage device. By means of the heat storage device, the μCHP device can be operated with a focus on following the electric load while storing the simultaneously produced heat. In general, the trend does not go towards large storages but rather in direction of small local storages, such as local hot water storages within households. Beyond that, it could be expected that within the next

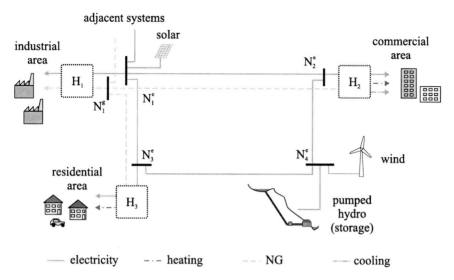

Figure 10.1: Sketch of a system of three interconnected energy hubs.

20 years, a huge amount of small and cheap energy storage units will be available, provided by PHEVs (plug-in hybrid electric vehicles).

Recently, research has addressed the *integrated* control of combined electricity and natural gas systems, e.g., in [2, 3, 21, 26]. While [21, 26] analyze the impact of natural gas infrastructures on the operation of electric power systems, [2, 3] directly address the integrated natural gas and electricity optimal power flow.

Figure 10.1 illustrates an exemplary hub based energy system supplied and interconnected by natural gas and electricity networks. The electricity network comprises four network nodes (N_1^e–N_4^e), whereas the natural gas network only features one network node N_1^g. Three hubs are present in the system, where each hub interfaces the natural gas and electricity distribution networks with the corresponding supply area. This illustration represents the supply of a town that is divided into industrial (hub H_1), commercial (hub H_2), and private/residential load (hub H_3) supply areas. The internal structure of each hub depends on the specific loads present at that hub. For example, hubs may contain electrical transformers, gas turbines, furnaces, heat exchangers, etc., but also storage devices such as heat storages or batteries. In the depicted system, both natural gas and electricity is exchanged with adjacent systems via network nodes N_1^g and N_1^e. Furthermore, a solar power plant is connected to the electric network node N_1^e as a power generation source outside the hub. Besides that, N_4^e connects the system with hydro and wind power plants.

10.1.2 Control of energy hubs

To determine the optimal operation of a multi-carrier energy system, an optimal power flow problem has to be solved. An optimal power flow problem is a general

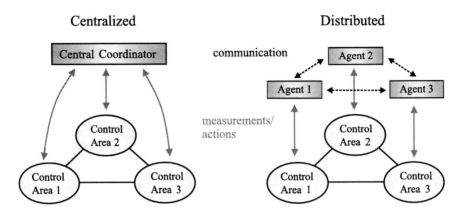

Figure 10.2: Sketch of a centralized and distributed control architecture for a system of three interconnected control areas. The solid arrows refer to measurements/actions between the physical system and the control unit(s). Information exchange between control units is indicated by dashed arrows.

optimization problem, which is formulated as an objective to be minimized, subject to system constraints to be satisfied. In particular, the power flow equations of the different energy carriers are part of these system constraints. By solving this optimal power flow problem, the optimal operational set-points of the system, i.e., of the energy generation units, converters and storage devices, can be determined.

In the considered model storage devices with dynamic behavior are present. Since these storage devices cause a dependency between consecutive time steps, optimization over multiple time steps is required. Therefore, for the optimal operation of the system, actions have to be determined taking the expected future behavior of the system into account. For optimizing the operation over multiple time steps, we propose to use model predictive control (MPC) [6, 19]. MPC is widely used in different application areas, since system dynamics, data forecasts, and operational constraints (system constraints) can be taken into account explicitly. In our case, we use MPC to determine the actions for the individual energy hubs that give the best predicted behavior, e.g., minimal energy costs, based on characteristics of the transmission infrastructures, the dynamics of the storage devices, and the load and price profiles. By using this predictive approach, the energy usage can be adapted to expected fluctuations in the energy prices, as well as to expected changes in the load profiles.

In an ideal situation, a centralized, supervisory controller can measure all variables in the network and determines actions for all actuators. This centralized controller solves at each decision step one optimization problem to determine actions for the entire system. The centralized control architecture is shown on the left-hand side in Figure 10.2, where a central coordinator supervises three interconnected control areas, e.g., hubs. Although for small-scale systems, centralized control may work well, for large-scale systems, a high amount of data needs to be transferred along the whole system and large optimization problems will have to be solved, resulting in high computational requirements. In addition, for large-scale systems, it may simply not be possible to have a single controller controlling all areas, in the case that these areas are owned by different parties. These difficulties could be overcome by implementing a distributed approach as explained in the following.

When solving the optimization problem in a distributed manner, each control area is controlled by its own respective control authority. Applying distributed control, the overall optimization problem is divided into subproblems which are solved in an iterative procedure. In order to guarantee the energy supply of the entire system, the control authorities have to coordinate their actions among one another (Figure 10.2, right-hand side).

The differences between centralized and distributed control in terms of supervision, synchronization, and type of optimization problem are summarized in Table 10.1. Distributed control has several advantages over centralized control. Distributed control is better suited for a distributed power generation infrastructure like the one considered in this chapter, since in distributed control the sometimes conflicting objectives of the individual hubs can explicitly be taken into account. Furthermore, distributed control has the potential to achieve higher robustness, since if the agent of one area fails, only this specific area is not controlled anymore, while other areas are still controlled. Furthermore, shorter computation times arise in distributed control, particularly for larger-scale systems. The control problems of the individual controllers are smaller in size and these local control problems can often be solved in parallel. The challenge is to design efficient coordination and communication among the individual controllers that provides overall system performance comparable to a centralized control authority.

Several approaches have been proposed for distributed control over the last decades, enabling coordination within a multi-area system. In [25] a variety of distributed MPC approaches applied to different application areas is summarized. The main approaches adopted are reviewed and a classification of a number of decentralized, distributed, and hierarchical control architectures for large-scale systems is proposed. Particular attention is paid to design approaches based on MPC. In [1] a decentralized MPC approach for linear, time-invariant discrete systems where the subproblems are solved in a noniterative way is proposed. In [22] a distributed MPC scheme based on decompositions of augmented Lagrangians is proposed for control

Table 10.1: Centralized versus distributed control.

	Centralized	**Distributed**
Supervision	Central coordinator supervises all areas	Each area is supervised by its agent only
Synchronization	Areas send data to a central coordinator	Agents exchange data among each other
Optimization problem	Central coordinator performs overall optimization problem	Overall optimization problem is decomposed into subproblems

of interconnected linear systems. In [28] a distributed MPC algorithm is presented where each local controller tends to move towards a Nash equilibrium by means of game theory considerations. This algorithm is based on discrete linear time invariant systems, too.

Work on distributed control that is not specifically addressed at MPC, that uses static models, but with nonlinear equations, is, e.g., [8, 14, 15, 23]. In [14, 15] coordination is achieved by adjustment of common variables at an existing of fictitious border bus between the areas. In [8, 23] coordination is carried out by specified constraints, referred to as coupling constraints, that contain variables from multiple control areas. For both decomposition procedures, the controllers do not need to know the information of the whole system. Only peripheral data of each control area need to be exchanged between the controllers. The approach of [14, 15] has as drawback that it requires appropriate tuning of weighting factors in order to obtain adequate convergence speed. The approach of [8, 23] has as drawback that the coupling constraints for enforcing the coordination are not arrangeable for every type of system. It depends on the physical constraints of the network nodes at the border of each area, i.e., if they depend on neighboring network nodes or not.

Since the systems that we consider are governed by nonlinear equations, and since is it possible to set up coupling constraints for these systems, we propose here a distributed MPC approach, based on the work for static systems described in [8], that does explicitly take into account dynamics.

10.1.3 Outline

This chapter is outlined as follows. In Section 10.2 the concept of energy hubs is discussed in detail. A model representing producers, transmission infrastructures, energy hubs, and consumers is presented in Section 10.3. Section 10.4 introduces a centralized, MPC formulation for controlling energy hub systems, and in Section 10.5 we propose our distributed MPC approach. Simulation results applying the centralized as well as the distributed scheme to a three-hub system are presented in Section 10.6. Section 10.7 provides conclusions and directions for future research.

10.2 Energy hub concept

Combining infrastructures means coupling them at certain nodes or branches, thereby enabling exchange of power between previously separated systems. As already mentioned, these couplings can be described by means of the energy hub concept. From a system point of view, an energy hub provides the functions of input, output, conversion, and storage of multiple energy carriers. An energy hub can thus be seen as a generalization or extension of a network node in an electrical network. An example of an energy hub is presented in Figure 10.3. Electricity, natural gas, district heat, and wood chips are consumed at the hub input and electricity, heating, and cooling is provided at the output port. For internal conversion

ENERGY HUB

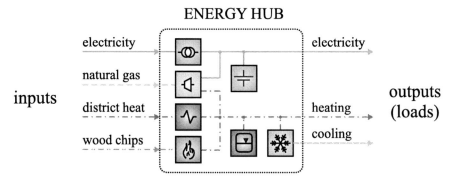

Figure 10.3: Example of an energy hub containing a transformer, μCHP, heat exchanger, furnace, absorption chiller, hot water storage, and battery.

and storage, the hub contains an electric transformer, a μCHP device, a furnace, an absorption chiller, a battery, and a hot water storage.

Energy hubs contain three basic elements: direct connections, converters, and storage devices. Direct connections deliver an input carrier to the hub output without converting it into another form or without significantly changing its quality (e.g., voltage, pressure). Examples of this type of elements are electric cables and overhead lines as well as gas pipelines. Besides that, converter elements are used to transform an input energy carrier into another output carrier. Examples are steam and gas turbines, combustion engines, electric machines, fuel cells, etc. Compressors, pumps, transformers, power electronic inverters, heat exchangers, and other devices may be used for conditioning, i.e., for converting power into desirable qualities and quantities to be consumed by loads. Storage devices are incorporated within the hubs in order to store energy and to use it at a later instant or in order to preserve excessive heat produced by a μCHP device. Examples are batteries for storing electric energy and hot water storages for conserving heat power.

The energy hub concept enables the integration of an arbitrary number of energy carriers and products (such as conversion and storage units) and thus provides high flexibility in system modeling. Co- or trigeneration power plants, industrial plants (paper mills, refineries), big building complexes (airports, hospitals, shopping malls), as well as supply areas like urban districts or whole cities can all be modeled as energy hubs. In [12] the energy hub approach has been applied to a hydrogen network that includes converters (electrolyzer and fuel cell), storage, and demand. This hydrogen network is part of an integrated energy system with electricity, gas and heat production and demand.

Combining and coupling different energy carriers in energy hubs provides a number of potential benefits:

- **Flexibility of supply**: Load flexibility is increased, since redundant paths within the hub offer a certain degree of freedom in satisfying the output demand. This offers the potential for optimization.

- **Increased reliability**: Since the loads do not depend on one single infrastructure, the reliability of energy supply is increased [16].
- **Synergy effects**: Synergy effects among various energy carriers can be exploited by taking advantage of their complementary characteristics. E.g., electricity can be transmitted over long distances with relatively low losses. Chemical energy carriers such as natural gas can be stored using relatively simple and cheap technologies.

10.3 Modeling multi-carrier systems

Multi-carrier energy systems are modeled as an interconnection of several interconnected hubs. Accordingly, two cases are distinguished concerning the modeling. First, the equations for power flow *within* the hubs are presented. These equations incorporate the power conversion and the energy storage of the various energy carriers. Then, the equations concerning energy transmission *between* the hubs are given. Finally, the equations for the hub and the transmission network model are combined resulting in a complete model description.

10.3.1 System setup

In the system under study (Figure 10.4), each energy hub represents a general consumer, e.g., a household, that uses both electricity and gas. Each of the hubs has its own local electrical energy production (G_i, with electric power production $P_{e,i}^G$, for $i \in \{1,2,3\}$). Hub H_1 is connected to a large gas network N_1, with gas infeed $P_{g,1}^G$. In addition, hub H_2 can obtain gas from a smaller gas network N_2 with limited capacity, modeled as gas infeed $P_{g,2}^G$. Each hub consumes electric power $P_{e,i}^H$ and gas $P_{g,i}^H$, and supplies energy to its electric load $L_{e,i}$ and its heat load $L_{h,i}$. The hubs contain converter and storage devices in order to fulfill their energy load requirements. For energy conversion, the hubs contain a μCHP device and a furnace. The μCHP device couples the two energy systems as it simultaneously produces electricity and heat from natural gas. All hubs additionally comprise a hot water storage device. Compressors (C_{ij}, for $(i,j) \in \{(1,2),(1,3)\}$) are present in the gas network within the pipelines originating from hub H_1. The compressors provide a pressure decay and enable the gas flow from the large gas network to the surrounding gas sinks. As indicated in Figure 10.4, the entire network is divided into three control areas (grey circles), where each area (including hub and corresponding network nodes) is controlled by its respective control agent. A more detailed description of the control areas follows in Section 10.5.

Depending on the prices and load profiles, the μCHP device is utilized differently. At high electricity prices, the μCHP device is mainly operated according to the electric load. The heat produced simultaneously is then either used to supply the thermal load or stored in the heat storage device. At low electricity prices, the

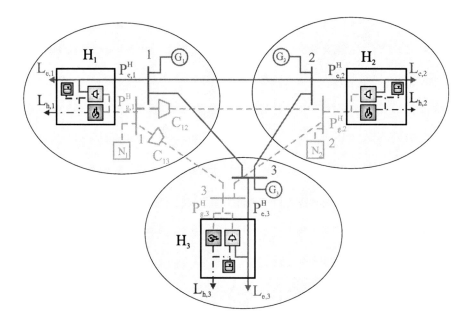

Figure 10.4: System setup of three interconnected energy hubs. Active power is provided by generators G_1, G_2, G_3. Hubs H_1 and H_2 have access to adjacent natural gas networks N_1, N_2.

electric load is preferably supplied directly by the electricity network and the gas is used for supplying the thermal load via the furnace. Hence, there are several ways in which electric and thermal load demands can be fulfilled. This redundancy increases the reliability of supply and at the same time provides the possibility for optimizing the input energies, e.g., using criteria such as cost, availability, emissions, etc. [10, 16].

Since the operation of the system is examined over a longer time duration, the model is based on discrete time steps $k = 0, 1, \ldots$, where a discrete time step k corresponds to the continuous time kT, where T corresponds to one hour.

10.3.2 Energy hub model

Here, the model of an energy hub is formalized, divided into a first part, describing the energy conversion, and into a second part, defining the energy storage models. The presented model is generic and can be applied to any configuration of converter and storage elements. The model is based on the assumption or simplification that within energy hubs, losses occur only in converter and storage elements. Furthermore, unidirectional power flows from the converter input to the converter output are implied. As an example, the hub equations for the energy hub depicted in Figure 10.4 are given.

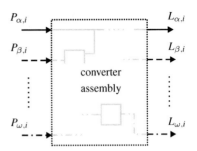

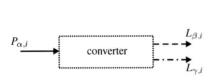

(a) Converter with a single input and two outputs.

(b) Converter arrangement with multiple inputs and multiple outputs.

Figure 10.5: Model of power converters with inputs $P_{\alpha,i}, P_{\beta,i}, \ldots, P_{\omega,i}$ and outputs (loads) $L_{\alpha,i}, L_{\beta,i}, \ldots, L_{\omega,i}$.

10.3.2.1 Energy conversion

Within an energy hub i, power can be converted from one energy carrier α into another energy carrier β. We consider a single-input multiple-output converter device, as it is commonly the case in practical applications. Figure 10.5(a) illustrates a converter with two outputs, such as a micro turbine or gas turbine, producing electricity and heat by means of gas. The input power $P_{\alpha,i}(k)$ and output powers $L_{\beta,i}(k)$, $L_{\gamma,i}(k)$ are at every time step k coupled as

$$L_{\beta,i}(k) = c_{\alpha\beta,i}(k)P_{\alpha,i}(k) \tag{10.1}$$
$$L_{\gamma,i}(k) = c_{\alpha\gamma,i}(k)P_{\alpha,i}(k), \tag{10.2}$$

where $c_{\alpha\beta,i}(k)$ and $c_{\alpha\gamma,i}(k)$ characterize the *coupling factors* between the input and output powers. In this case, the coupling factors correspond to the converter's steady-state energy efficiencies, denoted by $\eta_{\alpha\beta,i}$ and $\eta_{\alpha\gamma,i}$, respectively. More accurate converter models show non-constant efficiencies including the efficiency's dependency of the converted power level. This dependency can be incorporated by expressing the according coupling factor as a function of the converted power, i.e., $c_{\alpha\beta,i} = f_{\alpha\beta,i}(P_{\alpha,i}(k))$. As mentioned above, unidirectional power flows within the converters are assumed, i.e., $P_{\alpha,i}(k) \geq 0$, $P_{\beta,i}(k) \geq 0$, $P_{\gamma,i}(k) \geq 0$. Considering the entire hub (Figure 10.5(b)), various energy carriers and converter elements can be included, leading to the following relation:

$$\underbrace{\begin{bmatrix} L_{\alpha,i}(k) \\ L_{\beta,i}(k) \\ \vdots \\ L_{\omega,i}(k) \end{bmatrix}}_{\mathbf{L}_i(k)} = \underbrace{\begin{bmatrix} c_{\alpha\alpha,i}(k) & c_{\beta\alpha,i}(k) & \cdots & c_{\omega\alpha,i}(k) \\ c_{\alpha\beta,i}(k) & c_{\beta\beta,i}(k) & \cdots & c_{\omega\beta,i}(k) \\ \vdots & \vdots & \ddots & \vdots \\ c_{\alpha\omega,i}(k) & c_{\beta\omega,i}(k) & \cdots & c_{\omega\omega,i}(k) \end{bmatrix}}_{\mathbf{C}_i(k)} \underbrace{\begin{bmatrix} P_{\alpha,i}(k) \\ P_{\beta,i}(k) \\ \vdots \\ P_{\omega,i}(k) \end{bmatrix}}_{\mathbf{P}_i(k)}, \tag{10.3}$$

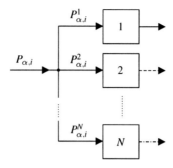

Figure 10.6: Dispatch of total input power $P_{\alpha,i}$ to converters $c = 1, 2, \ldots, N$.

which expresses how the input powers $\mathbf{P}_i(k) = [P_{\alpha,i}(k), P_{\beta,i}(k), \ldots, P_{\omega,i}(k)]^{\mathsf{T}}$ are converted into the output powers $\mathbf{L}_i(k) = [L_{\alpha,i}(k), L_{\beta,i}(k), \ldots, L_{\omega,i}(k)]^{\mathsf{T}}$. Matrix $\mathbf{C}_i(k)$ is referred to as the coupling matrix and is directly derived from the hub's converter structure and the converter's efficiency characteristics. Equation (10.3) illustrates a general formulation of a multi-input multi-output converter device. In reality, not every energy carrier is occurring at the input as well as at the output port. Moreover, the number of inputs and outputs do not have to coincide.

As the input powers $\mathbf{P}_i(k)$ can be distributed among various converter devices, so-called *dispatch factors* specify how much power goes into the corresponding converter device. Figure 10.6 outlines the concept, where the input carrier $P_{\alpha,i}(k)$ is divided over N converter devices as input carriers $P_{\alpha,i}^c$ ($c = 1, \ldots, N$),

$$P_{\alpha,i}^c(k) = \nu_{\alpha,i}^c(k) P_{\alpha,i}(k). \tag{10.4}$$

The conservation of power introduces the constraints

$$0 \leq \nu_{\alpha,i}^c(k) \leq 1 \qquad \forall \alpha, \forall c \tag{10.5}$$

$$\sum_{c=1}^{N} \nu_{\alpha,i}^c(k) = 1 \qquad \forall \alpha. \tag{10.6}$$

Hence, the coupling factors $c_{\alpha\beta,i}(k)$ for converters without explicitly preassigned inputs are defined as the product of dispatch factor and converter efficiency, i.e., $c_{\alpha\beta,i}(k) = \nu_{\alpha,i}^c(k) \eta_{\alpha\beta,i}$.

As long as the converter efficiencies are assumed to be constant, (10.3) represents a linear transformation. Including the power dependency as $c_{\alpha\beta,i}(k) = f_{\alpha\beta,i}(P_{\alpha,i}(k))$ results in a nonlinear relation. In either case, different inputs powers $\mathbf{P}_i(k)$ can can be found that fulfill the load requirements $\mathbf{L}_i(k)$ at the output, since the dispatch factor $\nu(k)$ is variable. This reflects the degrees of freedom in supply which are used for optimization.

Application example

The hub equations for power conversion are now derived for the exemplary hubs in
Figure 10.4. The electrical load $L_{e,i}(k)$ and the heat load $L_{h,i}(k)$ at a time step k are
related to the electricity $P_{e,i}^H(k)$ and gas hub input $P_{g,i}^H(k)$ as follows:

$$\underbrace{\begin{bmatrix} L_{e,i}(k) \\ L_{h,i}(k) \end{bmatrix}}_{\mathbf{L}_i(k)} = \underbrace{\begin{bmatrix} 1 & \nu_{g,i}(k)\eta_{ge,i}^{CHP} \\ 0 & \nu_{g,i}(k)\eta_{gh,i}^{CHP} + (1-\nu_{g,i}(k))\eta_{gh,i}^F \end{bmatrix}}_{\mathbf{C}_i(k)} \underbrace{\begin{bmatrix} P_{e,i}^H(k) \\ P_{g,i}^H(k) \end{bmatrix}}_{\mathbf{P}_i(k)}, \qquad (10.7)$$

for $i = 1, 2, 3$, where $\eta_{ge,i}^{CHP}$ and $\eta_{gh,i}^{CHP}$ denote the gas-electric and gas-heat efficiencies
of the μCHP device and where $\eta_{gh,i}^F$ denotes the efficiency of the furnace. The variable $\nu_{g,i}(k)$ ($0 \le \nu_{g,i}(k) \le 1$) represents the dispatch factor that determines how the
gas is divided between the μCHP and the furnace. The term $\nu_{g,i}(k)P_{g,i}^H(k)$ defines the
gas input power fed into the μCHP, and according to (10.6) the part $(1-\nu_{g,i}(k))P_{g,i}^H(k)$
defines the gas input power going into the furnace. (Since the gas dispatch involves
only two converter devices, the superscript c indicating the correspondent converter,
is omitted.)

10.3.2.2 Energy storage

The storage device is modeled as an ideal storage in combination with a storage
interface [9](Figure 10.7). The relation between the power exchange $Q_{\alpha,i}(k)$ and the
effectively stored energy $E_{\alpha,i}(k)$ at time step k is defined by the following equation:

$$\begin{aligned}
Q_{\alpha,i}(k) &= \frac{\dot{E}_{\alpha,i}}{e_{\alpha,i}} = \frac{1}{e_{\alpha,i}}\frac{dE_{\alpha,i}}{dt} \approx \frac{1}{e_{\alpha,i}}\frac{\Delta E_{\alpha,i}}{\Delta t} \\
&= \frac{1}{e_{\alpha,i}}\left(\frac{E_{\alpha,i}(k)-E_{\alpha,i}(k-1)}{\Delta t} + \dot{E}_{\alpha,i}^{stb}\right),
\end{aligned} \qquad (10.8)$$

with

$$e_{\alpha,i} = \begin{cases} e_{\alpha,i}^+ & \text{if } Q_{\alpha,i}(k) \ge 0 \quad \text{(charging/standby)} \\ 1/e_{\alpha,i}^- & \text{else} \qquad\qquad\quad \text{(discharging)}, \end{cases} \qquad (10.9)$$

where $e_{\alpha,i}^+$, $e_{\alpha,i}^-$ are the charging and discharging efficiencies of the heat storage
device, respectively, including the efficiency of the storage interface, converting the
energy carrier exchanged with the system $Q_{\alpha,i}(k)$ into the carrier stored internally
$\widetilde{Q}_{\alpha,i}(k)$, according to $\widetilde{Q}_{\alpha,i}(k) = e_{\alpha,i}Q_{\alpha,i}(k)$. The storage energy at time step k is
denoted by $E_{\alpha,i}(k)$, and $\dot{E}_{\alpha,i}^{stb}$ represents the standby energy losses of the heat storage
device per period ($\dot{E}_{\alpha,i}^{stb} \ge 0$).

Depending on which side of the converter the storage device is located, the following power flow equations result. Figure 10.8 illustrates the situation. If the
storage is located at the input side of the converter devices the power flow equations

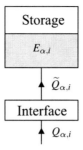

Figure 10.7: Storage element exchanging the power $Q_{\alpha,i}$; internal power $\widetilde{Q}_{\alpha,i}$, stored energy $E_{\alpha,i}$.

are described by

$$\widetilde{P}_{\alpha,i}(k) = P_{\alpha,i}(k) - Q_{\alpha,i}(k), \tag{10.10}$$

and if the storage is placed at the output side of to the converter device, the equations are given by

$$\widetilde{L}_{\beta,i}(k) = L_{\beta,i}(k) + M_{\beta,i}(k), \tag{10.11}$$

where $M_{\beta,i}(k)$ denotes the storage flow of a storage device at the output side of a converter, analogously to $Q_{\alpha,i}(k)$. Examples of storages before the converter devices are gas storages before a μCHP device or hydrogen storages before fuel cells. The hydrogen storage is filled by an electrolyzer, converting electricity into hydrogen. Storage examples after converters are heat storages after heat exchangers or μCHP devices or the above mentioned hydrogen storages after electrolyzers.

When merging all power flows, the inputs and outputs of the entire hub are then described by

$$\left[\mathbf{L}_i(k) + \mathbf{M}_i(k) \right] = \mathbf{C}_i(k) \left[\mathbf{P}_i(k) - \mathbf{Q}_i(k) \right], \tag{10.12}$$

where $\mathbf{Q}_{\alpha,i}(k)$ and $\mathbf{M}_{\alpha,i}(k)$ state all input-side and output-side storage power flows.

Here, we assume the converter efficiencies to be constant, i.e., to be independent of the converted power level, which results in a constant coupling matrix $\mathbf{C}_i(k)$ for each time step k. We can then apply superposition and summarize all storage flows in an equivalent output storage flow vector

$$\mathbf{M}_i^{\mathrm{eq}}(k) = \mathbf{C}_i(k) \mathbf{Q}_i(k) + \mathbf{M}_i(k). \tag{10.13}$$

With (10.8) and (10.13), the storage flows and the storage energy derivatives are related by

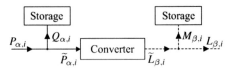

Figure 10.8: $\alpha\beta$-converter with α-storage at the input and β-storage at the output.

$$
\underbrace{\begin{bmatrix} M_{\alpha,i}^{\mathrm{eq}}(k) \\ \vdots \\ M_{\omega,i}^{\mathrm{eq}}(k) \end{bmatrix}}_{\mathbf{M}_i^{\mathrm{eq}}(k)} = \underbrace{\begin{bmatrix} s_{\alpha\alpha,i}(k) & \cdots & s_{\omega\alpha,i}(k) \\ \vdots & \ddots & \vdots \\ s_{\alpha\omega,i}(k) & \cdots & s_{\omega\omega,i}(k) \end{bmatrix}}_{\mathbf{S}_i(k)} \underbrace{\begin{bmatrix} \dot{E}_{\alpha,i}(k) \\ \vdots \\ \dot{E}_{\omega,i}(k) \end{bmatrix}}_{\dot{\mathbf{E}}_i(k)}, \tag{10.14}
$$

where the *storage coupling matrix* $\mathbf{S}_i(k)$ describes how changes within the storage energies affect the output flows, i.e., how the storage energy derivatives are mapped into equivalent output-side flows. According to (10.8), the storage energy derivatives correspond to

$$
\dot{\mathbf{E}}_i(k) = \mathbf{E}_i(k) - \mathbf{E}_i(k-1) + \dot{\mathbf{E}}_i^{\mathrm{stb}}. \tag{10.15}
$$

Adding the storage equation 10.14 to the general hub equation yields the following flows through an energy hub:

$$
\mathbf{L}_i(k) + \mathbf{S}_i(k)\dot{\mathbf{E}}_i(k) = \mathbf{C}_i(k)\mathbf{P}_i(k). \tag{10.16}
$$

Application example

For each hub depicted in Figure 10.4, hot water storage devices are implemented. Equation (10.7) is therefore completed with additional storage power flows, which are collected in a vector $\mathbf{M}_i(k)$:

$$
\underbrace{\begin{bmatrix} L_{\mathrm{e},i}(k) \\ L_{\mathrm{h},i}(k) + M_{\mathrm{h},i}(k) \end{bmatrix}}_{\mathbf{L}_i(k) + \mathbf{M}_i(k)} = \underbrace{\begin{bmatrix} 1 & \nu_{\mathrm{g},i}(k)\eta_{\mathrm{ge},i}^{\mathrm{CHP}} \\ 0 & \nu_{\mathrm{g},i}(k)\eta_{\mathrm{gh},i}^{\mathrm{CHP}} + (1-\nu_{\mathrm{g},i}(k))\eta_{\mathrm{gh},i}^{\mathrm{F}} \end{bmatrix}}_{\mathbf{C}_i(k)} \underbrace{\begin{bmatrix} P_{\mathrm{e},i}^{\mathrm{H}}(k) \\ P_{\mathrm{g},i}^{\mathrm{H}}(k) \end{bmatrix}}_{\mathbf{P}_i(k)}. \tag{10.17}
$$

10.3.3 Transmission model

As introduced above, we consider here a system where the hubs are interconnected by two types of transmission systems, an electricity and a natural gas network. However, district heating systems or hydrogen systems are also possible transmission systems for interconnecting hubs. For the transmission networks of both the electricity network and the gas pipeline network, power flow models based on nodal power balances are implemented.

10.3.3.1 AC electricity network

Electric power flows are formulated as nodal power balances of the complex power, according to the normal power flow equations [17]. At node m, the complex power balance at time step k is stated as

$$S_m(k) - \sum_{n \in N_m} S_{mn}(k) = 0, \qquad (10.18)$$

where $S_m(k)$ is the complex power injected at node m, and $S_{mn}(k)$ denotes the power flow to all adjacent nodes n of node m, summarized in the set N_m. The line flows are expressed by the voltage magnitudes $V(k)$ and angles $\theta(k)$ and the line parameters:

$$S_{mn}(k) = y_{mn}^* V_m(k) e^{j\theta_m(k)} (V_m(k) e^{-j\theta_m(k)} - V_n(k) e^{-j\theta_n(k)}) - j b_{mn}^{\mathrm{sh}} V_m^2(k), \qquad (10.19)$$

where the superscript * denotes the conjugate complex of the value. The line is modeled as a π-equivalent with the series admittance y_{mn} and the shunt susceptance b_{mn}^{sh} [17].

10.3.3.2 Pipeline network

Figure 10.9 shows the model of a gas pipeline composed of a compressor and a pipeline element. The volume flow balance at node m at time step k is defined as

$$F_m(k) - \sum_{n \in \mathcal{N}_m} F_{mn}(k) = 0, \qquad (10.20)$$

where $F_m(k)$ is the volume flow injected at node m, $F_{mn}(k)$ denotes the line flow between nodes m and n, and $\mathcal{N}_m$ denotes the set of neighboring nodes of node m, i.e., the nodes connected to node m through a pipeline. The line flow $F_{mn}(k)$ can be calculated as

$$F_{mn}(k) = k_{mn} s_{mn} \sqrt{s_{mn}(p_m^2(k) - p_n^2(k))}, \qquad (10.21)$$

where $p_m(k)$ and $p_n(k)$ denote the upstream and downstream pressures, respectively, and k_{mn} identifies the line constant. The variable s_{mn} indicates the direction of the gas flow as

$$s_{mn} = \begin{cases} +1 & \text{if } p_m(k) \geq p_n(k) \\ -1 & \text{otherwise.} \end{cases} \qquad (10.22)$$

The pipeline flow equation (10.21) is for most purposes a good approximation for all types of isothermal pipeline flows (liquid and gaseous). For obtaining more precise results for specific fluids and flow conditions a number of modified equations are available in [20].

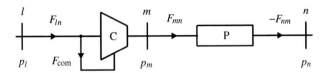

Figure 10.9: Model of a gas pipeline with compressor (C) and pipeline (P). Compressor demand is modeled as additional gas flow F_{com}.

To maintain a certain pressure level a compressor is needed. Here, the compressor is driven by a gas turbine which is modeled as additional gas flow

$$F_{\text{com}}(k) = k_{\text{com}} F_{mn}(k)(p_m(k) - p_l(k)),\qquad (10.23)$$

where $p_l(k)$ and $p_m(k)$ denote the pressures at the compressor input and output side, respectively, and k_{com} is a compressor constant. Basically, the amount of power consumed by the compressor depends on the pressure added to the fluid and on the volume flow rate through it. The resulting gas flow into the pipeline $F_{mn}(k)$ is therefore determined by

$$F_{mn}(k) = F_{ln}(k) - F_{\text{com}}(k).\qquad (10.24)$$

The pressure at the compressor output $p_m(k)$ is determined by

$$p_m(k) = p_{\text{inc}}(k)p_l(k),\qquad (10.25)$$

where $p_{\text{inc}}(k)$ defines the pressure amplification of the compressor. Depending on the required line flow $F_{mn}(k)$, $p_{\text{inc}}(k)$ is adjusted accordingly. For the purpose of this study, these simplified compressor models provide sufficient accuracy. More advanced compressor equations taking into account changing fluid properties are given in [20].

The volume flow rate $F_{mn}(k)$ corresponds to a power flow $P_{\text{g},mn}(k)$. The relation between volume and power flow is described by

$$P_{\text{g},mn}(k) = c_{\text{GHV}} F_{mn}(k),\qquad (10.26)$$

where c_{GHV} is the gross heating value of the fluid. The gross heating value depends on the fluid and is given in MWh/m^3. Values of different fluids can be found in [20].

10.3.4 Complete model description

The combined hub and transmission network model is obtained by combining the power flow models stated above. The system setup in Figure 10.4 serves again as example. For each time step k, the following three vectors are defined:

- algebraic state vector $\mathbf{z}(k)$: The algebraic state vector includes the variables for which no explicit dynamics are defined:

$$z(k) = [\mathbf{V}^T(k), \boldsymbol{\theta}^T(k), \mathbf{p}^T(k), \mathbf{p}_{inc}^T(k), (\mathbf{P}_e^H)^T(k), (\mathbf{P}_g^H)^T(k)]^T, \qquad (10.27)$$

where

- $\mathbf{V}(k) = [V_1(k), V_2(k), V_3(k)]^T$ and $\boldsymbol{\theta}(k) = [\theta_1(k), \theta_2(k), \theta_3(k)]^T$ denote the voltage magnitudes and angles of the electric buses, respectively,
- $\mathbf{p}(k) = [p_1(k), p_2(k), p_3(k)]^T$ denotes the nodal pressures of all gas buses,
- $\mathbf{p}_{inc}(k) = [p_{inc,1}(k), p_{inc,2}(k)]^T$ denotes the pressure amplification of the compressors,
- $\mathbf{P}_e^H(k) = [P_{e,1}^H(k), P_{e,2}^H(k), P_{e,3}^H(k)]^T$ denotes the electric inputs of the hubs, and
- $\mathbf{P}_g^H(k) = [P_{g,1}^H(k), P_{g,2}^H(k), P_{g,3}^H(k)]^T$ denotes the gas inputs of the hubs.

• dynamic state vector $\mathbf{x}(k)$: The dynamic state vector includes variables for which dynamics are included:

$$\mathbf{x}(k) = \mathbf{E}_h(k), \qquad (10.28)$$

where

- $\mathbf{E}_h(k) = [E_{h,1}(k), E_{h,2}(k), E_{h,3}(k)]^T$ denotes the energy contents of the heat storage devices.

• control vector $\mathbf{u}(k)$: The control variables include the operational set-points of the system:

$$\mathbf{u}(k) = \left[(\mathbf{P}_e^G)^T(k), (\mathbf{P}_g^G)^T(k), \boldsymbol{\nu}_g^T(k) \right]^T, \qquad (10.29)$$

where

- $\mathbf{P}_e^G(k) = [P_{e,1}^G(k), P_{e,2}^G(k), P_{e,3}^G(k)]^T$ denotes the active power generation of all generators,
- $\mathbf{P}_g^G(k) = [P_{g,1}^G(k), P_{g,2}^G(k)]^T$ defines the natural gas imports and
- $\boldsymbol{\nu}_g(k) = [\nu_{g,1}(k), \nu_{g,2}(k), \nu_{g,3}(k)]^T$ describes the dispatch factors of the gas input junctions.

Now, the model that we use to represent the multi-carrier network, including the hub equations with the dynamics, can be written in compact form as

$$\mathbf{x}(k+1) = \mathbf{f}(\mathbf{x}(k), \mathbf{z}(k), \mathbf{u}(k)) \qquad (10.30)$$
$$\mathbf{0} = \mathbf{g}(\mathbf{x}(k), \mathbf{z}(k), \mathbf{u}(k)). \qquad (10.31)$$

Equation (10.30) represent the difference equations describing the dynamics in the system, i.e., the dynamics in the storage devices. The equality constraints (10.31) represent the static, instantaneous relations in the system, i.e., the transmission and energy conversion components of the system.

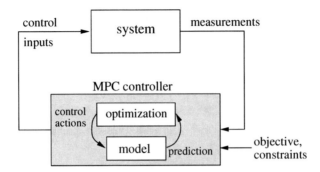

Figure 10.10: Illustration of model predictive control.

10.4 Centralized model predictive control

One way to determine the actions that yield the optimal operation of the system is by using centralized control. In centralized control, a centralized controller measures all variables in the network and determines actions or set-points for all actuators, i.e., the energy generation units, converters, and storage devices. We propose to use a model-based predictive control (MPC) scheme to determine the control variables $\mathbf{u}(k)$ in such a way that the total operational costs of the system are minimized while satisfying the system constraints. Below, we explain the basic idea of MPC. Then, the MPC problem for the considered hub system is formulated for centralized control.

10.4.1 Principle of operation

MPC [6, 19] is an optimization-based control strategy where an optimization problem is solved at each discrete decision step. This optimization problem uses an internal prediction model to find those actions that give the best predicted system behavior over a certain prediction horizon with length N. In this optimization operational constraints are also taken into account. MPC operates in a receding horizon fashion, meaning that at each time step new measurements of the system and new predictions into the future are made and new control actions are computed. By using MPC, actions can be determined that anticipate future events, such as increasing or decreasing energy prices or changes within the load profiles. MPC is suited for control of multi-carrier systems, since it can adequately take into account the dynamics of the energy storage devices and the characteristics of the electricity and gas networks.

In Figure 10.10 the operation of an MPC scheme is illustrated schematically. At each discrete control step k, an MPC controller first measures the current state of the system, $\mathbf{x}(k)$. Then, it computes which control input $\mathbf{u}(k)$ to be provided to the system, by using (numerical) optimization to determine the actions that give the best predicted performance over a prediction horizon of N time steps as defined by an objective function. The control variables computed for the first prediction step are then applied to the physical system. The system then transitions to a new state, $\mathbf{x}(k+1)$, after which the above procedure is repeated.

10.4.2 Problem formulation

In the MPC formulation the central controller determines the inputs $\mathbf{u}(k)$ for the network by solving the following optimization problem:

$$\min_{\tilde{\mathbf{u}}(k)} J(\tilde{\mathbf{x}}(k+1), \tilde{\mathbf{z}}(k), \tilde{\mathbf{u}}(k)) \tag{10.32}$$

subject to

$$\tilde{\mathbf{x}}(k+1) = \tilde{\mathbf{f}}(\tilde{\mathbf{x}}(k), \tilde{\mathbf{z}}(k), \tilde{\mathbf{u}}(k)) \tag{10.33}$$

$$\tilde{\mathbf{g}}(\tilde{\mathbf{x}}(k), \tilde{\mathbf{z}}(k), \tilde{\mathbf{u}}(k)) = \mathbf{0} \tag{10.34}$$

$$\tilde{\mathbf{h}}(\tilde{\mathbf{x}}(k), \tilde{\mathbf{z}}(k), \tilde{\mathbf{u}}(k)) \leq \mathbf{0}, \tag{10.35}$$

where the tilde over a variable represents a vector with the values of this variable over a prediction horizon of N steps, e.g., $\tilde{\mathbf{u}}(k) = [\, \mathbf{u}^{\mathrm{T}}(k), \ldots, \mathbf{u}^{\mathrm{T}}(k+N-1)\,]^{\mathrm{T}}$.

For the system setup under consideration, i.e., the system in Figure 10.4, the control objective is to minimize the energy costs, i.e., the costs for electricity energy and natural gas. The following objective function will be used in this minimization, in which costs of the individual energy carriers are modeled as quadratic functions of the corresponding powers:

$$J = \sum_{l=0}^{N-1} \sum_{i \in G} q_i^{\mathrm{G}}(k+l)(P_{\mathrm{e},i}^{\mathrm{G}}(k+l))^2 + q_i^{\mathrm{N}}(k+l)(P_{\mathrm{g},i}^{\mathrm{G}}(k+l))^2, \tag{10.36}$$

where G is a set of generation unit indices, i.e., the three generators and the two natural gas providers. The prices for active power generation $q_i^{\mathrm{G}}(k)$ and for natural gas consumption $q_i^{\mathrm{N}}(k)$ can vary throughout the day.

The equality constraints (10.33) and (10.34) represent the dynamic and static relations of the prediction model of the system. They correspond to equations (10.30) and (10.31), formulated over the prediction horizon N. The inequality constraints (10.35) comprise limits on the voltage magnitudes, active and reactive power flows, pressures, changes in compressor settings, and dispatch factors. Furthermore, power limitations on hub inputs and on gas and electricity generation are also incorporated into (10.35). Regarding the storage devices, limits on storage contents and storage flows are imposed.

The optimization problem (10.32)–(10.35) is a nonlinear programming problem [4], which can be solved using solvers for nonlinear programming, such as sequential quadratic programming [4]. In general, the solution space is nonconvex and therefore finding a global optimum cannot be guaranteed. Unless a multi-start ap-

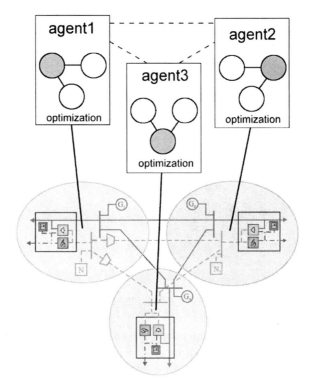

Figure 10.11: Three-hub system controlled by three communicating agents.

proach with a sufficient number of starts is used, a local optimum is returned by the numerical optimization.

10.5 Distributed model predictive control

Although a centralized controller could in theory give the best performance, practical and computational limitations prevent such a centralized controller from being useful in practice. The overall network may be owned by different entities, and these different entities may not want to give access to their sensors and actuators to a centralized authority. Even if they would allow a centralized authority to take over control of their part of the network, this centralized authority would have computational problems solving the resulting centralized control problem due to its large size. In that case, it has to be accepted that several different MPC controllers are present, each controlling their own parts of the network, e.g., their own households.

Figure 10.11 shows the introduced three-hub system controlled by three agents. Each agent, or controller, solves its own local MPC problem using the local model of its part of the system. However, the solution of a local MPC problem depends

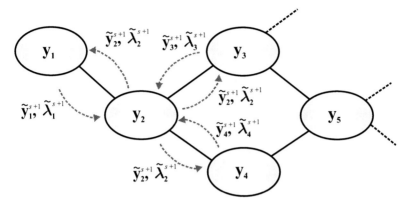

Figure 10.12: Coordination procedure between multiple interconnected areas by exchanging system variables **y** and Lagrangian multipliers **λ**.

on the solution of the MPC problems of the surrounding MPC controllers, since the electricity and gas networks interconnect the hubs. Therefore, the MPC problems of the controllers have to be solved in a cooperative way by allowing communication between the agents (dashed lines in Figure 10.11). This is not only to ensure that the controllers choose feasible actions, but also to allow the controllers to choose actions that are optimal from a system-wide point of view.

In our application, the MPC subproblems are based on nonlinear dynamic models. We therefore propose an extension of the static distributed control scheme in [8] that does take into account dynamics. Hence, the method is extended for optimization over multiple time steps in an MPC way. We then obtain an approach based on a combination of MPC and Lagrangian relaxation.

10.5.1 Principle of operation

Here, we explain the mathematical concept to decompose a general MPC optimization problem into several subproblems for individual distributed controllers. The procedure is presented on an interconnected multi-area system depicted in Figure 10.12. The areas $a = 1, 2, \ldots, A$ are interconnected in an arbitrary way. The system variables of each area a comprise the algebraic state vector $\tilde{\mathbf{z}}_a(k)$ and dynamic state vector $\tilde{\mathbf{x}}_a(k)$ as well as the control variables $\tilde{\mathbf{u}}_a(k)$, i.e.,

$$\tilde{\mathbf{y}}_a(k) = [\tilde{\mathbf{x}}_a(k), \tilde{\mathbf{z}}_a(k), \tilde{\mathbf{u}}_a(k)]^{\mathsf{T}} \qquad \text{for } a = 1, \ldots, A. \qquad (10.37)$$

The overall, centralized MPC optimization problem can then be defined as

$$\min_{\tilde{\mathbf{u}}_a(k)} \sum_{a=1}^{A} J_a(\tilde{\mathbf{y}}_a(k)) \qquad (10.38)$$

$$\text{subject to } \tilde{\mathbf{g}}(\tilde{\mathbf{y}}_a(k)) = \mathbf{0} \qquad \text{for } a = 1, \ldots, A \qquad (10.39)$$

$$\tilde{\hat{\mathbf{g}}}_a(\tilde{\mathbf{y}}_1(k),\dots,\tilde{\mathbf{y}}_a(k),\dots,\tilde{\mathbf{y}}_A(k)) = \mathbf{0} \qquad \text{for } a = 1,\dots,A, \qquad (10.40)$$

where only equality constraints are included for the sake of demonstration. Inequality constraints are handled analogously. The constraints are classified into two types of constraints. Constraints that involve only the local system variables are collected in (10.39). Besides these purely local constraints, so-called *coupling constraints* (10.40) (marked by a hat) are present, containing variables form multiple control areas. These coupling constraints are related to multiple areas and thus prevent the controllers of each subsystem from operating independently of each other. These constraints are the reason why coordination between the controllers is necessary.

10.5.1.1 Decomposition methodology

For decomposing this centralized MPC optimization problem into optimization problems for the controllers of the individual control areas, both the objective and the equality constraints are separated and assigned to a responsible control agent.

The constraints (10.39) with only local variables are assigned to the corresponding controller of each area. The coupling constraints (10.40) can in principle be assigned arbitrarily to the controllers. However, they are assigned to the area that contains the majority of the coupling variables. Coupling variables are the variables of the peripheral buses, also referred to as border buses, which are buses that are directly connected to buses of another area.

The subproblems for the individual controllers are now obtained by relaxing some of the coupling constraints and adding them to the objectives of the different controllers. Conventional Lagrangian relaxation is based on relaxing the own coupling constraints of each controller by incorporating them into their objective functions [15], weighted by Lagrangian multipliers. The obtained subproblems are then solved in a series of iterations, where each local optimization problem is solved with fixed values for the variables of the other controllers. After each iteration the Lagrangian multipliers are updated with a sub-gradient method. To avoid this update, which requires appropriate tuning of the update parameters, an advanced method establishes the subproblems by relaxing the coupling constraints assigned to the foreign areas (modified Lagrangian relaxation procedure [8]).

The resulting subproblem for each area $a = 1,\dots,A$ is then formally written as

$$\min_{\tilde{\mathbf{u}}_a(k)} \; J_a(\tilde{\mathbf{y}}_a(k)) + \sum_{b=1,b\neq a}^{A} (\tilde{\boldsymbol{\lambda}}_b^s)^{\mathsf{T}} \tilde{\hat{\mathbf{g}}}_b(\tilde{\mathbf{y}}^a(k)) \qquad (10.41)$$

$$\text{subject to } \; \tilde{\mathbf{g}}_a(\tilde{\mathbf{y}}_a(k)) = \mathbf{0}, \qquad (10.42)$$

$$\tilde{\hat{\mathbf{g}}}_a(\tilde{\mathbf{y}}^a(k)) = \mathbf{0}, \qquad (10.43)$$

where $\tilde{\mathbf{y}}^a(k) = [\tilde{\mathbf{y}}_1^s(k),\dots,\tilde{\mathbf{y}}_a(k),\dots,\tilde{\mathbf{y}}_A^s(k)]$ represents the system variables of all neighboring areas of area a. $\boldsymbol{\lambda}$ are the Lagrangian multipliers which will be explained below. The superscript s indicates the iteration step. As mentioned above, the optimization problems of the individual control agents are solved in an iterative procedure, keeping the variables of the neighboring areas constant. Both, the objective and the coupling constraints depend on variables of the foreign areas, referred to foreign variables, indicated by the superscript s.

The objective function of each controller consists of two parts. The first term expresses the main objective originating from the overall objective function (10.38). The second term is responsible for the coordination between the agents and consists of the coupling constraints introduced above. As indicated in (10.41) - (10.43), the coupling constraints of the own area are kept explicitly as hard constraints of the constraint set of the own controller (10.43) and are then added as soft constraints to the main objective of the other controllers. This follows the principle of the modified Lagrangian relaxation procedure [24]. The weighting factors of the soft constraints are the Lagrangian multipliers obtained from the optimization problem of the neighboring controllers.

10.5.1.2 Solution scheme

Both the objectives and the coupling constraints depend on variables of multiple controllers. To handle this dependency, the optimization problems of the controllers are solved in an iterative procedure:

- At each iteration step s, the MPC optimization problems of all control agents are solved independently of each other, while keeping the variables of the the other controllers constant.
- After each iteration, the controllers exchange the updated values of their variables, i.e., the variables $\tilde{\mathbf{y}}_i^{s+1}(k)$ and the Lagrange multipliers $\tilde{\boldsymbol{\lambda}}_i^{s+1}(k)$, where i refers to the corresponding control area. Figure 10.12 indicates the dependencies between area 2 and its surrounding areas. Only the variables between two directly connected areas need to be exchanged. Thus, area 5 does not need to send its variables to area 2.
- Convergence is achieved when the exchanged variables do not change more than a small tolerance γ_{tol} in two consecutive iterations.

Note that not the whole set of the updated system variables needs to be exchanged between the areas. Only the updated coupling variables have to be exchanged. For the sake of clarity of notation, the system variables and the effectively exchanged variables are not distinguished in the notation. In contrary to conventional Lagrangian relaxation procedures, a faster convergence is achieved as the weighting factors are represented by the Lagrangian multipliers of the neighboring optimization problem [24].

10.5.2 Application

We next apply the decomposition procedure to our three-hub system, as depicted in Figure 10.4. It is noted that although here we only consider three hubs, the presented decomposition procedure is also suited for large-scale systems. The considered three-hub network is divided into three control areas, according to the hubs. Each of the control areas has a controller for determining the local control actions.

10.5.2.1 Local variables

The controller of a particular hub considers as its variables the hub variables and the system variables of the nodes connected to it. For example, for the first controller, the state and control vectors for each time step k are defined as

$$\mathbf{x}_1(k) = E_{h,1}(k) \tag{10.44}$$

$$\mathbf{z}_1(k) = [V_1(k), \theta_1(k), p_1(k), p_{inc,1}(k), p_{inc,2}(k), P^H_{e,1}(k), P^H_{g,1}(k)]^T \tag{10.45}$$

$$\mathbf{u}_1(k) = [P^G_{e,1}(k), P^G_{g,1}(k), \nu_{g,1}(k)]^T. \tag{10.46}$$

The state and control vectors for the second and third controller are defined similarly according to Figure 10.4.

10.5.2.2 Objective functions

Each individual controller has its own control objective. In particular, the objective functions of the three controllers are:

$$J_1 = \sum_{l=0}^{N-1} q^G_1(k+l)(P^G_{e,1}(k+l))^2 + q^N_1(k+l)(P^G_{g,1}(k+l))^2 \tag{10.47}$$

$$J_2 = \sum_{l=0}^{N-1} q^G_2(k+l)(P^G_{e,2}(k+l))^2 + q^N_2(k+l)(P^G_{g,2}(k+l))^2 \tag{10.48}$$

$$J_3 = \sum_{l=0}^{N-1} q^G_3(k+l)(P^G_{e,3}(k+l))^2. \tag{10.49}$$

10.5.2.3 Coupling constraints

The three optimization problems have to be coordinated by adding the respective coupling constraints to the individual objectives given above. Below, the coupling constraints for the electric power and for the gas transmission systems are presented. Then, the resulting objectives are formulated.

Electric power systems

For applying the procedure to electric power systems, the constraints are arranged in the following way. The power flow equations of all inner buses of a particular area are incorporated into the equality constraints $\tilde{\mathbf{g}}_A(\tilde{\mathbf{y}}_A(k)) = 0$, $\tilde{\mathbf{g}}_B(\tilde{\mathbf{y}}_B(k)) = 0$. Inner buses are those buses of an area that have at least one bus in between themselves and the buses of another area. Buses that are directly connected to buses of another area are referred to as peripheral buses or border buses.

Regarding the couplings, the electric power flow equations at the border buses serve as coupling constraints. A coupling between the areas is only enabled when these power flow equations comprehend variables of both areas. This implies that the constraints for the active and reactive power balance serve as coupling constraints, but not the equations regarding voltage magnitude and angle reference settings. Hence, having PQ buses (active and reactive power are specified [17]) at the common tie-lines results in two coupling constraints per peripheral bus. A less tight coupling is achieved with PV buses (active power and voltage magnitude are specified [17]), yielding only one coupling constraint. If the slack bus (voltage magnitude and voltage angle are specified [17]) is situated at one of the border buses, the procedure is not implementable, because only voltage magnitude and angle reference settings have to hold for these kind of buses. For the case of active power control, the slack bus is modeled as a PV bus with an additional angle reference in order to obtain enough coupling constraints. The inequality constraints are occurring with transmission limits on tie-lines belonging to both areas. To classify the inequality constraints into own and foreign constraints the tie-lines need to be allocated to one area, arbitrarily.

For the studied three-hub system, the active power balances of all nodes of the electricity system require coordination as they depend on the neighboring voltage magnitudes and angles. For each coupling constraint, the dependencies of the own and foreign system variables (marked by superscript s, which specifies the current iteration step) are indicated. Since each node serves as border bus of the respective control area, a coupling constraint is set up for each node. The following active power balances need to be fulfilled:

$$\triangle P_1(k) = P_{e,1}^G(k) - P_{12}(k) - P_{13}(k) - P_{e,1}^H(k) \tag{10.50}$$
$$= f_{P_1}(V_1(k), \theta_1(k), V_2^s(k), \theta_2^s(k), V_3^s(k), \theta_3^s(k)) = 0$$
$$\triangle P_2(k) = P_{e,2}^G(k) + P_{12}(k) - P_{13}(k) - P_{e,2}^H(k) \tag{10.51}$$
$$= f_{P_2}(V_1^s(k), \theta_1^s(k), V_2(k), \theta_2(k), V_3^s(k), \theta_3^s(k)) = 0$$
$$\triangle P_3(k) = P_{e,3}^G(k) + P_{12}(k) + P_{13}(k) - P_{e,3}^H(k) \tag{10.52}$$
$$= f_{P_3}(V_1^s(k), \theta_1^s(k), V_2^s(k), \theta_2^s(k), V_3(k), \theta_3(k)) = 0.$$

Pipeline networks

Implementing the decomposition procedure for natural gas systems, the constraints are arranged in the same way. The constraints $\tilde{\mathbf{g}}_A(\tilde{\mathbf{y}}_A(k)) = 0$, $\tilde{\mathbf{g}}_B(\tilde{\mathbf{y}}_B(k)) = 0$ comprise the volume flow equations of all inner buses as well as the pressure reference settings (slack bus). Coordination is required due to the nodal flow balances at the border buses, since the injected volume flows are dependent on the nodal pressures of the neighboring busses. Inequality constraints consist of pressure limits and compressor limits. No coupling inequality constraints are incorporated. Here, each node serves as border bus as well, thus, a coupling constraint is set up for each node. The following volume flow balances need to be fulfilled:

$$\triangle F_1(k) = P_{g,1}^G(k) - F_{12}(k) - F_{13}(k) - F_{com,12}(k) - F_{com,13}(k) - P_{g,1}^H(k) \qquad (10.53)$$
$$= f_{F_1}(p_1(k), p_2^s(k), p_3^s(k)) = 0$$
$$\triangle F_2(k) = P_{g,2}^G(k) + F_{12}(k) - F_{13}(k) - P_{g,2}^H(k) \qquad (10.54)$$
$$= f_{F_2}(p_1^s(k), p_2(k), p_3^s(k)) = 0$$
$$\triangle F_3(k) = F_{12}(k) + F_{13}(k) - P_{g,3}^H(k) \qquad (10.55)$$
$$= f_{F_3}(p_1^s(k), p_2^s(k), p_3(k)) = 0,$$

where $F_{com,12}(k)$ and $F_{com,13}(k)$ describe the gas flows into the compressors C_{12} and C_{13}, respectively. For combined electricity and natural gas networks, the constraints are merged. Summarizing, for each controller, there exists one coupling constraint for the electricity and one for the natural gas system.

Resulting objective functions

The resulting objective functions for the controllers are obtained by adding in each case the coupling constraints of the neighboring areas. These constraints are weighted with the corresponding Lagrangian multipliers, obtained by the correspondent neighboring area. For example, the objective of the first controller takes into account the constraints of the second and third controller which are weighted by the Lagrangian multipliers obtained at the previous iteration step. The Lagrangian multipliers related to the electricity system and gas system are referred to as λ_{el} and λ_{gas}, respectively. We then obtain the following objective functions:

$$J_1(\cdot) = \sum_{l=0}^{N-1} q_1^G(k+l)(P_{e,1}^G(k+l))^2 + q_1^N(k+l)(P_{g,1}^G(k+l))^2$$
$$+ \lambda_{el,23}^s(k)\begin{bmatrix}\triangle P_2(k)\\ \triangle P_3(k)\end{bmatrix} + \lambda_{gas,23}^s(k)\begin{bmatrix}\triangle F_2(k)\\ \triangle F_3(k)\end{bmatrix} \qquad (10.56)$$

$$J_2(\cdot) = \sum_{l=0}^{N-1} q_2^G(k+l)(P_{e,2}^G(k+l))^2 + q_2^N(k+l)(P_{g,2}^G(k+l))^2$$
$$+ \lambda_{el,13}^s(k)\begin{bmatrix}\triangle P_1(k)\\ \triangle P_3(k)\end{bmatrix} + \lambda_{gas,13}^s(k)\begin{bmatrix}\triangle F_1(k)\\ \triangle F_3(k)\end{bmatrix} \qquad (10.57)$$

$$J_3(\cdot) = \sum_{l=0}^{N-1} q_3^G(k+l)(P_{e,3}^G(k+l))^2$$
$$+ \lambda_{el,12}^s(k)\begin{bmatrix}\triangle P_1(k)\\ \triangle P_2(k)\end{bmatrix} + \lambda_{gas,12}^s(k)\begin{bmatrix}\triangle F_1(k)\\ \triangle F_2(k)\end{bmatrix}. \qquad (10.58)$$

10.6 Simulation results

Simulations are presented, applying the MPC scheme proposed above to the three-hub system shown in Figure 10.4. Note that the scheme is general according to the discussion above and not only valid or applicable for our illustrative three-hub system. Next the setup of the simulation is given. Then, simulation results in which the centralized and distributed MPC approach are applied are presented. As the considered optimization problems are nonconvex, finding the global optimum cannot be guaranteed when applying numerical methods. However, the values of the centralized problem serve as a reference of optimality and the simulation results obtained by distributed optimization are compared with these values in order to judge the performance of the distributed approach. The solver `fmincon` provided be the Optimization Toolbox of Matlab is used [27].

10.6.1 Simulation setup

Each hub has a daily profile of its load demand and the energy prices. Here, we assume that the price and load forecasts are known. However, in reality, there are always forecast errors. As a first study, we assume perfect forecasts and it is believed that the following results are representative also for small forecast errors since the storage devices are able to balance deviations within load forecasts. The given profiles are typical profiles for a household. The electricity and heat loads are assumed to be the same for all hubs and are depicted in Figure 10.13(a) in per unit (p.u.) values.

Regarding the prices, electricity generation at hubs H_2 and H_3 is twice as expensive as at hub H_1, as illustrated in Figure 10.13(b) in m.u./p.u.^{2} values, where m.u. refers to monetary units. The reason for choosing different electricity prices is to obtain three hubs with different setups. (Hub H_1 has a cheap access to electricity and gas, hub H_2 has an expensive electricity and a limited gas access, and hub H_3 has an expensive electricity access and no gas access.) Gas prices remain constant throughout the day.

Regarding the electricity network, bus 1 is modeled as slack bus, i.e., having the voltage angle and voltage magnitude fixed ($V_1(k)$ has a magnitude of 1 p.u. and an angle of $0°$). The other two buses are modeled as PV buses, for which the net active power and the voltage magnitude are specified. Also within the gas network bus 1 serves as slack bus, having a fixed pressure value of 1 p.u. The coefficients and simulation parameters used are listed in Table 10.2. Since hub H_2 is assumed to have only access to a network with limited capacity, a flow rate constraint of 2 p.u. is imposed on $P_{g,2}^G(k)$. The gas network is mainly supplied via the large gas network at bus 1, i.e., via $P_{g,1}^G(k)$, which delivers gas to the neighboring buses by means of the two compressors.

Based on the profiles, the total generation costs are minimized for a simulation period of $N_{sim} = 24$ steps, where one time step corresponds to 1 hour. To analyze the performance of the proposed control scheme, we vary the length of the prediction

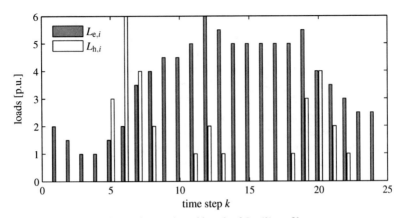

(a) Electricity $L_{e,i}(k)$ and heat load $L_{h,i}(k)$ profiles.

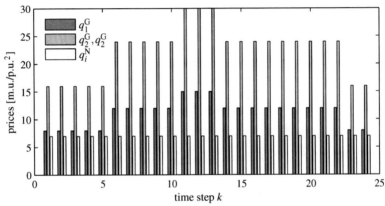

(b) Price profiles for electricity $q_i^G(k)$ and natural gas consumption $q_i^N(k)$.

Figure 10.13: Daily profiles used for simulation.

Table 10.2: Bounds and parameter values for the three-hub system in p.u.

variable	bounds	category	coefficients
V_i	$0.9 \leq |V_i| \leq 1.1$	μCHP	$\eta_{ge,i}^{CHP} = 0.3,\ \eta_{gh,i}^{CHP} = 0.4$
$P_{e,i}^G$	$0 \leq P_{e,i}^G \leq 10$	F	$\eta_{gh,i}^F = 0.75$
p_i	$0.8 \leq p_i \leq 1.2$	$E_{h,i}^{stb}$	$E_{h,i}^{stb} = 0.2$
$p_{inc,i}$	$1.2 \leq p_{inc,i} \leq 1.8$	$e_{h,i}$	$e_{h,i}^+ = e_{h,i}^- = 0.9$
ν_i	$0 \leq \nu_i \leq 1$		
$P_{g,i}^G$	$0 \leq P_{g,1}^G \leq 20,\ 0 \leq P_{g,2}^G \leq 2$		
E_i	$0.5 \leq E_i \leq 3$		
$M_{h,i}$	$-3 \leq M_{h,i} \leq 3$		

horizon N used between $N = 1$, i.e., no prediction, and $N = 24$, i.e., predicting for all 24 time steps at once.

10.6.2 Centralized control

First, the results for a specific prediction horizon are analyzed in more detail. Second, the performance of the control scheme operating with different prediction horizon lengths is compared. Finally, the operation costs are presented when comparing the operation of the μCHP device with and without heat storage support. Furthermore, the costs are compared with the decoupled operation mode, i.e., when the electricity and natural gas system are operated independently of each other, i.e., when no μCHP devices are in use.

10.6.2.1 Prediction horizon with length $N = 5$

The behavior of the system is illustrated for a prediction horizon with length $N = 5$. This length of prediction horizon is adequate for practical applications as it represents a proper trade-off between control performance on the one side and obtainable forecasts and computational effort on the other side, as is illustrated below in Section 10.6.2.2.

An optimization for 24 time steps is run, at each time step k implementing only the control variables for the current time step k and then starting the procedure again at time step $k + 1$ using updated system measurements. The operational costs for the entire simulation period $[0, 24]$ are $2.73 \cdot 10^4$ m.u. Figure 10.14 shows the evolution of the active power generation and natural gas import at the first hub. The electricity generation mainly corresponds to the electricity load pattern and the natural gas import evolves similar to the heat loads. However, natural gas is also used during time periods, in which no heat is required. During these periods gas is converted by the μCHP for supporting the electricity generation. The heat produced thereby is stored and used later for the heat supply.

In Figure 10.17, the content of all three storage devices over time is shown for $N = 1, 3, 5, 24$. The dotted line represents the storage behavior for a prediction horizon with length $N = 5$. In general, the storage devices are mainly discharged during the heat load peaks and charged when no heat is required. However, the heat storage devices are not only important for the heat supply but indirectly also for electricity generation, since the μCHP devices can be operated according to the electricity load requirements by means of the heat storage devices. At high electricity prices, electricity generation via μCHP is cheaper than via the generators, thus, the μCHP devices are preferably used for supplying the electricity demand while storing all excessive produced heat. This is also the reason why the storage contents of storages E_1 and E_2 rise again at the end of the simulation. Nevertheless, during the heat peak loads all gas is diverted into the furnaces because the thermal efficiencies of the μCHPs are not sufficient in order to supply the heat loads. During these time periods, the operational costs increase correspondingly.

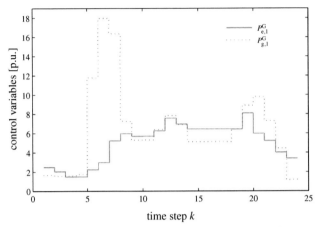

Figure 10.14: Active power generation $P^G_{e,1}(k)$ and natural gas import $P^G_{g,1}(k)$ of hub H_1 over the simulation horizon.

10.6.2.2 Comparison of different prediction horizon lengths

For showing the effect of prediction, prediction horizons with different lengths N are compared. In order to obtain a fair comparison, the prediction horizon is reduced towards the end of the simulation. Hence, in each case, the controller knows the same data, i.e., the measurements of the same 24 time steps. Figure 10.15 shows the total operation costs defined in (10.36) for different lengths of the prediction horizon N. Generally, the operation costs decrease with increasing prediction horizon. But this is not always the case. Depending on the input profiles, some prediction horizon lengths yield poorer results since the planned actions are suboptimal with respect to the whole simulation horizon. It should be noted that this conclusion is valid for this specific load profiles and that other load profiles might yield other results. As can be seen, a fast decay of the operation costs occurs within prediction horizon lengths $N = 1, \ldots, 5$. For longer prediction horizons, not much reduction of the cost is gained, except for optimizing for all 24 time steps at once ($N = 24$). Besides that, computational effort increases with increasing prediction horizon length. Figure 10.16 shows the computation time for different prediction horizon lengths. As can be seen, computational effort increases considerably for prediction horizon lengths larger than $N = 5$.

In Figure 10.17, the storage contents for different lengths of prediction horizons are presented. The horizontal lines indicate the storage limits ($0.5 \leq E_i(k) \leq 3$). At a prediction horizon with a length of $N = 1$ (dotted line) and $N = 3$ (solid line), the storage devices are filled up too late or are even emptied (time steps 1–3) because the controller sees the heat load peaks too late. With increasing N, the storage devices are filled up earlier. In fact, the optimization of the system would continuously proceed. For demonstration purposes, the optimization is stopped after 24 time steps. Therefore, no terminal constraint for the storage is imposed, such as requiring the storages to be half full at the end of the simulation period.

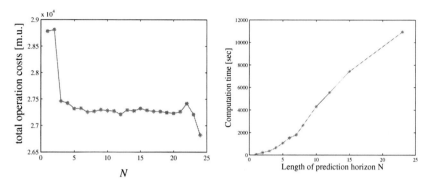

Figure 10.15: Total operation costs for different lengths of prediction horizon N.

Figure 10.16: Computation time for different lengths of prediction horizon N.

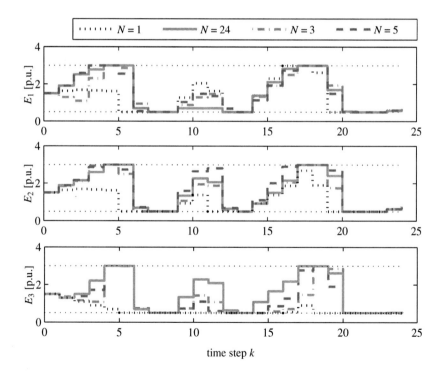

Figure 10.17: Storage evolution over simulation horizon. Comparison for different lengths of the prediction horizon, $N = 1, 3, 5, 24$.

When optimizing for all 24 time steps at once the most optimal behavior over the simulation horizon is obtained. The control variables for all next 24 time steps are determined and applied at time step k. But optimizing for all time steps at once is not applicable in practice since the data for the whole next day is normally not known in advance. Moreover, possibly occurring disturbances cannot be handled and computational effort becomes too high. Hence in practice, applying MPC with a properly chosen length of prediction horizon is the best choice. For the application example presented in this paper, a prediction horizon length of $N = 5$ yields the best results. In general, depending on the specifications, a trade-off between control performance and computational effort has to be made. Issues such as obtainable forecasts and size of possible disturbances also influence the choice of an adequate length of prediction horizon.

10.6.2.3 Comparison with decoupled mode

In the following the operation costs are compared for different system setups regarding the μCHP and the storage devices. The configuration with μCHP and storage devices serves as base case. In Table 10.3 the increase in costs for the different cases are presented, in each case the optimization is made with a prediction horizon length of $N = 5$. In the first two cases, the μCHP is utilized and the performance with and without heat storages is compared. Using the μCHP devices without the heat storages, total operation costs of $2.98 \cdot 10^4$ m.u. are obtained, corresponding to an increase of 9.2%. This is due to the fact that the μCHP devices cannot be utilized during periods without heat loads because the thereby produced heat cannot be dispensed. The second two cases present the costs obtained in decoupled operation mode, namely when the electricity and natural gas networks are optimized independently of each other. No power is converted by the μCHP devices in this mode. Running the optimization without μCHP usage but including the heat storages, total costs of $2.94 \cdot 10^4$ m.u. are obtained. Thus, by decoupling both infrastructures instead of operating them at once, generation costs are increased by 7.7%. Running the simulation with either the μCHP nor the storage devices yields total costs of $3.07 \cdot 10^4$ m.u., corresponding to an increase of 12.5%. Note that the combination of both devices, μCHP and storage device, have a higher effect on the total operation costs than each device itself. There exists an interplay between both devices which make both of them necessary.

Table 10.3: Comparison of operation costs, $N = 5$.

μCHP	storage	costs [m.u.]	increase
yes	yes	$2.73 \cdot 10^4$	base
yes	no	$2.98 \cdot 10^4$	9.2%
no	yes	$2.94 \cdot 10^4$	7.7%
no	no	$3.07 \cdot 10^4$	12.5%

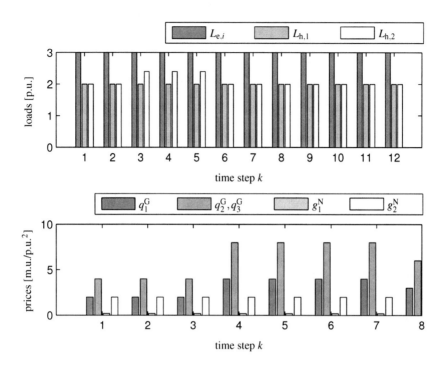

Figure 10.18: Profile for electricity $L_{e,i}(k)$ and heat loads $L_{h,i}(k)$ (upper plot) and prices for electricity $q_i^G(k)$ and natural gas consumption $q_i^N(k)$ (lower plot).

10.6.3 Distributed control

For the distributed case, again, as a preliminary case study, we assume a perfect forecast, in which no disturbances within the known profiles are occurring. The total generation costs are here minimized for a simulation horizon $N_{sim} = 10$. The length of the prediction horizon N is chosen as $N = 3$. Hence, an optimization over N time steps is run N_{sim} times, at each time step k implementing only the control variable for the current time step k and then starting a new optimization at time step $k+1$ with updated system measurements.

The price and load profiles of all hubs used in this study are shown in Figure 10.18. The electricity load $L_{e,i}$ and the gas import prices q_i^N remain constant over time. Variations are assumed only in the prices of the electric energy generation units $q_i^G(k)$ and in the heat load of hub H_2, $L_{h,2}$, in order to exactly retrace the storage behavior. In this study, only two storage devices E_1, E_2 are available for demonstrating the cooperative behavior. Control areas 1 and 2 are supposed to support control area 3 to fulfill its load requirements. Control area 3 has neither a gas access, nor a local heat storage, nor a cheap electricity generation possibility. The other system parameters are as given in Table 10.2.

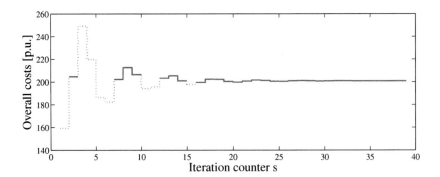

Figure 10.19: Intermediate solutions of the distributed algorithm applied to the system. Dotted lines represent infeasible solutions, solid lines are feasible solutions.

10.6.3.1 Single simulation step

Feasibility of distributed algorithm

In order to evaluate whether the solution determined by the distributed algorithm is feasible for the real system, the following simulation is run. The quality of the intermediate solutions in case that these would be applied to the system is shown in Figure 10.19. The distributed MPC optimization problem is is solved at time step $k = 1$, for $N = 3$. At each iteration counter s, the overall system costs are shown, when applying the control variables determined by the distributed algorithm to the system. The dotted values refer to the infeasible solutions. As the number of iterations increases, the distributed MPC algorithm converges, and, in fact, the solution obtained at the end of the iterations approaches the solution obtained by the centralized MPC approach (200.98 m.u.). After iteration 16, the values of all control variables are feasible. After 39 iterations, the algorithm converges.

Basically, the amount of backup energy provided by the storage devices determine whether the solution of the distributed MPC algorithm is feasible. Applying the solution to the system, the control variables are kept fixed, while the values of the storages are varied within their range attempting to fulfill the load requirements, i.e., to find an overall feasible solution. Hence, if the storage devices have not been operated close to their limits at the previous time step, a solution of the distributed algorithm may yield a feasible system solution, although the controller solution is considerably far away from a coordination between the individual control areas.

Convergence between control areas

Running the algorithm for the first simulation step with a prediction horizon length of $N = 3$ yields overall production costs of 200.77 m.u. Figure 10.20(a) shows the evolution of the objective values of all control areas as well as the total objective value. The costs of area 1 are higher since it contributes the highest amount of energy for overall system. The control variables are plotted in Figure 10.20(b).

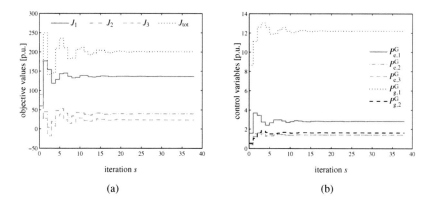

Figure 10.20: (a) Objective values of areas $1,2,3$ and total objective value; (b) control variables: active power generation and natural gas import.

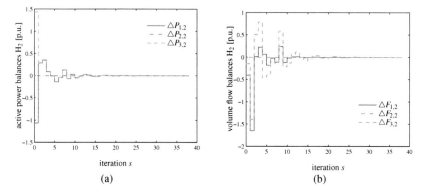

Figure 10.21: Convergence of coupling constraints at nodes 2: (a) active power balances and (b) volume flow balances.

Their steady state values adjust according to the prices for electricity generation and for the natural gas consumption, respectively.

For analyzing convergence between the control areas the evolution of the coupling constraints is plotted. In Figure 10.21, the coupling constraints obtained by the optimization of area 2 are presented. Figure 10.21(a) shows the active power balances obtained at all node of the electricity system and Figure 10.21(b) presents the volume flow balances at all nodes of the natural gas system. The active power balance and the volume flow balance as considered by node 2, denoted by $\triangle P_{2,2}$ and $\triangle F_{2,2}$, respectively, remain zero, i.e., the balances are always fulfilled, as they are implemented as hard constraints in the optimization problem of area 2. With increasing iterations, the coupling constraints decrease to zero, i.e., they are fulfilled, indicating that a successful coordination between the control agents has been achieved.

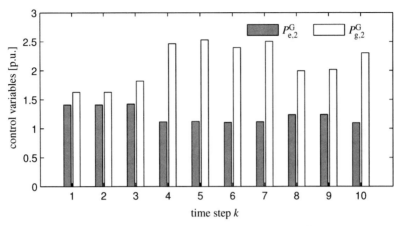

Figure 10.22: Active power generation $P_{e,2}^G$ and natural gas import $P_{g,2}^G$ of hub H_2 over time.

10.6.3.2 Simulation of multiple time steps

When minimizing the energy costs over the full simulation of N_{sim} time steps, a total cost of 850.62 m.u. is obtained for the load and price profiles given above. Applying centralized MPC, the overall costs are lower, 849.78 m.u., since, due to the imposed convergence tolerance γ_{tol} of the distributed algorithm, the centralized approach finds a slightly different solution at some iteration steps. In Figure 10.22 the active power generation and the natural gas import of hub H_2 are shown. As can be seen, active power generation is reduced at time steps with higher generation costs, i.e., time steps 4–7 and time step 10. During these time steps more gas is consumed. The electrical loads are now predominantly supplied by the μCHP devices in order to save costs. Most of the gas is diverted into the μCHP device and less into the furnace. For still supplying the heat load, the heat storage devices come into operation. Figure 10.23 shows the content of both storage devices evolving over the time steps. Both storage devices start at an initial level of 1.5 p.u. Since the heat load at hub H_2 is increased by 20% at time steps 3-5 (Figure 10.18), storage E_2 attempts to remain full before this increase and then operates at its lower limit during the heat load peaks. At the subsequent electricity price peaks (time steps 6, 7) both storages are recharged. The electrical loads are mainly supplied by the μCHP devices and all excessive heat produced during these time steps is then stored in the storage devices. Storage device E_1 is refilled more than E_2, as hub H_2 has a limited gas access.

If the controllers have a shorter prediction horizon than $N = 3$, the storage devices are filled up less and also later. With a prediction horizon length of $N = N_{sim}$, the storage devices are filled up earlier and the lowest costs are obtained, although calculation time becomes considerably longer and the system is insensitive to unknown changes in the load and price profiles.

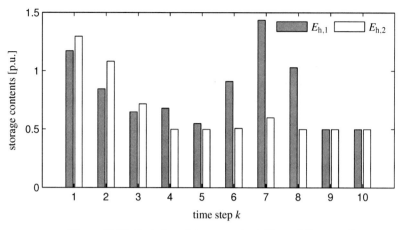

Figure 10.23: Evolution of storage contents $E_{h,1}$ and $E_{h,2}$ over time.

10.7 Conclusions and future research

In this chapter we have proposed the application of model predictive control to energy hub systems. The dynamics of storage devices, forecasts on energy prices and demand profiles, and operational constraints are taken into account adequately by the predictive control scheme, which is an effective control approach for this type of systems. The performance of different prediction horizons of varying length have been compared. With an increasing length of the prediction horizon the total operation costs decrease, but the computational effort increases accordingly. The analyses show that due to storage capability, even more flexibility on energy conversion inside the hubs is provided, which brings about more freedom in system planning and operation.

A distributed model predictive control (MPC) approach has been proposed for solving the overall optimization problem in a distributed way. In a case study, we have analyzed the quality of intermediate solutions obtained throughout the iterations of the proposed approach to ensure that applying the control to the real system yields feasible solutions. A cooperative behavior is shown, where the neighboring areas help to support the system wide objective.

Future research will address the incorporation of forecast errors into the scheme instead of assuming perfect forecasts. The goal is to implement different load forecast models and to analyze what size in forecast error the procedure can handle. Basically, the storage devices are the units that are able to balance a deviation within the load forecasts. Hence, the quality of the forecasts defines the size of the storage. The more appropriate forecasts are available, the smaller storage devices are needed to compensate the load prediction errors.

Furthermore, conditions and measures for guaranteeing convergence have to be investigated more precisely. As the solution spaces of the individual optimization problems are nonconvex, each control agent may have multiple choices for a local minimum. If neighboring agents each go for a local minimum that does not yield a system-wide feasible solution, the agents end up in a situation fighting against each other which results in a hunting or zig-zag behavior. If this zig-zag behavior is not too large in size, the distributed solution still yields a feasible system solution, as the storages can balance out a certain amount of disagreement. In these situations, it has to be decided which one is the dominant area, i.e., which one is decisive in the conflict. This decision is possibly based on economics or other relevant criteria. One aspect of future work will focus on determining which additional information has to be exchanged in order to make the algorithm more robust.

In addition, network operators, which could influence the energy exchanges between the hubs, are to be incorporated to the interconnected hub system. The final goal is to apply the developed procedures to larger systems with more than three hubs. Thereby, multiple hubs probably are taken into the same control area for enabling coordination within reasonable time. Then, firstly coordination *between* the different control areas and then coordination *within* the individual control areas is carried out.

Acknowledgements This research is supported by the project "Vision of Future Energy Networks" (VoFEN) of ABB, Areva T& D, Siemens, and the Swiss Federal Office of Energy, the BSIK project "Next Generation Infrastructures (NGI)", the Delft Research Center Next Generation Infrastructures, the European STREP project "Hierarchical and distributed model predictive control (HD-MPC)", and the project "Multi-Agent Control of Large-Scale Hybrid Systems" (DWV.6188) of the Dutch Technology Foundation STW.

References

1. A. Alessio and A. Bemporad. Decentralized model predictive control of constrained linear systems. In *Proceedings of the European Control Conference 2007*, pages 2813–2818, Kos, Greece, July 2007.
2. S. An, Q. Li, and T. W. Gedra. Natural gas and electricity optimal power flow. In *Proceedings of the 2003 IEEE PES Transmission and Distribution Conference*, pages 138–143, Dallas, Texas, September 2003.
3. M. Arnold, R. R. Negenborn, G. Andersson, and B. De Schutter. Distributed control applied to combined electricity and natural gas infrastructures. In *Proceedings of the International Conference on Infrastructure Systems*, Rotterdam, The Netherlands, November 2008.
4. D. P. Bertsekas. *Nonlinear Programming*. Athena Scientific, Beltmore, Massachusetts, 2003.
5. I. Bouwmans and K. Hemmes. Optimising energy systems: Hydrogen and distributed generation. In *Proceedings of the 2nd International Symposium on Distributed Generation: Power System Market Aspects*, pages 1–7, Stockholm, Sweden, October 2002.
6. E. F. Camacho and C. Bordons. *Model Predictive Control*. Springer-Verlag, New York, New York, 2004.
7. G. Chicco and P. Mancarella. A comprehensive approach to the characterization of trigeneration systems. In *Proceedings of the 6th World Energy System Conference*, Turin, Italy, July 2006.
8. A. J. Conejo, F. J. Nogales, and F. J. Prieto. A decomposition procedure based on approximate newton directions. *Mathematical Programming, Series A*, 93(3):495–515, December 2002.

9. M. Geidl and G. Andersson. Optimal coupling of energy infrastructures. In *Proceedings of PowerTech 2007*, pages 1398–1403, Lausanne, Switzerland, July 2007.
10. M. Geidl and G. Andersson. Optimal power flow of multiple energy carriers. *IEEE Transactions on Power Systems*, 22(1):145–155, 2007.
11. H. M. Groscurth, T. Bruckner, and R. Kümmel. Modeling of energy services supply systems. *Energy*, 20(9):941–958, January 1995.
12. A. Hajimiragha, C. Canizares, M. Fowler, M. Geider, and G. Andersson. Optimal energy flow of integrated energy systems with hydrogen economy considerations, August 2007. Presented at the IREP Symposium Bulk Power System Dynamics and Control – VII.
13. J. Hernandez-Santoyo and A. Sanchez-Cifuentes. Trigeneration: An alternative for energy savings. *Applied Energy*, 76(1–3):219–277, 2003.
14. B. H. Kim and R. Baldick. Coarse-grained distributed optimal power flow. *IEEE Transactions on Power Systems*, 12(2):932–939, May 1997.
15. B. H. Kim and R. Baldick. A comparison of distributed optimal power flow algorithms. *IEEE Transactions on Power Systems*, 15(2):599–604, May 2000.
16. G. Koeppel and G. Andersson. The influence of combined power, gas and thermal networks on the reliability of supply. In *Proceedings of 6th World Energy System Conference*, pages 646–651, Turin, Italy, July 2006.
17. P. Kundur. *Power System Stability and Control*. McGraw-Hill, New York, New York, 1994.
18. R. H. Lasseter and P. Piagi. Microgrid: A conceptual solution. In *Proceedings of the IEEE 35th Annual Power Electronics Specialists Conference*, pages 4285–4290, Aachen, Germany, June 2004.
19. J. M. Maciejowski. *Predictive Control with Constraints*. Prentice Hall, Harlow, England, 2002.
20. E. S. Menon. *Gas Pipeline Hydraulics*. Taylor & Francis, New York, New York, 2005.
21. M. S Morais and J. W. Marangon Lima. Natural gas network pricing and its influence on electricity and gas markets. In *Proceedings of the 2003 IEEE Bologna PowerTech Conference*, Bologna, Italy, June 2003.
22. R. R. Negenborn, B. De Schutter, and J. Hellendoorn. Multi-agent model predictive control for transportation networks: Serial versus parallel schemes. *Engineering Applications of Artificial Intelligence*, 21(3):353–366, April 2008.
23. A. J. Nogales, F. J. Prieto, and A. J. Conejo. A decomposition methodology applied to the multi-area optimal power flow problem. *Annals of Operations Research*, 120(1-4):99–116, April 2003.
24. F. J. Nogales, F. J. Prieto, and A. J. Conejo. A decomposition methodology applied to the multi-area optimal power flow problem. *Annals of Operations Research*, 120:99–116, April 2003.
25. R. Scattolini. Architectures for distributed and hierarchical model predictive control – A review. *Journal of Process Control*, 19(5):723–731, May 2009.
26. M. Shahidehpour, Y. Fu, and T. Wiedman. Impact of natural gas infrastrucutre on electric power systems. *Proceedings of the IEEE*, 93(5):1024–1056, 2005.
27. The Mathworks. Optimization Toolbox User's Guide, 2008.
28. A. N. Venkat, J. B. Rawlings, and S. J. Wright. Stability and optimality of distributed model predictive control. In *Proceedings of 44th IEEE Conference on Decision and Control*, pages 6680–6685, Seville, Spain, December 2005.

Part III
Road Traffic Infrastructures

Chapter 11
Model-based Control of Intelligent Traffic Networks

B. De Schutter, H. Hellendoorn, A. Hegyi, M. van den Berg, and S.K. Zegeye

Abstract Road traffic networks are increasingly being equipped and enhanced with various sensing, communication, and control units, resulting in an increased intelligence in the network and offering additional handles for control. In this chapter we discuss some advanced model-based control methods for intelligent traffic networks. In particular, we consider model predictive control (MPC) of integrated freeway and urban traffic networks. We present the basic principles of MPC for traffic control including prediction models, control objectives, and constraints. The proposed MPC control approach is modular, allowing the easy substitution of prediction models and the addition of extra control measures or the extension of the network. Moreover, it can be used to obtain a balanced trade-off between various objectives such as throughput, emissions, noise, fuel consumption, etc. Moreover, MPC also allows the integration and network-wide coordination of various traffic control measures such as traffic signals, speed limits, ramp metering, lane closures, etc. We illustrate the MPC approach for traffic control with two case studies. The first case study involves control of a freeway stretch with a balanced trade-off between total time spent, fuel consumption, and emissions as control objective. The second case study has a more complex layout and involves control of a mixed urban-freeway network with total time spent as control objective.

B. De Schutter
Delft University of Technology, Delft Center for Systems and Control & Marine and Transport Technology, Delft, The Netherlands, e-mail: b@deschutter.info

H. Hellendoorn, M. van den Berg, S.K. Zegeye
Delft University of Technology, Delft Center for Systems and Control, Delft, The Netherlands, e-mail: {j.hellendoorn,monique.vandenberg,s.k.zegeye}@tudelft.nl

A. Hegyi
Delft University of Technology, Department of Transport & Planning, Faculty of Civil Engineering and Geosciences, Delft, The Netherlands, e-mail: a.hegyi@tudelft.nl

R.R. Negenborn et al. (eds.), *Intelligent Infrastructures*, Intelligent Systems, Control and Automation: Science and Engineering 42, DOI 10.1007/978-90-481-3598-1_11,

11.1 Introduction

11.1.1 Positioning and relation of intelligent traffic networks with other networks

Each infrastructure has its own characteristics. Electricity networks are governed by the laws of Kirchhoff, which state that voltage and current distribute equally over the network. Water networks are governed by the law of gravity, which states that water is always flowing downwards, unless one makes use of pumping stations. Gas networks are determined by pressure, telecommunication networks by the behavior of the users. In all these cases the subjects that are transported are passive: electrons, water, gas molecules, and bits and bytes do not have an own will. Road networks form a special class, because the subjects, the drivers, are self-willed and do not always obey the proposed traffic measures. Nevertheless, the road network is too important to let it be uncontrolled. Economics and society depend heavily on efficient roads. For instance, in the European Union 44 % of all goods are moved by trucks over roads and 85 % of all persons are transported by cars, buses, or coaches on the roads.

There are several aspects of the road network that have to be mentioned. Firstly, the possible measures that can be taken and the limitations of these measures to control individual drivers. Secondly, the role of governments and other (non-government) parties in traffic control, in particular with regards to traffic jams and environmental issues. Thirdly, the road network as a critical infrastructure that is vital for the economy. And fourthly, the expectation of modern societies for flexible road networks without frequent construction works.

Until half a century ago streets and roads were passive infrastructures. The main function of the infrastructure was to facilitate comfortable and quick driving. But the number of vehicles has increased significantly over the last decades, which has led to traffic jams and dangerous situations on crossings. So traffic signals were introduced, first with a fixed-time scheme, and later on with computer programs connected to induction loops in the streets. Moreover, on highways variable speed limits and ramp metering were introduced, nowadays combined with Dynamic Route Information Panels (DRIPs). The behavior of drivers towards the road signs has changed: they now expect that traffic control measures can oversee the total traffic situation and, at the same time, have a good understanding of the needs of individual drivers, such that the drivers are not waiting seemingly meaningless for a traffic signal or a ramp metering installation, or have to drive slower than meaningful on a highway. Due to improved car mechanics, safer cars, and better roads, drivers have adapted their driving behavior, e.g., by keeping less distance to each other. In emergency cases this can lead to dangerous traffic situations and severe accidents.

Governments play a large role in defining targets for traffic management. Traffic safety has become a main political issue leading to new road constructions, separate lanes, and more roundabouts. Governments have clear targets to reduce the number of road casualties. They are also under severe pressure to reduce traffic jams, which leads, e.g., to the use of shoulder lanes in combination with camera surveillance during rush hours. Furthermore, legislation on environmental protection forces governments to take actions against pollution of CO_2, NO_x, HC, and particulate matters. Interest groups of truck drivers, automobile clubs, ecology movements, and neighboring residents and companies also play an important role in the public discussion about road constructions and traffic management.

Roads as well as electricity and drinking water networks have become critical infrastructures. They are of vital importance for the well-functioning of modern society. Traffic jams have large economical impact that may sum up to several percentages of the Gross Domestic Product (GDP), especially in (parts of) countries with a dense population where an accident on one of the main junctions can lead to a total collapse of the road network. In The Netherlands the length of traffic jams increased with 8.1 % in 2007, the number of traffic jams between 3.30 p.m. and 8.00 p.m. increased with 6.5 % in the same period, and between 10.00 a.m. and 3.30 p.m. the length of traffic jams increased by even 22.8 %. Worldwide, traffic incidents cost approximately 1–2 % of the GDP, i.e., approximately 65 billion dollar. Road congestion amounts to an average 1 % of GDP in the European Union, with Great Britain and France at 1.5 %. So reducing congestion contributes to a healthier economy.

Society demands more functionality, capacity, and quality of the road network. Construction works are necessary, but should influence the traffic flow as little as possible. Road networks should be flexible on the long term: it should be possible to quickly implement new political or societal desires. Extensions of the network or new driving concepts like adaptive cruise control or platooning should not lead to construction works that last many years.

There are several (partial) solutions to address all the issues discussed above, and one of them is the use of intelligent dynamic traffic management, which is the topic of this chapter.

11.1.2 The need for intelligent dynamic traffic management

As already indicated above, the need for mobility is increasing due to the growing number of road users as well as the increasing number of movements per user [36]. This leads to an increase in the frequency, length, and duration of traffic jams. These traffic jams cause large delays, resulting in higher travel costs and they also have a negative impact on the environment due to, e.g., noise and pollution.

To tackle these congestion problems there exist different solution approaches: constructing new roads, levying tolls, promoting public transport, or making more efficient use of the existing infrastructure. In this chapter we consider the last approach, implemented using dynamic traffic management and traffic control measures, such as on-ramp metering, dynamic speed limits, traffic signals, dynamic routing, provision of congestion information, etc., since this solution is effective on the short term, and inexpensive compared to constructing new roads. In addition, it is flexible enough to deal with the new infrastructures that will be constructed in the long term.

Current traffic control approaches usually focus on either urban traffic or free-way traffic. In urban areas traffic signals are the most frequently used control measures. Traditionally, they are controlled locally using fixed-time settings, or they are vehicle-actuated, meaning that they react on the prevailing traffic situation. Nowadays sophisticated, dynamic systems that aim at coordinating different available control measures in order to improve the total performance, are also making progress: systems such as SCOOT [46], SCATS [58], Toptrac [3], TUC [13], Mitrop [17], Motion [8], and UTOPIA/SPOT [43] use a coordinated control method to improve the urban traffic circulation, e.g., by constructing green waves. Control on freeways is done using different traffic control measures. Ramp metering is applied on on-ramps, using methods like ALINEA [40]. Overviews of ramp metering methods and results are given in [41, 50]. The use of variable speed limits on freeways is described in [2, 21, 32, 49], and the use of route guidance in [11, 13, 26]. Several authors have described methods for coordinated control of freeways using different traffic control measures [6, 19, 28, 29].

Usually these traffic control measures operate based on local data (occupancy, intensity, or speed measurements). However, considering the effect of the measures on the network level has many advantages compared to local control. E.g., solving a local traffic jam only may have as a consequence that the vehicles run faster into another (downstream) traffic jam, whereas still the same amount of vehicles have to pass the bottleneck (with a given capacity), and so the average travel time at the network level will still be the same. Another reason for considering the effects of control at the network level, is that in a dense network a local control measure can have effects on more distant parts of the network: an improved flow may cause congestion somewhere else in the network or a reduced flow may prevent congestion somewhere else in the network. Another source of degraded network performance is that congestion may block traffic flows on routes that do not pass the bottleneck (or incident location), such as a freeway with a congested off-ramp where the vehicles that want to leave the freeway block the mainstream traffic. Similar arguments also hold for urban and mixed urban-freeway traffic networks.

The traffic flows on freeways are often influenced by traffic flows on urban roads, and vice versa. Freeway control measures like ramp metering or speed limits allow a better flow and a larger throughput, but could lead to longer queues on on-ramps. These queues may spill back and block urban roads. On the other hand, urban traffic management policies often try to get vehicles on the freeway network as soon as possible, displacing the congestion toward neighboring freeways. The problems between the two road types are often increased by the fact that in several countries urban roads and freeways are managed by different traffic authorities, each with their own policies and objectives.

Hence, there is a clear need for coordinated and network-wide traffic management and control. Therefore, we present a coordinated control approach for mixed urban-freeway networks that provides an appropriate trade-off between the performance of the urban and freeway traffic operations, and that results in a significant improvement of the performance of the overall network.

As control method we use a model predictive control (MPC) approach [9, 34], adapted for traffic control. MPC is an online model-based predictive control approach in which a prediction model and (online) optimization are used to determine the control actions that optimize a given performance criterion over a given time horizon subject to given constraints. Using a receding horizon approach, only the first step of the computed control signal is applied, and next the optimization is started again with the prediction horizon shifted one time step further. MPC has already been applied to coordinated control of freeway networks in [6, 19, 29]. In this chapter, which collects and extends several of our previous results reported in [19–21, 52–54, 59, 60], we demonstrate how MPC can be used to obtain coordinated and integrated control of traffic networks containing both freeways and urban roads as well as a wide variety of traffic control measures, such that a balanced trade-off is obtained between various performance criteria and such that (hard) constraints are taken into account. Other publications that deal with MPC or MPC-like approaches for traffic control are [12, 16, 29, 43].

11.1.3 Overview of the chapter

This chapter is organized as follows. We first present an integrated traffic flow model for networks that contain both urban roads and freeways, as well as an emission and fuel consumption model in Section 11.2. Next, we describe MPC-based control for intelligent traffic networks in Section 11.3. The proposed approach is then illustrated in Section 11.4 via two simple case studies: one involving the balanced optimization of total time spent, total fuel consumption, and total emissions, and one involving a mixed urban-freeway network with total time spent as cost criterion. Section 11.5 concludes the chapter.

11.2 Traffic models

Traffic flow models can be distinguished according to the level of detail they use to describe the traffic. An overview of existing traffic models is given in [25].

In this chapter we use macroscopic traffic models. Macroscopic models describe traffic flows using aggregated variables such as flows and densities. They are suited very well for online control since these models give a balanced trade-off between accurate predictions and computational efforts. Indeed, the computation time for a macroscopic model does not depend on the number of vehicles in the network, which makes the model well-suited for online control, where the prediction should be performed online in an optimization setting, requiring that the model should run several times faster than real-time. Examples of macroscopic models are the LWR model [33, 45], the models of Helbing [22] and Hoogendoorn [24], and METANET [35].

We use an extended version of the METANET traffic flow model [35] to describe the freeway traffic, and a modified and extended model based on a queue length model developed by Kashani and Saridis [27] for the urban traffic. We also discuss how the freeway and the urban model have to be coupled, and we explain how the resulting traffic flow model can be linked with a model that describes emissions and fuel consumption. This leads to a macroscopic traffic model for mixed networks with urban roads and freeways, especially suited for an MPC-based traffic control approach.

Note that we will explicitly make a difference between the simulation time step T_f for the freeway part of the network, the simulation time step T_u for the urban part of the network, and the controller sample time T_c. We will also use three different counters: k_f for the freeway part, k_u for the urban part, and k_c for the controller. For simplicity, we assume that T_u is an integer divisor of T_f, and that T_f is an integer divisor of T_c:

$$T_f = M_{fu}T_u, \quad T_c = M_{cf}T_f = M_{cf}M_{fu}T_u, \tag{11.1}$$

with M_{fu} and M_{cf} integers. The value for T_f must be selected in such a way that no vehicle can cross a freeway segment in one time step, which results in a typical value of 10 s for freeway segments of length 0.5 km. The value of T_u is in general selected small enough to obtain an accurate description of the traffic, typically between 1 and 5 s, depending on the length of the roads. The control time step T_c should be large enough to allow the traffic controller to determine the new control signal, which depends on the required computation time, and short enough to deal with changing traffic conditions. Typical values for T_c are 1–5 min.

11.2.1 Freeway traffic flow model

For the prediction of the traffic flows on the freeway part of the network we use the destination-independent METANET model from [29, 30, 39]. We will briefly present the basic METANET model here. For a full description we refer the interested reader to [29, 30, 39].

The METANET model represents a network as a directed graph with the links (indicated by the index m) corresponding to freeway stretches. Each freeway link has uniform characteristics, i.e., no on-ramps or off-ramps and no major changes in geometry. Where major changes occur in the characteristics of the link or in the road geometry (e.g., at an on-ramp or an off-ramp), a node is placed. Each link m is divided into N_m segments (indicated by the index i) of length L_m (see Figure 11.1). Each segment i of link m is characterized by the *traffic density* $\rho_{m,i}(k_f)$ (veh/km/lane), the *mean speed* $v_{m,i}(k_f)$ (km/h), and the *traffic volume* or *outflow* $q_{m,i}(k_f)$ (veh/h), where k_f indicates the time instant $t = k_f T_f$, and T_f is the time step used for the simulation of the freeway traffic flow (typically $T_f = 10$ s).

The following equations describe the evolution of the network over time. The outflow of each segment is equal to the density multiplied by the mean speed and the number of lanes on that segment (denoted by λ_m):

$$q_{m,i}(k_f) = \rho_{m,i}(k_f)v_{m,i}(k_f)\lambda_m. \tag{11.2}$$

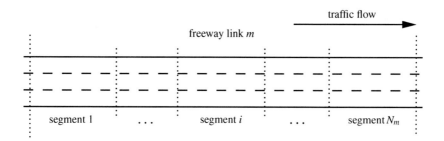

Figure 11.1: In the METANET model a freeway link is divided into segments.

The density of a segment equals the previous density plus the inflow from the up-stream segment, minus the outflow of the segment itself (conservation of vehicles):

$$\rho_{m,i}(k_f+1) = \rho_{m,i}(k_f) + \frac{T_f}{L_m \lambda_m}\left(q_{m,i-1}(k_f) - q_{m,i}(k_f)\right). \tag{11.3}$$

While (11.2) and (11.3) are based on physical principles and are exact, the equations that describe the speed dynamics and the relation between density and the desired speed are heuristic. The mean speed at step k_f+1 equals the mean speed at step k_f plus a relaxation term that expresses that the drivers try to achieve a desired speed V, a convection term that expresses the speed increase (or decrease) caused by the inflow of vehicles, and an anticipation term that expresses the speed decrease (increase) as drivers experience a density increase (decrease) downstream:

$$v_{m,i}(k_f+1) = v_{m,i}(k_f) + \frac{T_f}{\tau}\left(V\left(\rho_{m,i}(k_f)\right) - v_{m,i}(k_f)\right) +$$
$$\frac{T_f}{L_m}v_{m,i}(k_f)\left(v_{m,i-1}(k_f) - v_{m,i}(k_f)\right) - \frac{\vartheta T_f}{\tau L_m}\frac{\rho_{m,i+1}(k_f) - \rho_{m,i}(k_f)}{\rho_{m,i}(k_f) + \kappa}, \tag{11.4}$$

where τ, ϑ, and κ are model parameters. The expression for the desired speed V is given by

$$V\left(\rho_{m,i}(k_f)\right) = v_{\text{free},m}\exp\left[-\frac{1}{a_m}\left(\frac{\rho_{m,i}(k_f)}{\rho_{\text{crit},m}}\right)^{a_m}\right], \tag{11.5}$$

with a_m a model parameter, and where the free-flow speed $v_{\text{free},m}$ is the average speed that drivers assume if traffic is freely flowing, and the critical density $\rho_{\text{crit},m}$ is the density at which the traffic flow is maximal.

When a speed limit is active in the segment, (11.5) becomes

$$V\left(\rho_{m,i}(k_f)\right)$$
$$= \min\left(v_{\text{free},m}\exp\left[-\frac{1}{a_m}\left(\frac{\rho_{m,i}(k)}{\rho_{\text{crit},m}}\right)^{a_m}\right], (1+\alpha)v_{\text{control},m,i}(k_f)\right), \tag{11.6}$$

where $v_{control,m,i}(k_f)$ is the speed limit imposed on segment i of link m at time step k_f, and $1+\alpha$ is the non-compliance factor that expresses that drivers usually do not fully comply with the displayed speed limit and that their target speed is usually higher than what is displayed[1].

Origins are modeled with a simple queue model. The length of the queue equals the previous queue length plus the demand $d_o(k_f)$, minus the outflow $q_o(k_f)$:

$$w_o(k_f+1) = w_o(k_f) + T_f\big(d_o(k_f) - q_o(k_f)\big). \tag{11.7}$$

The outflow of the origin depends on the traffic conditions on the mainstream and, for the metered on-ramp, on the ramp metering rate[2] $r_o(k_f)$, where $r_o(k_f) \in [0,1]$. More specifically, $q_o(k_f)$ is the minimum of three quantities: the available traffic in time period k_f (queue plus demand), the maximal flow that could enter the freeway because of the mainstream conditions, and the maximal flow allowed by the metering rate:

$$q_o(k_f) = \min\left[d_o(k_f) + \frac{w_o(k_f)}{T_f}, \; Q_o r_o(k_f), \; Q_o\left(\frac{\rho_{max,m} - \rho_{m,1}(k_f)}{\rho_{max,m} - \rho_{crit,m}}\right)\right], \tag{11.8}$$

where Q_o is the on-ramp capacity (veh/h) under free-flow conditions and $\rho_{max,m}$ (veh/km/lane) is the maximum density of link m, and m is the index of the link to which the on-ramp is connected.

The above equations can be extended with terms that account for the speed drop caused by merging phenomena if there is an on-ramp, or for the speed reduction due to weaving phenomena when there is a lane drop, see [29].

The coupling equations to connect links are as follows. Every time there is a major change in the link parameters, like a junction or a bifurcation, a node is placed between the links. This node provides the incoming links with a virtual downstream density (required for the speed update equation (11.4)), and the leaving links with an inflow and a virtual upstream speed (required for the density update equation (11.3) and the speed update equation (11.4)). The flow that enters node n is distributed among the leaving links according to

$$Q_n(k_f) = \sum_{\mu \in I_n} q_{\mu,N_\mu}(k_f) \tag{11.9}$$

$$q_{m,0}(k_f) = \beta_{m,n}(k_f)Q_n(k_f), \tag{11.10}$$

where $Q_n(k_f)$ is the total flow that enters the node at step k_f, I_n is the set of links that enter node n, $\beta_{m,n}(k_f)$ expresses the turning rates (i.e., the fraction of the total flow through node n that leaves via link m), and $q_{m,0}(k_f)$ is the flow that leaves node n via link m.

---

[1] Data from the Dutch freeways show that when the speed limits are not enforced the average speed is approximately 10 % higher than what is displayed ($\alpha = 0.1$), and when they are enforced the average speed is approximately 10 % lower than what is displayed ($\alpha = -0.1$).

[2] For an unmetered on-ramp we can also use (11.8) by setting $r_o(k_f) \equiv 1$.

When node n has more than one leaving link, the virtual downstream density $\rho_{m,N_m+1}(k_f)$ of entering link m is given by

$$\rho_{m,N_m+1}(k_f) = \frac{\sum_{\mu \in O_n} \rho_{\mu,1}^2(k_f)}{\sum_{\mu \in O_n} \rho_{\mu,1}(k_f)}, \qquad (11.11)$$

where O_n is the set of links leaving node n.

When node n has more than one entering link, the virtual upstream speed $v_{m,0}(k_f)$ of leaving link m is given by

$$v_{m,0}(k_f) = \frac{\sum_{\mu \in I_n} v_{\mu,N_\mu}(k_f) q_{\mu,N_\mu}(k_f)}{\sum_{\mu \in I_n} q_{\mu,N_\mu}(k_f)}. \qquad (11.12)$$

Extensions to the basic METANET model are presented in [19–21].

11.2.2 Urban traffic flow model

Now we present a macroscopic model that describes the evolution of the traffic flows in the urban part of the network. This model is based on the Kashani model [27], but with the following extensions:

1. Horizontal, turning-direction-dependent queues;
2. Blocking effects, represented by maximal queue lengths and a flow constraint on flows that want to enter the blocked link, so no vehicle will be able to cross a blocked intersection;
3. A shorter time step[3], to get a more accurate description of the traffic flows.

The main variables used in the urban model are shown in Figures 11.2(a) and 11.2(b). The most important variables are the queue length x expressed in number of vehicles, the number of arriving vehicles m_{arr}, and the number of departing vehicles m_{dep}. Using these variables, the model is formulated as follows.

The number of vehicles that intend to leave the link $l_{o_i,s}$, connecting origin o_i and intersection s, toward destination d_j at time $t = k_u T_u$ is given by:

$$m_{dep,int,o_i,s,d_j}(k_u) =$$

$$\begin{cases} 0 & \text{if } g_{o_i,s,d_j}(k_u) = 0, \\ \min \left(x_{o_i,s,d_j}(k_u) + m_{arr,o_i,s,d_j}(k_u), \right. \\ \qquad \left. S_{s,d_j}(k_u), T_u Q_{cap,o_i,s,d_j} \right) & \text{if } g_{o_i,s,d_j}(k_u) = 1, \end{cases} \qquad (11.13)$$

where T_u is the urban step with k_u as counter, $x_{o_i,s,d_j}(k_u)$ is the queue length consisting of vehicles coming from origin o_i and going to destination d_j at intersection s,

---

[3] The original Kashani model of [27] uses the cycle time as time step, which restricts the model to effects that take longer than the cycle time. For MPC-based traffic control the other effects can also be relevant, and one might also want to control the cycle times.

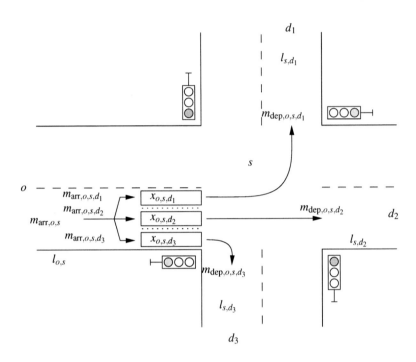

(a) Variables for an urban intersection.

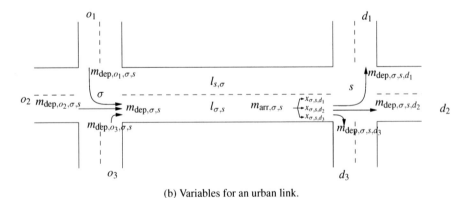

(b) Variables for an urban link.

Figure 11.2: Overview of the urban network variables.

$m_{\mathrm{arr},o_i,s,d_j}(k_\mathrm{u})$ is the number of vehicles arriving at the end of this queue, $S_{s,d_j}(k_\mathrm{u})$ is the free space in the downstream link expressed in number of cars, $Q_{\mathrm{cap},o_i,s,d_j}$ is the saturation flow[4], and $g_{o_i,s,d_j}(k_\mathrm{u})$ is a binary signal that is 1 when the specified traffic direction has green, and zero otherwise. This means that $g_{o_i,s,d_j} = 0$ corresponds to a red traffic signal, and $g_{o_i,s,d_j} = 1$ to a green one.[5]

The free space $S_{\sigma,s}$ in a link $l_{\sigma,s}$ expresses the maximum number of vehicles that can enter the link. It can never be larger than the length $L_{\sigma,s}$ of the link expressed in number vehicles, and is computed as follows:

$$S_{\sigma,s}(k_\mathrm{u}+1) = S_{\sigma,s}(k_\mathrm{u}) - m_{\mathrm{dep},\sigma,s}(k_\mathrm{u}) + \sum_{d_j \in D_s} m_{\mathrm{dep},\sigma,s,d_j}(k_\mathrm{u}), \qquad (11.14)$$

where $m_{\mathrm{dep},\sigma,s}(k_\mathrm{u})$ is the number of vehicles departing from intersection σ towards link $l_{\sigma,s}$, and D_s is the set of destinations connected to intersection s.

The number of vehicles departing from intersection s towards link l_{s,d_j} can be computed as

$$m_{\mathrm{dep},s,d_j}(k_\mathrm{u}) = \sum_{o_i \in O_s} m_{\mathrm{dep},o_i,s,d_j}(k_\mathrm{u}). \qquad (11.15)$$

These vehicles drive from the beginning of the link l_{s,d_j} toward the tail of the queue waiting on the link. This gives a time delay $\delta_{s,d_j}(k_\mathrm{u})$ which is approximated as:

$$\delta_{s,d_j}(k_\mathrm{u}) = \mathrm{ceil}\left(\frac{S_{s,d_j}(k_\mathrm{u}) L_{\mathrm{av,veh}}}{v_{\mathrm{av},s,d_j}} \right), \qquad (11.16)$$

where $L_{\mathrm{av,veh}}$ is the average length of a vehicle, and v_{av,s,d_j} the average speed on link l_{s,d_j}.

The time instant at which the vehicle enters the link and the vehicle's delay on the link result in the time instant at which the vehicle will arrive at the end of the queue. It can happen that vehicles that have entered the link at different instants reach the end of the queue during the same time step. To take this into account the variable $m_{\mathrm{arr},s,d_j}(k_\mathrm{u})$ that describes the vehicles arriving at the end of the queue is updated accumulatively every time step. This results in:

$$m_{\mathrm{arr},s,d_j}(k_\mathrm{u}+\delta_{s,d_j}(k_\mathrm{u}))_{\mathrm{new}} = m_{\mathrm{arr},s,d_j}(k_\mathrm{u}+\delta_{s,d_j}(k_\mathrm{u}))_{\mathrm{old}} + m_{\mathrm{dep},s,d_j}(k_\mathrm{u}), \qquad (11.17)$$

where $m_{\mathrm{arr},s,d_j}(k_\mathrm{u}+\delta_{s,d_j}(k_\mathrm{u}))$ is the number of vehicles arriving at the end of the queue at time $k_\mathrm{u}+\delta_{s,d_j}(k_\mathrm{u})$, and $m_{\mathrm{dep},s,d_j}(k_\mathrm{u})$ is the number of vehicles entering link l_{s,d_j}.

---

[4] The saturation flow is the maximum flow that can cross the intersection under free-flow conditions.

[5] The computed green time is the effective green time. The exact signal timing including the amber time can easily be derived from this effective green time.

The traffic flow reaching the tail of the queue in link l_{s,d_j} divides itself over the subqueues according to the turning rates $\beta_{o_i,s,d_j}(k_u)$:

$$m_{\mathrm{arr},o_i,s,d_j}(k_u) = \beta_{o_i,s,d_j}(k_u) m_{\mathrm{arr},o_i,s}(k_u). \qquad (11.18)$$

The subqueues are then updated as follows:

$$x_{o_i,s,d_j}(k_u+1) = x_{o_i,s,d_j}(k_u) + m_{\mathrm{arr},o_i,s,d_j}(k_u) - m_{\mathrm{dep},o_i,s,d_j}(k_u). \qquad (11.19)$$

The total flow entering a destination link consists of several flows from different origins. The available space in the destination link should be divided over the entering flows, since the total number of vehicles entering the link may not exceed the available space. We divide this available space equally over the different entering flows. When one flow does not fill its part of the space, the remainder is proportionally divided over the rest of the flows. For a detailed description of this process we refer to [54].

11.2.3 Interface between the freeway and the urban traffic flow models

The urban part and the freeway part are coupled via on-ramps and off-ramps. In this section we present the formulas that describe the evolution of the traffic flows on these on-ramps and off-ramps. The main problems are the different simulation time steps T_f and T_u and the boundary conditions that the models create for each other. We assume that the time steps are selected such that $T_f v_{\mathrm{free},m} < L_m$.

11.2.3.1 On-ramps

Consider an on-ramp r that connects intersection s of the urban network to node p of the freeway network, as shown in Figure 11.3(a). The number of vehicles that enter the on-ramp from the urban network is given by $m_{\mathrm{arr},s,r}(k_u)$. These vehicles have a delay $\delta_{s,r}(k_u)$ similar to (11.16). The evolution of the queue length is first described with the urban model. At the end of each freeway time step, the queue length as described in the urban model is then translated to the queue length for the freeway model as explained below.

Consider the freeway time step k_f corresponding to the urban time step $k_u = M_{fu}k_f$ (recall that $T_f = M_{fu}T_u$). In order to get a consistent execution of the urban and freeway models the computations should be done in the following order:

1. Determine the on-ramp departure flow $q_{r,p}(k_f)$ during the period $[k_f T_f, (k_f+1)T_f]$ using (11.8).
2. Assume that these departures spread out evenly over the equivalent urban simulation period $[k_u T_u, (k_u + M_{fu})T_u]$. Compute the departures for each urban time step in this period using $m_{\mathrm{dep},s,r,p}(k) = \dfrac{q_{r,p}(k_f)T_f}{M_{fu}}$ for $k = k_u, \ldots, k_u + M_{fu} - 1$.

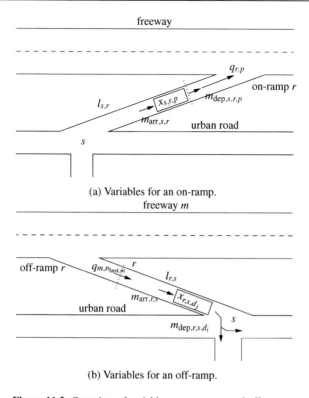

Figure 11.3: Overview of variables on on-ramps and off-ramps.

3. The number of arriving vehicles, the free space, and the queue length $x_{s,r,p}$ at link $l_{s,r}$ can now be computed using the equations for the urban traffic model given in Section 11.2.2.

4. When the queue length $x_{s,r,p}(k_u + M_{fu})$ is computed, we set $w_o(k_f + 1) = x_{s,r,p}(k_u + M_{fu})$. It is easy to verify that this is equivalent to (11.7).

11.2.3.2 Off-ramps

The evolution of the traffic flows on an off-ramp r is computed for the same time steps as for the on-ramp, starting at time step $k_u = M_{fu}k_f$. The variables are displayed in Figure 11.3(b). The following steps are required to simulate the evolution of the traffic flows, in order to get a consistent execution of the urban and freeway models:

1. Determine the number of departing vehicles from link $l_{r,s}$ at intersection s during the period $[k_u T_u, (k_u + M_{fu})T_u]$ using the urban traffic flow model.

2. Compute the maximal allowed flow $q_{r,1}^{max}(k_f)$ that can leave the freeway and enter the off-ramp in the period $[k_f T_f, (k_f + 1)T_f]$ based on the available storage space in the link $l_{r,s}$ at the end of the period:

$$q_{r,1}^{max}(k_f) = \frac{1}{T_f} S_{r,s}(k_u) + \sum_{k=k_u}^{k_u+M_{fu}-1} \sum_{d_j \in D_s} m_{dep,r,s,d_j}(k). \qquad (11.20)$$

The effective outflow $q_{m,n_{last,m}}(k_f)$ of freeway link m between node p and off-ramp r is then given by

$$q_{m,n_{last,m}}(k_f) = \min\left(q_{m,n_{last,m}}^{normal}(k_f), q_{r,1}^{max}(k_f) \right), \qquad (11.21)$$

where $q_{m,n_{last,m}}^{normal}(k_f)$ is the flow that would have entered the freeway if the off-ramp would not have been blocked.

3. Now the METANET model can be updated for simulation step k_f+1.
4. We assume that the outflow of the off-ramp is distributed evenly over the period $[k_f T_f, (k_f+1)T_f]$ such that

$$m_{arr,r,s}(k+\delta_{r,s}) = \frac{q_{m,n_{last,m}}(k_f)T_f}{M_{fu}} \quad \text{for } k = k_u,\dots,k_u+M_{fu}-1. \qquad (11.22)$$

The corresponding urban queue lengths $x_{r,s,d_j}(k)$ for $k = k_u+1,\dots,k_u+M_{fu}$ can be updated using the urban traffic flow model.

11.2.4 Emission and fuel consumption model

In this section we present the dynamic emission and fuel consumption model VT-micro and we show how it can be integrated with the macroscopic METANET traffic flow model of Section 11.2.1 (see also [59, 60]). A similar approach can also be used to integrate VT-micro with the urban model of Section 11.2.2.

11.2.4.1 Emission and fuel consumption models

Traffic emission and fuel consumption models calculate the emissions produced and fuel consumed by vehicles based on the operating conditions of the vehicles. The main inputs to the models are the operating conditions of the vehicle (such as speed, acceleration, engine load) [23]. These models can be either *average-speed-based* or *dynamic*. Average-speed-based emission and fuel consumption models estimate or predict traffic emission and fuel consumption based on the trip-based average speed of traffic flow [37]. These models can also be used with second-by-second speeds to take some of the variation of the speeds into account [7]. On the contrary, dynamic (or also called microscopic) emission and fuel consumption models use the second-by-second speed *and* acceleration of individual vehicles to estimate or predict the emissions and the fuel consumption. Such models provide better accuracy than average-speed-based models. Therefore, we will consider a dynamic emission and fuel consumption model and integrate it with the macroscopic traffic flow model of Section 11.2.1.

VT-micro [1] is a microscopic dynamic emission and fuel consumption model that yields emissions and fuel consumption of one individual vehicle using second-by-second speed and acceleration. The model has the form

$$E_x(k_f) = \exp\left(\tilde{\mathbf{v}}^{\mathrm{T}}(k_f)\mathbf{P}_x\tilde{\mathbf{a}}(k_f)\right), \qquad (11.23)$$

where E_x is the estimate or prediction of the emission or fuel consumption variable $x \in \{CO, NO_x, HC, fuel\}$, with $\tilde{v}(k_f) = [1 \ \ v(k_f) \ \ v(k_f)^2 \ \ v(k_f)^3]^T$ where $v(k_f)$ is the speed of the vehicle at freeway time step k_f, with $\tilde{a}(k_f) = [1 \ \ a(k_f) \ \ a(k_f)^2 \ \ a(k_f)^3]^T$ where $a(k_f)$ is the acceleration of the vehicle at freeway time step k_f, and with $\mathbf{P}_x$ the model parameter matrix for the variable x. The values of the entries of P_x for $x \in \{CO, NO_x, HC, fuel\}$ can be found in [1].

The VT-micro emission model does not yield estimates of CO_2 emission. But since there is almost an affine relationship between the fuel consumption and the CO_2 emission [38], we can compute the CO_2 emission as

$$E_{CO_2}(k_f) = \delta_1 + \delta_2 E_{fuel}(k_f), \qquad (11.24)$$

where δ_1 and δ_2 are model parameters, the values of which can be found in [38].

Figure 11.4 depicts the CO_2 and fuel consumption versus the vehicle speed for three acceleration values using the equations presented above.

11.2.4.2 Integrating METANET with VT-micro

The VT-micro model is a microscopic traffic emission and fuel consumption model while METANET is a macroscopic traffic flow model. Thus, these two different models are required to be integrated in such a way that VT-micro can get speed and acceleration inputs of the traffic flow from the METANET model at every simulation time step. The speed of the traffic flow can be easily obtained from (11.4). However, the computation of the acceleration is not as straightforward. In the sequel we show how to obtain the acceleration from the METANET model.

Since the METANET model is discrete in both space and time, there are two acceleration components involved in the model. The first is the "temporal" acceleration of the vehicle flow within a given segment. The second component is the "spatial" acceleration of the vehicles flowing from one segment into another in one simulation time step (see Figure 11.5).

Temporal acceleration

The temporal acceleration of vehicles in a segment i of link m at time step k_f is given by:

$$a_{m,i}(k_f) = \frac{v_{m,i}(k_f+1) - v_{m,i}(k_f)}{T_f}. \qquad (11.25)$$

This equation is the same for all segments of all links.

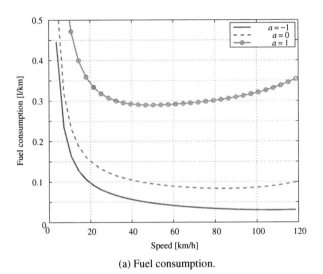

(a) Fuel consumption.

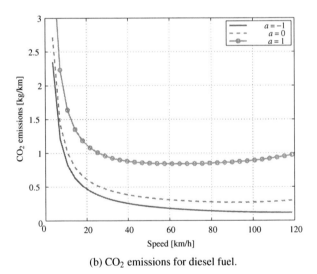

(b) CO_2 emissions for diesel fuel.

Figure 11.4: Fuel consumption and CO_2 emission curves of vehicles as a function of the speed for accelerations $a \in \{-1, 0, 1\}$ m/s.

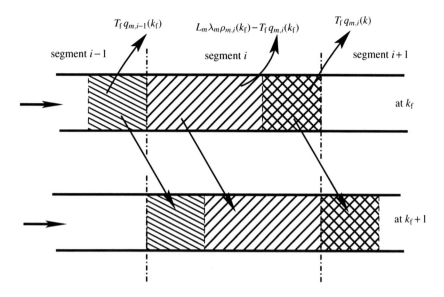

Figure 11.5: Illustration of traffic flow in METANET.

Let us now determine how many vehicles are subject to this temporal acceleration from time step k_f to time step $k_f + 1$. At time step k_f the number of vehicles in segment i of link m is equal to $L_m \lambda_m \rho_{m,i}(k_f)$ and from time step k_f to $k_f + 1$ (i.e., in the time period $[k_f T_f, (k_f + 1)T_f]$) the number of vehicles leaving segment i is $T_f q_{m,i}(k_f)$ (see Figure 11.5). Therefore, the number of vehicles accelerating at the temporal acceleration provided in (11.25) is

$$n_{m,i}(k_f) = L_m \lambda_m \rho_{m,i}(k_f) - T_f q_{m,i}(k_f). \tag{11.26}$$

This equation is also the same for all segments of all links.

Spatial acceleration

The spatial acceleration depends on the geometry of the traffic network. It is different for a link, an on-ramp, an off-ramp, merging links, and splitting links. Here we present the spatial accelerations for a link and an on-ramp. Similar equations can be derived for off-ramps and for splitting and merging links (see [59]).

Link

In the simulation time step from k_f to $k_f + 1$, the spatial acceleration of the vehicles leaving segment $i - 1$ of link m and going to segment i of link m, and the corresponding number of the vehicles are respectively described by (see also Figure 11.5):

$$a_{m,\Delta i}(k_f) = \frac{v_{m,i}(k_f + 1) - v_{m,i-1}(k_f)}{T_f} \tag{11.27}$$

$$n_{m,\Delta i}(k_f) = T_f q_{m,i-1}(k_f).$$ (11.28)

On-ramp

Consider an on-ramp connected to segment i of link m. We assume the initial speed on the on-ramp to be v_{on}. If the inflow of the on-ramp is given by $q_{on,m,i}(k_f)$, the spatial acceleration of the vehicles as their speed is changing from v_{on} to $v_{m,i}(k_f+1)$ and the corresponding number of vehicles are respectively

$$a_{on,m,\Delta i}(k_f) = \frac{v_{m,i}(k_f+1) - v_{on}}{T_f}$$ (11.29)

$$n_{on,m,\Delta i}(k_f) = T_f q_{on,m,i}(k_f).$$ (11.30)

11.2.4.3 VT-macro

Recall that as input for the VT-micro model we have to specify speed-acceleration pairs. In all the derivations above both the temporal and spatial accelerations have the form $a_y = \frac{v_2 - v_1}{T_f}$. The corresponding speed input is then taken to be v_1. Furthermore, since the speed-acceleration pair holds for a number of vehicles, the emissions and fuel consumption obtained for the given pair have to be multiplied by the corresponding number of vehicles in order to obtain the total emissions and fuel consumption (as the VT-micro model describes the emissions or fuel consumption of one individual vehicle).

This results in a new traffic emission and fuel consumption model

$$E_{x,m,i}^{total}(k_f) = n^{temp}(k_f)E_{x,m,i}^{temp}(k_f) + n^{spat}(k_f)E_{x,m,\Delta i}^{spat}(k_f),$$ (11.31)

where the superscripts "temp" and "spat" in n and E refer to the temporal or spatial variables, n denotes the number of vehicles, and $E_{x,m,i}$ and $E_{x,m,\Delta i}$ denote the emission or fuel consumption for segment i of link m with respectively the temporal and spatial speed and acceleration inputs. More specifically, $J_{x,m,i}$ and $J_{x,m,\Delta i}$ are computed as in (11.23) or (11.24).

We call this new model the VT-macro emission and fuel consumption model.

11.3 Model-based predictive traffic control

In the previous section we have developed a model that describes traffic networks that contain both urban roads and freeways. This model forms the basis for our model-based predictive control method.

11.3.1 Motivation

To find the optimal combination of traffic control measures (control inputs) we apply a model predictive control (MPC) framework [9, 15, 34]. MPC is an optimal control method applied in a rolling horizon framework. Optimal control has successfully been applied in [29–31] to coordinate or integrate traffic control measures. Both optimal control and MPC have the advantage that the controller generates control signals that are optimal according to a user-supplied objective function. However, MPC has some important advantages over traditional optimal control. First, optimal control has an *open-loop* structure, which means that the disturbances (in our case: the traffic demands) have to be completely and exactly known in advance, and the traffic model has to be very accurate to ensure sufficient precision for the whole period of operation. MPC operates in *closed-loop*, which means that the traffic state and the current demands are regularly fed back to the controller, and the controller can take disturbances (here: demand prediction errors) into account and correct for prediction errors resulting from model mismatch. Second, adaptivity is easily implemented in MPC, because the prediction model and/or its parameters can be updated during the operation of the controller. This may be necessary when traffic behavior changes (e.g., in case of incidents, changing weather conditions, lane closures for maintenance). Third, for MPC a shorter prediction horizon is usually sufficient, which reduces complexity, and makes the real-time application of MPC feasible.

11.3.2 Principle of operation

In MPC a discrete-time model is used to predict the future behavior of the traffic network. During a control sampling interval the control signals are taken to be constant. The goal of the controller is to find the control signals that result in an optimal behavior of the traffic flows. To express performance an objective function is defined and the control signals that optimize this function are found via (numerical) optimization.

The control is applied in a rolling-horizon scheme: at each control step k_c (corresponding to the time instant $t = k_c T_c$ with T_c the control time step (typically in the range of 1 to 5 min for traffic network control)), a new optimization is performed over the prediction horizon $[k_c T_c, (k_c + N_p) T_c]$, and only the first value of the resulting control signal (the control signal for time instant k_c) is applied to the process (see Figure 11.6). At the next control step $k_c + 1$ this procedure is repeated. To reduce complexity and improve stability often a control horizon N_c ($\leq N_p$) is introduced, and after the control horizon has been passed the control signal is taken to be constant.

So there are two loops in the MPC scheme: the rolling-horizon loop and the optimization loop inside the controller. The loop inside the controller of Figure 11.6 is executed as many times as required to find the optimal control signals at control time step k_c, for given N_p, N_c, traffic state, and expected demand. The loop connect-

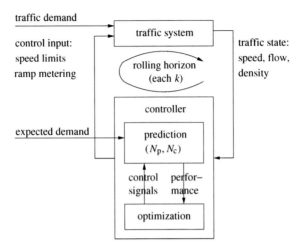

Figure 11.6: Schematic view of the model predictive control (MPC) structure.

ing the controller and the traffic system is performed once for each k_c and provides the state feedback to the controller. Recall that this feedback is necessary to correct for (the ever present) prediction errors, and disturbance rejection (compensation for unexpected traffic demand variations). Another advantage of this rolling-horizon approach is that it results in an online adaptive control scheme that allows us to take changes in the system or in the system parameters into account by regularly updating the model of the system.

When optimizing large networks, the computational complexity may become too high. In these cases the network should be separated into some subnetworks that are controlled by separate MPC controllers. The subnetworks should be chosen such that the interaction between them is as small as possible. To handle the remaining interactions (e.g., occasionally overflowing queues) adequately, structures such as hierarchical control or agent-based control can be used (see [47, 55, 56]).

Objective function

The MPC algorithm has to determine the control signals **c** (such as the ramp metering rates, dynamic speed limits, traffic signal settings, etc.) that minimize a given objective function over the period $[k_c T_c, (k_c + N_p)T_c]$. Possible performance criteria are the total time spent (TTS) in the freeway and the urban part of the network, emission levels, and fuel consumption. Let us now derive the expressions for these performance criteria.

The TTS in the freeway part of the network is given by

$$J_{\text{TTS,f}}(k_c) = T_f \sum_{j \in \mathcal{K}_f(k_c, N_p)} \left(\sum_{(m,i) \in M_{\text{ls,f}}} \rho_{m,i}(j) L_m \lambda_m + \sum_{o \in O_f} w_o(j) \right), \qquad (11.32)$$

where $\mathcal{K}_f(k_c, N_p)$ is the set of freeway time steps k_f that correspond to the considered prediction period $[k_c T_c, (k_c + N_p) T_c]$, $M_{ls,f}$ is the set of pairs (m, i) of link indices m and segment indices i of the freeway part of the network, and O_f is the set of all mainstream origins and on-ramps at the boundaries of the mixed network (note that on-ramps connecting the urban and the freeway part of the network are considered in the expression for the urban TTS given next).

To compute the TTS for the urban part of the network the number of vehicles in each urban link $n_{veh, l_{\sigma,s}}$ is required:

$$n_{veh, l_{\sigma,s}}(k_u) = L_{\sigma,s} - S_{\sigma,s}(k_u), \tag{11.33}$$

where $L_{\sigma,s}$ is the maximum number of vehicles that the link can contain. Using this equation the number of vehicles for all urban links, on-ramps, and off-ramps can be determined. The TTS in the urban part of the network is then given by

$$J_{TTS,u}(k_c) = T_u \sum_{l \in \mathcal{K}_u(k_c, N_p)} \left(\sum_{l_{o_j,s} \in I_u} n_{veh, l_{o_j,s}}(l) + \sum_{l_{s,r} \in R_{on} \cup R_{off}} n_{veh, l_{s,r}}(l) \right.$$
$$\left. + \sum_{o \in O_u} n_{veh,o}(l) \right), \tag{11.34}$$

where $\mathcal{K}_u(k_c, N_p)$ is the set of urban time steps k_u that correspond to $[k_c T_c, (k_c + N_p) T_c]$, I_u is the set of all urban links, O_u is the set of all urban origins o, R_{on} is the set of urban links $l_{r,s}$ connected to the on-ramps, and R_{off} is the set of urban links $l_{s,r}$ connected to the off-ramps.

The expression for the emissions and the fuel consumption over the prediction period $[k_c T_c, (k_c + N_p) T_c]$ is given by

$$J_x(k_c) = \sum_{j \in \mathcal{K}_f(k_c, N_p)} \sum_{(m,i) \in M_{ls,f}} E^{total}_{x,m,i}(j) \tag{11.35}$$

for $x \in \{CO, NO_x, HC, fuel\}$, where $E^{total}_{x,m,i}$ is defined by (11.31).

In order to get smoother control signals, one often also imposes a penalty on the temporal variation of the control signal $\mathbf{c}$:

$$J^{temp}_{var}(k_c) = \sum_{j=k_c}^{k_c+N_c-1} \|\mathbf{c}(j) - \mathbf{c}(j-1)\|^2. \tag{11.36}$$

In a similar way, one could also define a penalty on the spatial variation for speed limits $v_{control,m,i}(j)$ defined on control time steps j:

$$J^{spat}_{var}(k_c) = \sum_{j=k_c}^{k_c+N_c-1} \sum_{(m,i) \in V_c} (v_{control,m,i}(j) - v_{control,m,i-1}(j))^2, \tag{11.37}$$

where V_c is the set of pairs of pairs (m, i) of link indices m and segment indices i such that a speed limit is active on both segments i and $i - 1$.

The overall objective function used in MPC is then a weighted sum of the above partial objective functions:

$$J_{\text{MPC}}(k_c) = \sum_{i=1}^{n} \frac{\gamma_i}{J_{\text{nominal},i}} J_i(k_c), \tag{11.38}$$

where $J_{\text{nominal},i}$ is the nominal value of partial objective function J_i (for normalization purposes) and γ_i is a weighting factor.

Operational constraints

The constraints may contain upper and lower bounds on the control signal, but also linear or nonlinear equality and inequality constraints on the states of the system. The constraints are used, e.g., to keep the system working within safety limits, or to avoid unwanted situations.

Optimization algorithms

At each control step the MPC controller computes an optimal control sequence over a given prediction horizon. In general, this optimal control sequence is the solution of a nonlinear, non-convex optimization problem in which the objective function is minimized subject to the model equations and the constraints. To solve this optimization problem different numerical optimization techniques can be applied, such as multi-start sequential quadratic programming (SQP) (see, e.g., [42, Chapter 5]) or pattern search (see, e.g., [44]) for real-valued problems, and genetic algorithms [10], and tabu search [18] or simulated annealing [14] for mixed-integer problems that arise when discrete control measures (e.g., lane closures) are included.

Tuning of N_p and N_c

The tuning rules to select appropriate values for N_p and N_c that have been developed for conventional MPC cannot be applied straightforwardly to the traffic flow control framework presented above. However, based on a heuristic reasoning we can determine an initial guess for these parameters.

The prediction horizon N_p should be larger than the typical travel time from the controlled segments to the exit of the network, because if we take the prediction horizon N_p shorter than the typical travel time in the network, the effect of the vehicles that are influenced by the current control measure and — as a consequence — have an effect on the network performance before they exit the network, will not be taken into account. Furthermore, a control action may affect the network state (by improved flows, etc.) even when the actually affected vehicles have already exited the network. On the other hand, N_p should not be too large because of the computa-

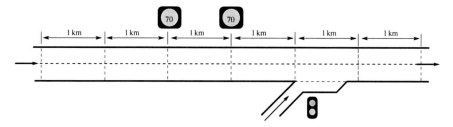

Figure 11.7: A 6 km freeway with metered on-ramp and two dynamic speed limit control.

tional complexity of the MPC optimization problem. So based on this reasoning we select N_p to be about the typical travel time in the network.

For the control horizon N_c we select a value that represents a trade-off between the computational effort and the performance.

11.4 Case studies

In order to illustrate the control framework presented above we will now apply it to two case studies: a simple one involving a freeway stretch and a more complex one involving a mixed urban-freeway network.

11.4.1 MPC for the reduction of emissions, fuel consumption and travel time on freeways

11.4.1.1 Network and scenario

Since we want to focus on the relevant points of the approach presented in this chapter, the benchmark network (see Figure 11.7) for this experiment was chosen as simple as possible. The network consists of one mainstream freeway link with two speed limits, and one metered on-ramp. The on-ramp is located at a distance of 4 km from the mainstream origin of the freeway link, and it has a capacity of 2000 veh/h. The mainstream freeway link has two lanes with a capacity of 2100 veh/h each. Segments 3 and 4 of the freeway are equipped with a variable message sign where speed limits can be displayed. The outflow at the end of the freeway is considered to be unrestricted. We assume that the queue length at the on-ramp may not exceed 100 vehicles, in order to prevent spill-back to a surface street intersection.

We use the network parameters as found in [30]: $T_f = 10$ s, $\tau = 18$ s, $\kappa = 40$ veh/km/lane, $\vartheta = 60$ km²/h, $\rho_{\max} = 180$ veh/km/lane, $a = 1.867$, and $\rho_{\text{crit}} = 33.5$ veh/km/lane. Furthermore, we assume that the desired speed is 10% higher than the displayed speed limit, that is $\alpha = 0.1$.

The controller sampling time T_c is chosen to be 1 min. Moreover, in order to include relevant dynamics of the system states, the prediction horizon and the control horizon are selected to be respectively $N_p = 15$ and $N_c = 7$.

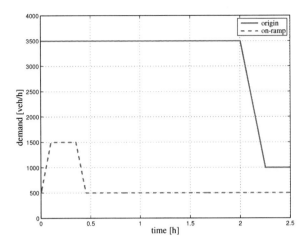

Figure 11.8: On-ramp demand (dashed line) and mainstream demand (solid-line) profiles used in the simulations for the first case study.

To examine the effect of the combination of variable speed limits and ramp metering typical demand profiles are considered for the mainstream origin and the on-ramp (see Figure 11.8). The mainstream demand has a constant, relatively high level and a drop after 2 h to a low value in a time span of 15 min. The demand on the on-ramp increases to near capacity, remains constant for 15 min, and decreases finally to a constant low value. For the given demand profiles one uncontrolled (Case 1) and four controlled (Cases 2 to 5) situations are compared. When no control is applied to the system (Case 1), the speed limit is set to a constant value of 102 km/h and the ramp rate is constant and equal to 1. We consider this case as a benchmark to compare the results of the simulations when an MPC controller is implemented. For the controlled cases the objective of the MPC controller is to optimize the following performance criteria (all deemed equally important within each case, i.e., $\gamma_i = 1$ for all i):

Case 2: total time spent,

Case 3: total fuel consumption and total time spent,

Case 4: total NO_x emissions and total time spent, and

Case 5: total fuel consumption, total NO_x emissions, and total time spent.

In this experiment the MPC optimization problem is a non-convex, nonlinear problem with real-valued optimization variables (the dynamic speed limits and the ramp metering rates). To solve this optimization problem we have selected multi-start SQP as optimization algorithm. The SQP algorithm is implemented in the `fmincon` function of the Matlab optimization toolbox [51].

Table 11.1: Simulation results for the freeway case study.

Control objectives	Simulation results		
	TTS (veh·h)	Total NO$_x$ emissions (kg)	Total fuel consumption (l)
Case 1: Uncontrolled	1459	8.719	6108
Case 2: TTS	1247 (-14.6%)	8.288 (-4.9%)	5274 (-13.7%)
Case 3: TTS + fuel	1257 (-13.9%)	8.147 (-6.6%)	4934 (-19.2%)
Case 4: TTS + NO$_x$	1412 (-3.2%)	7.654 (-12.2%)	5290 (-13.4%)
Case 5: TTS + fuel + NO$_x$	1336 (-8.5%)	7.786 (-10.7%)	5088 (-16.7%)

11.4.1.2 Results

In Table 11.1 we present the simulation results of the uncontrolled simulation (Case 1) and the controlled simulations (Case 2 to Case 5).

When an MPC controller is implemented (Case 2 to Case 5) the values of all the performance indicators are reduced by a certain amount compared to the uncontrolled situation. But, the reduction of the respective performance indicators is dependent on the objective of the controller. As can be seen in the table, when the objective of the controller is to reduce the TTS (Case 2), the TTS is reduced with 14.6 %. Moreover, the total NO$_x$ emissions and the total fuel consumption are reduced with 4.9 % and 13.7 % respectively. This indicates that under the given traffic demand and traffic scenario, reducing the TTS can also help in reducing the total NO$_x$ emissions and the total fuel consumption.

When the objective of the controller includes the total fuel consumption as well as the TTS (Case 3), the results for TTS differ slightly compared to Case 2, while the total fuel consumption shows a significant reduction. So for Case 3 more fuel is saved by making a small sacrifice in the TTS.

In Case 4 the objective of the controller is to reduce the total NO$_x$ emissions and the TTS where both criteria are weighted equally. In this case, the TTS does not show a significant improvement compared to the uncontrolled case. On the other hand, the total NO$_x$ emissions decrease significantly. Moreover, we can see that the total fuel consumption is reduced by 13.4 %, which is less than in Case 3. Thus, although the combination of a larger TTS and more fuel consumption indicates inefficient driving behavior in terms of the non-environmental performance indicators (travel time and energy consumption), there is a positive impact on the emissions.

Case 5 encompasses the concerns regarding the travel time, energy consumption, and the environment, as it addresses all three performance indicators by weighting them equally. The simulation results in Table 11.1 show that in this case the MPC controller achieves a balanced trade-off between total time spent, the total fuel consumption, and the total NO$_x$ emissions.

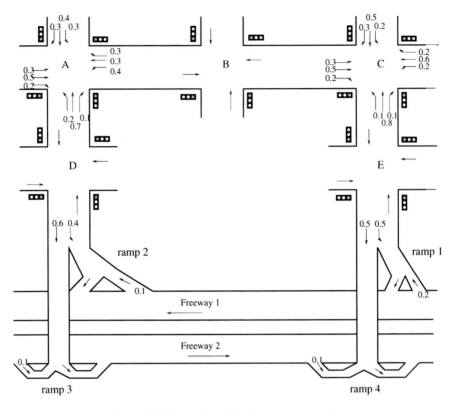

Figure 11.9: Network used in the second case study.

11.4.2 MPC for mixed urban-freeway networks

11.4.2.1 Set-up and scenarios

For this case study a simple network is used, as shown in Figure 11.9. The network consists of two two-lane freeways (freeway 1 and 2) each with two on-ramps and two off-ramps (ramp 1 to 4). Furthermore, there are two urban intersections (A and C), which are connected to the freeway and to each other. Between these intersections and the freeways there are some crossing roads (B, D, and E), where there is only crossing traffic that does not turn into other directions, e.g., pedestrian traffic or bicycles. We have selected this network because it contains the most essential elements from mixed networks. There are freeways with on-ramps and off-ramps and controlled intersections not far away from the freeways, resulting in a strong relation between the traffic on the two types of road. The network is small enough to use intuition to analyze and interpret the results, but large enough to make the relevant effects visible.

We will consider four different traffic scenarios all of which are created starting from a "basic" scenario. This basic scenario has a demand of 3600 veh/h for freeway origins and 1000 veh/h for urban origins, and turning rates as shown in Figure 11.9.

Each of the scenarios is a variation on this basic scenario, with one variable or parameter changed, or with a constraint added. The total simulated time is 30 min. These are the four scenarios:

Scenario 1: Congestion on the freeway: A traffic jam exists at the downstream end of freeway 1. This congestion grows into the upstream direction and blocks the on-ramps, causing a spill-back leading to urban queues. The congestion is created by imposing a downstream density of 65 veh/km/lane for the last segment of the freeway.

Scenario 2: Blockage of an urban intersection: On intersection D an incident has occurred, and the whole intersection is blocked. The queues spill back into neighboring intersections, and also block the off-ramps of the freeways. This incident is simulated by setting the saturation flow of all links leaving the intersection to 0 veh/h.

Scenario 3: Rush hour: In this scenario the demand at the origins becomes larger during a short period. We have selected a flow of 500 veh/h with a peak of 2000 veh/h for the urban origins, and a flow of 2000 veh/h with a peak of 4000 veh/h for freeway origins. The duration of the peak is 10 min.

Scenario 4: Maximum queue length: Here, the queue on the link from intersection A toward intersection B may not become longer than 20 vehicles. This can be a traffic management policy, e.g., when the link is in a residential area.

For all control systems the implementation of the simulations and the controller is completely done in Matlab. We use the traffic flow model described in Section 11.2 both as the real-world model and as the prediction model. With this set-up we can give a proof of concept of the developed control method, without introducing unnecessary side effects.

In our case study the MPC optimization problem is a non-convex, nonlinear problem with real-valued optimization variables. To solve this optimization problem we have applied multi-start SQP, using the SQP algorithm implemented in the fmincon function of the Matlab optimization toolbox [51].

As cost function we select the total time spent (TTS) in the urban and the freeway part of the network. The model parameters are selected as follows. The parameters of the METANET model are selected according to [30]: $v_{\mathrm{free},m} = 106$ km/h, $\rho_{\mathrm{crit},m} = 33.5$ veh/km/lane, $\rho_{\mathrm{max},m} = 180$ veh/km/lane, $Q_{\mathrm{cap},m} = 4000$ veh/h, $\tau = 18$ s, $\kappa = 40$ veh/km/lane, $\vartheta = 65$ km^2/h, and $a_m = 1.867$. The parameters of the urban model are: $Q_{\mathrm{cap},o,s,d} = 1000$ veh/h, $L_{\mathrm{av,veh}} = 6$ m, and $v_{\mathrm{av},l_{s,d}} = 50$ km/h.

We have selected the following time steps: $T_{\mathrm{c}} = 120$ s, $T_{\mathrm{f}} = 10$ s, and $T_{\mathrm{u}} = 1$ s. There are three parameters that can be tuned for the MPC controller. We have selected $N_{\mathrm{p}} = 8$ and $N_{\mathrm{c}} = 3$ as horizons, and $\gamma_1 = \gamma_2 = 1$ as weights for the urban and freeway partial performance criteria in the cost function (without normalization in this case).

11.4.2.2 Alternative control methods

Dynamic traffic control systems have already been implemented in the real world. Some examples of these systems are SCATS [58], Toptrac [3], SCOOT [46], UTOPIA/SPOT [43], MOTION [8], and IN-TUC [13].

Here we will use SCOOT and UTOPIA/SPOT to make a comparison between the developed MPC control method and some existing systems. We have selected these methods because they are good representatives of this kind of dynamic traffic control systems. However, these systems are commercial systems, meaning that real specifications are not publicly available. This means that we can only approximate their functioning (see [54] for details). Note that both systems only target the urban traffic. So they optimize the intersections independently of the neighboring freeway.

The main difference between the MPC-based system proposed in this paper and the existing systems like SCOOT and UTOPIA/SPOT is that MPC-based traffic control takes the influences and interactions between the urban and freeway parts of the network into account. By simulating the effect of one measure on both kinds of roads, control settings can be found that provide a trade-off between improving traffic conditions on the freeway and delaying traffic on the urban roads, and vice versa.

Furthermore, the MPC-based system we have developed can handle hard constraints on both the control signals and the states of the traffic network. All control systems can handle constraints that are directly linked to the control signals, e.g., maximal and minimal green times, or maximal cycle times. But the MPC-based system can also handle more indirect constraints such as maximum queue lengths, maximum delays, etc. These constraints are included as hard constraints in the MPC optimization problem, which is subsequently solved using a constrained optimization algorithm. In the other systems such a constraint (like maximum queue lengths) is implemented by adding a penalty term that penalizes the constraint violation to the performance function. This penalty term must become relatively large when the maximum queue length is reached. This results in a very high value of the cost when the maximum queue length is violated. While the purpose of the control is to minimize the cost function, a trade-off will have to be made between minimizing the original cost and violating the queue length constraint. This can lead to either satisfying the constraints with a degraded performance, or violating the constraints and obtaining a better performance.

11.4.2.3 Results

We have applied SCOOT, UTOPIA/SPOT, and MPC to the case study network for each of the four scenarios. The results are listed in Table 11.2. This table shows the TTS for the freeway part of the network, for the urban part, and for the whole network. The last column of the table shows the improvement of the MPC method compared to SCOOT (first number) and to UTOPIA/SPOT (second number). This makes it possible to determine in which part of the network the largest improvements are obtained. For the fourth scenario the largest attained queue length is also shown.

The first two scenarios show that the MPC method can improve the performance for the urban as well as for the freeway part of the network when a problem arises in one of the two. The immediate negative effects of such a problem are reduced as well as the negative influence on the rest of the network.

The third scenario shows that the MPC method can control the traffic slightly better that SCOOT and UTOPIA/SPOT when a large peak in the demand occurs. In this scenario the trade-off between the freeway and urban parts of the network can clearly be seen. A reduction of the performance on the urban network can lead to an improvement of the performance on the freeway network, and vice versa. This can be used to obtain a better performance for the total network.

The maximum queue length constraint is implemented in SCOOT and UTOPIA by adding an extra penalty term in the cost function. This term has a relative weight that allows a trade-off between the performance of the network and the importance of the maximum queue length constraints. When the weight is high the queue length constraint is satisfied but the performance is low, as shown in the first simulations done for the fourth scenario. In the second set of simulations the weighting term for the queue constraint is low, resulting in a better performance, but now the maximum queue length is exceeded. The values for MPC are the same for both simulation sets because the queue length constraint is implemented as a hard constraint for the optimization algorithm[6].

11.5 Conclusions and future research

In this chapter we have shown how model predictive control (MPC) can be used to obtain network-wide coordination and integration of various control measures (e.g., variable speed limits, ramp metering, and traffic signals) as well as a balanced trade-off between various performance criteria (such as the total time spent, emissions, and fuel consumption). As models are an important component of MPC we have presented an integrated traffic flow model for mixed urban-freeway networks. We have also proposed an emissions and fuel consumption model that can be interfaced with the combined urban-freeway traffic flow model. Finally, we have presented two case studies that illustrate the proposed control approach.

The MPC-based traffic control approach presented in this paper can be extended in various directions. One important topic is the issue of scalability and computational complexity, which could be addressed using a distributed, hierarchical, or multi-agent approach [47, 55, 56]. Other issues include the further development of fast, but accurate traffic flow and emission models for large-scale traffic networks. Moreover, the approach could also be extended to include additional control mea-

---

[6] The MPC-based method violates the constraint with 1 vehicle at the start of the simulation. This is due to infeasibility problems during the optimization, related to the initial state of the network at the start of the simulation. This issue can be solved by increasing the horizons N_p and N_c.

Table 11.2: Results for the second case study: total time spent for the freeway part of the network, for the urban part, and for the total network. The improvement of the MPC-based method compared to SCOOT and UTOPIA/SPOT is also listed.

Scenario 1: Congestion on the freeway.

	SCOOT	UTOPIA/SPOT	MPC	improvement
freeway	595.4	565.1	563.9	5.3 / 0.3%
urban	313.6	335.7	305.7	3.0 / 9.0%
total	909.0	900.8	869.6	4.4 / 3.5%

Scenario 2: Blockage of an urban intersection.

	SCOOT	UTOPIA/SPOT	MPC	improvement
freeway	498.0	526.2	495.0	0.7 / 6.0%
urban	665.9	672.3	620.3	6.9 / 7.8%
total	1163.9	1198.5	1115.3	4.2 / 7.0%

Scenario 3: Rush hour.

	SCOOT	UTOPIA/SPOT	MPC	improvement
freeway	244.6	280.1	253.3	-3.5 / 9.6%
urban	409.0	383.5	386.8	5.5 / -1.6%
total	653.6	663.6	640.1	2.1 / 3.5%

Scenario 4a: Maximum queue length of 20 vehicles, with a large weight.

	SCOOT	UTOPIA/SPOT	MPC	improvement
freeway	367.2	510.3	373.9	-1.8 / 26.8%
urban	309.7	435.4	264.4	15.7 / 39.3%
max. queue	19	19	21	
total	676.9	945.7	638.3	6.8 / 32.6%

Scenario 4b: Maximum queue length of 20 vehicles, with a small weight.

	SCOOT	UTOPIA/SPOT	MPC	improvement
freeway	367.1	428.1	373.9	-1.8 / 13.7%
urban	303.0	360.5	264.5	13.8 / 26.7%
max. queue	93	43	21	
total	670.1	788.6	638.3	5.8 / 19.1%

sures, including those arising in the context of automated highway systems and intelligent vehicle highway systems [4, 5, 48, 57].

The proposed approach is not limited to traffic networks only but could — with proper modifications — also be applied to other types of networks, such as water networks, electricity networks, railway networks, etc. (see also the other chapters of this book).

Acknowledgements Research supported by the BSIK programs "Next Generation Infrastructures (NGI)" and "Towards Sustainable Mobility (TRANSUMO)", the European 7th framework STREP project "Hierarchical and Distributed Model Predictive Control (HD-MPC)" (contract number ICT-INFSO-223854), the European COST Action TU0702, the project "Multi-agent control of large-scale hybrid systems" (DWV.6188) of the Dutch Technology Foundation STW, the NWO-CONNEKT project 014-34-523 "Advanced multi-agent control and information for integrated multi-class traffic networks (AMICI)", the Shell/TU Delft Sustainable Mobility program, the Transport Research Centre Delft, and the Delft Research Center Next Generation Infrastuctures.
We would also like to thank the reviewers for their careful reading of the original manuscript and for their useful feedback and suggestions.

References

1. K. Ahn, A.A. Trani, H. Rakha, and M. Van Aerde. Microscopic fuel consumption and emission models. In *Proceedings of the 78th Annual Meeting of the Transportation Research Board*, Washington, DC, January 1999. CD-ROM.
2. A. Alessandri, A. Di Febbraro, A. Ferrara, and E. Punta. Nonlinear optimization for freeway control using variable-speed signaling. *IEEE Transactions on Vehicular Technology*, 48(6):2042–2052, November 1999.
3. TPA Traffic & Parking Automation. Toptrac; verkeersafhankelijke regeling voor netwerken. Technical report, TPA Traffic & Parking Automation, NHTV Breda, The Netherlands, 2002. In Dutch.
4. L. D. Baskar, B. De Schutter, and H. Hellendoorn. Hierarchical traffic control and management with intelligent vehicles. In *Proceedings of the 2007 IEEE Intelligent Vehicles Symposium (IV'07)*, pages 834–839, Istanbul, Turkey, June 2007.
5. L.D. Baskar, B. De Schutter, and H. Hellendoorn. Model predictive control for intelligent speed adaptation in intelligent vehicle highway systems. In *Proceedings of the 17th IEEE International Conference on Control Applications*, pages 468–473, San Antonio, Texas, September 2008.
6. T. Bellemans. *Traffic Control on Motorways*. PhD thesis, Faculty of Applied Sciences, K.U.Leuven, Leuven, Belgium, May 2003.
7. P. G. Boulter, T. Barlow, I. S. McCrae, S. Latham, D. Elst, and E. van der Burgwal. Road traffic characteristics, driving patterns and emission factors for congested situations. Technical report, TNO Automotive, Department Powertrains-Environmental Studies & Testing, Delft, The Netherlands, 2002. OSCAR Deliverable 5.2.
8. F. Busch and G. Kruse. MOTION for SITRAFFIC – A modern approach to urban traffic control. In *Proceedings of the IEEE Conference on Intelligent Transportation Systems*, pages 61–64, Oakland, California, 2001.
9. E. F. Camacho and C. Bordons. *Model Predictive Control in the Process Industry*. Springer-Verlag, Berlin, Germany, 1995.

10. L. Davis, editor. *Handbook of Genetic Algorithms*. Van Nostrand Reinhold, New York, New York, 1991.

11. F. P. Deflorio. Evaluation of a reactive dynamic route guidance strategy. *Transportation Research Part C*, 11(5):375–388, October 2003.

12. C. Diakaki, V. Dinopoulou, K. Aboudolas, and M. Papageorgiou. Signal management in real time for urban traffic networks. Technical report, Technical University of Crete – IST programme, Chania, Greece, July 2002.

13. C. Diakaki, M. Papageorgiou, and T. McLean. Integrated traffic-responsive urban corridor control strategy in Glasgow, Scotland: Application and evaluation. *Transportation Research Record*, (1727):101–111, 2000.

14. R. W. Eglese. Simulated annealing: A tool for operations research. *European Journal of Operational Research*, 46(3):271–281, June 1990.

15. C. E. García, D. M. Prett, and M. Morari. Model predictive control: Theory and practice — A survey. *Automatica*, 25(3):335–348, May 1989.

16. N. H. Gartner. Development of demand-responsive strategies for urban traffic control. In P. Thoft-Christensen, editor, *Proceedings of the 11th IFIP Conference on System Modelling and Optimization*, pages 166–174. New York, New York: Springer-Verlag, 1984.

17. N. H. Gartner, J. D. C. Little, and H. Gabbay. Simultaneous optimization of offsets, splits, and cycle time. *Transportation Research Record*, (596):6–15, 1976.

18. F. Glover and M. Laguna. *Tabu Search*. Kluwer Academic Publishers, Boston, Massachusetts, 1997.

19. A. Hegyi. *Model Predictive Control for Integrating Traffic Control Measures*. PhD thesis, Delft University of Technology, Delft, The Netherlands, February 2004. TRAIL Thesis Series T2004/2.

20. A. Hegyi, B. De Schutter, and H. Hellendoorn. Model predictive control for optimal coordination of ramp metering and variable speed limits. *Transportation Research Part C*, 13(3):185–209, June 2005.

21. A. Hegyi, B. De Schutter, and J. Hellendoorn. Optimal coordination of variable speed limits to suppress shock waves. *IEEE Transactions on Intelligent Transportation Systems*, 6(1):102–112, March 2005.

22. D. Helbing, A. Hennecke, V. Shvetsov, and M. Treiber. Micro- and macro-simulation of freeway traffic. *Mathematical and Computer Modelling*, 35(5–6):517–547, March 2002.

23. J. Heywood. *Internal Combustion Engine Fundamentals*. McGraw-Hill, New York, New York, 1988.

24. S. P. Hoogendoorn and P. H. L. Bovy. Generic gas-kinetic traffic systems modeling with applications to vehicular traffic flow. *Transportation Research Part B*, 35(4):317–336, May 2001.

25. S. P. Hoogendoorn and P. H. L. Bovy. State-of-the-art of vehicular traffic flow modelling. *Proceedings of the Institution of Mechanical Engineers, Part I: Journal of Systems and Control Engineering*, 215(4):283–303, August 2001.

26. A. Karimi, A. Hegyi, B. De Schutter, J. Hellendoorn, and F. Middelham. Integrated model predictive control of dynamic route guidance information systems and ramp metering. In *Proceedings of the 7th International IEEE Conference on Intelligent Transportation Systems (ITSC 2004)*, pages 491–496, Washington, DC, October 2004.

27. H. R. Kashani and G. N. Saridis. Intelligent control for urban traffic systems. *Automatica*, 19(2):191–197, March 1983.

28. A. Kotsialos and M. Papageorgiou. Motorway network traffic control systems. *European Journal of Operational Research*, 152(2):321–333, January 2004.

29. A. Kotsialos, M. Papageorgiou, M. Mangeas, and H. Haj-Salem. Coordinated and integrated control of motorway networks via non-linear optimal control. *Transportation Research Part C*, 10(1):65–84, February 2002.

30. A. Kotsialos, M. Papageorgiou, and A. Messmer. Optimal coordinated and integrated motorway network traffic control. In *Proceedings of the 14th International Symposium of Transportation and Traffic Theory (ISTTT)*, pages 621–644, Jerusalem, Israel, July 1999.
31. A. Kotsialos, M. Papageorgiou, and F. Middelham. Optimal coordinated ramp metering with advanced motorway optimal control. In *Proceedings of the 80th Annual Meeting of the Transportation Research Board*, Washington, DC, 2001. Paper no. 01-3125.
32. H. Lenz, R. Sollacher, and M. Lang. Nonlinear speed-control for a continuum theory of traffic flow. In *Proceedings of the 14th IFAC World Congress (IFAC'99)*, volume Q, pages 67–72, Beijing, China, January 1999.
33. M. J. Lighthill and G. B. Whitham. On kinematic waves: I. Flood movement in long rivers. *Proceedings of the Royal Society of London*, 299A:281–316, May 1955.
34. J. M. Maciejowski. *Predictive Control with Constraints*. Prentice Hall, Harlow, UK, 2002.
35. A. Messmer and M. Papageorgiou. METANET: A macroscopic simulation program for motorway networks. *Traffic Engineering and Control*, 31(8/9):466–470, August/September 1990.
36. Ministry of Transport, Public Works and Watermanagement. Mobiliteitsonderzoek Nederland 2005, tabellenboek. Technical report, AVV transport research center, Rotterdam, The Netherlands, 2006. In Dutch.
37. L. Ntziachristos and Z. Samaras. Speed-dependent representative emission factors for catalyst passanger cars and influencing parameters. *Atmospheric Environment*, 34(27):4611–4619, March 2000.
38. M. T. Oliver-Hoyo and G. Pinto. Using the relationship between vehicle fuel consumption and CO_2 emissions to illustrate chemical principles. *Journal of Chemical Education*, 85(2):218–220, February 2008.
39. M. Papageorgiou, J. M. Blosseville, and H. Haj-Salem. Modelling and real-time control of traffic flow on the southern part of Boulevard Périphérique in Paris: Part II: Coordinated on-ramp metering. *Transportation Research Part A*, 24(5):361–370, September 1990.
40. M. Papageorgiou, H. Hadj-Salem, and J. M. Blosseville. ALINEA: A local feedback control law for on-ramp metering. *Transportation Research Record*, (1320):58–64, 1991.
41. M. Papageorgiou and A. Kotsialos. Freeway ramp metering: An overview. *IEEE Transactions on Intelligent Transportation Systems*, 3(4):271–280, December 2002.
42. P. M. Pardalos and M. G. C. Resende, editors. *Handbook of Applied Optimization*. Oxford University Press, Oxford, UK, 2002.
43. Peek Traffic. *UTOPIA/SPOT - Technical Reference Manual*. Peek Traffic Scandinavia, January 2002.
44. D. A. Pierre. *Optimization Theory with Applications*. Dover Publications, New York, New York, 1986.
45. P. I. Richards. Shock waves on the highway. *Operations Research*, 4(1):42–57, January–February 1956.
46. D. I. Robertson and R. D. Bretherton. Optimizing networks of traffic signals in real time - the SCOOT method. *IEEE Transactions on Vehicular Technology*, 40(1):11–15, February 1991.
47. R. Schleiffer, ed. Special Issue on Intelligent Agents in Traffic and Transportation. *Transportation Research Part C*, 10(5–6):325–527, October–December 2002.
48. S. E. Shladover, C. A. Desoer, J. K. Hedrick, M. Tomizuka, J. Walrand, W. B. Zhang, D. H. McMahon, H. Peng, S. Sheikholesham, and N. McKeown. Automatic vehicle control developments in the PATH program. *IEEE Transactions on Vehicular Technology*, 40(1):114–130, February 1991.
49. S. Smulders. Control of freeway traffic flow by variable speed signs. *Transportation Research Part B*, 24(2):111–132, April 1990.
50. H. Taale and F. Middelham. Ten years of ramp-metering in The Netherlands. In *Proceedings of the 10th International Conference on Road Transport Information and Control*, pages 106–110, London, UK, April 2000.
51. The MathWorks, Natick, Massachusetts. *Optimization Toolbox 4 — User's Guide*, March 2009. Revised for Version 4.2.

52. M. van den Berg, B. De Schutter, A. Hegyi, and J. Hellendoorn. Model predictive control for mixed urban and freeway networks. In *Proceedings of the 83rd Annual Meeting of the Transportation Research Board*, Washington, DC, January 2004. Paper 04-3327.

53. M. van den Berg, A. Hegyi, B. De Schutter, and J. Hellendoorn. A macroscopic traffic flow model for integrated control of freeway and urban traffic networks. In *Proceedings of the 42nd IEEE Conference on Decision and Control*, pages 2774–2779, Maui, Hawaii, December 2003.

54. M. van den Berg, A. Hegyi, B. De Schutter, and J. Hellendoorn. Integrated traffic control for mixed urban and freeway networks: A model predictive control approach. *European Journal of Transport and Infrastructure Research*, 7(3):223–250, September 2007.

55. R. T. van Katwijk, P. van Koningsbruggen, B. De Schutter, and J. Hellendoorn. Test bed for multiagent control systems in road traffic management. *Transportation Research Record*, (1910):108–115, 2005.

56. R.T. van Katwijk. *Multi-Agent Look-Ahead Traffic Adaptive Control*. PhD thesis, Delft University of Technology, Delft, The Netherlands, January 2008. TRAIL Thesis Series T2008/3.

57. P. Varaiya. Smart cars on smart roads: Problems of control. *IEEE Transactions on Automatic Control*, 38(2):195–207, February 1993.

58. B. Wolshon and W.C. Taylor. Analysis of intersection delay under real-time adaptive signal control. *Transportation Research Part C*, 7(1):53–72, February 1999.

59. S.K. Zegeye, B. De Schutter, H. Hellendoorn, and E. Breunesse. Integrated macroscopic traffic flow and emission model based on METANET and VT-micro. In *Proceedings of the International Conference on Models and Technologies for Intelligent Transportation Systems*, Rome, Italy, June 2009.

60. S.K. Zegeye, B. De Schutter, H. Hellendoorn, and E. Breunesse. Reduction of travel times and traffic emissions using model predictive control. In *Proceedings of the 2009 American Control Conference*, St. Louis, Missouri, June 2009.

Chapter 12
Intelligent Road Network Control

J.L.M. Vrancken and M. dos Santos Soares

Abstract An important challenge in road traffic control (RTC) consists in the transition from the current mainly local and reactive control to network-wide, proactive control. This chapter presents a distributed architecture (DTCA, Distributed Traffic Control Architecture) and real-life test bed aimed at this challenge. DTCA consists of the addition of multi-agent control to an existing architecture for RTC, which, at the network-level, features only single-agent, top-down control. The fact that, in Traffic Engineering, it is still largely unknown how to do network-level control, results in a need for highly adaptable systems that can accommodate new insights into network-level control. This requirement has been addressed by the application of advanced Software Engineering (SE). An implementation of a control system based on DTCA, in the Dutch city of Alkmaar, has thus resulted in a real-life test bed for combined single- and multi-agent RTC.

J. Vrancken, M. dos Santos Soares
Delft University of Technology, Faculty Technology, Policy and Management, Delft, The Netherlands, e-mail: {j.l.m.vrancken,m.dossantossoares}@tudelft.nl

R.R. Negenborn et al. (eds.), *Intelligent Infrastructures*, Intelligent Systems, Control and Automation: Science and Engineering 42, DOI 10.1007/978-90-481-3598-1_12,
© Springer Science+Business Media B.V. 2010

12.1 Introduction

In many countries, road traffic is both the most important and the most problematic means of transportation. The problems are well-known: insufficient capacity resulting in frequent congestion, high fatality rates and serious environmental damage. Addressing the lack of capacity by building more infrastructure is increasingly difficult due to high costs, environmental concerns and lack of space. As a consequence, many countries have resorted to traffic management in order to maximize the capacity of the existing infrastructure. Traffic management helps. For instance, the effectiveness of traffic signals is beyond debate and they continue to be improved. Since the seventies of the last century, many more traffic management measures have been developed and deployed. Since then, measures have been characterized by increasing levels of automation and traffic-responsiveness [7, 14, 26].

Traffic management consists of a number of activities supporting and regulating traffic, such as Road Traffic Control (RTC), incident handling and pricing. This chapter is solely about RTC, which deals with influencing traffic flows by means of visual signals to drivers, for the purposes of increasing capacity and safety and reducing environmental damage.

There have been mainly two approaches to RTC: a technology oriented approach, embedded within ITS (Intelligent Transportation Systems [11, 12, 21]), and a traffic engineering approach based on mathematical modeling and simulation of traffic flows [10, 13]. Here, we will combine these two approaches.

Current RTC is, from the traffic engineering point of view, mainly local, with limited network-wide coordination, and reactive [22]: it tries to solve congestion after it occurs. Technically, it is implemented by a wide variety of legacy systems, developed in various periods in the past by a variety of manufacturers. They feature the usual characteristics of legacy systems: hard to maintain, hard to change and hard to communicate with [18].

Besides further improvements of local measures and the development of new ones, an important challenge in RTC consists in the transition from local to network-wide and from reactive to proactive. Local control has an obvious limitation: it cannot prevent congestion and solving congestion locally is often just shifting the problem to a nearby location. Proactive control, which means preventing rather than solving problems, requires looking into the future. In the case of traffic, this implies that the network context has to be taken into account: the near-future of one location in a network is to a large extent determined by the current situation elsewhere in the network. In addition, society's goals in traffic management, such as improved travel time reliability and reduced total delay in traffic, are network related goals. It is unlikely that purely local measures will achieve the optimum for these goals.

All RTC by traffic management authorities is still done by influencing drivers by means of locally displayed signals. Other means of communication, such as cell phone or radio broadcast communication, is currently still unable to convey mandatory instructions. Although these means of communication have an increasing influence on traffic, they are left to private parties. Non-mandatory instructions can only support a per user optimum, not the network optimum that traffic management authorities aim at. This dependence on locally displayed visual signals means that network-level control does not replace local measures but is solely about co-ordinating local measures. It must be added that network-level control, seen as the coordination of local measures, is still largely uncharted territory [30]. Causes for this are, first, that network-level control is essentially more complex than local control: a sizeable network has far more elements and interactions between elements than for instance a single crossing or a single on-ramp. A second reason is that the current legacy systems were not designed for network-level control and, due to their legacy properties, are hard to adapt to this new way of using them.

The approach here is new in several aspects. First of all, we chose a multi-disciplinary approach, by combining the two aforementioned traditional approaches and by adding the application of advanced Software Engineering (SE). Second, we chose to apply a hybrid control strategy at the network-level, by combining single-agent, top-down control with multi-agent, bottom-up control. Third, we combined simulation of traffic with real-life implementation.

From ITS, we inherit the architecture-based approach to system development [12], which is also in accordance with current practice in SE [2, 19]. In The Netherlands, there is a well developed and widely used architecture for RTC: the Traffic Control Architecture (TCA). It was originally developed for motorways, but later it was extended to so-called area-oriented control, which means that it aims at the control of all types of networks in a given area, not only motorways but also urban and provincial or rural networks. At the network-level, this architecture features only single agent, top-down control: all coordination of local measures in an area is done via the traffic management center of the area and the pertinent traffic operator with his/her support tools. There is no direct communication between local measures. Taking the successful TCA as a starting point, we added multi-agent control, resulting in the Distributed Traffic Control Architecture (DTCA), which was applied in a real-life control system.

There are two aspects of network-level RTC that will not be addressed in this chapter. First of all, network-level control is often multi-level control. A large network can be considered at different levels. For RTC, this has been addressed elsewhere [31]. Here we consider only one level, the so called focal areas level [31], i.e., specific problem areas in networks such as belt roads. Second, we do not consider the problem of measure interaction and consistency. The target areas of measures may overlap. Then it is not always the case that all activated measures have synergic and consistent effects [28] in this overlap. For the control measures applied in the Alkmaar system, this happens to be no problem, but in more diverse sets of measures, this problem should not be ignored.

In this chapter, we will first formulate the research objectives in a number of research questions, followed by the research approach in more detail and an overview of related work. Then, we give a description of TCA, followed by a description of its extension to DTCA. Following this, we describe the implementation of a control system based on DTCA on the belt road of Alkmaar in The Netherlands. The chapter is concluded by some ideas for future research and by the answers we can currently give to the research questions.

12.2 Research questions

The objectives of the research described here can be expressed in the following research questions:

1. How to do network-level RTC?
2. How to implement systems for network-level RTC?

Within an architecture based approach, these questions lead to two additional questions:

3. How does an architecture for network-level RTC look like?
4. What are the SE requirements for network-level control systems?

For the evaluation of the proposed architecture we have the questions:

5. How to measure the effectiveness of network-level RTC?
6. To what extent do the architecture-based control systems comply with the SE requirements?

Definite answers to these questions, especially the first one, should not be expected. We made a step towards answering these questions and we will substantiate that the steps we took are necessary.

12.3 Approach

The chosen approach takes the existing architecture TCA, described below in Section 5, as starting point. To this architecture, multi-agent, bottom-up control has been added, not to replace top-down control but in order to add a wide range of new control options. This addition at the RTC layer (Figure 12.1) leads to a number of SE requirements resulting in extensions of the technical layers of TCA. Finally, a real-life control system has been implemented that serves both as a test case for the implementability of the architecture and as a test bed for future experiments with hybrid control strategies in RTC. A real-life test bed is on the one hand essential, as any result based on simulation only leaves open the question as to the value of the result in real traffic. There is no other way to answer this question than via real-life implementation. On the other hand, it brings in serious extra problems: one has to deal with existing legacy systems in the target area, with the interests of all the parties involved in real traffic, and with the circumstance that reality is not repeatable and much less suitable to play with than simulation. The ideal here seems to be the combination of simulation and real-life implementation (see Future Research in Section 8). The typical problems of real-life implementations are an important reason for the prominent role of SE in this study.

12.4 Related work

In order to convey an impression of the current status of network management and of multi-agent approaches to network management, a number of examples of recent publications are briefly described below.

Trafik Stockholm [8] is a system for network management applied on the belt road around Stockholm in Sweden. The system integrates several existing, more locally oriented control systems, but is not itself an automated control system. Its main function is to support human operators who take all the control decisions at this level of coordination.

Most approaches to network management still deal with the coordination of traffic signals. Two recent examples are [15] and [4]. The first deals with the MOTION system in the German city of Muenster. This system updates the signal plans of a number of traffic signals throughout the city every 5 to 15 minutes, on the basis of data from a city-wide traffic monitoring system. In this way, signal behavior is adapted to current traffic patterns. The second example deals with the experimental BALANCE system, applied at 46 crossings in the German town of Ingolstadt. This system is somewhat more advanced than the one in Muenster. It applies a genetic algorithm for traffic lights optimization. In both cases it is still too early for hard numerical evidence of the effectiveness of the systems involved.

[3] is an example of an approach to traffic management with a somewhat different goal than usual, namely reducing the environmental damage by traffic. It shows that from the control point of view, not much changes, but that still quite different sensors and quite different modeling techniques for monitoring the proper quantities are needed.

A comprehensive description of multi-agent systems, independent of any application domain, is given by Vidal [27]. He deals with various agent activities relevant in control situations, such as learning, negotiating, bidding, voting, and game playing.

In [25] a test bed is described with some similarity to the system described in this chapter. The main differences are that the test bed mentioned is not real-life, it does not deal with hybrid forms of control, it deals only with traffic signals and it has no ambitions in applying advanced SE. Katwijk's thesis [24] offers a thorough treatment of a specific case of network-level control: small groups of traffic signaling systems, coordinated by model predictive, multi-agent control.

In [1] a multi-agent approach to modeling traffic has been applied in which the vehicles are the agents and not the network elements as in our case.

[16] illustrates the multi-agent model predictive control in a quite different domain: electric power systems. Although the author the potential of this way of controlling power generation, his proposal has only been tested in simulation. Moreover, he abstains from the problems of real-life implementation in an environment with an abundance of legacy systems.

Summarizing, network management is still most often restricted to coordinating traffic signals. Many partial aspects of the problem of multi-agent control have been dealt with in previous studies, but the multi-disciplinary, real-life approach

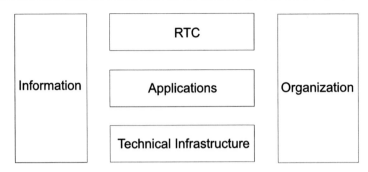

Figure 12.1: TCA Framework.

to network-level RTC described here seems to be new. The more advanced agent activities described by Vidal [27] will play a role in future research.

12.5 TCA: The architecture for road traffic control

We give a short description of the relevant parts of TCA [20, 22, 29], the architecture that will be extended towards DTCA in the next section.

The TCA framework, showing the main parts of the architecture, is depicted in Figure 12.1. The three horizontal layers are most relevant here. The RTC layer describes the TCA approach to RTC, in typical traffic engineering terms. This approach is typically single-agent, top-down control. All coordination of local measures is done via the traffic management centres (TMCs) and ultimately via the desk and the tools of human traffic operators. The Applications and Technical Infrastructure layers form the technical or ITS parts of the architecture. The Applications layer describes the applications to implement the RTC features mentioned in the RTC layer, such as the control systems for each of the local measures, and the operator's user interface and support tools in the TMC. This layer includes roadside sensors and actuators used in control systems. The Technical Infrastructure is about the data communications network and the processing platforms along the roads and in the TMCs. This layer also comprises a vitally important middleware layer that facilitates high level communication between application components. The RTC layer can be expanded into the layered model, depicted in Figure 12.2.

The purpose of this model is to show how societal goals in traffic management, such as improved travel time reliability and less delay, can be translated, in five steps, into the signals shown to drivers. Only the lower three layers are of relevance here. The Control Scenarios layer is the layer at which operators in TMCs work. Control scenarios are integrated programs of control measures that cover the area under control of the TMC. Their execution is triggered by and responds to certain recurring patterns in traffic, such as the morning rush hour. In principle, each recurring pattern found in traffic has its own control scenario. Control measures are the usual local measures, such as traffic signals, ramp metering, speed instructions

Figure 12.2: The RTC layer expanded.

and warning signals. All coordination of local measures takes place at the Control Scenarios layer. In TCA, local measures are not supposed to communicate with each other. Finally, the Signals layer deals with the various visual signals shown to drivers.

TCA has a number of limitations, primarily caused by its top-down, single-agent nature. First of all, single-agent (also called centralized or top-down) control is known to have serious limitations in handling complex situations, in addition to limited scalability, communication overhead and limited computational capacity. Traffic seldom fits one of the covered patterns exactly. Control scenarios are not very responsive to deviations from known patterns. This makes it a rather rough means of coordination. The limited scalability is due to the fact that the controlled area cannot easily be extended: that would soon overload the operator.

12.6 The DTCA architecture

In order to cope with the limitations of TCA, DTCA consists of a number of additions to TCA, first of all in the RTC layer (Figure 12.1). From this addition, and the reasons behind it, a number of non-functional SE requirements [19] can be derived that give guidance for the additions and choices in the technical layers.

In TCA's RTC layer, multi-agent control is added as an extra measure in the Control Measures sub-layer (Figure 12.2). This measure entails that network elements, such as crossings and road segments between crossings, but also routes and origin-destination pairs, become active agents, measuring their own traffic state and communicating with other network elements, mostly the adjacent ones (in the case of crossings and road segments). The communication consists of requests for information about traffic states in nearby elements and of requests to take certain measures, such as reducing the inflow to the requesting element. The bottom-up control is intended as an addition to and fine-tuning of the top-down control, certainly not as its replacement.

On itself, it is not obvious why the addition of multi-agent control could be useful. A multi-agent system is usually not very predictable in its behavior, but there are reasons why, in this case, this addition looks promising. Due to the many local measures, control of traffic is already strongly distributed. The essential addition here is that we make local measures communicate with each other. More precisely, measures are assigned to network elements and the network elements communicate with each other about the measures. This means that no options are lost and a wide realm of new control options is added. Moreover, such local, direct influencing of network elements among themselves gives a much quicker, real-time response to local changes in traffic and a much shorter control loop than via the TMC. In addition, multi-agent control, by its local nature, can easily scale to any size of network.

The inevitability of multi-agent control can be argued as follows. This argument is described in more detail in [31]. No decision making entity has infinite capacity. Actually, the capacity of both automated and of human decision making is rather limited. This means that each control entity has a limited geographical extent; in practice this may be as small as a single crossing. These control areas touch upon each other at their borders, which means that all the ingredients for a typical multi-agent control setting are present: many entities, each with its own control area, and with mutual influence because the areas have common borders.

12.6.1 SE requirements

For the technical layers, the addition of multi-agent control has a number of consequences for which we first formulate a set of SE requirements:

- **Flexibility**: Above, it was mentioned that the way to do network-level control is still largely to be discovered, by simulation but also by real-life experiments. This means that control systems should allow frequent adaptations, which makes flexibility (more precisely adaptability) a prime requirement.

- **Scalability**: The scalability of multi-agent control should not be hampered by any technical aspects, so the technical part of control systems should also be scalable.

- **Future-proofness of roadside equipment**: An important aspect of RTC systems is that they require equipment along roads. The roadside equipment, i.e., the technical infrastructure (data communications network and computing platforms) and the sensors and actuators, constitute by far the most expensive part of a RTC system. This equipment should therefore be future proof to some degree. Ideally, it should last much longer than the specific applications (software) running on it and it should accommodate several generations of control applications.

- **Real-time properties**: RTC is a form of process control. Therefore, as mentioned above, real-time properties of applications are a major concern.

- **Reliability**: Traffic involves human life. This means that applications and technical infrastructure should feature high reliability, as to not jeopardize human life by system malfunctioning.

- **Interoperability**: When experimenting with different forms of multi-agent control, applications should be easy to connect to and operate with, which means that interoperability between RTC-applications is important.

- **Interoperability with legacy systems**: Finally, the presence of legacy systems along the roads, and the circumstance that it would be too costly to ignore the existing systems, implies that interoperability with legacy systems is an important requirement.

12.6.2 DTCA technical layers

Given the addition of multi-agent control in the RTC layer, and the ensuing SE requirements, we come to the following additions to TCA in the technical layers.

In the Applications and Components layer, components are added that represent the communicating network elements, such as road segments, crossings and routes. This is an immediate consequence of the choice to add multi-agent control to the RTC layer.

In the Technical Infrastructure layer of TCA, an explicit choice is made for asynchronous Publish/Subscribe middleware. This is the most appropriate kind of middleware, given the requirements of real-time properties and scalability [5, 6, 9]. In addition, it is known for the efficient use of network bandwidth. This kind of middleware can accommodate both the top-down and the bottom-up control mechanisms, in addition to the communication needed for data collection.

Summarizing, the DTCA consists of the TCA architecture with a number of additions for multi-agent control. The network elements, such as crossings and road segments, become control agents.

12.7 Experimental implementation

The DTCA has been applied in an experimental first implementation: the HARS system (Figure 12.3. HARS: *Het Alkmaar Regelsysteem*, the Alkmaar control system). Alkmaar (Figure 12.4) is a medium sized Dutch city with serious traffic prob-

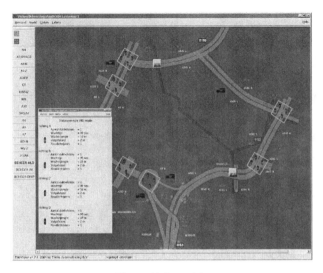

Figure 12.3: HARS.

lems on its belt road, which is part of a provincial north-south corridor. It consists of motorways, as well as provincial and urban roads.

A consortium of commissioners (the city council of Alkmaar, the provincial authorities of the province of North Holland and the motorways traffic management authority Rijkswaterstaat) sent out a call for tenders in 2005, explicitly asking for an innovative solution. The tender by the offering consortium of Trinité Automatisering B.V., specialized in the construction of traffic management systems, and Goudappel Coffeng, a traffic engineering consultancy firm, was given the job to build an RTC system for the belt road. The system was built and installed mainly in 2006 and is currently in operation.

The goal of HARS was therefore first of all to tackle the traffic problems in Alkmaar. As an implementation of DTCA, the goal of the system was to prove the concepts behind DTCA, especially the assumption that in RTC, a dual control strategy is stronger than just top-down control. HARS can be considered successful if it has a noticeable, positive effect on traffic in Alkmaar, if the multi-agent control plays a substantial role in this and if the system turns out to meet the SE requirements mentioned in Section 6.

HARS takes the existing local RTC measures and their systems, such as traffic signals at crossings and VMSs (Variable Message Signs, also called DRIPs: Dynamic Route Information Panels), as starting point, and adds a software overlay layer to these local measures. It is operated from the existing regional traffic management center for the province of North-Holland. It does not have any sensors or actuators of its own. At the technical infrastructure layer, the proprietary Publish/Subscribe middleware of Trinité, called DSS (Data Subscription System), has been applied, together with a component framework, called PDA (Programmable Distributed Architecture) by the same company [23].

Figure 12.4: The belt road in Alkmaar.

The top-down control part is in accordance with TCA. It consists of control scenarios, defined on the basis of regularly occurring traffic patterns, such as the morning rush hour or the weekend exodus, made proactive by a predictive traffic simulation model, called MaDAM [17]. This simulation model, built by Goudappel Coffeng, generates traffic states half an hour ahead, in steps of 5 minutes. MaDAM also has a role in supplying data at locations where the system assumes a sensor for traffic data collection, but no physical sensor happens to be present. The bottom-up control is implemented by means of communicating road segments and crossings. Each element E offers, to each adjacent network element F downstream of E, the service of reducing inflow into F. This service is implemented in various ways, according to the available local measures in the network element: a road segment may implement it by reducing maximum speed or closing one or more of its lanes; a crossing with traffic signals may implement it by reducing green times of the pertinent signal groups. Bottom-up control also applies look ahead by means of the simulation model. A reduce-inflow request is issued when this look ahead indicates overloading of the network element in the near future. The elements receiving the request try to fulfill it by using their own available capacity. When this is not enough, reduce-inflow requests will be issued further upstream. In this way, reserve capacity in the network is used in order to avoid overloading individual crossings or segments.

HARS is currently in operation. Several interruptions of the system due to maintenance, and the simultaneous deterioration of the traffic condition, showed that it already has a positive effect on network performance. The top-down control part works according to expectations. The dual control strategy has a distinctive practical advantage. Just a bottom-up strategy would probably be deemed too innovative to be accepted by commissioners for real-life implementation, besides the functional limitations that one would then have. In order to measure the contribution of bottom-up control, a more refined network performance monitoring system is needed than is currently available.

12.8 Future challenges

On the basis of our experiences with HARS, we see the following challenges in order to achieve the goal of effective network-wide RTC:

- **Measuring network performance**: Measuring network performance is essential in order to be able to discern successful approaches to network-level control from ineffective or even counterproductive approaches. Measuring network performance is far from trivial. First of all, the definition of network performance should be given due attention. Most likely, the proper definition depends on the societal goals to be achieved. Second, the measuring process involves deriving an overall picture of network performance from a set of noisy and inhomogeneous data streams produced by the various kinds of sensors in the network.

- **Tuning multi-agent control**: Once it is possible to measure the appropriate quantities for network performance, one can start to test various local control programs for the agents and compare results. The challenge here is that in order to make results from different control programs comparable, one should have the same traffic load at different times. In practice this happens only rarely. What is needed here is a simulation of both traffic and the control system. This is in essence a copy of the real control system, with its inputs and outputs connected to a traffic simulator. The presence of a real-life system is essential in order to make the simulated setup behave realistically.

- **Finding guiding principles**: Once we have a network performance measuring system and a simulated setup of control system and traffic, we will face the problem of finding our way in this vast solution space. In principle any information item in the system can be used in an infinity of ways in the control programs of large numbers of agents. Without guiding principles, one may easily get lost in this space. We hope to find guiding principles by studying examples of successful cases of network control elsewhere and by studying approaches to multi-agent control in other areas.

- **Model-driven software engineering**: In order to further improve the adaptability of RTC systems, it is envisaged to apply Model-Driven Software Engineering techniques in order to be able to generate RTC systems from a model of the road network under consideration and its traffic characteristics.

- **Self-learning control systems**: The self-learning property is often associated with multi-agent systems [27]. The feedback resulting from measuring network performance leads to adaptations in system and agent configuration. Some of these adaptations may very well allow automation. In this way it is feasible, in the somewhat more remote future, to make DTCA-based RTC systems self-learning.

12.9 Conclusions and future research

An important challenge in RTC consists of the transition from local, reactive control to network-level proactive control, in order to optimize network performance for the traffic policy in force. In practice, this amounts to effectively coordinating all the local control measures in a given area. Our approach to this challenge is an architecture-based, multi-disciplinary approach, combining the decade old TCA architecture with simulation models from traffic engineering and techniques from ITS and Software Engineering. The addition of multi-agent control to TCA leads to DTCA, the Distributed Traffic Control Architecture. A real-life implementation of a DTCA-based control system has been realized on the belt road of the Dutch city of Alkmaar, The Netherlands. This implementation can serve as a real-life test bed for multi-agent RTC. Thus in the HARS experiment, some important steps were taken towards answering the research questions in Section 2, steps that in several aspects went beyond what already was available. On the basis of the experiences until now, we can give the following preliminary answers to the research questions:

- **How to do network-level RTC?** The method known as area-oriented RTC, combined with a predictive on-line traffic model, has been shown to have a noticeable effect on network performance. To support this statement numerically and to measure the degree by which the bottom-up control contributes to this effect, remains to be done.

- **How can network-level control systems be implemented?** The DTCA offers an effective way of implementing such control systems, in the sense that the SE requirements mentioned in Section 6 can be met. The layeredness of DTCA contributes to the flexibility requirement. The choice for Publish/Subscribe middleware contributes to flexibility, scalability, real-time properties and the interoperability with legacy systems. HARS illustrated that DTCA-based control systems can be deployed on the existing technical infrastructure and existing sensors and actuators, thereby contributing to the future-proofness of these parts. Only the interoperability with other DTCA-based control systems has not been demonstrated, but this is most probably easier than interoperability with legacy systems and is also taken care of by the middleware.

This answers Questions 3, 4 and 6. Question 5, about network-level monitoring, will be addressed in future research.

References

1. J. L. Adler, G. Satapathy, V. Manikonda, B. Bowles, and V. J. Blue. A multi-agent approach to cooperative traffic management and route guidance. *Transportation Research Part B: Methodological*, 39(4):297–318, May 2005.
2. L. Bass, P. Clements, and R. Kazman. *Software Architecture in Practice*. Addison-Wesley Professional, April 2003.
3. S. Boerma, N. Rodrigues, and W. Broeders. Deploying environmental dynamic traffic management. In *Proceedings of the 7th European Congress and Exhibition on Intelligent Transport Systems and Services*, Geneva, Switzerland, June 2008.
4. R. Braun, C. Kemper, F. Weichenmeier, and J. Wegmann. Improving urban traffic in Ingolstadt. In *Proceedings of the 7th European Congress and Exhibition on Intelligent Transport Systems and Services*, Geneva, Switzerland, June 2008.
5. P. T. Eugster, P. A. Felber, R. Guerraoui, and A-M. Kermarrec. The many faces of publish/-subscribe. *ACM Computing Surveys*, 35(2):114–131, June 2003.
6. L. Fiege, M. Cilia, G. Muhl, and A. Buchmann. Publish-subscribe grows up: Support for management, visibility control, and heterogeneity. *IEEE Internet Computing*, 10(1):48–55, January/February 2006.
7. P. B. Hunt, D. I. Robertson, R. D. Bretherton, and R. I. Winton. SCOOT - A traffic responsive method of coordinating signals. Technical Report 1014, Transport and Road Research Laboratory, Berkshire, 1981.
8. T. Julner. Unique integrated traffic management using decision support at trafik stockholm. In *Proceedings of the 7th European Congress and Exhibition on Intelligent Transport Systems and Services*, pages 1–7, Geneva, Switzerland, June 2008.
9. D. Lea, S. Vinoski, and W. Vogels. Guest Editors' introduction: Asynchronous middleware and services. *IEEE Internet Computing*, 10(1):14–17, January/February 2006.
10. W. Leutzbach. *Introduction to the theory of traffic flow*. Springer Verlag, Berlin, Germany, 1988.
11. R. P. Maccubbin, B. L. Staples, M. R. Mercer, F. Kabir, D. R. Abedon, and J. A. Bunch. Intelligent transportation systems benefits, costs and lessons learned. Technical Report FHWA-OP-05-002; FHWA-JPO-05-002, Mitretek Systems and Federal Highway Administration, Washington, DC, 2005.
12. B. McQueen and J. McQueen. *Intelligent Transportation Systems Architectures*. Artech House Publishers, Norwood, Massachusetts, 1999.
13. W. R. McShane, R. P. Roess, and E. S. Prassas. *Traffic Engineering*. Prentice-Hall, Upper Saddle River, New Jersey, 1998.
14. P. Mirchandani and F-Y. Wang. RHODES to intelligent transportation systems. *IEEE Intelligent Systems*, 20(1):10–15, January/February 2005.
15. J. Mueck. The German approach to adaptive network control. In *Proceedings of the 7th European Congress and Exhibition on Intelligent Transport Systems and Services*, Geneva, Switzerland, June 2008.
16. R. R. Negenborn. *Multi-agent Model Predictive Control with Applications to Power Networks*. PhD thesis, Delft University of Technology, Delft, The Netherlands, December 2007.
17. OmniTRANS. OmniTRANS - offering the best of both worlds. *Traffic Engineering & Control*, 47(9):1–3, October 2006.

18. R. C. Seacord, D. Plakosh, and G. A. Lewis. *Modernizing Legacy Systems: Software Technologies, Engineering Processes, and Business Practices*. Addison-Wesley Professional, 2003.
19. I. Sommerville. *Software Engineering*. Addison-Wesley Longman Publishing Co., Inc., Boston, Massachusetts, 2007.
20. H. Stoelhorst and F. Middelham. State of the art in regional traffic management planning in The Netherlands. In *Proceedings of the 11th IFAC Symposium on Control in Transportation Systems*, pages 451–456, Delft, The Netherlands, August 2006.
21. R. Stough. *Intelligent Transport Systems: cases and policies*. Edward Elgar Publishing, 2001.
22. H. Taale, M. Westerman, H. Stoelhorst, and D. van Amelsfort. Regional and sustainable traffic management in The Netherlands: Methodology and applications. In *Proceedings of the European Transport Conference*, Strasbourg, France, October 2004.
23. H. ter Doest, , F. Ottenhof, H. Eertink, and J. L. M. Vrancken. When theory meets practice, Building traffic control systems made easy. *Traffic Technology International*, September 2003.
24. R. T. van Katwijk. *Multi-Agent Look-Ahead Traffic-Adaptive Control*. PhD thesis, Delft University of Technology, Delft, The Netherlands, 2008.
25. R. T. van Katwijk and P. van Koningsbruggen. Coordination of traffic management instruments using agent technology. *Transportation Research Part C*, 10(5–6):455–471, October-December 2002.
26. S. Venglar and T. Urbanik. Evolving to real-time adaptive traffic signal control. In *Proceedings of the 2nd World Congress on Intelligent Transport Systems*, pages 10–15, Yokohama, Japan, November 1995.
27. J. M. Vidal. *Fundamentals of Multiagent Systems*. 2007.
28. J. L. M. Vrancken. Layered models in IT standardization. In *Proceedings of the 2006 IEEE International Conference on Systems, Man and Cybernetics*, pages 3862–3865, Taipei, Taiwan, October 2006.
29. J. L. M. Vrancken, V. A. Avontuur, M. Westerman, and J. C. Blonk. Architecture development for traffic control on the Dutch motorways. In *Proceedings of the ITS World Congress*, Seoul, South Korea, October 1998.
30. J. L. M. Vrancken and O. C. Kruse. Intelligent control in networks: The case of road traffic management. In *Proceedings of the 2006 IEEE International Conference on Networking, Sensing and Control*, pages 308–311, Fort Lauderdale, Florida, April 2006.
31. J. L. M. Vrancken and M. S. Soares. Multi-level control of networks, the case of road traffic control. In *Proceedings of the 2007 IEEE International Conference on Systems, Man and Cybernetics*, pages 1741–1745, Montréal, Canada, October 2007.

Chapter 13
An Integrated Dynamic Road Network Design Approach with Stochastic Networks

H. Li, M.C.J. Bliemer, and P.H.L. Bovy

Abstract Stochastic supply and fluctuating travel demand in transport systems leads to stochastic travel times and travel costs for travelers. This chapter will establish a dynamic road network design approach considering stochastic capacity and its influence on travelers' choice behavior under uncertainty. This chapter will firstly work on the modeling of travelers' departure time/route choice behavior under stochastic capacities. A reliability-based dynamic network design approach is proposed and formulated of which numbers of lanes on all the potential links are the design variables. A combined road network-oriented Genetic Algorithm and set evaluation algorithm is proposed to solve the dynamic network design problem. The proposed reliability-based dynamic network design approach is applied to a hypothetical network, and its solutions are compared to a corresponding static network design approach. It is concluded that the static network design approach may lead to inferior designs. In general static traffic assignment underestimates the overall total network travel time and total network travel costs. Dynamic network design approach appears to derive a fairly good allocation of road capacity over space and makes the best utilization of the network capacity over time.

H. Li, M.C.J. Bliemer, P.H.L. Bovy
Delft University of Technology, Faculty of Civil Engineering and Geosciences, Delft, The Netherlands, e-mail: {h.li,m.c.j.bliemer,p.h.l.bovy}@tudelft.nl

R.R. Negenborn et al. (eds.), *Intelligent Infrastructures*, Intelligent Systems, Control and Automation: Science and Engineering 42, DOI 10.1007/978-90-481-3598-1_13,

13.1 Introduction

The economy of a nation or region depends heavily upon an efficient and reliable transportation system to provide accessibility and to promote the safe and efficient movement of people and goods. As road authorities/designers, it is aimed to design road networks to provide efficient and reliable services to the travelers, while travelers react to the design of the networks by adapting their departure time and route choices. Road networks are characterized by uncertainty of traffic conditions on the networks over space and time dimensions. Many sources contribute to the uncertainty, ranging from irregular and random incidents such as earthquakes, floods, adverse weather, traffic accidents, breakdowns, signal failures, and road works to regular fluctuations of travel demand in times of day, days of the week, and seasons of the year [44]. All these sources lead to variations in network capacities or fluctuations in travel demand, which cause travel times and travel costs for travelers to vary on the same network in the same time period from day to day and to vary from time to time within a day [15, 45, 46].

Travel time reliability has been defined and widely adopted as a concept reflecting the properties of the travel time stochasticity. We refer to [3, 9, 11, 45] for categories of reliability definitions and measures of travel time reliability for different purposes. On the one hand, travel time reliability might be adopted for purpose of modeling travelers' choice behavior in response to stochastic and uncertain travel times. On the other hand, it is also applicable as a measure of the service quality that a road network provides to its users. Therefore the travel time reliability related research could be generally grouped into reliability-based choice behavior modeling under uncertainty and reliable network performance assessment or network designs.

Modeling travelers' choice behavior under uncertainty, for instance route choice and departure time choice behavior, has gained increasing attention [1, 5, 48]. Especially in the context of uncertainty, modeling departure time adaptation is fairly important since travelers attempt to arrive on time at their destination with a high probability and to reduce the schedule delay late by departing earlier [30, 32]. At the same time, this time adaptation appears to be one of the major responses of travelers to changing congestion conditions on their trips. How to model travelers' choice behavior under uncertainty and whether to model departure time choice behavior under uncertainty will have significant impact on the traffic assignment and thus lead to different network performance evaluations and different network designs.

Currently, in most road network design studies where stochastic networks or travel demand are considered, choice behavior under uncertainty is not modeled [33, 36, 41]. In most cases with stochastic networks or travel demand, static network design problems are proposed and formulated in which the lower level traffic assignment cannot capture the dynamics, flow propagations and spillback effects. Departure time choice modeling is in most cases neglected, which is however a very important aspect of travel behavior that needs to be modeled in the context of uncertainty [48]. In several dynamic network design studies, for instance [21], deterministic networks and travel demand are considered. Therefore, this chapter aims to design reliable road network infrastructures to provide reliable and efficient services to road users considering stochastic properties of road infrastructures. This chapter, other that earlier network design studies as aforementioned, develops a dynamic road network design approach with stochastic capacities with modeling travelers' route/departure time choice behavior under uncertainty. Dynamic traffic assignment is adopted, which is able to model the dynamics in flow propagations and spillback effects and represents the reality better. Travel time reliability is explicitly taken into account on both the network design level and the individual choices. The structure of the network and numbers of lanes on a set of potential links are aimed to be designed.

This chapter is structured as follows: firstly, travelers' departure time and route choice behavior under travel time uncertainty is modeled, based on which a reliability-based dynamic network design problem is proposed and formulated with stochastic capacities. Then the solution approaches of modeling the long term user equilibrium with stochastic capacities and of solving the proposed network design problem is presented. Thirdly, an application to a hypothetical road network is given with comparative results from our dynamic and traditional static network design approaches. Finally, conclusions are drawn and future research is discussed.

13.2 Modeling travelers' departure time/route choice behavior

The network performance evaluation for network designs is based on travelers' choice modeling. Traveler's choice behavior under uncertainty, for instance route choice, departure time choice and mode choice, has recently gained increasing attention. Especially, departure time adaptation appears even more significant than route choice adaptation for the sake of attempting to arrive on time at work and to reduce the schedule delay, and minimizing the travel costs by departing earlier. Thus, the impacts of travel time unreliability on traveler's choice behavior, especially on departure time choice, are badly needed to be investigated. Modeling travelers' departure time/route choices is crucial in network modeling, network evaluation and network design under uncertainty.

We work in the framework of utility theory. It is assumed that travelers make their departure time/route decisions in order to maximize their utility (i.e., minimize the disutility). Therefore, a utility function is defined to model travelers' choice behavior. In tradition with deterministic demand and capacities, the disutilities for a traveler making a trip are for instance travel time, schedule delay early and late, tolls, fuel costs, etc. A typical utility function in the deterministic case for modeling travelers' departure time/route choice, widely adopted, is for example:

$$u_p^{od}(t) = -\alpha \tau_p^{od}(t) - \gamma_1 \left(PAT - \left(t + \tau_p^{od}(t) \right) \right)^+ - \gamma_2 \left(t + \tau_p^{od}(t) - PAT \right)^+ \quad (13.1)$$

where $u_p^{od}(t)$ denotes the travel (dis)utility on route p between origin-destination (OD) (o, d) when departing at time instant t. $\tau_p^{od}(t)$ denotes travel time on route p between OD pair (o, d) when departing at time instant t. PAT denotes the preferred arrival time. The second and the third components in the utility function denote the schedule delay early and late costs respectively. Function $(x)^+$ is equivalent to $\max\{0, x\}$, since there is either early schedule delay or late schedule delay on a specific day. It can never occur that travelers experience both delay costs at the same trip. Parameters α, γ_1, and γ_2 denote value of travel time, values of schedule delay early and late, respectively. Homogeneous travelers are assumed in this utility function with an identical PAT and the same parameter values. This travel cost function assumes linearity in schedule delay costs which simplifies the reality. It is assumed that travelers have perfect information about the traffic conditions they may encounter at all potential departure times. Travelers make trade-offs between all the disutility components.

In transport networks, due to the stochastic supply and fluctuating travel demand, travel times and travel costs experienced by traveler's within-day and between-days are stochastic. The travel time for a specific departure time instant follows some probability distribution. The influence of travel time variability on travelers' choice behavior under uncertainty has gained increasing attention. A lot of studies [47] have been carried out analyzing the impacts of travel time variability on travelers' departure time/route choice behavior and on how to model their choice behaviors under uncertainty. Several empirical studies, e.g., [1, 5–8, 27], suggest that travelers are interested not only in travel time savings but also reduction of travel time variability. Most travelers prefer reliable travel times.

Different hypotheses have been proposed and different models have been assumed to model travelers' choice behavior under uncertainty. A hypothesis of a 'safety margin' being selected by travelers has once been specified [26], which assumed that travelers make their choice decisions by considering the expected travel time and adding an extra time budget, the so-called safety margin, to cope with the uncertainties. Travelers might also trade off their expected travel time and travel time variability based on their past experiences. However with these hypotheses, the impacts of travel time variability on scheduling costs and travelers' reaction on the scheduling cost uncertainty are not explicitly considered, which turns out to be very important in representing travelers' choice behavior under uncertainty [42].

A mean-variance approach has been proposed [23], in which the individual travel cost function is composed of expected travel time and travel time variance (or standard deviation of travel time). It hypothesizes that travelers are interested not only in travel time savings but also a reduction of travel time variability. Variability of travel time is usually measured by the standard deviation of travel time [43].

Due to the stochastic properties of travel times, a traveler departing everyday at the same time instant t may arrive early, late or on time as preferred, largely relying on the traffic situations on that specific day. Travel time variability directly leads to uncertainty in arrival times at the destination. Departure time shifts are very important reactions of travelers to travel time unreliability, since travel time unreliability causes uncertainty of their arrival time, which cause punishment for delays. Travelers attempt to depart earlier in order to arrive at the destination on time and reduce the schedule delay costs. Several empirical studies show that scheduling costs play a major role in the timing of departures [42].

A scheduling approach based on the expected utility theory [40] has originally been specified [38], which hypothesizes that scheduling delay cost plays a very important role in the timing of departures under uncertainty. However, it is found that scheduling delay costs cannot capture the travel time unreliability completely [37, 47]. Besides a scheduling effect, travel time unreliability appears to be a separate source of travel disutility. Travelers may not only consider the expected travel time (i.e., duration of their trip), the expected schedule delay costs, but also the variability in travel times as an indicator representing travelers' perceived uncertainty.

Based on the research [29], it is proven that the mean-variance approach and the scheduling approach are equivalent conditional to departure times. A generalized utility function has been proposed to model travelers' departure time and route choice behavior under uncertainty, given by:

$$u_p^{od}(t) = -\alpha E\left[\tilde{\tau}_p^{od}(t)\right] - \gamma_1 \left(PAT - \left(t + E\left[\tilde{\tau}_p^{od}(t)\right]\right)\right)^+$$
$$- \gamma_2 \left(t + E\left[\tilde{\tau}_p^{od}(t)\right] - PAT\right)^+ - \beta std\left[\tilde{\tau}_p^{od}(t)\right], \quad \forall (o,d), p, t, \quad (13.2)$$

where $u_p^{od}(t)$, $\tilde{\tau}_p^{od}(t)$ denote the individual's travel (dis)utility and stochastic travel times on route p between OD pair (o, d) departing at t under uncertainty respectively. In this chapter all stochastic variables are denoted with a tilde. $E\left[\tilde{\tau}_p^{od}(t)\right]$ and $std\left[\tilde{\tau}_p^{od}(t)\right]$ denote the expectation of travelers' experienced travel times and the standard deviation of travel times on route p between OD pair (o, d) departing at t respectively. Again the function $(x(t))^+ = \max\{0, x(t)\}$, and PAT denotes the preferred arrival time. The second and the third component are the schedule delay costs of being early and late based on expected travel time respectively. Parameter β is the value of reliability. With this utility function, it is assumed that travelers make trade-offs between expected travel time, schedule delay early and late based on expected travel time and the travel time variability, and then make departure time/route decisions. It is assumed that travelers are fully aware of the distributive properties of their experienced travel times based on the past cumulative experiences under uncertainty. The utility function (13.2) will be adopted in this chapter to model travelers' departure time and route choices under uncertainty caused by stochastic capacities from day to day.

13.3 Reliability-based dynamic network design

The network design problem (NDP) has been recognized as one of the most difficult and challenging problems in the transport field. This section proposes an integrated reliable network design problem with two main actors involved, namely road authorities/designers on the network level and individual travelers on the lower level. In the circumstance with uncertainties, travelers make their travel decisions considering travel costs and travel time reliability. Road authorities/designers also pay increasing attentions to the reliability of the services provided to road users. Thus the network design problem is a typical bi-level optimization problem. Joint departure time and route choices are modeled on the lower level under uncertainty, aiming to capture the dynamics in the network with stochastic capacities and to evaluate the network performance more reasonably. Reliability will be modeled on both the network design level and individual choice level.

A long term equilibrium departure pattern is achieved by assuming that travelers make their strategic departure time choice based on their past experiences. A deterministic departure pattern will result in which no traveler can increase his/her long term future trip utilities (defined by utility function (13.2)) by unilaterally changing his/her departure time. With this deterministic long term departure pattern, the within-day traffic situation is, however, not necessarily an equilibrium due to the varied capacities from day to day. It is assumed that travel time variability is only caused by stochastic capacities representing day-to-day dynamics, while within-day capacity is assumed constant.

A reliable road network design problem constrained to a reliability-based long term user equilibrium with joint departure time/route choice modeling is proposed and formulated as following. Time will be discretized into time intervals $k \in K$. A vector $\mathbf{L}$ of numbers of lanes on a set of potential links is the design variable. The design problem is a Mathematical Program with Reliability-based Dynamic Equilibrium Constraints (MPRDEC). As indicated by [28, 34], an MPEC (Mathematical Program with Equilibrium Constraints) is a special case of a bi-level programming problem:

$$
\bar{\mathbf{L}} = \underset{\mathbf{L} \in \Lambda}{\arg\min} \left(\vartheta E \left(\sum_{(o,d)} \sum_p \sum_k \tilde{\tau}_p^{od} \left(k \,|\, \mathbf{L}, \bar{\mathbf{f}} \right) \bar{f}_p^{od} \left(k \,|\, \mathbf{L} \right) \right) \right.
$$

$$
\left. + (1-\vartheta)\, std \left(\sum_{(o,d)} \sum_p \sum_k \tilde{\tau}_p^{od} \left(k \,|\, \mathbf{L}, \bar{\mathbf{f}} \right) \bar{f}_p^{od} \left(k \,|\, \mathbf{L} \right) \right) \right), \qquad (13.3)
$$

where the objective function (13.3) aims to derive the optimal lane strategies $\bar{\mathbf{L}}$ to minimize a weighted sum of the average total network travel time (TNT) and network travel time unreliability (for short, objective mean+std of TNT). $\bar{f}_p^{od}(k)$ is the equilibrium flow departing at time interval k on route p between OD pair (o, d). $\bar{\mathbf{f}}$ is the equilibrium flow matrix on all routes in all departure time intervals. $\tilde{\tau}_p^{od} \left(k \,|\, \mathbf{L}, \bar{\mathbf{f}} \right)$ is the stochastic travel time at time interval k on route p between OD pair (o, d) given the equilibrium flow pattern $\bar{\mathbf{f}}$. The total network travel time distribution at long term equilibrium varies according to the given design strategy $\mathbf{L}$.

The design problem is subject to reliability-based dynamic long term equilibrium constraints under stochastic capacities (13.4), the budget constraint (13.6), and integer solution space limitations (13.7):

$$\sum_{o,d} \sum_{p\in P^{od}} \sum_k \bar{F}_p^{od}(k\,|\bar{\mathbf{f}},\mathbf{L}) \left(f_p^{od}(k) - \bar{f}_p^{od}(k) \right) \geq 0, \quad \forall \mathbf{f} \in \Omega_d, \qquad (13.4)$$

where

$$\bar{F}_p^{od}(k) = \left(\bar{f}_p^{od}(k) - \hat{f}_p^{od}(k) \right) \frac{\partial \bar{c}_p^{od}(k)}{\partial f_p^{od}(k)}, \qquad (13.5)$$

$$I(\mathbf{L}) \leq \mathbf{B}, \qquad (13.6)$$

$$\mathbf{L} \in \Lambda, \quad \Lambda = \{\mathbf{L}|L_a = 0,1,2,\ldots,L_a^{\max}\}, \qquad (13.7)$$

where $\hat{f}_p^{od}(k)$ is the intermediate departure flow, calculated by the OD demand for (o, d) multiplied by the proportion of travelers choosing route p among all route alternatives for an OD pair (o, d) and departure time interval k. The proportion can be calculated by any probabilistic model such as Path-Size logit model, C-logit model, Path Size Correction Logit (PSC-Logit) [6], which account for the spatial overlap of routes. Ω_d is the set of feasible dynamic route flow matrices $\mathbf{f}$, defined by the flow conservation (13.8) and non-negativity (13.9) constraints, respectively:

$$\sum_{p\in P^{od}} \sum_k f_p^{od}(k) = q^{od}, \quad \forall (o,d), \qquad (13.8)$$

$$f_p^{od}(k) \geq 0, \quad \forall (o,d), p, k, \qquad (13.9)$$

where q^{od} is the total travel demand between (o, d).

Expression (13.4) is the finite-dimensional VI formulation of the long term probabilistic user equilibrium with stochastic networks in a discrete-time version. The equilibrium travel costs and equilibrium departure flows are dependent on the design strategies. Expression (13.6) gives the budget inequality constraints. B is the available budget for the construction. The construction cost, denoted by $I(\mathbf{L})$, is a function of lane design $\mathbf{L}$ and the lengths corresponding to $\mathbf{L}$. The construction cost is computed as a sum of the construction costs over all the links. The construction cost on link a, denoted by I_a, can be formulated as a function of L_a and the length of link a, denoted as d_a:

$$I_a = x(L_a, d_a) L_a d_a, \qquad (13.10)$$

where $x(L_a,d_a)$ is the average construction cost per kilometer per lane, which firstly decreases and then increases with the number of lanes. Expression (13.7) is called the domain constraint with a restriction on the maximum number of lanes $L_a^{\max}$ for each link. Note that zero-lane is also an option, which means not building a certain road segment. The problem must be solved for the values of the variables L_a that satisfy the restrictions and meanwhile minimize the objective function. A vector $\mathbf{L} \in \Lambda$ satisfying all the constraints is called a feasible solution to the optimization problem. The collection of all feasible solutions forms a feasible region.

Given the number of lanes for a link in the potential network, the link capacity $\tilde{C}_a$ is assumed a stochastic continuous variable, following a distribution P, expressed as:

$$\tilde{C}_a(L_a) \sim P\left(\mu_a(L_a), \sigma_a^2(L_a)\right), \tag{13.11}$$

where $\mu_a(L_a)$ and $\sigma_a(L_a)$ are the expectation and standard deviation of the link capacity distribution. Clearly the characteristics of the link capacity distribution are functions of the numbers of lanes. The standard deviation of the link capacities is a function of the mean capacity depending on the number of lanes, expressed as:

$$\sigma_a = f(\mu_a(L_a)). \tag{13.12}$$

The larger the number of lanes of a link, the smaller its relative capacity variation. For instance, with a one-lane road, the standard deviation of its capacity is about 10% of the mean capacity [10], while with a two-lane road, the standard deviation is 8% of the corresponding mean capacity.

13.4 Solution approach

13.4.1 A simulation-based approach for modeling the long term user equilibrium

A simulation-based approach is developed and a framework is established to model the long term user equilibrium under uncertainty, which is depicted in Figure 13.1. The simulation-based approach is able to model stochastic/deterministic networks, dynamic/static user equilibria, with/without departure time choices.

The inputs of the model are static OD demand, the travel cost function with parameter values and *PAT*, and the infrastructure network with link capacities. There are two loops in the simulation-based approach: 1) an inner loop with DNL (Dynamic Network Loading), which propagates the traffic flows through the network and models the within-day dynamics. Each iteration loop represents a hypothetical day with a randomly drawn link capacity vector by repeating the within-day dynamics with varied link capacity vectors; 2) an outer loop with simultaneous departure time/route choices. This is a typical process with simulation-based approaches.

The applied DNL model version simulating horizontal queue and spillback is taken from Bliemer [4]. The reason to choose this DNL model is that it does not rely on the unrealistic restrictions made by other models.

To well represent the capacity distribution, a quasi-Monte Carlo approach using Halton draws [20] is adopted to generate stochastic capacities.

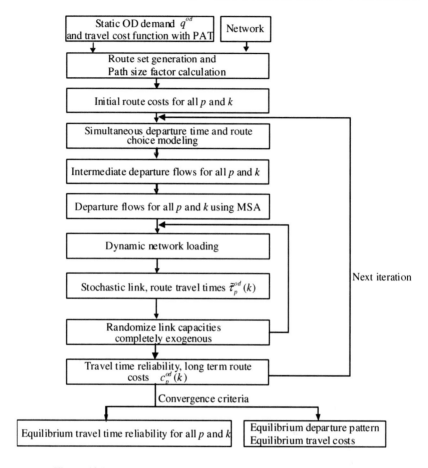

Figure 13.1: Framework of the iterative simulation-based approach.

13.4.2 A combined set evaluation and genetic algorithm

The formulated network design problem is a constrained combinatorial optimization problem. Well known algorithms for solving the optimization problem can be classified as either exact or approximate algorithms. An exact algorithm is the one which always produces an optimal solution if it exists. For finding the exact optimal solution for our formulated dynamic network design problem, a systematic search in the solution space is needed, for instance the exhaustive search and full enumeration of all possible designs. However such algorithm involves exponential asymptotical computational complexity. Therefore approximate or heuristic algorithms, which sacrifice the guarantee of finding the optimal solution and produce a feasible and hopefully very good solution in polynomial-time, are adopted frequently for finding a sufficiently good solution [16]. Promising feasible heuristic algorithms to solve the combinatorial problem are among others: local search (LS), simulated annealing (SA) [25], genetic algorithm (GA) [17, 39], ant colony optimization (ACO) [12, 13], tabu search (TS) [18, 19], particle swarm optimization (PSO) [14, 24], etc. SA, GA, ACO and PSO are naturally inspired metaheuristics. In this chapter, a combined genetic algorithm and set evaluation approach is adopted to solve the network design problem.

13.4.2.1 Optimization of the solution space

Prior to the optimizations, the set of lane designs firstly is filtered. Such a filtering is especially useful in a transport network design context where trips between origin and destination have to be served. A universal set of lane designs is defined, which contains all possible combinations of the number of lanes on all the potential links in the network satisfying the domain constraint with L_a^{max} for each link. The size of the universal set can be easily computed as a product of the possibility for link a over all links, expressed as: $\prod_\mathbf{a} \left(\mathbf{L_{a,max}} + 1 \right)$. Since zero-lane is also a possibility, the possible numbers of lanes for link a therefore is $\mathbf{L_{a,max}} + 1$. Each lane design in this universal set can be represented as a point in a multidimensional space where the links are the axes and the numbers of lanes are the values on these axes.

The master set is defined as the subset of the universal set (i.e., a subset of points in the multidimensional space), which contains all feasible, connected, and logical lane designs. A design satisfying the budget constraint is defined as a feasible design. An unconnected design is defined as a lane vector $\mathbf{L}$, which disables the connectivity between any OD pair in the network. Illogical designs are defined by lane vectors with which the roads constructed will never be used. The size of the master set is strongly dependent on the network structure. By eliminating the infeasible, unconnected, and illogical lane designs, the solution space can be reduced substantially, which saves computation time considerably. See for more details [29, 31].

13.4.2.2 Road network-oriented genetic algorithm

Upon the derived master set, a genetic algorithm is applied to find a fairly good or even the optimal design. A Genetic algorithm was firstly developed by Holland in 1975 [22]. GA is used to produce and evolve the lane designs from the master set toward a fairly good or the optimal solution for the network design problem. The fitness of each lane design is evaluated based on the user's long term equilibrium considering travel time reliability in departure time and route choices, which is simulated by the DTA model. GA iteratively evolves the lane designs until reaching the predefined stopping criteria. An initial set of lane designs, namely a population, is randomly selected from the master set. All the designs are different in the population. The size of the population is predefined according to the number of design variables.

There are three GA operators: crossover, mutation, and reproduction, with which the new lane designs (so-called offspring) are created from the designs from the previous generation (so-called parents). The behavior of GA is characterized by a balance between exploitation and exploration in the search space. The balance is strongly affected by several parameters such as size of the design set at each generation, coefficient of reproduction selectivity (a parameter involved in the fitness function), probability of crossover, probability of mutation, and maximum generations. The mutation probability p_m is decreased gradually along with the elapse of generations to facilitate the exploration in the whole solution space at the beginning and convergence at the end using the following equation (similar function, see [17]):

$$p_m = 0.5 - 0.4 \frac{g}{G_{\max}}, \qquad (13.13)$$

where g denotes the current generation number and $G_{\max}$ denotes the maximum number of generations. The mutation probability will decrease from 0.5 to 0.1 as the number of generations increases to $G_{\max}$.

At each generation, the best two designs from the previous generation (the so-called parents) are kept in the set for current generation (the so-called offspring). Each design is characterized by its fitness derived from the simulation with the DTA model. The fitness of each design L is defined as the inverse of the objective function in the optimization problem. The fitness function in some literature is calculated as a negative exponential transformation of the objective function. According to the fitness, each design has a probability of being selected to generate new designs. The higher the fitness, the higher the possibility it will be selected. Whether to perform crossover or not and the selection of designs are random processes. The selection process used in this study is the so-called 'roulette wheel' [39]. Crossover point and mutation position are all randomly chosen.

All the evaluated designs after each generation are stored in a pool with the fitness, all the network cost components (for instance, average total network travel time at equilibrium, standard deviation of total network travel time, total schedule delay early and late, construction costs, etc.) and the duality gap. At each generation, the newly generated designs are checked before the evaluation to make sure that whether it has already been evaluated in previous generations or not. If so, the fitness can be directly retrieved from the pool to avoid repeated evaluations. In case an infeasible, or unconnected or illogical lane design is generated, a new design from the master set is randomly selected to take place. A real value encoding is adopted.

The convergence of GA is considered to be achieved when the average of the set performance will not be significantly improved for M successive generations, or a maximum number of N generations has been performed.

13.4.2.3 A combined genetic and set evaluation approach

In order to find a fairly good solution or the optimal solution, more generations are needed with GA. However, it takes very long computation time due to the time consuming simulations to reach the dynamic user equilibrium with randomized capacities. Therefore, a set evaluation approach is proposed and combined with GA. This means that firstly a full evaluation of all the designs in the master set is performed with a reliability-based static assignment with stochastic capacities. Then, a number of the top designs in terms of network performance are selected and eval-

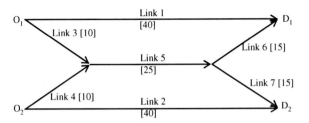

Figure 13.2: Network description.

uated by the reliability-based dynamic traffic assignment model. The best design is derived with the set evaluation. The GA is also applied. If the solution derived from GA is found in the top design set from static network design approach, then we think the set is sufficiently large and the set evaluation derives a better solution than GA. Set evaluation is quite efficient in terms of computation time and enables a better solution for the dynamic network design problem.

13.5 Case study

This section will present an application of the proposed dynamic network design approach to a hypothetical network for a complete network design. All the potential links in the network are designed with the possibility to have zero lanes. The dynamic network design approach will be compared to the static network design approach. The purpose of this chapter is to show the feasibility and the merits of the proposed reliability-based dynamic network design approach.

13.5.1 Network description

The proposed reliability-based dynamic discrete network design approach is applied to a hypothetical network with 7 potential links, as depicted in Figure 13.2, given the land use patterns. The length of each link is given in square brackets with a unit of kilometer. In total 6 routes are potentially available between the two origins and destinations. Routes are numbered from 1 to 6 with the route-link table given as in Table 13.1. The total travel demand for each OD pair is given in Table 13.2 in vehicles.

Each link capacity is assumed to follow a normal distribution. Fifty Halton draws are used to replicate the link capacity distribution in the quasi-Monte Carlo process. The domain constraint is defined as $L_{max} = 3$, restricting a maximum number of lanes for each potential link to three. It is assumed that the standard deviation is equal or less than 10% of the mean capacity, depending on the number of lanes. The mean

Table 13.1: Link incidence matrix for the potential complete network.

		Link 1	Link 2	Link 3	Link 4	Link 5	Link 6	Link 7
(O_1, D_1)	Route 1	1	0	0	0	0	0	0
(O_1, D_1)	Route 2	0	0	1	0	1	1	0
(O_1, D_2)	Route 3	0	0	1	0	1	0	1
(O_2, D_1)	Route 4	0	0	0	1	1	1	0
(O_2, D_2)	Route 5	0	0	0	1	1	0	1
(O_2, D_2)	Route 6	0	1	0	0	0	0	0

Table 13.2: OD demand (unit in vehicles).

	D_1	D_2
O_1	5000	1000
O_2	2000	5000

capacities for one-lane, two-lane, and three-lane are 2150, 4300, and 6900 pcu/hour, and the standard deviations are 215, 350, and 500, respectively.

The path-size factor depends on the network structure and the route sets in this potential network. Different designs will result in different network structures and route sets, thus changed path-size factors. If all the potential links are built, the path-size factors are computed for all the 6 routes sequentially as [0, 1.0397, 1.0397, 1.0397, 1.0397, 0]. For general networks, path-size factors need to be recomputed for different designs.

According to the definition of unconnected designs, links 3, 4, 5, 6, 7 cannot have zero lanes to preserve the connectivity between OD pairs (O_1, D_2) and (O_2, D_1) in the network. Illogical routes do not exist for this hypothetical network. Therefore the path size factors will remain the same regardless of the network structures and designs for this hypothetical network. For this hypothetical network, PSC-logit is applied for the traffic assignment.

A value of time of 10 euros/hour has been adopted. Relative parameters of 1.2, 0.8, and 1.2 are utilized in the travel cost function for travel time reliability, schedule delay early and schedule delay late respectively. The impact of the different parameter settings on the design solution will be investigated in future work. A time period of 90 minutes is modeled with $PAT = 80$ [min]. The maximum speed on each link is 100 km/hour. A time step of 30 seconds is adopted in the dynamic network loading, while the minimum link travel time in the network is 6 minutes. A departure time period of 1 minute is used in modeling the departure flows.

A construction cost of about 210,000 euros/km/lane/year is taken from [35]. Lack of cost information forced us to use only rough cost factors. Since we work on a constrained network design, for this hypothetical network the construction cost calculation does not influence the designs since the budget and the construction costs are scalable and not a component in the objective function. The demand modeled in this case is assumed 20% of the daily total demand, 5 days a week and 52 weeks per year are accounted for. Thus the total travel cost for a year is computed by multiplying the total travel cost derived from the simulation for 90 minutes by 1300.

With the genetic algorithm, a decreasing mutation probability from 0.5 to 0.2 is taken. The convergence of GA is considered to be achieved with $M=10$ and $N=50$.

13.5.2 Design solutions

This section presents the design solution from the reliability-based dynamic road network design approach. The purpose is to investigate the effects of dynamic network modeling and departure time choice on network designs. The static network design approach is applied as well to compare different network design approaches.

A sufficient set of the top 100 designs from static designs is evaluated with dynamic traffic assignment model. The GA is also applied to the reliability-based dynamic network design problem. The solution given by GA appears to be one of the designs in the 100 designs, which means that the evaluation of the set of designs will lead to a better solution than the GA algorithm.

Table 13.3 presents the weighted sum of the mean and standard deviation of total network travel time and total network travel costs by components of the top 10 designs. However with a static network design approach, the top 10 designs are different from the dynamic network design approach, as seen in Table 13.4.

It can be seen that $\bar{L}_d$ (denotes the design solution from dynamic network design) of [3 2 2 2 2 1 1] is different from the solution $\bar{L}_s$ of [2 3 1 1 2 2 1] from the static network design approach. It appears that the static solution $\bar{L}_s$ from the static network design is not even a good solution in the top 10 best designs from dynamic network design approach. Because of its poor travel time estimations, a static traffic assignment may lead to poor approximation of the network performance, thus to poor network designs.

It is also found that in general the estimated total network travel times from dynamic traffic assignment expectedly are much higher than from static assignment since the dynamics and spillbacks are captured. It is also noticed that larger variations of total network travel time (the coefficient of variation of TNT is about 3–4%) appear than in the static network design case (about 2.2%). Static traffic assignment significantly underestimates the total network travel time and the variability of total network travel time, thus also underestimates the value of the investments.

Travelers make a trade-off between the route travel times, route travel time reliability, schedule delay costs at all available departure time intervals and make their route and departure time choice accordingly. For comparison, Figure 13.3 and Figure 13.4 present the equilibrium departure patterns on all routes in the network with the dynamic design $\bar{L}_d$ and the static design $\bar{L}_s$.

Table 13.3: Dynamic network design evaluations on a yearly basis.

Lane designs	Weighted mean+std of TNT (Hours/ year)	Total travel costs (Millions of Euros/ year)	Mean of TNT (Hours/ year)	Std of TNT (Hours/ year)	Total travel time costs (Millions of Euros/ year)	Total re-liability costs (Millions of Euros/ year)	Construc-tion cost (Millions of Euros/ year)
[3 2 2 2 2 1 1]	7040815	158,10	11489484	367812	114,89	7,72	66.68
[3 2 1 2 2 1 1]	7116073	160,94	11615972	366226	116,16	8,21	64.59
[2 2 3 2 2 2 1]	7238355	170,79	11736390	491302	117,36	8,86	63.55
[2 2 3 3 2 2 1]	7238411	170,49	11735793	492339	117,36	8,91	65.63
[2 2 2 3 2 3 1]	7288140	171,65	11816261	495960	118,16	8,92	66.68
[2 2 2 3 2 2 2]	7293935	171,69	11827147	494116	118,27	8,94	66.68
[2 2 1 3 2 2 2]	7295356	171,56	11829379	494322	118,29	8,96	64.59
[2 2 2 2 2 1 2]	7295545	172,85	11848845	465595	118,49	9,25	61.47
[2 2 2 2 2 1 3]	7302294	172,93	11857264	469840	118,57	9,29	64.59
[2 2 3 2 2 3 1]	7314904	171,98	11857998	500262	118,58	8,92	66.68

Table 13.4: Static network design evaluations on a yearly basis.

Lane designs	Weighted mean+std of TNT (Hours/ year)	Total travel costs (Millions of Euros/ year)	Mean of TNT (Hours/ year)	Std of TNT (Hours/ year)	Total travel time costs (Millions of Euros/ year)	Total re-liability costs (Millions of Euros/ year)	Construc-tion cost (Millions of Euros/ year)
[2 3 1 1 2 2 1]	5041101	85.77	8278582	184878.5	82.79	2.98	65.63
[3 2 2 2 2 1 1]	5044122	85.87	8288688	177274.3	82.89	2.98	66.68
[3 2 1 3 2 1 1]	5046498	85.92	8289389	182162.3	82.89	3.02	66.68
[3 2 1 2 2 1 1]	5049941	85.96	8291679	187334.8	82.92	3.04	64.60
[3 2 1 1 2 2 1]	5097991	86.96	8354704	212922.3	83.55	3.4	65.63
[3 2 1 1 2 1 2]	5117852	87.36	8391531	207333.1	83.92	3.45	65.63
[2 3 2 2 2 1 1]	5118624	87.43	8386537	216755.3	83.87	3.58	66.68
[3 2 3 1 2 1 1]	5123687	87.44	8399110	210553.1	83.99	3.45	66.68
[2 3 1 3 2 1 1]	5123973	87.54	8391606	222524.6	83.92	3.63	66.68
[3 2 2 1 2 1 1]	5124443	87.50	8397241	215245.5	83.97	3.53	64.60

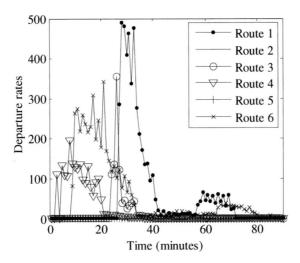

Figure 13.3: Dynamic equilibrium departure patterns with $\bar{L}_d$ [3 2 2 2 2 1 1].

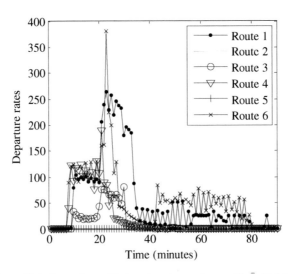

Figure 13.4: Dynamic equilibrium departure patterns with $\bar{L}_s$ [2 3 1 1 2 2 1].

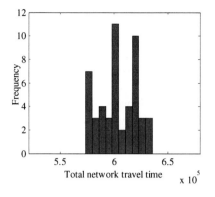
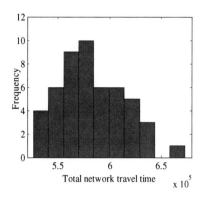

(a) Dynamic solution $\bar{\mathbf{L}}_d$ based on the dynamic assignment.

(b) Static solution $\bar{\mathbf{L}}_s$ based on the dynamic assignment.

Figure 13.5: Total network travel time distribution.

Different equilibrium departure patterns over time appear for both designs, which directly influence the network performance. With designs for instance $\bar{\mathbf{L}}_s$, departures from different OD pairs start and finish within almost the same time region whereas with designs for instance $\bar{\mathbf{L}}_d$, the departures from different OD pairs appear relatively separate over time. The dynamic design solution $\bar{\mathbf{L}}_d$ leading to relatively separate departures among the four OD pairs, explains the best network performance. A wider distribution of departures over time among different OD pairs will generate lower total network travel time and total network travel costs. The distribution of departures is a result of the designs. Therefore the design solution, leading to the best network performance, makes the most efficient utilization not only of the physical capacity itself, but also the efficiency over time. Modeling departure time choice is very crucial for network designs.

A static assignment model cannot capture the dynamics over time, even not to mention the departure dynamics over time which has significant influence on network performance. A static network design approach only attempts to utilize the physical capacity, not the capacity over time. Utilizing network capacity over the time dimension can reduce congestions, thus result in a better network performance. That's the reason why some research has suggested that a wider distribution of *PAT* from the employer may help alleviate the congestions [2]. A wider distribution of *PAT* will lead to relatively separate departures from different OD pairs, which has been shown to lead to lower network costs and better network performance.

For comparison, Figure 13.5(a) and Figure 13.5(b) present the total network travel time distributions based on the dynamic assignment caused by the stochastic link capacities with the design solutions $\bar{\mathbf{L}}_d$ and $\bar{\mathbf{L}}_s$. It is found that with the dynamic solution $\bar{\mathbf{L}}_d$, the total network travel time at long term equilibrium is narrowly distributed, while with static solution $\bar{\mathbf{L}}_s$ the total network travel time exhibits much larger variability. It means the design $\bar{\mathbf{L}}_d$ is much more reliable than $\bar{\mathbf{L}}_s$. In this sense also a static network design approach may lead to a poor network design.

13.6 Conclusions and future research

This chapter aimed to design road network infrastructures in order to provide reliable and efficient services to road users considering the stochastic property of the road infrastructures (i.e., stochastic capacities of road network infrastructures) and its impact on travelers' choice behavior.

A reliability-based network design approach is developed, for which reliability is modeled on both the network design level and the individual choice level. Dynamic traffic assignment is performed to evaluate the network performances under uncertainty. A simulation-based approach is established to model the long term user equilibrium with stochastic capacities.

The numbers of lanes on all potential links are the design variables. A combined GA and set evaluation algorithm has been developed to solve the dynamic network design problem. The solution space is filtered prior to the optimization by eliminating the infeasible, unrealistic and illogical designs to save computation time.

The proposed dynamic network design approach has been applied to a hypothetical network for a complete design. For comparison, a static network design approach was applied as well. The proposed design approach appears feasible and enables identification of the best network designs.

It was found that static traffic assignment may lead to poor approximations of the network performance, thus poor network infrastructure design. Static traffic assignment underestimates the total network travel time, the unreliability of the total network travel time, and total network travel costs.

Modeling departure time choice is crucial when dealing with the network design problem. A dynamic network design approach with departure time choice aims to derive a fairly good or even the optimal allocation of the capacity over space and makes the best utilization of the network over time. Relatively separate departures among different OD pairs, as a result of the designs, lead to better network performances. This effect cannot be captured by a static network design approach.

In conclusion, reliability of the service that a road network can provide to its users should be considered and incorporated in the network infrastructure design procedure, since network capacities, travel demand, and travel costs experienced by travelers are stochastic. Modeling travel time reliability in travelers' choice behavior has significant influence on the network performance evaluations, thus the network designs. Road network infrastructure designs should be based on dynamic network modeling, instead of the traditional static modeling.

Road network infrastructure design is different from other infrastructure designs, for instance water networks, power networks, etc., because travel behavior of human being is involved. Travel behaviors of different travelers are very random. However, the principle of designing road networks, water networks, and power networks is similar, which aims to design the network capacities, for instance the number of lanes, the diameter of the water pipe, etc., to satisfy the demand, for instance travel demand, water demand, power demand, etc. in an efficient and reliable manner. For solving the network design problem, the proposed approach in this chapter to filter and to reduce the solution space prior to the optimization can in future research be used for any other network infrastructure design, to save computation time substantially.

Acknowledgements This publication is supported by the research programs of the Transport Research Centre Delft, The Dutch foundation "Next Generation Infrastructures", and TRAIL Research School.

References

1. M. Abdel-Aty, M. R. Kitamura, and P. Jovanis. Investigating effect of travel time variability on route choice using repeated measurement stated preference data. *Transportation Research Record*, 1493:39–45, 1996.
2. R. Arnott, T. Rave, and R. Schöb. *Alleviating Urban Traffic Congestion*. MIT Press, 2005.
3. Y. Asakura and M. Kashiwadani. Road network reliability caused by daily fluctuation of traffic flow. In *Proceedings of the 19th PTRC Summer Annual Meeting*, pages 73–84, Brighton, UK, July 1991.
4. M. C. J. Bliemer. Dynamic queuing and spillback in an analytical multiclass dynamic network loading model. *Transportation Research Record*, 2029:14–21, 2007.
5. E. A. I. Bogers and H. J. Van Zuylen. The importance of reliability in route choice in freight transport for various actors on various levels. In *Proceedings of the European Transport Conference*, Strasbourgh, France, October 2004.
6. P. H. L. Bovy, S. Bekhor, and C. G. Prato. The factor of revisited path size: Alternative derivation. *Transportation Research Record*, 2076:132–140, 2008.
7. D. Brownstone, A. Ghosh, T. F. Golob, C. Kazimi, and D. Van Amelsfoort. Drivers' willingness-to-pay to reduce travel time: Evidence from the San Diego I-15 congestion pricing project. *Transportation Research A*, 37(4):373–387, 2003.
8. D. Brownstone and K. A. Small. Valuing time and reliability: Assessing the evidence from road pricing demonstrations. *Transportation Research Part A: Policy and Practice*, 39(4):279–293, 2005.
9. C. Cassir, M. G. H. Bell, and Y. Iida. Introduction. In M. G. H. Bell and C. Cassir, editors, *Reliability of Transport Networks*, pages 1–10. Research Studies Press Ltd, Newcastle, UK, 2000.
10. AVV Transport Research Center. Handboek capaciteitswaarden infrastuctuur autosnelwegen (Capacity handbook of motorway infrastructure). Technical report, AVV Transport Research Center, Rotterdam, The Netherlands, 1999.
11. A. Chen, H. Yang, H. K. Lo, and W. H. Tang. Capacity reliability of a road network: An assessment methodology and numerical results. *Transportation Research Part B: Methodological*, 36(3):225–252, 2002.
12. A. Colorni, M. Dorigo, and V. Maniezzo. Distributed optimization by ant colonies. In *Proceedings of the European Conference on Artificial Life*, Paris, France, December 1991.
13. M. Dorigo, V. Maniezzo, and A. Colorni. The ant system: Optimization by a colony of cooperating agents. *IEEE Transactions on Systems, Man, and Cybernetics, Part B*, 26(1):29–42, 1996.
14. R. C. Eberhart and Y. Shi. Particle swarm optimization: Developments, applications and resources. In *Proceedings of the IEEE Congress on Evolutionary Computation*, Seoul, South Korea, May 2001.
15. E. B. Emam and H. Al-Deek. A new methodology for estimating travel time reliability in a freeway corridor. In *Proceedings of the 84th Transportation Research Board*, Washington, DC, January 2005.

16. J. R. Evans and E. Minieka. *Optimization Algorithms For Networks And Graphs*. Marcel Dekker Inc, 1992.
17. M. Gen and R.W. Cheng. *Genetic algorithms and engineering optimization*. John Wiley & Sons Ltd, 2000.
18. F. Glover. Tabu search – Part I. *ORSA Journal on Computing*, 1(3):190–206, 1989.
19. F. Glover. Tabu search – Part II. *ORSA Journal on Computing*, 2:4–32, 1990.
20. J. Halton. On the efficiency of certain quasi-random sequences of points in evaluating multi-dimensional integrals. *Numerische Mathematik*, 2:84–90, 1960.
21. B. Heydecker. Dynamic equilibrium network design. In *Proceedings of the 15th International Symposium on Transportation and Traffic Theory*, pages 349–370, Adelaide, Australia, July 2002.
22. J. H. Holland. *Adaptation in natural and artificial systems*. University of Michigan Press, Ann Arbor, Michigan, 1975.
23. B. W. Jackson and J. V. Jucker. An empirical study of travel time variability and travel choice behavior. *Transportation Science*, 6(4):460–475, 1981.
24. J. Kennedy and R. C. Eberhart. Particle swarm optimization. In *Proceedings of the IEEE International Conference on Neural Networks*, Piscataway, New Jersey, November 1995.
25. S. Kirkpatrick, C. D. Gelatt Jr., and M. P. Vecchi. Optimization by simulated annealing. *Science*, 220:671–680, 1983.
26. T. E. Knight. An approach to the evaluation of changes in travel time unreliability: A "Safety Margin" hypothesis. *Transportation*, 3:393–408, 1974.
27. T. Lam. *The Effect of Variability of Travel Time on Route and Time-of-day Choice*. PhD thesis, University of California, Los Angeles, California, 2000.
28. S. Lawphonegpanich and D. W. Hearn. A mpec approach to second-best toll pricing. In *Proceedings of the Triennial symposium on Transportation Analysis (TRISTAN) V*, Le Gosier, France, June 2004.
29. H. Li. *Reliability-based Dynamic Network Design with Stochastic Networks*. PhD thesis, Delft University of Technology, Delft, The Netherlands, 2009.
30. H. Li, M. C. J. Bliemer, and P. H. L. Bovy. Strategic departure time choice in a bottleneck with stochastic capacities. In *Proceedings of the 87th Transportation Research Board*, Washington, DC, January 2008.
31. H. Li, M. C. J. Bliemer, and P. H. L. Bovy. Reliability-based dynamic discrete network design with stochastic networks. In *Proceedings of the 18th International Symposium on Transportation and Traffic Theory*, Hong Kong, July 2009.
32. H. Li, P. H. L. Bovy, and M. C. J. Bliemer. Departure time distribution in the stochastic bottleneck model. *International Journal of ITS Research*, 6(2):79–86, 2008.
33. Y. Y. Lou, Y. F. Yin, and S. Lawphonegpanich. A robust approach to discrete network designs with demand uncertainty. In *Proceedings of the 88th Transportation Research Board*, Washington, DC, January 2009.
34. Z. Q. Luo, J. S. Pang, and D. Ralf. *Mathematical programs with equilibrium constraints*. Cambridge University Press, New York, New York, 1996.
35. A. M. H. Meeuwissen. Exchange between congestion cost and cost for infrastructure. In *Proceedings of the 19th Dresden Conference on Traffic and Transport Sciences*, Dresden, Germany, Septembeer 2003.
36. M. W. Ng and S. T. Waller. Reliable system optimal network design: a convex mean-variance type model with implicit chance constraints. In *Proceedings of the 88th Transportation Research Board*, Washington, DC, January 2009.
37. R. B. Noland and J. W. Polak. Travel time variability: A review of theoretical and empirical issues. *Transport Reviews*, 22(1):39–54, 2002.
38. R. B. Noland and K. A. Small. Travel-time uncertainty, departure time choice, and the cost of the morning commutes. *Transportation Research Record*, 1493:150–158, 1995.
39. M. Obitko. Genetic algorithms, 1998. URL: http://www.obitko.com/tutorials/genetic-algorithms/.

40. J. Polak. Travel time variability and departure time choice: A utility theoretic approach. Technical Report Discussion paper 15, Transport Studies Group, Polytechnic of Central London, 1987.
41. S. Sharma, S. V. Ukkusuri, and T. V. Mathew. A pareto optimal multi-objective optimization for the robust transportation network design problem. In *Proceedings of the 88th Transportation Research Board*, Washington, DC, January 2009.
42. K. A. Small. The scheduling of consumer activities: Work trips. *American Economic Review*, 72:467–479, 1982.
43. K. A. Small, R. B. Noland, X. Chu, and D. Lewis. Valuation of travel time savings and predictability in congested conditions for highway user-cost estimation. Technical Report NCHRP report 431, Transportation Research Board, National Research Council, 1999.
44. M. A. P. Taylor. Dense network traffic models, travel time reliability and traffic management: II application to reliability. *Journal of Advanced Transportation*, 33(2):235–244, 1999.
45. H. Tu. *Monitoring Travel Time Reliability on Freeways*. PhD thesis, Delft University of Technology, Delft, The Netherlands, 2008.
46. H. Tu, J. W. C. van Lint, and H. J. Van Zuylen. The impact of traffic flow on travel time variability of freeway corridors. *Transportation Research Record*, 1993:59–66, 2007.
47. D. Van Amelsfort, P. H. L. Bovy, M. C. J. Bliemer, and B. Ubbels. Travellers' responses to road pricing: Value of time, schedule delay and unreliability. In E. T. Verhoef, M. C. J. Bliemer, L. Steg, and B. van Wee, editors, *Pricing in Road Transport*. Edward Elgar Publishing Limited, 2008.
48. D. H. Van Amelsfort. *Behavioral Responses and Network Effects of Time-Varying Road Pricing*. PhD thesis, Delft University of Technology, Delft, The Netherlands, 2008.

Chapter 14
Dealing with Uncertainty in Operational Transport Planning

J. Zutt, A. van Gemund, M. de Weerdt, and C. Witteveen

Abstract An important problem in transportation is how to ensure efficient operational route planning when several vehicles share a common road infrastructure with limited capacity. Examples of such a problem are route planning for automated guided vehicles in a terminal and route planning for aircraft taxiing at airports. Maintaining efficiency in such transport planning scenarios can be difficult for at least two reasons. Firstly, when the infrastructure utilization approaches saturation, traffic jams and deadlocks may occur. Secondly, incidents where vehicles break down may seriously reduce the capacity of the infrastructure and thereby affect the efficiency of transportation. In this chapter we describe a new approach to deal with congestion as well as incidents using an intelligent infrastructure. In this approach, infrastructural resources (road sections, crossings) are capable of maintaining reservations of the use of that resource. Based on this infrastructure, we present an efficient, context-aware, operational transportation planning approach. Experimental results show that our context-aware planning approach outperforms a traditional planning technique and provides robustness in the face of incidents, at a level that allows application to real-world transportation problems.

J. Zutt, A.J.C. van Gemund, M.M. de Weerdt, C. Witteveen
Delft University of Technology, Department of Software Technology, Delft, The Netherlands,
e-mail: {j.zutt,a.j.c.vangemund,m.m.deweerdt,c.witteveen}@tudelft.nl

R.R. Negenborn et al. (eds.), *Intelligent Infrastructures*, Intelligent Systems, Control and
Automation: Science and Engineering 42, DOI 10.1007/978-90-481-3598-1_14,
© Springer Science+Business Media B.V. 2010

14.1 Introduction

Transportation is one of the strongest growing activities in our society, and as a result, there is an increasing need for improving the transportation process of public as well as freight transportation. One way to meet these efficiency demands is to automate significant parts of the transportation process. For example, in 1988, European Container Terminals realized the first ever robotized container terminal where completely controlled automated guided vehicles drive around 24 hours a day. Another initiative is the to be realized fully automated underground logistic system at Amsterdam airport, connecting the flower market of Aalsmeer, the airport, and a new rail terminal to one another. Also in public transportation automation has been applied. Already in 1939, General Motors presented a vision of "driver-less" vehicles moved under automated control at that year's World's Fair in New York. A fully automated highway was initially examined by General Motors during the late 1970s. Due to the advances in computing technologies, microelectronics, and sensors in the 1980s, the University of California program Partners for Advanced Transit and Highways has carried out significant research and development efforts in highway automation since the 1980s. In Europe, automation has been applied in public transportation as well. For example, in Paris, the underground railway is partly automated. Other examples of automation are car navigation systems that not only are capable to advise and assist, but also, maybe in the near future, will be used to monitor and partially direct human driver behavior.

One of the main promises of automation in public and freight transportation is to help in detecting and resolving possible *conflicts* (e.g., collisions) between large amounts of vehicles moving on a rather restricted infrastructure (highways, railways, or air corridors). For example, on-board computers help in detecting obstacles, assist the driver if sight is limited, and give advice if accidents are about to occur.

Another main feature of automation, however, is that it might also assist in *preventing* such conflicts to occur by careful *planning* of transportation activities, where route choices are determined and communicated in advance. One significant property of planning in transportation is that it occurs in a *distributed* way, without centralized control. Since, usually, every vehicle in a transportation process is planning its transportation independently from the other, conflicts have to be detected locally and solved in real-time. The traditional solution to preventing these conflicts between plans is as taken in our everyday traffic: a set of operational conflict-resolution rules such as traffic rules (keep right), traffic lights (semaphores), and dynamic traffic guidance systems ensure effective conflict resolution when the plans are *executed*, i.e., when the previously planned route choices are followed.

The problem with this traditional approach is that, first of all, due to the high load on road section resources, in reality travel times as planned are nearly unpredictable, even if no incidents occur. This means that, normally, the quality of plans cannot be guaranteed as vehicles are dependent upon the plans of others. In addition, operational conflict rules might not always be successful: the high load may easily lead to deadlocks, where a number of vehicles is waiting for one another without any progress. This, for example, might happen if four vehicles approach an intersection at the same time from different directions. Furthermore, as a consequence of this type of conflict resolution, the infrastructure is often used inefficiently: sometimes vehicles are waiting, while a short detour could have prevented this.

To solve these problems, we need to deal with conflict resolution not during the *execution* of the individual transportation plans, but during the *planning* phase. Therefore, a more sophisticated approach would be to split the problem into a first phase, in which routes are found for each of the vehicles individually, and a second phase, in which feasibility of the collection of routes is ensured, *before* these individual plans are actually executed. In fact, this form of plan conflict resolution can be considered as a form of *plan repair*, modifying individual plans if they are in conflict. As a representative of this approach, Broadbent et al. [3] employ a simple shortest path algorithm to find a set of initial routes. In case of *catching-up* conflicts, some vehicles are slowed down; for *head-on* conflicts, an alternative route is found that does not make use of the road at which the conflict occurred. Broadbent's algorithm can be used both on unidirectional and bidirectional infrastructures, but in the latter case it need not find the optimal solution. Like Broadbent et al., the approach proposed by Hatzack and Nebel [8] can also be considered as such a two-step approach. Compared with earlier approaches, however, Hatzack and Nebel use a more refined model of the infrastructure by considering parts of the infrastructure (such as lane segments and intersections) as resources having a limited capacity. Once the individual routes have been chosen, conflict resolution is then modeled as a *job-shop scheduling* problem with blocking to ensure that the capacity constraints on the resources are not violated.

Instead, however, of using such a two-step approach to distributed transportation planning, the approach in this chapter aims at a full *integration* of the route planning and the conflict-resolution process. To this end, we assume the existence of an automated, distributed, intelligent infrastructure that can be used by planners to realize a personal but so-called *context-aware* traffic plan. The basic idea here is that infrastructural resources, like road sections and crossings, maintain a list of reservations (time windows) of the use of that resource. While making their transportation plans, vehicles query a resource for its availability during some time interval and make time-window reservations for the use of that resource. Hence, individual vehicles are capable of planning a route in such a way that the influence of all plans of other vehicles have been taken into account, i.e., they are able to construct a *context-aware* route plan.

In this chapter, firstly, we elaborate upon the above idea for context-aware planning, because such a method can guarantee that traffic plans do not suffer from inherent unpredictability, since context-aware planning methods take into account conflicts in the planning phase instead of the execution phase. In particular, we not only review different existing variants of context-aware planning methods that may be used given such an intelligent infrastructure, we also present an algorithm that is optimal for one vehicle, given the reservations of the others, and computationally more efficient than any of its predecessors. We compare the performance of this improved context-aware planning method to a traditional planning method that solves conflicts in the plan execution phase.

Secondly, we address the distributed transportation problem when incidents *do* occur. In our case, we consider incidents to be events that cannot be anticipated in advance and that have a negative influence on the planned activities, such as a collision, or a vehicle with a motor that is malfunctioning. Note that if incidents occur, even our context-aware route planning algorithm again cannot guarantee conflict-free execution of individual transportation plans and we are forced, as in the traditional approach, to resolve conflicts in the execution phase. As we will show, however, we are able to revise plans to take the consequences of incidents into account by making some adaptations of the context-aware route planning algorithm. We show that also under incident conditions, a context-aware route planning approach performs better than resolving conflicts in the execution phase.

These contributions are presented in this chapter as follows. We first introduce our model of an infrastructure, vehicles, and requests, and a description of the types of solutions we expect (Section 14.2). Then we discuss some existing work on context-aware approaches, and introduce our improved method in Section 14.3. In Section 14.4 we propose two different ways to extend context-aware methods to deal with incidents. Thereafter, we evaluate these methods by running a number of simulations in a synthetic grid infrastructure, as well as in a realistic network of the Dutch national airport near Amsterdam.

14.2 A framework for distributed operational transport planning

To discuss the context-aware route planning algorithms that we compare with the traditional approaches, we first need to specify the building blocks of our approach to distributed transportation planning. In our framework, we distinguish a *transportation network* consisting of resources, *transportation requests (tasks)*, and *transport agents* who make *transportation plans* for vehicles to serve the transportation requests using the transportation network.

Transportation network A transportation network, or *infrastructure* is a tuple (R, E, K, D, S). Here, each of the n (infrastructure) *resources* $r \in Rr$ represents space that can be occupied by the transport agents, e.g., a road, a road segment, part of an intersection or a parking space. The directed *connectivity* relation $E \subseteq R \times R$ defines which infrastructure resources a transport agent can traverse to from a given infrastructure resource. The *distance* function $D : R \to \mathbb{R}^+$ gives, for each resource $r \in R$, the positive non-zero distance of traversing r.

For all infrastructure resources $r \in R$, the *capacity* function $K : R \to \mathbb{N}$ specifies the number $K(r)$ of agents that can use resource r simultaneously, and the *speed* function $S : R \to \mathbb{R}^+$ specifies the maximum allowed driving speed $S(r)$ at resource

r. Note that this maximum is defined independently from the vehicle using the resource.

Transportation requests A transportation request or *task* j is modeled by a tuple $(f_j, s_j, \tau_j^s, d_j, \tau_j^d, \pi_j)$ and specifies the request to pick up some amount of freight f_j to a certain source location $s_j \in R$ within a specified time window $\tau_j^s = [t_{j,1}^s, t_{j,2}^s]$ and to deliver it at a specified destination location d_j within a time window $\tau_j^d = [t_{j,1}^d, t_{j,2}^d]$. Associated with each request j there is a *reward function* $\pi_j : W \times W \to \mathbb{R}$, where W denotes a set of time windows. This reward is maximized if the request is executed within its time windows, and will typically be (much) smaller if one or both of the time windows of the transportation request are violated.[1]

Transport agents The transportation agents A are the transport planners. Each transport agent owns exactly one vehicle. The task of a transport agent is to compute a plan for the tasks (transportation requests) it accepted and then to execute this plan. Intuitively, each agent will try to maximize the rewards associated to its transportation requests, while minimizing the (traversal) costs. We assume that similar to infrastructure resources, for agents A there is a a *capacity* function $K : A \to \mathbb{N}$, specifying the maximum load capacity. If J_t is the set of transportation requests loaded by an agent $a \in A$ at time t, the constraint $\sum_{j \in J_t} f_j \le K(a)$ (i.e., the sum of all loaded freight is smaller than the capacity of the vehicle) always holds. Likewise, the *speed* function $S : A \to \mathbb{R}$ specifies the maximum driving speed $S(a)$ of the vehicle of agent a. Furthermore, at each point $t \in T$ in time, each transport agent $a \in A$ claims exactly the infrastructure resource r at which agent a resides at time t. If r_1 and r_2 are infrastructure resources with $(r_1, r_2) \in E$, and agent a holds a claim at infrastructure resource r_1, it should claim resource r_2 after having traversed resource r_1, but before (or when) releasing its claim on resource r_1.

Transportation plans Each agent $a \in A$ plans a route to execute the transportation requests it has been assigned. Such a route $Rt_a = (r_1, r_2, \ldots, r_k)$ for agent a is represented as a sequence of resources such that resources r_i and r_{i+1} are connected to each other, i.e., $(r_i, r_{i+1}) \in E$ for $1 \le i < k$. Accompanying this route Rt_a, the schedule Sd_a of agent a provides information on when each of these resources in Rt_a are claimed. Schedule $Sd_a = (t_1, t_2, \ldots, t_k)$ is a sequence of time points, where t_i specifies the time at which agent a claims resource r_i. This implies that agent a uses r_i during the time window $[t_i, t_{i+1})$ for $1 \le i < k$ and uses resource r_k during time window $[t_k, \infty)$. Obviously, at any time, the route and schedule have the same length, i.e., $\forall a \in A : |Rt_a| = |Sd_a|$.[2]

Although every individual agent plan might be feasible, it may be the case that the set of all agent plans is not feasible, because some of these plans may be in *conflict*.

---

[1] Allen's interval algebra [1] defines that a time-interval τ' is during time-interval τ if and only if time window τ' does not start before time window τ and does not end at a later time.

[2] We use $|S|$ to denote the cardinality of a set S.

Somewhat simplifying[3] we say that there is a conflict in the set of all agent plans if there is a subset $A' \subseteq A$, a resource $r \in R$, and a time t, such that at time t, every agent $a \in A'$ has a claim on resource r and $K(r) < |A'|$. If there is no such conflict, the set of agent plans is said to be conflict free.

The ultimate goal of solving a distributed transportation planning problem based upon this framework is to come up with a set of conflict-free route plans before these plans are executed. We describe such a route planning method in the next section.

14.3 Operational transport planning methods

In this section, we first briefly discuss some traditional approaches to operational transport planning. Then, we give the state-of-the art in *context-aware* routing, where conflicts between agents are removed during planning and before execution. At the end of this section, we propose a more efficient context-aware routing algorithm that we will also use in our experiments.

14.3.1 Conventional transportation planning approaches

The traditional solution to operational transport planning is to leave the planning to the individual agents, but to constrain plan execution in a way similar to the everyday traffic regulation approach: use a set of operational conflict-resolution rules, such as traffic rules (keep right), traffic lights, and dynamic traffic guidance systems, to ensure effective conflict resolution in the operational stage.

Using such an approach in our framework, the simplest way to determine the route for an agent, neglecting the presence of other agents on a given infrastructure, is to have each agent plan a shortest path from its current resource location to its destination resource. A basic *shortest-path algorithm*, such as Dijkstra [6], can be used by the agents to achieve this.

The resolution of operational conflicts is then done by defining resource usage rules that prioritize vehicles when they enter an intersection at the same time. These resource usage rules can be based on static aspects related to the importance of the task, or on dynamic aspects, such as who arrives first. Examples of such static resource usage rules are:

- **Task-priority**: when the rewards for executing tasks are not constant, agents that are executing tasks with a high reward should be given priority. For example, an ambulance, police, or fire brigade in action could precede regular traffic;

- **Vehicle-priority**: each type of vehicle is assigned its own priority level. For example, buses and trucks can be assigned higher priority than personal vehicles.

Examples of dynamic resource usage rules are:

---

[3] To define all types of conflicts that might occur is somewhat more involved. The reader is referred to the PhD thesis of Zutt [20] for a complete overview.

- **First-come-first-served**: for example, in the USA at a four-way-stop road crossing, the vehicle that enters a crossroad first gains the highest priority;

- **Longest-waiter-first**: during plan execution, agents possibly have to wait at several occasions. This rule is similar to the previous one, but sums the waiting times of multiple crossroads;

- **Urgent-deadline-first**: when agents are executing tasks with delivery deadlines, the agent that has the least slack should go first.

As remarked above, the problem with these traditional approaches is that travel times become almost *unpredictable*: an agent must at least have some knowledge of what the other agents are doing to know how this will affect its own plan. Even more important is the possibility that *deadlocks* occur, which means that the agents will not even be able to execute their plans. Finally, more knowledge about each others actions can improve the agent's decision making, which in turn may improve the performance of all agents.

More in general, the traditional approach can be compared with a pure *generate-and-test* approach to solving transportation planning problems: first, the routes are generated and then a test approach is followed, which detects conflicts during execution (e.g., the resource requested is already occupied) and then finds a solution to the problem (by using priorities). The resource-usage rules described above are much more effective, however, if one could push the test approach into the route generator. This implies that agents should use more conflict information available during the generation stage and hence make better decisions. This is the essence of the *context-aware routing* approach, which we propose as an alternative to the traditional approach that, from now on, will be used as a *baseline approach* (UNINFORMED) in comparison with the context-aware approach.

14.3.2 Context-aware route planning

Context-aware routing can be viewed as a multi-agent extension to single-agent shortest-path algorithms such as Dijkstra [6]. Typically, however, shortest-path algorithms do not take into account the plans of other agents while searching for a shortest time path from a source to a destination location. Taking into account the effect of other plans has some important consequences. For example, when a path to a node has been found in Dijkstra's shortest path algorithm, it is known that the current path to this node is the shortest, and the algorithm does not need to consider any other paths leading to this node. In context-aware routing, however, the first (and shortest) path to reach a resource is not necessarily the one that will result in the shortest path from the source to the destination, via the current resource. Consider the example in Figure 14.1. From the first (and only) free time window (a period without any reservations) on start resource r_s, we can reach both free time windows on resource r_1 (which is on a direct path to the destination resource r_d). However, from the first free time window on r_1, $f_{1,1} = [0, 2]$, we cannot reach any free time window on r_d, because on the destination resource there is a reservation

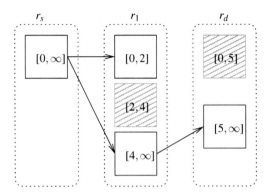

Figure 14.1: This graph represents the time windows at three resources. The shaded boxes denote time intervals reserved by other agents. The traversal time for each resource is one. The first arrival at resource r_1, at time 1, will not lead to a shortest path to destination resource r_d (in fact, a path to r_1 should first return to r_s since r_1 has been reserved from 2 to 4). Instead, we must consider the path that visits r_1 during its second free time window, which starts from 4.

until time 5. Hence, we must go from r_s to r_1 at time 4 (assuming travel times of 1 for all resources, by time 4 we will have had enough time to traverse r_s). Then, we can leave r_1 at time five, entering r_d at time 5, at the start of the free time window on the destination resource. This clearly shows that in context-aware routing we sometimes need to consider more than one visit to a resource due to reservations made by other agents.

Hence, a context-aware routing algorithm has to take into account not only whether a next resource is reachable from the current one, but also whether there is a suitable *free time window* on it, in which no reservations have been made by other agents. Consequently, for each location a set of disjoint free time windows is computed from all known reservations that exactly specifies the time windows at which the load of the location is smaller than its capacity (see Figures 14.2 and 14.3). Furthermore, a reachability relation is defined that specifies which free time windows at one location can be reached from which free time windows at another location (similar to arcs in a normal graph). When a shortest-path algorithm is used in this graph, each individual plan is optimal given that the reservations of all other agents do not change.

To illustrate this approach, let us look at the example depicted in Figure 14.2. The task is to route from source location r_s to destination location r_d starting at time $t = 0$, either by traveling via location r_1 or via location r_2, while taking into account the specified reservations of other agents. At first sight, it seems quicker to traverse via location r_1, because location r_2 is already reserved up to time $t = 6$. However, from location r_2 the journey cannot continue because of the reservation $[0, 5)$ in location r_d, and the agent is not allowed to wait in location r_1 due to the reservation $[2, 8)$ over there. Hence, the agent must wait in resource r_s until time $t = 8$ if it desires to travel the upper route. The free time-window graph gives somewhat more information. There is only an arc from location r_1 to location r_d from the free time window $[8, \infty)$, so one can immediately infer that the upper route has costs 9, and thus the lower route with costs is quicker with costs 7 (assuming the all locations have a traversal time of 1).

This single-agent source-destination routing computes a shortest path (in time) for an agent to traverse from its initial location to a destination location. However, in transportation planning, agents have to look further. They can be assigned multiple tasks, and hence have to create a plan to visit multiple locations one after another. The sequence of loading and unloading locations that an agent has to travel to is referred to as the *visiting sequence* of that agent. The route for a visiting sequence can be found by computing a shortest path between any two locations using context-aware routing, and by including the loading or unloading time at each location.

Unfortunately, finding a globally optimal plan (minimizing the sum of the costs) for all visiting sequences of all vehicles is NP-hard [13], so we cannot expect to find an efficient (polynomial time) algorithm for this problem. We will therefore restrict ourselves to an efficient algorithm that does not find optimal solutions. The idea is to apply a context-aware routing method each time an agent takes up a new transportation request. For a transportation request from a loading to an unloading location all possible ways of inserting these two locations in a visiting sequence are considered as long as the loading takes place before the unloading and the capacity of the vehicle is not exceeded. The agents insert such transportation requests in the order in which they arrive and without interleaving the planning with other agents. This causes the resulting plans to be usually sub-optimal. We will illustrate this sub-optimality by two examples. First, due to the arbitrary ordering: suppose that agent a_1 plans earlier than agent a_2 in this ordering of agents, and that agents a_1 and a_2 share some infrastructure resources in their routes. If an optimal plan requires that agent a_2 precedes agent a_1 in at least one of these shared infrastructure resources, this plan might not be found (in the case that agent a_1 reaches the infrastructure resource earlier than agent a_2). Second, due to the absence of interleaved planning: if agent a_1 first has to precede agent a_2, but later has to take priority in any optimal plan, then such a plan also cannot be found, because the agents create a complete plan for all of their transportation requests at once when it is their turn.

We would like to point out that there are a number of approaches related to this approach, in which route planning problems consist of finding a free path of resources from the origin to a destination, taking into account reservations that have been made by other agents using the same infrastructure. For example, the algorithm proposed by Huang et al. [9] finds a path through the (graph of) free time windows on the resources, rather than directly through the graph of resources. Huang et al.'s algorithm assumes unit capacity for all resources and is optimal both for unidirectional and bidirectional networks. Their algorithm runs in $O((n+m)^2 \log(n+m))$ for one vehicle, with n being the number of resources, and m the number of connections between the resources. Fujii et al. [7] combine the search through free time windows with a heuristic that calculates the shortest path from the current resource to the destination resource, assuming no other traffic. The solution method proposed should result in an optimal, polynomial-time algorithm, but the description

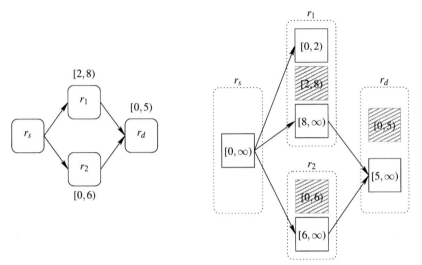

Figure 14.2: Transport network. Arcs represent connections between locations. The time intervals are reservations of other agents.

Figure 14.3: Free time-window graph. Connections specify the reachability relation between free intervals.

of the algorithm is not entirely correct, and the time complexity of this algorithm is unknown. The work of Kim and Tanchoco [10] is similar to the work of Fujii et al., but their treatment of the problem and the analysis of their algorithm is more comprehensive. Kim and Tanchoco's algorithm finds the (individually) optimal solution for both uni- and bidirectional networks, and they give an $O(v^4 n^2)$ time complexity for their algorithm, where v is the number of vehicles in the system, and n is the number of resources in the infrastructure. Because of this relatively high run-time complexity (especially given the limited computational resources of an early 90s PC), Taghaboni-Dutta and Tanchoco [17] developed an approximation algorithm that decides at every intersection to which resource to go next, based on the estimated traffic density of the resources from the current intersection to the destination. The authors show only a small loss of plan quality, and they claim that the algorithm consumes significantly fewer computational resources. However, the run-time complexity of this approximation algorithm is unknown, and we have not found any quantitative comparisons.

Finally, considering the approach by Hatzack and Nebel [8] in more detail, we see that their approach consists of two phases. In the first phase each agent computes a context-unaware shortest path from its current location to its destination location. In the second phase, the agents schedule their routes sequentially (each agent completes its reservations before the next agent), while preventing conflicts with of the other agents. At the end, all agents have a plan and the joint plan is guaranteed to be free of conflicts.

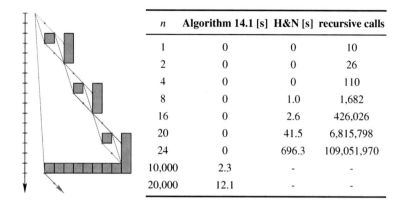

n	Algorithm 14.1 [s]	H&N [s]	recursive calls
1	0	0	10
2	0	0	26
4	0	0	110
8	0	1.0	1,682
16	0	2.6	426,026
20	0	41.5	6,815,798
24	0	696.3	109,051,970
10,000	2.3	-	-
20,000	12.1	-	-

Figure 14.4: The figure on the left illustrates how these difficult instances for the algorithm of Hatzack and Nebel are created. The vertical axis represents the progress of time. The horizontal axis represents the resources on the path from the source (on the left) to the destination (on the right). The rectangles represent already reserved time windows. The table on the right gives the time required to find a plan for different problem sizes n, and the number of recursive calls used by the algorithm of Hatzack and Nebel.

A drawback of their approach is that their algorithm uses backtracking. Since they do not make use of the idea that a free time window needs to be considered at most once, it is possible to construct examples in which the algorithm keeps backtracking through the same paths of time windows. Figure 14.4 depicts such an example in which this worst-case behavior is realized. This figure illustrates the reservations on the sequence of resources (horizontally from source resource s via $1, 2, 3, \ldots$ to destination resource d); vertically, the progress of time is shown. Each resource is assumed to have a traversal time of 1. To create such an instance where the algorithm needs $2^n + 1$ updates, no more than the following $5n$ reservations are needed in a route $s, r_1, r_2, \ldots, r_{3n}, d$ of $3n + 2$ resources:

- resources r_{3i-2} are reserved during $[5i-3, 5i-2)$ for $1 \leq i \leq n$,
- resources r_{3i} are reserved during $[5i-3, 5i)$ for $1 \leq i \leq n$,
- resources r_i are reserved during $[5n, 5n+1)$ for $1 \leq i \leq 3n$.

The table shown in Figure 14.4 shows that if such a structure of reservations occurs, the algorithm of Hatzack and Nebel will not be able to solve the instance within acceptable time. In the next section we propose an algorithm (Algorithm 14.1) that has no problem with these instances at all.

14.3.3 A more efficient context-aware routing algorithm

One of the disadvantages of the context-aware planning approach discussed above is its rather high run-time complexity; especially when the number v of vehicles is growing, this will prohibit an efficient context-aware route finding process. We therefore propose an alternative context-aware single-agent routing algorithm that is more efficient. This algorithm is embedded in the same approach for multiple agents discussed above, but has a time complexity that is much better than all of the above-mentioned existing methods (cf. [18]). This context-aware routing method is based on Dijkstra's shortest path method.

Let us consider Dijkstra's algorithm for computing a shortest path from a given source node r_s to all other nodes in order of their distance from r_s. The nodes for which the shortest path to s is already known are maintained in a list R_s. In each iteration one of the neighbors j of the nodes in R_s is added to R_s. The shortest path to each neighbor j using only nodes in R_s is simply the minimum of the distance to a node i in R_s plus the length of the edge from i to j. These neighbors are stored in a priority queue sorted on this distance. For the neighbor with the minimum distance in this queue, it is known that there is no shorter path using nodes outside of R_s, because all these nodes are further away. This node is then added to R_s, and the distances of its neighbors in the priority queue are updated if they can be reached through this node.

Our context-aware routing algorithm differs from Dijkstra's shortest path method in a number of aspects. First, we apply this method on a network in which the costs are in the resources, not in the connections between the resources. Second, we apply this method to search for a shortest path in the free time-window graph, and we make sure that only the relevant part of this time-window graph is represented. Third, we ensure that all constraints on the neighbor (connectivity) relation of two infrastructure resources are explicitly checked. Fourth, we sort the priority queue on the exit time of a certain resource (instead of the shortest distance to a node). Fifth, for an agent a we use the time distance function $D'(r) = \frac{D(r)}{\min(S(a),S(r))}$ instead of just the length of the edges between nodes. This then boils down to Algorithm 14.1.

Let us briefly go over the main steps in this algorithm. In Line 2, we initialize the priority queue Q of free time windows (with priority on exit time) to the start resource and the earliest exit time of this resource. In Line 5, we retrieve the time window (r_i, t_i) with the lowest cost exit time. To expand the current free time window, we consider in Line 9 and 10 all (resource, free time window) pairs that are reachable from (r_i, t_i). In Line 11, we check whether it is really possible for the agent to go from r_i with earliest exit time t_i through resource r_j within the interval $[t_j, t'_j)$. This check involves checking whether there is an exit time $t \leq t_i$ from r_i that is later than t_j and for which also $t + D(r_j) \leq t'_j$. Furthermore, it is ensured that this step does not involve a conflict with the reservations of other agents (see Section 14.2). If all constraints are met, this option is added to the priority-queue. Finally, we remove the free time windows up to t_j from resource r_j's set of free time windows, since we have already found the shortest path to this time window (since the exit time of r_i was the earliest). This is an important step, as it guarantees that we do not consider any free time window for expansion more than once.

To analyze the run-time complexity of this algorithm, we first place a bound on the number of time windows, using n for the number of resources in the infrastructure, and v for the number of vehicles. Assuming that vehicles visit each resource only a constant number of times, there are at most $O(nv)$ free time win-

Require: start resource s, destination resource d, start time t.
Ensure: exit time from d for the shortest path from s to d.
1: **if** s is free from t to $t+D'(s)$ **then**
2:      $Q \leftarrow \{(s,t+D'(s))\}$
3: **end if**
4: **while** $Q \neq \varnothing$ **do**
5:      $(r_i,t_i) \leftarrow \mathrm{pop}(Q)$
6:      **if** $r_i = d$ **then**
7:          **return** (r_i,t_i)
8:      **end if**
9:      **for all** neighboring resources r_j **do**
10:         **for all** free time windows $[t_j,t'_j)$ of r_j **do**
11:             **if** possible to go to this window from (r_i,t_i) **then**
12:                 $Q \leftarrow \mathrm{insert}\big(Q,\big(r_j,\max\{t_i,t_j\}+D'(r_j)\big)\big)$
13:                 remove all time windows up to t_j from the list of free time windows of r_j
14:             **end if**
15:         **end for**
16:     **end for**
17: **end while**

Algorithm 14.1: Context-aware routing.

dows, since for each free time window there must be a reservation as well. Based on this observation, we can now bound the run-time complexity of this algorithm by $O(mv+nv\log(nv))$ where m is the number of connections in the infrastructure. This can be seen as follows. Each free time window is considered at most once in an iteration of the while loop (Line 4). Because there are $O(nv)$ free time windows in total, the while loop is executed at most nv times. Since Q is a priority queue, removing the smallest element from the list takes $O(\log nv)$ time. Lines 1–7 therefore contribute $O(nv\log(nv))$ to the complexity of the algorithm. Rather than looking at lines 9–13 in the context of the while loop, we observe that over the whole run of the algorithm, these lines will be executed at most once for each time window of each neighbor in the time window graph, i.e., $O(mv)$. Similarly, regarding line 12, we can see that this line will be executed at most once per time window, i.e., $O(nv)$. Since this inserting in a priority queue takes $O(\log(nv))$, the total run-time for this part of the algorithm is $O(mv+nv\log(nv))$. Hence, Algorithm 14.1 has a run-time complexity of $O(mv+nv\log(nv))$.

Thus, compared to earlier work, our approach is more efficient. For example, the run-time complexity of the (single-agent) free time-window graph routing algorithm of Kim and Tanchoco [10] is $O(v^4n^2)$, while our algorithm has a run-time of $O(mv+nv\log(nv))$. The reason for this difference is that their conflict detection procedure is inefficient. Kim and Tanchoco [10] did not make use of the fact that it is enough to consider only the direct successor and predecessor to check for catching-up conflicts (instead, they iterated through all present vehicles). Furthermore, they used a different framework in which they both had to check for conflicts in the locations, as well as on lanes. In our framework lanes are also modeled as resources with the same properties as locations.

14.4 Dealing with uncertainty

Thus far we have considered the transportation planning problem as a static problem. This is, of course, in reality not the case. Several sources of uncertainty can be distinguished [4]. Uncertainty can be caused by incidents, such as communication failures between automated guided vehicles and the system maintaining reservations, break-down of a mobile entity (engine failure), or failures in the transport network (e.g., due to traffic accidents). Uncertainty can also be caused by a modification of the transportation requests. For example, the arrival of a new transportation request renders a current plan infeasible.

Uncertainty, and especially incidents can be dealt with pro-actively or reactively. Pro-active methods attempt to create robust schedules, while reactive methods recover from incidents at the moment they occur. A typical pro-active approach is to insert slack in plans such that for example delays have (almost) no consequences, and new requests can easily be inserted, see, e.g., [5]. However, in case nothing unexpected happens, these plans take much longer than strictly required. Therefore, in the following, we focus on approaches that deal with uncertainty reactively.

In principle, uncertainty does not offer any additional problem to the traditional approach, since the operational conflict resolution methods are usually sufficient to handle conflicts due to new arrivals or incidents as well. For context-aware route planning systems, however, such unexpected changes are a serious threat. Of course we could add an operational conflict resolution system to context-aware routing to remove any conflicts during execution. However, then it easily might turn out that such systems are not better than the traditional route planning approaches regarding efficiency and predictability. Therefore, we propose to extend context-aware routing with the ability to deal with such conflicts by reconsidering the plans of the involved vehicles (and possibly some of the others) each time an unexpected event occurs.

We consider the following two possible improvements. In the *revising-priorities* method a number of selected vehicles reconsider only the timing of their reservations in order of their priority, which is determined by a heuristic function. The *revising-routes* method also examines alternative routes for the vehicles. After all, it might be the case that a vehicle is better off taking a detour if the heuristic function determines that this vehicle should wait for other vehicles.

14.4.1 Revising priorities

In context-aware routing, reservations are (in principle) permanent, and later requests take existing reservations into account. Consequently, the performance of the system depends on the order in which the vehicles request reservations. The idea of the REVISING-PRIORITIES method is to improve the performance by re-evaluating reservations at certain moments, for instance, when a vehicle has accepted a new transportation request, a transportation request has been modified, or when a vehicle is bothered by an incident on its path. This method usually leads to a revised ordering of the vehicles' reservations of certain infrastructural resources, hence the

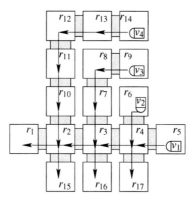

Figure 14.5: Initially the plan of v_1 is $(r_5, [0,2), r_4, [2,4), r_3, [4,6), r_2, [6,7), r_1, [7,\infty))$. It thus has to wait at every resource on its route. With REVISING-PRIORITIES vehicle v_2 can give way to v_1. The plan of v_1 then is much more efficient, while v_2 has to wait only one additional time step.

name. Below we describe a centralized version of this method where one scheduler controls the order in which the vehicles can make new reservations. Typically, the vehicle requesting rescheduling determines a group of involved vehicles, such as all vehicles that share at least one infrastructure resource with the requesting vehicle. This set of vehicles can quickly be determined by considering reservations for infrastructure resources in the plan of the requesting vehicle.

When the request to revise priorities is granted by the scheduler, all of the participating vehicles dispose of their current reservations, but maintain their route. In turn, these vehicles can request new reservations for the first part (block) of their remaining route. A heuristic function is used to determine the order in which this should be done. Examples of such heuristic functions are given in Table 14.1. This algorithm terminates when each vehicle in the group has a new schedule for each complete route.

Table 14.1: List of heuristics used by the REVISING-PRIORITIES and REVISING-ROUTES methods.

Heuristic	Description
random	for baseline comparison, chooses a random agent to go first
delays	agent with highest sum of expected delays goes first
deadlines	agent with least amount of slack before the deadline goes first
profits	agent with lowest expected profits goes first
wait	agent that waits longest to enter its current location goes first
task	agent that is assigned the task that has the highest reward goes first
inv task	the reversed ordering of the task heuristic, used to see the effect of bad versus good heuristics

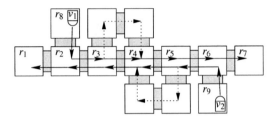

Figure 14.6: When the route of v_1 is r_2, r_3, r_4, r_5, r_6 and the route of v_2 exactly the opposite, one of these vehicles will have to wait until the other one has completed its route. With REVISING-PRIORITIES one of them can take a small detour instead, resulting in a more efficient global plan.

To see that this method cannot only be used in case of incidents, but also has some potential to improve plans under regular operating conditions, consider the following example in Figure 14.5. If all resources traversals take one time step, vehicle v_1 has to wait an additional time step at each resource on its route for another vehicle. If v_1 is given priority at only one of these resources, there is no need for v_1 to wait at later resources anymore, while it will cost another vehicle only one more time step.

14.4.2 Revising routes

Besides reconsidering the priorities of vehicles when making reservations, the REVISING-ROUTES method also allows vehicles to reserve a completely new route. In principle, this method is very similar to REVISING-PRIORITIES: first a selection of vehicles dispose of their reservations, and then each vehicle in turn requests new reservations for a part of their route. The difference is that in REVISING-ROUTES, a vehicle now also disposes of its route, and plans a new route when it is its turn. When planning a new route, vehicles do not consider a re-ordering of the visiting sequence, because finding an optimal visiting sequence is a very hard problem in itself, and takes too much time to compute.[4] However, reconsidering the routes between each pair of locations in a visiting sequence of a group of vehicles still gives much more flexibility than only reconsidering the reservations.

For example, when two vehicles plan to use the same sequence of resources, but in the opposite direction, a context-aware routing method lets one of the vehicles wait before entering this sequence (see Figure 14.6). REVISING-PRIORITIES can improve the average efficiency by a heuristic that determines which vehicle should wait. The REVISING-ROUTES method can further improve the resulting plans by allowing vehicles to take a detour. Especially when a certain resource is blocked for some time due to an incident, being able to compute a new route is a great advantage. We therefore expect this approach to result in better plans than REVISING-PRIORITIES, but the computation costs (in CPU time) are also higher, since the

---

[4] This problem is similar to the traveling salesman problem, which is NP-complete.

context-aware routing algorithm is now called as a subroutine for each pair of locations of each of the agents' vehicle sequences. In the next section an experimental evaluation is presented to indicate under which conditions which method is the best alternative.

14.5 Experimental evaluation

To evaluate the proposed context-aware routing algorithms, we perform simulations in two settings. First, we compare the context-aware routing (INFORMED) and its two improvements (REVISING-PRIORITIES and REVISING-ROUTES) to the baseline approach in which no intelligent infrastructure with reservations is used (UNINFORMED). This evaluation is done in a synthetic grid infrastructure. In this setting we also evaluate the effect of incidents on the context-aware methods that we propose. Second, we study the effect of using our improved context-aware routing method in a realistic situation, i.e., taxiing at Amsterdam airport. Here we approximate current practice with the reservation-based planning method of Hatzack and Nebel [8] and compare the resulting plans to the ones produced by our proposed approach.

14.5.1 Comparing context-aware methods on a grid network

For these experiments an 8×8-grid network is used with 32 vehicles and 192 requests. The problem instances are randomly generated in such a way that we know that an optimal solution exists, i.e., a solution for which all requests are served within the given deadlines.

In this section we consider the effect of the planning method and several incident settings (these are the independent variables). The *dependent variables* (or performance indicators) are the relative system reward and the CPU time required to finish the particular simulation. The reader is referred to [20] for a study on the influence of the request load, the size and the topology of the transport network, and the number of vehicles. The planning methods are divided into four categories: *(i)* baseline approach (also called UNINFORMED), *(ii)* context-ware routing (also called INFORMED), *(iii)* revising priorities, and *(iv)* revising routes.

For each request j, the agents receive a reward defined by the reward function π_j (see Section 14.2). The relative system reward is the ratio of the realized reward, divided by the maximum reward for all transportation requests. A relative reward of 1 means that the maximum possible reward is obtained for all of the transportation requests. This maximum reward cannot always be achieved, especially not if there are many requests or incidents.

All results presented have been obtained by making use of the transport planning simulator TRAPLAS, see [20]. The simulations are run on one processor, and a central controller was used. However, in this simulator, each vehicle and each resource has its own separate light-weight thread, which should make it relatively easy to

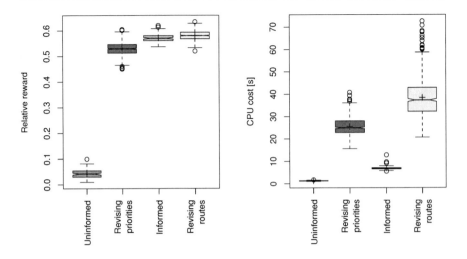

Figure 14.7: The performance of the four methods. With one exception, all methods execute all transportation requests (but not all on time). The UNINFORMED method, however, executes only 48.5% on average. The '+' symbol indicates the mean (often coinciding with the median in the figure) and the 'o' symbol is used for outliers.

re-use the source code for a distributed implementation on a real infrastructure. A free software environment for statistical computing and graphics called R [16] has been used to combine the output of TRAPLAS, to plot the graphs, and for all further data analysis. All experiments were done on the Distributed ASCI Supercomputer (DAS-3 [2]).

14.5.1.1 Comparing the planning methods

Figure 14.7 shows a box-and-whisker plot [19] for the four different planning-method categories. The UNINFORMED and INFORMED plots are based on 100 samples (10 simulations and 10 different sets of 192 transportation requests); the REVISING-PRIORITIES and REVISING-ROUTES categories consist of 7 times more samples to test each of the 7 different heuristics listed in Table 14.1.

An analysis-of-variance study shows that the differences in the mean values between the methods are statistically significant. The first thing to note is that the UNINFORMED method has a much lower relative reward. This can be explained by the fact that this is the only method that was not able to execute all transportation requests. This is due to deadlocks that occur without any coordination between the agents. On average, the UNINFORMED method executed 48.5% of the transportation requests.

The INFORMED context-aware planning method makes use of an intelligent infrastructure that manages reservations. This method produces much better results. All transportation requests are executed successfully. The plot on the right-hand side in Figure 14.7 shows the sole strength of the UNINFORMED method; it is the cheapest in terms of CPU cost. In general, it can be seen that the CPU cost required by the planning methods increases if more information is considered. Although this might be an advantage for very large transportation instances, the INFORMED method is usually fast enough for practical instances.

A further observation is that the performance of the INFORMED method is in between the REVISING-PRIORITIES and REVISING-ROUTES methods. This is a bit surprising, as we introduced REVISING-PRIORITIES as an improvement over the standard INFORMED context-aware planning method. However, this can be explained by the results for the seven heuristics (see Figure 14.8).

Most of the heuristics that we study perform no better than the random heuristic. The results for these heuristics significantly influence the average relative reward for REVISING-PRIORITIES in Figure 14.7. In addition, we can see that both the REVISING-PRIORITIES and the REVISING-ROUTES method have the best performance when using the *wait* heuristic (in this grid network with a relatively small number of requests and no incidents).

The good performance of the *wait* heuristic can be understood by appreciating the intuitive resemblance with the first-come-first-served heuristic [11], which works so well in scheduling. The *wait* heuristic aims to minimize the waiting time of transport resources, by giving priority to the longest waiting vehicle. Hence, the *wait* heuristic attempts to increase the throughput of transportation requests, which increases the performance of the agents.

14.5.1.2 Studying the influence of incidents

As discussed in Section 14.4, there are many types of uncertainty in operational transport planning. The focus regarding uncertainty in this chapter is on uncertainty caused by incidents. We model such incidents by a time period (the repair time) during which a certain resource or vehicle endures a slow-down between 0 (no slow-down) and 1 (full stop). This model can capture all kinds of real-life situations ranging from regular traffic jams (delays of infrastructural resources for certain periods of the day) to unexpected vehicle break-downs (full stop for some time until repaired). To evaluate the effect of incidents on the proposed context-aware methods, we generate situations ranging from a few incidents with a low impact to many incidents with long repair times and a high impact. An incident is generated in analogy to the concept of mean-time-between-failure, which is in our case based on an exponential distribution with a failure probability between 0 and 0.2, and the repair time is drawn from a normal distribution with mean 400 and standard deviation 50. We have chosen to create a relatively small number of incidents, but with quite a significant repair time, because this brings out the differences between planning methods more explicitly [14].

The impact of all incidents in a generated problem instance is expressed by the *incident level*. This incident level is the sum over all incidents of the impact of that incident times the duration of the incident, and varies in our case between 0 and 323981 seconds (about half a week). For six scenarios within this range we compute the relative reward of the INFORMED context-aware routing method, REVISING-

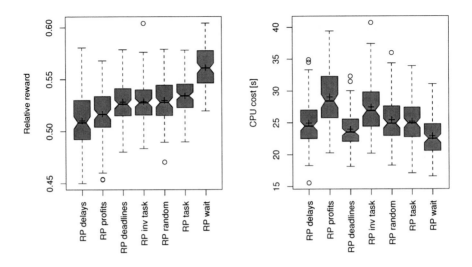

Figure 14.8: The performance of each of the heuristics in combination with the REVISING-PRIORITIES method. The '+' symbol indicates the mean and the 'o' symbol is used for outliers.

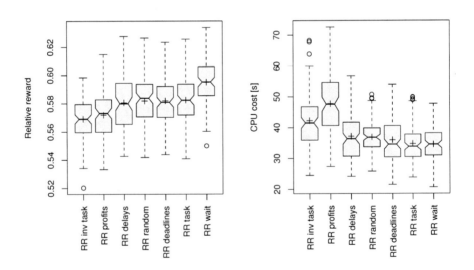

Figure 14.9: The performance of each of the heuristics in combination with the REVISING-ROUTES methods. The '+' symbol indicates the mean and the 'o' symbol is used for outliers.

PRIORITIES, and REVISING-ROUTES. For the latter two we use the *wait* heuristic, since this heuristic performs best according to our earlier experiments.

Figure 14.10 shows that the two methods that have been superior so far, i.e., REVISING-PRIORITIES and REVISING-ROUTES, now clearly outperform the standard context-aware method when the incident level is high, with a relative reward that is almost twice as high. The performance of all methods degrades when the incident level is increased. The differences of the means of the performance of the INFORMED and the other two methods are statistically significant for all six scenarios.

Next, let us consider the CPU time and the simulation time of these runs (see Figure 14.11). As expected, we can again clearly observe that REVISING-ROUTES requires significantly more computation time than REVISING-PRIORITIES, and both are significantly outperformed in this respect by INFORMED. However, the execution of the plans takes much longer for INFORMED in cases with a high incident level, since some vehicles may encounter multiple incidents on their planned route, without being able to plan around them (either in time or in space).

14.5.2 Taxiing at Amsterdam airport

Besides these experiments based on a synthetically generated test set, we evaluate our methods also on a real-life transport network. For this, we conduct experiments on the Amsterdam airport network (Schiphol) in the Netherlands consisting of 1016 infrastructure resources (see Figure 14.12 for a map of the center part of this network).[5] In this transportation network, the taxiing problem of aircrafts (on the ground) plays an a crucial role in the performance of the whole airport [18]. The usual sequence of an airplane after touch-down is to first taxi to a gate, then wait for services, such as cleaning, boarding, safety checks, and possibly a visit to a de-icing station. Currently approximately 300 airplanes per day go through this process (a number that is increasing), and thus efficient and robust routing methods are required.

The goal of the airplane taxiing experiments is to compare the current practice in a realistic setting to our context-aware routing method. Current practice is comparable to the approach of Hatzack and Nebel [8], in which first routes are planned, and then the use of resources is scheduled in order to prevent conflicts.

To make a fair comparison between these two approaches, the current practice as well as two of our context-aware routing methods are used to compute a route for the same start-destination pair, given the *same set of prior reservations* on the infrastructure. For each of such a set of reservations, we measure the average time and the average quality of finding a conflict-free path for 20 randomly chosen start-destination pairs, one of which is actually reserved for the next round. We repeat this 3000 times, reserving 3000 routes in the end. First, we study a setting where the sets of prior reservations are constructed using the algorithm of Hatzack and Nebel

---

[5] The network model of Amsterdam airport was kindly provided by the National Aerospace Laboratory (NLR).

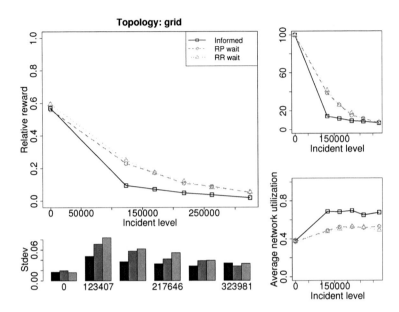

Figure 14.10: The relative reward for three planning methods on a 8×8 grid network decreases when the incident level increases (with a fixed number of 192 requests).

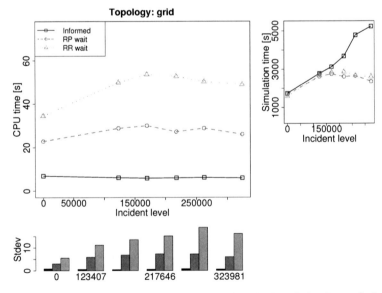

Figure 14.11: The CPU costs (in time) required by the selection of planning methods on grid networks; the incident level is increasing and the request load is fixed to 192 requests.

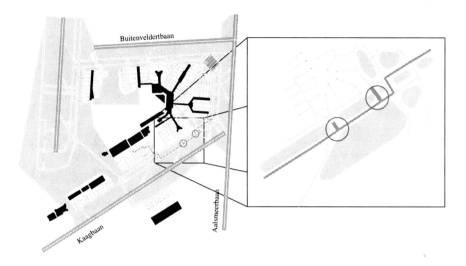

Figure 14.12: If cycles are allowed and there are many reservations of other agents present in the network, the context-aware algorithm often produces plans with a cycle. An example of a produced shortest path is highlighted by the thick line, and the side-steps are indicated by the two circles. Such plans are not considered by the acyclic version of the routing algorithm.

[8]. Then, we do the same measurements for sets of reservations based on the plans obtained by our context-aware approach.

Running our experiments, we have discovered that the context-aware routing algorithm (Algorithm 14.1) considered visiting the same part of the infrastructure twice, for example to step aside to let another airplane pass. This is illustrated in Figure 14.12. Since this is considered undesirable, we also introduce a simple modification of the context-aware routing algorithm in which such cycles are not allowed. For this *acyclic* variant of the algorithm, in Line 11 of Algorithm 14.1, the time windows to be considered do not only need to be possible to go to, but they also may not already occur in the plan of the agent. This thus requires some additional administration as well.

In the following experiments, we thus compare not only the context-aware routing algorithm (INFORMED) to the plans produced by the two-stage method of Hatzack and Nebel [8], but also the context-aware routing method that excludes cycles in the plans ("No cycles").

From Figure 14.13, it can be concluded that the context-unaware approach of the algorithm of Hatzack and Nebel is so fast, the context-aware algorithms look slow by comparison. A closer look reveals that the context-aware algorithms are still quite fast, as a solution (for one additional request) is found on average within two tenths of a second. Also, the 95% confidence intervals are reasonably small, so this performance is reasonably stable. With regard to the different variants of the context-aware routing methods, it can be seen that the no-cycles variant is significantly faster than the other, despite the fact that checking for absence of cycles is not

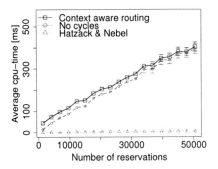

Figure 14.13: Planning for one additional request takes more CPU time when there are more resource reservations in case these reservations have been made using Hatzack and Nebel [8].

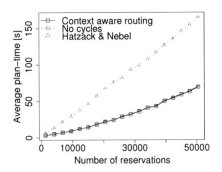

Figure 14.14: The average end-times of plans are later when there are more resource reservations, but there is a significant difference between context-aware routing and Hatzack and Nebel [8].

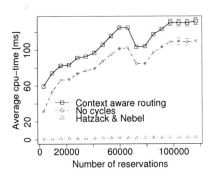

Figure 14.15: Planning for one additional request takes also more CPU time when there are more resource reservations in case these reservations have been made using context-aware routing.

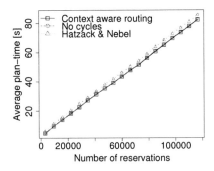

Figure 14.16: The average end-times of plans are later when there are more resource reservations (confidence intervals too small to display).

a very cheap operation. This is due to the fact that the no-cycles variant can ignore all routes where a resource is visited twice.

In Figure 14.15, a clear notch is visible, just before 80,000 reservations, which shows the CPU cost (in time) is not growing monotonically for an increasing number of reservations in the transport network. Intuitively, the more reservations there are, the more difficult it is to search a path through the transport network. However, after adding certain reservations, the search might become easier all of a sudden, because a "difficult" part of the network does not have to be searched through anymore (due to a reservation now prohibiting this). The exact position of such a notch depends on the transport network and the order in which agents create the reservations for traveling to their desired target locations.

Looking at the end-times of the generated plans, the plans generated by the no-cycles variant are of equal quality as those generated by Algorithm 14.1. When the set of reservations is produced by Hatzack and Nebel [8], the context-aware methods both succeed in finding a much more efficient plan by routing around bottlenecks (Figure 14.14). When the set of reservations of all other agents is produced by the context-aware routing method, however, the plan made by the algorithm of Hatzack and Nebel [8] for one additional agent is only slightly longer (in time), i.e., about 3 to 4% (Figure 14.16). In this case, there are apparently much shorter queues for bottle-necks than in the case when the route for all agents is fixed on forehand (as done by Hatzack and Nebel).

14.6 Conclusions and future research

In this chapter we have presented an alternative to the traditional approach to operational transport planning where conflicts are prevented during execution. The proposed method relies on an infrastructure that maintains reservations of its use at specific times, and can thereby resolve conflicts during planning. This ensures predictable travel times and thereby significantly reduces the uncertainty in transportation. Based on this intelligent infrastructure, we proposed three new context-aware route planning algorithms that either have a better run-time complexity than earlier work, or are much more sophisticated in finding alternatives in case of incidents or bottlenecks. These algorithms have been evaluated in a synthetic grid network (with and without incidents), and in an airplane taxiing simulation of Amsterdam airport. We have shown that methods using the reservation information of infrastructure resources outperform local (traffic) coordination rules, and that performance increases even more when vehicles' reservations or even parts of their routes are rescheduled in case of incidents. This latter method can deal with incidents and is still fast enough (in the airport-taxiing experiments finding an additional route can be done with less than a second), and gives the best results in terms of plan quality.

Apart from studying the behavior of this method experimentally in other realistic infrastructures and under other incident conditions, there are quite a number of other interesting directions for future work. For example, in the context of multiple requests and the possibility of incidents, a method in which all agents optimize the length of their own plan (by reserving required resources as early as possible) may not always lead to the best result. Possibly introducing some slack in the plan may lead to better overall results. This idea can definitely be exploited when agents plan to arrive too early at their destination. The proposed method does not take this into account and just lets the agent wait at the last resource until it is the right time to deliver. Secondly, the context-aware method introduced in this chapter allows for a distributed implementation to potentially allow for better scalability and reliability. It would be interesting to evaluate whether these promises can be fulfilled by performing some additional experiments. Another direction is to optimize not just the length of an agent's plan, but to allow for multiple objectives. For this, the work presented could be integrated with a multi-objective routing package called Samcra [12, 15]. Finally, we believe there is still some room for improving the solution quality of the resulting operational transport plans, since this is an NP-hard problem and our algorithms are polynomial. Repeatedly (re)running the context-aware routing algorithm for a set of agents can thus not always give optimal results, even though our context-aware routing algorithm for a single agent is both optimal and very efficient.

References

1. J. F. Allen. Maintaining knowledge about temporal intervals. *Communications of the ACM*, 26(11):832–843, November 1983.
2. H. Bal, R. Bhoedjang, R. Hofman, C. Jacobs, T. Kielmann, J. Maassen, R. van Nieuwpoort, J. Romein, L. Renambot, T. Rühl, and others. The distributed ASCI supercomputer project. *ACM SIGOPS Operating Systems Review*, 34(4):76–96, October 2000.
3. A. J. Broadbent, C. B. Besant, S. K. Premi, and S. P. Walker. Free ranging AGV systems: Promises, problems and pathways. In *Proceedings of the 2nd International Conference on Automated Guided Vehicle System*, pages 221–237, June 1985.
4. A. J. Davenport and J. C. Beck. A survey of techniques for scheduling with uncertainty. Unpublished manuscript, 2000.
5. A. J. Davenport, C. Gefflot, and J. C. Beck. Slack-based techniques for robust schedules. In *Proceedings of the Sixth European Conference on Planning*, pages 7–18, September 2001.
6. E. W. Dijkstra. A note on two problems in connexion with graphs. *Numerische Mathematik*, 1:269–271, 1959.
7. S. Fujii, H. Sandoh, and R. Hozaki. A routing control method of automated guided vehicles by the shortest path with time-windows. In M. Pridham and C. O'Brien, editors, *Production Research: Approaching the 21st Century*, pages 489–495. Taylor & Francis, London, UK, 1991.
8. W. Hatzack and B. Nebel. Solving the operational traffic control problem. In *Proceedings of the 6th European Conference on Planning*, Toledo, Spain, September 2001.
9. J. Huang, U. S. Palekar, and S. G. Kapoor. A labelling algorithm for the navigation of automated guides vehicles. *Journal of Engineering for Industry*, 115(3):315–321, 1993.
10. C. W. Kim and J. M. A. Tanchoco. Conflict-free shortest-time bidirectional AGV routing. *International Journal on Production Research*, 29(12):2377–2391, 1991.
11. R. L. Kruse. *Data Structures & Program Design*. Prentice-Hall, Inc., Upper Saddle River, New Jersey, 1984.
12. F. A. Kuipers and P. Van Mieghem. Conditions that impact the complexity of QoS routing. *IEEE/ACM Transaction on Networking*, 13(4):717–730, August 2005.
13. J. K. Lenstra and A. H. G. R. Kan. Complexity of vehicle routing and scheduling problems. *Networks*, 11(2):221–227, 1981.
14. S. Maza and P. Castagna. A performance-based structural policy for conflict-free routing of bi-directional automated guided vehicles. *Computers in Industry*, 56(7):719–733, September 2005.
15. H. De Neve P. Van Mieghem and F. A. Kuipers. Hop-by-hop quality of service routing. *Computer Networks*, 37(3-4):407–423, November 2001.
16. R Development Core Team. *R: A language and environment for statistical computing*. R Foundation for Statistical Computing, Vienna, Austria, 2007. URL http://www.R-project.org.
17. F. Taghaboni-Dutta and J. M. A. Tanchoco. Comparison of dynamic routing techniques for automated guided vehicle system. *International Journal of Production Research*, 33(10): 2653–2669, 1995.

18. A. W. Ter Mors, J. Zutt, and C. Witteveen. Context-aware logistic routing and scheduling. In *Proceedings of the 17th International Conference on Automated Planning and Scheduling*, pages 328–335, Providence, Rhode Island, September 2007.

19. J. W. Tukey. *Exploratory Data Analysis*. Addison Wesley, Boston, Massachusetts, 1977.

20. J. Zutt. *Operational Transport Planning in a Multi-Agent Setting*. PhD thesis, Delft University of Technology, 2009. Forthcoming.

Chapter 15
Railway Dynamic Traffic Management in Complex and Densely Used Networks

F. Corman, A. D'Ariano, D. Pacciarelli, and M. Pranzo

Abstract This chapter is the first thorough assessment of a full implementation of the concept of dynamic traffic management in combination with advanced optimization tools. In the last years, several studies on partial implementations of this concept have been carried out reporting promising results. The development of new strategies for railway traffic control experienced an increasing interest due to the expected growth of traffic demand and to the limited possibilities of enhancing the infrastructure, which increase the needs for efficient use of resources and the pressure on traffic controllers. Improving the efficiency requires advanced decision support tools that accurately monitor the current train positions and dynamics, and other operating conditions, predict the potential conflicts and reschedule trains in real-time such that consecutive delays are minimized. We carry on our study using an innovative computerized railway traffic management system, called ROMA (Railway traffic Optimization by Means of Alternative graphs). An extensive computational study is carried out, based on two complex and busy dispatching areas of the Dutch rail network. We study practical size instances and different types of disturbances, including train delays and blocked tracks. Our results show the high potential of ROMA as a support tool to improve punctuality through intelligent use of the rail infrastructure and efficient use of the available transport capacity.

F. Corman
Delft University of Technology, Department of Transport and Planning, Delft, The Netherlands,
e-mail: f.corman@tudelft.nl

A. D'Ariano (corresponding author), D. Pacciarelli
Università degli Studi Roma Tre, Dipartimento di Informatica e Automazione, Roma, Italy,
e-mail: {a.dariano,pacciarelli}@dia.uniroma3.it

M. Pranzo
Università degli Studi di Siena, Dipartimento di Ingegneria dell'Informazione, Siena, Italy,
e-mail: pranzo@dii.unisi.it

R.R. Negenborn et al. (eds.), *Intelligent Infrastructures*, Intelligent Systems, Control and Automation: Science and Engineering 42, DOI 10.1007/978-90-481-3598-1_15,
© Springer Science+Business Media B.V. 2010

15.1 Introduction

The aim of railway traffic control is to ensure safety, regularity, reliability of service and punctuality of train operations. Railway business strongly needs to improve the quality of service and to accommodate growth while reducing the costs. The punctuality analysis represents an important measure of rail operation performance and is often used as standard performance indicator. As reported in Goverde [9], in the autumn of 2001 the punctuality of the Dutch railway system decreased to below 80% (percentage of trains arriving at scheduled stops with a delay < 3 min). In 2003, a report by four major companies operating in The Netherlands [21] indicated a punctuality level of 95% as a target to reach within year 2015, despite the expectation of a significant increase of traffic intensity and the limited budget available to build new rail infrastructure. This chapter addresses such challenging target that can only be achieved through intelligent use of the existing rail infrastructure and efficient use of the available transport capacity. We describe traffic management strategies and dispatching support systems that can be used to improve punctuality of railway operations under disturbed traffic conditions.

Performance management is usually achieved by railway managers by carefully designing an off-line timetable and operating in real-time with strict adherence to it. However, train operations are intrinsically stochastic and traffic needs to be dynamically managed. When the scheduled railway traffic is disturbed, decisions have to be taken that modify the plan of operations in order to reduce delay propagation.

In The Netherlands, new pro-active approaches have been proposed to construct and manage the timetable. Two currently adopted strategies are the following:

- Development of robust timetables including specific running time supplements and buffer times to handle minor disturbances. This strategy requires to develop reliable estimation of delay propagation (see, e.g., [29]) in such a way that the amount of disturbances absorbed by the time reserves is increased.

- Development of dynamic traffic management strategies in which less details are fixed during the planning of activities and more control decisions are left to the dispatchers (see, e.g., [24]). This approach requires to develop effective traffic control procedures to fully exploit the enlarged degree of freedom.

To a certain extent these two strategies are complementary, since they both aim to improve punctuality of train operations at a microscopic level. In fact, minor train delays represent the vast majority of all delays and their influence on the level of train service can be minimized by careful management, provided that detailed and reliable information is available. This chapter focuses on the potential of the latter strategy and evaluates the performance of given timetables for varying the degree of freedom and the traffic control procedures.

Dynamic traffic management strategies offer an interesting possibility to improve railway services by operating flexible timetables in which each train has to fit in a time window of arrival at a given set of feasible platforms/passing tracks. A specific platform and the exact arrival/departure times are then defined in real-time by the traffic controllers, which are therefore required to perform more actions with respect

Figure 15.1: A train dispatcher at the traffic control center (Source: ProRail).

to non-flexible timetables. Under this strategy, traffic controllers have enhanced possibilities to react to unexpected events by adapting the timetable to the status of the rail network. Due to the strict time limits for computing a new timetable in case of disturbances, they usually perform manually a few modifications, such as adjustments of train routes, orders and speeds, while the efficiency of the chosen measures is often unknown [14, 31].

Dispatching systems support human dispatchers to manage traffic flow (as shown in Figure 15.1). Existing support systems compute rescheduling solutions on the basis of local information, i.e., they operate only "on the spot" and "now" and may implement simple dispatching rules. More advanced traffic management systems take into account the whole traffic in a larger area, detecting future conflicts affecting train movements, scheduling automatically trains in the whole area by using global information and suggesting possible changes of train orders or routes to the dispatcher, as well as displaying advisory speeds to train drivers. A comprehensive review on the related literature is reported in D'Ariano [4].

Most of the computerized decision support systems developed, so far, can provide fairly good solutions for small instances and simple perturbations. However, they cannot deal with heavy disturbances in larger networks as the actual train delay propagation is only roughly estimated and does insufficiently take into account interactions among trains in the whole network. Therefore, extensive control actions are necessary to obtain globally feasible solutions.

In this chapter we compare a simple dispatching procedure with an advanced traffic management system. The simple procedure is a first come first served rule, a common practice in railway real-time management. The advanced system is the recently developed software, ROMA (Railway traffic Optimization by Means of Alternative graphs), that makes use of an optimization tool based on global information on the future evolution of the train traffic. This tool can be applied to various types of disturbances (such as multiple delayed trains, dwell time perturbations, block sections unavailability, and others) within a short computation time. The mathematical model and algorithms will be briefly described in Section 15.4. For more information, we refer the interested reader to, e.g., [2, 4–7].

The innovative scientific contribution of ROMA is characterized by a combination of blocking time theory (see, e.g., [12, 22]) for the recognition of timetable conflicts in case of disturbances and a general discrete optimization model, based on the alternative graph formulation of [15], for the real-time evaluation of train reordering and rerouting in rail networks, while the costs of the different options are measured in terms of maximum and average delays at stations and other relevant points within the investigated network.

Computational experiments are based on two complex and densely occupied dispatching areas of the Dutch rail network, namely the Schiphol bottleneck area [13] and the Utrecht station area. The former is a dispatching area subject to high frequency passenger traffic, while the latter consists of a complex set of routes, heterogeneous traffic and less dense traffic conditions. A large set of disturbances are proposed for increasing values of train delays, multiple track blockage and different time horizons of traffic prediction. For each perturbed situation we generate several feasible schedules by using different configurations of the ROMA system. This allows us to quantify the effects of different traffic management strategies, in terms of train delays and computation time.

The next section presents the traffic management problem and related terminology. Section 15.3 describes the basic strategies for dynamic traffic management. Section 15.4 deals with the architecture of the ROMA system, while Section 15.5 reports on our computational experiments. The last section discusses the main achievements and gives some future research directions.

15.2 Problem description

In its basic form a rail network is composed of stations, links and block sections separated by signals. For safety reasons, the signals control the train traffic on the routes, and impose a minimum distance headway between consecutive trains. Signals, interlocking and Automatic Train Protection (ATP) control the train traffic by imposing a minimum safety separation between trains, setting up conflict-free routes and enforcing speed restrictions on running trains. The minimum safety distance and time headways between two consecutive trains (see the example of Figure 15.2) depend on their speeds (running time of the involved trains), the braking rate of the second train (considered in the approaching time to the next signal), the train length of the first train (clearing time), the signal spacing (including sight and reaction time) and the swithing time. In case of technical or human failures, ATP

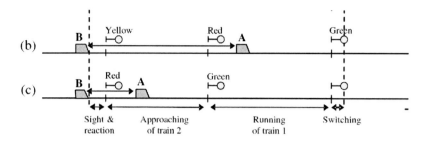

Figure 15.2: Track occupation of two trains (A and B) in case of different signal aspects.

ensures safe rail operations. Specifically, ATP causes automatic braking if the train ignores the valid speed restrictions. Signals are located before every junction as well as along the lines and inside the stations. A block section is a track segment between two main signals, that governs the train movements, and may host at most one train at a time.

The standard feature of a railway signaling system is characterized by the three-aspect fixed block signaling. For fixed block signaling systems, a train may enter a block section only after the train ahead has completely cleared the block section and is protected by a stop signal. A signal aspect may be red, yellow or green. A red signal aspect means that the subsequent block section is either out of service or occupied by another train (see Figure 15.2(c)), a yellow signal aspect means that the subsequent block section is empty but the following block section is still occupied by another train (see Figure 15.2(b)), and a green signal aspect indicates that the next two block sections are empty (see Figure 15.2(a)). A train is allowed to enter the next block section if the signal aspect is either green or yellow, but the latter requires deceleration and stop before the next signal if this remains red. A detailed description of different aspects of railway signaling systems and traffic control regulations can be found, e.g., in [8, 9, 12].

The passage of a train through a particular block section is called an *operation*. A *route* of a train is a sequence of operations to be processed in a track yard during a *service* (train run). At any time a route is *passable* if all its block sections are available and the corresponding block signal is green or yellow, i.e., there are no blocked tracks. The *timing* of a route specifies the starting time t_i of each operation in the route. Each operation requires a traveling time, called *running time*, which depends on the actual speed profile followed by the train while traversing the block section. A speed profile is furthermore constrained by the rolling stock characteristics (maximum speed, acceleration and braking rates), physical infrastructure characteristics (maximum allowed speed and signaling system) and driver behavior (coasting, braking and acceleration profiles when approaching variable signals aspects). The running time includes the time needed to accelerate (or decelerate) a train due to a scheduled stop, as well as speed variations between two consecutive speed signs. Furthermore, the running time is known in advance since all trains

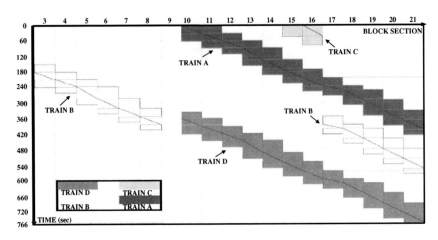

Figure 15.3: A timetable with four trains (A, B, C and D) running on a network of 21 block sections.

travel at their scheduled speed, which usually contains some margins for recovery. In yards or complex station interlocking areas, the routes for the individual trains need to be setup before entering and cleared after leaving, which takes a certain switching time.

The running time of a train on a block section starts when its head (the first axle) enters the block section. Safety regulations impose a minimum distance separation between the trains running in the network, which translates into a minimum *setup time* (time headway) between the exit of a train from a block section and the entrance of the subsequent train into the same block section. This time takes into account the time between the entrance of the train head in a block section and the exit of its tail (the last axle) from the previous one, plus additional time margins to release the occupied route and sighting distance (see, e.g., [20]). Railway timetable design usually includes *recovery times* and *buffer times* between the train routes. The recovery time is an extra time added to the travel time of a train between two stations, which corresponds to planning train speed smaller than maximum. Recovery times can be utilized in real-time to recover from delays by running trains at maximum speed. The buffer is an extra time inserted in the timetable between consecutive train paths besides the minimum headway, which prevents or reduces the propagation of train delays in the network.

We consider a timetable (see, e.g., the illustrative example of Figure 15.3) which describes the movements of all trains running in the network during a given time period of traffic prediction, specifying, for each train, the planned arrival/passing times at a set of relevant points along its route (e.g., stations, junctions, and the exit point of the network). At stations, a train is not allowed to depart from a platform stop before its scheduled departure time and is considered late if arriving at the platform later than its scheduled arrival time. At a platform stop, the scheduled stopping time of each train is called *dwell time*. Additional practical constraints related to pas-

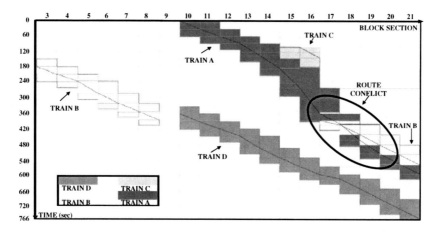

Figure 15.4: A route conflict between trains *A* and *B* caused by train *C* (see timetable in Figure 15.3).

senger satisfaction are to be considered, such as minimum transfer times between connected train services. This is the time required to allow passengers to alight from one train, move to another platform track and board the other train. Constraints due to rolling stock circulation must also be taken into account. In fact, a train completes a number of round-trips during the service of a line and its length may be changed by (de)coupling. For this reason, railway timetables include a *turn-around time* at terminal stations, which is a time margin between the arrival of the train and the start of a new train service in the opposite direction using the same rolling stock. Similarly, crew scheduling constraints impose a minimum time elapsed between the arrival of a train carrying (part of) the crew of another train and the departure of the latter train. In case of severe disturbances a scheduled train may be delayed from the beginning because of the unavailability of rolling stock or train personnel.

Timetables are designed to satisfy all traffic regulations. However, unexpected events occur during operations, which cause delays with respect to the operations scheduled in the timetable. The delay may propagate causing a domino effect of increasing disturbances. In Figure 15.4, the delay of one train (train *C*) propagates to the other trains running in the railway area (trains *A* and *B*). We define an *entrance perturbation* as a set of train delays at the entrance in a dispatching area, due to the propagation of delays from previous dispatching areas. A *disruption* is the modification of some infrastructure characteristics, such as the temporary unavailability of one or more block sections, which causes alterations in the train travel times and routes. Running time prolongation may occur because of headway conflicts between consecutive trains or technical failures, route changes are due to some block section being unavailable for a certain amount of time and dwell time perturbations are due to traffic delays at stations.

A *conflict* occurs when two or more trains claim the same block section simultaneously, and a decision on the train ordering has to be taken (see, e.g., the disturbed

Vertrek	Naar / Opmerkingen	Spoor	Trein
00:10	**Groningen/Leeuwarden** via Nijkerk, Harderwijk, Zwolle	2	⇝ Intercity **+ 30 minuten**
00:22	**Amersfoort Vathorst** via	5b	⇝ Sneltrein **+ 5 minuten**
00:52	**Utrecht Centraal** via Den Dolder,	6	⇝ Stoptrein **+ 5 minuten**
04:40	**Amsterdam Centraal** via Baarn, Hilversum, Naarden-B	7	⇝ Sneltrein
05:14	**Amsterdam Centraal** via Baarn, Hilversum, Naarden-B	7	⇝ Sneltrein
05:44	**Alkmaar** via Baarn, Hilversum, Amsterdam C	7	⇝ Sneltrein
05:46	**Utrecht Centraal** via Den Dolder,	4b	⇝ Stoptrein

Figure 15.5: Display of train delays for a disturbed traffic situation (Source: Nederlandse Spoorwegen).

traffic situation of Figure 15.4). A set of trains cause a *deadlock* when each train in the set claims a block section ahead which is not available, due either to a disruption or to the occupation/reservation for another train in the set.

Real-time railway traffic management copes with temporary infeasibility by adjusting the timetable of each train, in terms of routing and timing, and/or by resequencing the trains at the entrance of each merging/crossing point. The railway traffic is predicted over a given time horizon of traffic prediction. The task of dispatchers is to regulate traffic in a given dispatching area with the main objective of minimizing train delays in such a way that the new schedule is compliant with rail operating rules and with the entrance position of each train. The latter information is taken into account in the computation of the *release time* of each train that is the expected time, with respect to the starting time t_0 of traffic prediction, at which the current train enters its first block section in the area under study. The *total delay* is the difference between the estimated train arrival time and the scheduled time at a relevant point in the network (see the train delays shown in Figure 15.5), and can be divided into two parts. The *initial delay* (primary delay) is caused by original failures and disturbances and can only be recovered by exploiting available running time reserves, i.e., trains traveling at maximum speed. The *knock-on delays* (consecutive or secondary delays) are caused by the hinder from other trains or dispatching measures like late setting-up of train routes.

The real-time railway traffic management problem is the following: given a railway network, a set of train routes and passing/stopping times at each relevant point in the network, and the position and speed of each train being known at time t_0, find a deadlock-free and conflict-free schedule, compatible with their initial positions and such that the selected train routes are not blocked, each train enters the network at its release time, no train departs from a relevant point before its minimum scheduled departure time, rolling stock constraints and connected train services are respected, and trains arrive at the relevant points with the smallest possible knock-on delay.

15.3 Dynamic traffic management strategies

The standard practice in railway traffic management consists of the off-line construction of a timetable and the real-time control of trains with strict adherence to the timetable [1]. This rigid control practice makes timetable development a complex problem in which a compromise between capacity utilization and timetable robustness has to be provided [30].

Schaafsma [24] suggested the new concept of *Railway Dynamic Traffic Management* (RDTM) for improving railway system robustness (i.e., the resilience to disturbance in operation), without decreasing the capacity of the lines (i.e., the maximum number of trains which may be operated through a line per time period). The basic idea is to keep train traffic flowing in the bottleneck by avoiding unnecessary waiting time. This can be achieved by relaxing some of the timetable specifications, such as train routing, arrival/departure times and sequencing. This concept has been further refined within the Dutch railway undertakings in the last years (see, e.g., [17, 18, 25, 26, 28]). The resulting RDTM principles consist basically of the following dynamic information for the management of congested areas:

1. Strict arrival/departure times are replaced by time windows of [minimum, maximum] arrival/departure times at each platform and relevant timetable points of the network. A large time window corresponds to having more flexibility. In this case, the operational timetable, used by railway managers, includes both minimum and maximum arrival and departure times, while the public timetable, available to passengers, includes the maximum arrival time and the minimum departure time only. The longer travel times would be compensated by a higher reliability of train services, i.e., low variability of travel times and connections. This would allow a greater possibility to control traffic.

2. The scheduled order of trains at overtakes and junctions may be provisionally, or even partially defined in the operational timetable and finally determined in real-time. In the latter case, the timetable might contain conflicts to be solved during operations. Enabling the change/specification of train orders in real-time would allow to reduce delays if the dispatcher can find good schedules within strict time limits of computation.

3. The default platform/passing track for a train at a station is replaced by a set of feasible platform/passing tracks, leaving the final choice to traffic control. In this case, an additional dynamic information system would guide the passengers to their trains. The operational timetable might also specify a set of routing options for each train that would use the infrastructure with more flexibility.

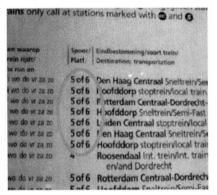

Figure 15.6: Routing flexibility by static (left side) and dynamic (righ side) information (Source: [26]). Note that "5 of 6" is Dutch and translates to "5 or 6" in English.

With an RDTM strategy, the plan of operations is only partially defined off-line in the timetable and then fixed in real-time, based on the actual status of the network and on the current train positions. The increasing degree of freedom left to real-time control leads to a larger workload for traffic controllers who must take several real-time actions. Thus, computerized traffic management systems are necessary to support dispatchers to exploit at best the opportunity offered by RDTM concepts.

Middelkoop and Hemelrijk [17] investigate the possible effects of introducing the RDTM principles in real-time traffic management by using the macroscopic simulation tool SIMONE. In a follow up paper, Middelkoop and Loeve [18] report on a computational analysis using the microscopic simulation tool FRISO.

Schaafsma and Bartholomeus [26] describe the first implementation of some of the RDTM principles at the Schiphol bottleneck of the Dutch rail network (see the static and dynamic information systems of Figure 15.6). The authors limit the assessment to routing flexibility and to train resequencing, using the first come first served rule at some specific railway junctions to resequence train movements.

D'Ariano et al. [7] compute optimal schedules with ROMA to resequence train movements, and assess the benefits of flexible departure times and flexible train sequencing based on the Schiphol area but using different, more challenging, timetables than in [26] and [27].

Up to now the three principles of dynamic traffic management have been evaluated separately and on a single case study. Hence, there is a need to assess the full implementation of the three RDTM principles on different networks and timetables in order to evaluate their potential and limitations more in general. This need motivates the present chapter. In the next section we illustrate the decision support system used to carry on this assessment.

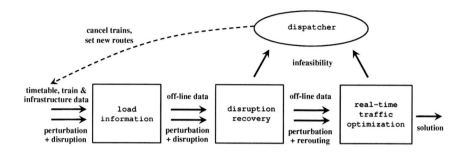

Figure 15.7: Architecture of the decision support system.

15.4 Decision support system

This section describes the implementation of the decision support system called ROMA (Railway traffic Optimization by Means of Alternative graphs) for railway traffic management. The ROMA system is implemented in C++ language, is compatible with Linux and Windows platforms and uses the *AGLibrary* developed by the "Aut.Or.I." Research Group of Roma Tre University. Given a disturbed timetable, the real-time railway traffic management problem is divided into three subproblems: *(i)* data loading and exchange of information with the field, *(ii)* assignment of a passable route to each train in order to avoid blocked tracks and *(iii)* definition of optimal train routes, orders and specification of the exact arrival and departure times at stations as well as at a set of relevant points in the network, such as specific junctions and passing points. The ROMA system addresses the resolution of the three subproblems.

Figure 15.7 presents the ROMA system architecture, which is composed of inter-related procedures. A human dispatcher can interact with the decision support system by adding/removing constraints or changing the timetable. We now describe the function of each procedure and how the three introduced subproblems are solved:

- Data loading (subproblem *(i)*): Collect data from the field such as the current infrastructure status, the existing timetable, the actual position and speed of all trains, and forecast the time needed to complete the next scheduled operations (e.g., entrance delay of a train in the network, dwell time perturbations, etc.). We assume that the exact speed and location of each train are updated in real-time. Hence, the impact of inaccurate train data is supposed to be negligible.
- Disruption recovery (subproblem *(ii)*): Given a default route and a prioritized set of rerouting options, find a passable route for each train by avoiding the blocked tracks in the area.
- Real-time traffic optimization (subproblem *(iii)*): Given a set of dynamic traffic management strategies, i.e., flexible orders, routes and departure times, find a new deadlock-free and conflict-free schedule by rescheduling and/or rerouting trains with the aim of minimizing train delays in the network.

The real-time railway traffic management system must be able to efficiently detect and solve the conflicts arising in the rail network during perturbed operations. The traffic optimization procedure identifies the potential headway and route conflicts with a high level of accuracy (considering the physical characteristics of the rolling stock and infrastructure used by each train) while considering all trains simultaneously. The potential headway and route conflicts are determined by predicting the future location of trains based on information about the actual state of the rail network. The traffic optimization procedure allows to take train rescheduling and rerouting decisions considering all the train speed profiles fixed as scheduled (each train travels with the minimum traversing time on each block section). To alter the original timetable as little as possible, we focus here on the development of conflict resolution actions in the dispatching area under study. Large timetable modifications and cancelation of train routes are among the possible dispatching measures but they are not performed automatically. The decision support system solution is suggested to the traffic controller before its actual implementation. After checking the suggested solution, the traffic controller may either confirm the proposed actions or choose other dispatching measures. The dispatcher's decision would then be communicated to the interlocking system, that sets the interconnected switches and signals, and to the drivers in the train cabins by means of radio data transmission. Consequently, we suppose that trains are equipped with on-board computers for automatic train control.

Due to the synchronization time, i.e., the time to react to changing conditions, speed and location modifications may happen while the decision support system is computing a solution. However, since the decision support system is able to compute a feasible solution in a few seconds, depending on the time period of traffic prediction, we assume that such real-time variations would not affect the principal validity of the rescheduling solution.

The next three subsections address each procedure separately, and in a fourth subsection we point out the limitations and approximations of the proposed approach.

15.4.1 Load information

The *data loading* procedure periodically collects all the information from the field, which is required by the other procedures (subproblem *(i)*). The primary condition for calculating the future train movements is the availability of a detailed and accurately updated data set. Precisely, running times and setup times for each operation are computed in accordance with the actual speed and position of each train at its entrance of the network, the current infrastructure status (e.g., track layout, speed limits), the timetable data and the rolling stock characteristics. We next describe *real-time* and *off-line* data.

We consider real-time data gathered from the field that can change or be decided during real-time operations. Clearly, a continuous and reliable communication with the trains is assumed, i.e., a real-time data processing unit on-board and in the traffic control center is necessary. Among the real-time data, the current operating situation has to be included, i.e., actual position and speed of the running trains at the beginning of the considered time horizon (i.e., time t_0). The expected entrance time/route for each incoming train, time windows of availability for all block sections/platforms, possible temporary speed limits occurring at some block sections, and additional scheduled stops on open tracks and stations with their scheduled arrival and departure times are real-time data which the traffic controller has to set in the decision support system before execution. All this information is stored before the other procedures start.

Off-line (planning) data consists of detailed information about the infrastructure, timetable and rolling stock characteristics. The timetable contains a list of arrival/departure times (time windows of minimum/maximum arrival/departure times) for a set of relevant points in the network, including all the station platform tracks visited by each train. The infrastructure consists of a set of available block sections delimited by signals. For each block section the status, length, grade, speed limitations, traversing directions and maximum speed are given. The route release and switching times are also known off-line. The specific technical characteristics of the rolling stock of each train (power, train length and weight, maximum speed and friction rates between rails and wheels) are recorded off-line in order to enable a re-calculation of the required minimal running time. The data associated with each train also includes a prioritized list of routing options, chosen by the dispatcher. The (mean) acceleration and braking rates are to be calculated on the basis of traction force/speed diagrams and scheduled maximum speeds. Here we apply speed profiles on the basis of standard acceleration and braking tables used by the Dutch infrastructure manager ProRail. We assume that the drivers follow standard braking and acceleration profiles. The weather condition, the train load (number of passengers) and weight are assumed to be a priori defined. Although some of these data may be different from day to day, for the purpose of rescheduling they are computed as off-line data. In case of substantial real-time variability of these factors, a more accurate estimation of the trains speeds and movements should be considered.

We distinguish between scheduled speed profiles (used during the timetable planning phase) and operational speed profiles (adopted in the conflict detection and resolution phase). Operational speed profiles, used in the rescheduling process, suppose that trains travel at their maximum speed according to the train characteristics, infrastructure speed restrictions and adopted standard drivers' behavior, and they are obtained by using off-line data.

After the completion of the loading phase, the other ROMA procedures are executed assuming that real-time variations of these data would not affect the principal validity of the rescheduling solution. We also adopt ROMA to predict railway traffic for several time horizons. The solutions obtained are thus applicable in real-time only for those instances solved within a few minutes, depending on the prevailing traffic conditions. However, the proposed decision support system is a laboratory version tested on a real-world off-line data set and does not include transmission of actual train monitoring data and data communication protocols between the system and the trains' on-board units.

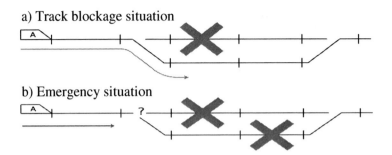

Figure 15.8: Disrupted railway corridors.

15.4.2 Disruption recovery

The *disruption recovery* procedure checks if there are unavailable block sections in the network (i.e., track blockage situation of Figure 15.8 (a)), which make some train route unpassable. This activity corresponds to the resolution of subproblem *(ii)*. For each train, this procedure discards disrupted routes, sorts the passable routing options on the basis of a priority list (given by traffic controllers) and then assigns the one with the highest priority, called the *default route*. The default route of each train and the set of remaining passable routes are then given to the real-time railway traffic optimization procedure of the ROMA system.

Since the ROMA system is only allowed to select the route of each train from a given set, when no passable route is available for a train, the system requires external support by the human dispatcher (e.g., emergency situation of Figure 15.8 (b)). In case of a heavy disruption, the human dispatcher authorizes ROMA to use an emergency timetable in which train routes are strongly modified, e.g., enabling a specific train to reverse its running direction according to a specific movement authority given by the traffic controllers.

15.4.3 Real-time traffic optimization

The *real-time railway traffic optimization* procedure is the decisional kernel of the decision support system. This procedure is responsible for detecting and solving train conflicts while minimizing the train delay propagation. Given all the necessary information by the data loading procedure and (at least) a passable route for each train by the disruption recovery procedure, the conflict detection and resolution problem (i.e., subproblem *(iii)*) is addressed as follows. A conflict detection procedure checks whether the timetable is deadlock-free and detects potential conflicting train paths in a given period of traffic prediction (e.g., 15 minutes ahead). Given the actual train delays and predicted conflicts, a conflict resolution procedure computes a new feasible schedule (i.e., deadlock-free and conflict-free) compatible with the status of the network, by defining routes, orders and times for all trains.

The conflict detection and resolution problem can be formulated as a job shop scheduling problem with no-store and no-swap constraints. Mascis and Pacciarelli [15] show that the alternative graph is a suitable model for this job shop problem and several real-world constraints can be easily modeled by it. The real-time railway traffic optimization procedure uses an alternative graph formulation of the conflict detection and resolution problem. This graph represents the routes of all trains in a given control area along with their precedence constraints (minimum headways). Since a train must traverse the block sections in its route sequentially, a route is modeled in the alternative graph with precedence constraints and a chain of associated nodes. This formulation requires that a passable routing for each train is given and a fixed traversing time for each block section is known in advance, except for a possible additional waiting time between operations in order to solve train conflicts. A train schedule therefore corresponds to the set of the starting time of each operation. Since a block section cannot host two trains at the same time, a potential conflict occurs whenever two or more trains require the same block section. In this case, a passing order must be defined between the trains which is modeled in the graph by introducing a suitable pair of alternative arcs for each pair of trains traversing a block section. A deadlock-free and conflict-free schedule is next obtained by selecting one of the two alternative arcs from each pair, in such a way that there are no positive length cycles in the graph (i.e., deadlock). In other words, the alternative arcs represent operational choices such as the train order at a crossing or merging section. In order to evaluate a schedule, we use the maximum consecutive delay as performance indicator of a solution, which is the maximum delay introduced when solving conflicts in the dispatching area. This is caused by the propagation of the input delays of late trains to the other trains in the railway area. In general, other train operators' objectives could also be considered in the problem formulation, such as dynamic train priorities including intercity, local and freight trains, passengers' dissatisfaction due to extra running times or change of platform stops, et cetera.

We now present a small railway network with two trains running at different speeds. Figure 15.9 shows the studied infrastructure with four block sections (denoted as 1, 2, 3 and 9), a simple station with two platforms (6 and 7) and three junctions (4, 5 and 8). In this example, we only show the location of the most relevant block signals. However, each block section has, clearly, the capacity of one train at a time. At the starting time t_0, there are two trains in the network. Train A is a slow train running from block section 3 to block section 9 and stopping at platform 6. Train A can enter a block section only if the signal aspect is yellow or green. Train B is a fast train running from block section 1 to block section 9 through platform 7 without stopping. Train B can enter a block section at high speed only if the signal aspect is green. At t_0, we therefore assume that train B requires two empty block sections.

Figure 15.10 presents an alternative graph formulation for the traffic situation of Figure 15.9. For the sake of clarity, a node of the graph can be identified by a pair (train, block section), by a pair (train, scheduled stop) or by a pair (train, exit point), except for the dummy nodes 0 and n. Each pair of alternative arcs is associated

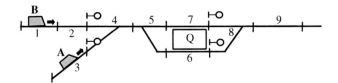

Figure 15.9: Example of a small railway network with two trains.

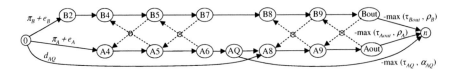

Figure 15.10: Alternative graph formulation of the example in Figure 15.9.

with the usage of a common block section by two trains (i.e., with two conflicting operations), and is represented by connecting the two paired arcs with a small circle. For simplicity, the length of fixed and alternative arcs representing running and setup constraints is not depicted. Since trains A and B share block sections 4, 5, 8 and 9, there are four pairs of alternative arcs. The values $\pi_A + e_A$ and $\pi_B + e_B$, depicted respectively on arcs $(0, A4)$ and $(0, B2)$, represent the time at which the heads of trains A and B are scheduled to reach the end of their current block sections. The values π_A and π_B are their scheduled entrance times while the values e_A and e_B are their entrance delays. The length d_{AQ} of arc $(0, A8)$ is the scheduled departure time of train A from the scheduled stop Q, while the length of arc $(AQ, A8)$ (not depicted in Figure 15.10) is its scheduled dwell time.

We next discuss the objective function for the formulation of Figure 15.10. Let τ_{AQ} be the earliest possible arrival time of train A at Q computed according to its initial position and speed, its assigned route and following a maximum speed profile in the empty network (i.e., by disregarding the presence of other trains). Let α_{AQ} be the scheduled arrival time of a train A at Q in the timetable, which can be infeasible in case of real-time disturbances. The length $-\max\{\tau_{AQ}, \alpha_{AQ}\}$ is the *modified due date* of train A at the scheduled stop. Let ρ_A and ρ_B be the scheduled exit times of trains A and B, the modified due dates at the exit of the network ("out") are $-\max\{\tau_{Aout}, \rho_A\}$ and $-\max\{\tau_{Bout}, \rho_B\}$. The resulting makespan $l^S(0, n)$ corresponds to the maximum consecutive delay computed for all trains at their relevant points.

The main value of the alternative graph is the detailed but flexible representation of the network topology at the level of railway signal aspects and operational rules. In case of fixed block signaling each block signal corresponds to a node in the alternative graph and the arcs between nodes represent blocking times or headway times. Moreover, other constraints relevant to the railway practice can be included into the alternative graph model (see, e.g., [4–7]).

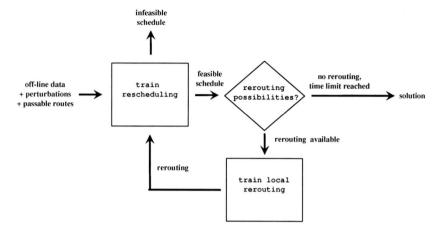

Figure 15.11: Architecture of the real-time traffic optimization procedure.

The real-time optimization procedure is in charge of computing a first feasible schedule and then is looking for better solutions in terms of delay minimization. Its architecture is described in Figure 15.11. Given a timetable, a set of passable routes associated with each train and the current status of the network, the train scheduling procedure returns a feasible schedule for each train, i.e., defines its entrance time on each block section. Specifically, the first run of this procedure considers the default routings defined by the disruption recovery procedure. If no feasible schedule is found within a predefined time limit of computation, the human dispatcher is in charge of avoiding deadlocks by taking decisions that are forbidden to the automated system, such as the cancelation of a train connection and movement authorities. When a feasible schedule is found, the train rerouting procedure verifies whether local rerouting options may lead to better solutions. For each changed route, the running times and setup times are modified accordingly. Whenever some route is replaced, the train scheduling procedure computes a new deadlock-free and conflict-free timetable by thoroughly rescheduling the train movements. The combined scheduling and rerouting procedure returns the best solution found when a time limit of computation is reached or no local rerouting improvement is possible. We next introduce the algorithms used by ROMA.

Since the resolution of train conflicts has direct impact on the level of punctuality, this chapter compares two scheduling algorithms:

- **Branch and Bound (BB)**: This is an exhaustive algorithm that explores all the reordering alternatives and chooses the one minimizing the maximum consecutive delay. Here we consider a truncated branch and bound [6] that returns near-optimal schedules for practical size problems within a short computation time.

- **First Come First Served (FCFS)**: This is a well-known dispatching rule which gives precedence to the train arriving first at a block section. This rule requires no dispatching action since trains pass at merging or crossing points on the basis of their actual order of arrival and not necessarily as in the timetable.

We also implemented rerouting algorithms based on advanced heuristics, i.e., local search and tabu search. The heuristics analyze the alternative routes of each train, searching for a train route potentially leading to a better schedule. Whenever a better schedule is found, the new route is set as default route and the search is repeated. Specifically, the effectiveness of extensive rerouting strategies is explored by incorporating the local search for new routes in a novel Tabu Search (TS) algorithm, in order to escape from local minima [2]. Since the combinatorial structure of the conflict detection and resolution problem is similar to that of the job shop scheduling problem with routing flexibility, we focus on the tabu search approach that achieved very good results with the latter problem [16].

If the real-time railway traffic optimization procedure is unable to find deadlock-free and conflict-free schedules, the dispatcher has to carry out other types of timetable modifications such as introduction of new train routes, application of short-turning of trains in case of track blockage or even cancelation of train services at some stations (e.g., connections between passenger trains).

15.4.4 Discussion

The ROMA system is designed to help the dispatchers to cope with disturbances in the traffic flow by suggesting adjustments to the timetable of each train in terms of routing and timing, and by resequencing the trains at the entrance of each merging/crossing point. A strong point of ROMA is the simultaneous management of all trains running in an area, which allows maximization of punctuality and best exploitation of the available rail infrastructure (subproblem *(iii)*). ROMA is able to optimize railway traffic flow also in case of severe traffic disturbances, including the presence of blocked tracks and routes (subproblem *(ii)*), i.e., when emergency timetables are required and dispatchers need more support in their task.

The proposed decision support system is still a laboratory tool whose input/output interface is still limited to the loading of static (infrastructure and timetable) and dynamic (train positions and speeds) data gathered from the field (subproblem *(i)*). Besides, train speed profiles are computed separately from the scheduling problem in a preprocessing step. In order to incorporate ROMA within an advanced traffic management system, technical implementation issues concerning the practical operation, such as data transmission, communication of delays and the realization of the proposed dispatching measures, have still to be implemented, as well as the computation of new train dynamics when changing the train orders at conflict points. These issues are outside the scope of this chapter (we refer the reader to, e.g., [11, 19, 23]).

We are aware that the required information may not always be readily available, but the current railway signaling systems are fitted with intermittent and continuous automatic train protection systems. The existing train describer system records automatically the occupation and release of each signal block and the train number and its actual passing time at the critical block signals of the network. This information is used by many railways for comparison with the scheduled passing times. Thus, the difference between the scheduled and measured times is computed and could be used, too, as input data for the proposed decision support system (see, e.g., [3, 10]).

15.5 Computational experiments

This section presents the experiments performed to evaluate the ROMA system over a large sample of real-life instances. The aim of the study is to assess to which extent train delays could be minimized by choosing suitable dispatching actions and dynamic traffic management strategies. We compare the solutions obtained by simple and advanced algorithms. ROMA runs on a PC equipped with a processor Intel Pentium D (3 Ghz), 1 GB RAM and Linux operating system. Each run of the BB algorithm is truncated after 10 seconds of computation (the best-known solution is often found during the first few seconds of computation), while the FCFS algorithm takes less than one second of computation. The whole time allowed to the real-time optimization procedure to compute a solution is limited to 60 seconds in order to be compatible with real-time operations.

15.5.1 Description of the test cases

This subsection describes the two dispatching areas under study. Both infrastructures offer interesting possibilities for train reordering and rerouting. Each train has a default route and a set of local rerouting options. Rerouting options can be applied along corridors or within station areas, where trains may be allowed to stop at different nearby platforms. There are several potential conflict points in each dispatching area that are merging and crossing points along each traffic direction.

15.5.1.1 Schiphol dispatching area

The dispatching area around Schiphol tunnel is shown in Figure 15.12. The network consists of 86 block sections, 16 platforms and two traffic directions. The rail infrastructure is around 20 km long and consists mainly of four tracks, divided into two pairs for each traffic direction. Trains enter/leave the network from/to ten access points: the High Speed Line (HSL), the station of Nieuw Vennep, the shunting yard of Hoofddorp station, and two stations in Amsterdam, namely Amsterdam Lelylaan and Amsterdam Zuid WTC. The two traffic directions are largely independent except around Amsterdam Lelylaan station and at the border of Hoofddorp shunting yard. There are two intermediate stations: Hoofddorp and Schiphol.

We use an experimental timetable for passenger trains, designed to face the expected increase in traffic through this bottleneck area in the next years. This challenging timetable is very close to capacity saturation of this area, thus making it an

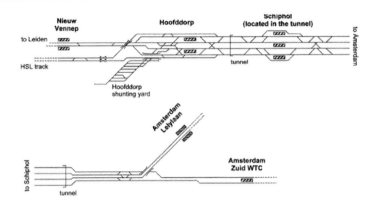

Figure 15.12: Schiphol dispatching area (Source: [13]).

interesting test case for our study. The timetable is cyclic with a period length of one hour and contains 27 trains per direction, for a total of 54 trains running each hour. This is a timetable with a limited amount of time reserves to recover delays, due to the high number of trains which is not far from the network capacity saturation. It is worthwhile observing that the actual number of trains per hour scheduled at Schiphol during year 2007 was 20 trains per direction [13]. In 2009, the hourly timetable of Schiphol is expected to include 24 trains per direction [27]. We chose the more challenging timetable with 27 trains per direction in order to assess the effectiveness of ROMA under even more dense traffic conditions.

For some trains in the timetable, we consider alternative platform stops at Schiphol station. This flexibility is only applied to nearby platforms in order to limit passengers' discomfort. In total, there are 111 routes available for train rerouting.

Table 15.1 describes the disturbances for the Schiphol dispatching area. The first column indicates the number of trains delayed at their entrance in the network (we only delay the trains that enter the studied dispatching area in the first 15 minutes of traffic prediction), the second column the maximum values of entrance delay (in seconds), the third column the number of blocked tracks in the network, plus the corresponding percentage of passable train routes. The chosen blocked tracks are unavailable platforms at Schiphol station, causing some trains to be rerouted on another available platform in order to continue their trip. Finally, the fourth column of Table 15.1 reports the length of the time horizon of traffic prediction (in minutes).

Table 15.1: Description of the Schiphol instances.

Number of delayed trains	Maximum entrance delay	Number of blocked trains	Time horizon length
1 / 3 / 5	100 / 300 / 600	0 (100%) / 1 (82%) / 2 (67%)	15 / 30 / 60

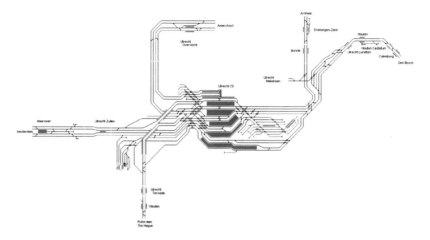

Figure 15.13: Utrecht dispatching area (Source: ProRail).

Three instances are generated for each value of columns 1, 2, 3 and 4, yielding a total of 81 entrance delay configurations.

15.5.1.2 Utrecht dispatching area

The railway network around Utrecht Central station is shown in Figure 15.13. Five main lines converge to Utrecht, connecting the North and South regions of The Netherlands to the lines to the West and the East. The network considered is delimited by the following stations: Utrecht Overvecht on the line to Amersfoort, Driebergen-Zeist on the line to Arnhem, Culemborg on the line towards Den Bosch, Vleuten on the line to Rotterdam and the Hague plus Maarssen on the line towards Amsterdam. In total, the diameter of the dispatching area is around 20 km long.

Utrecht Central station is one of the most complex railway areas in The Netherlands, including more than 600 block sections and a very complicated and densely occupied interlocking area, defining a large amount of inbound and outbound routes. Most of the trains have a scheduled stop at one of the 20 platform tracks. The total amount of travelers at Utrecht Central Station is around 150.000 per day.

We use a provisional 2008 timetable that is cyclic with a cycle length of one hour. The trains are mostly for passenger services, operated by NS (Nederlandse Spoorwegen), except for a few freight trains. The timetable schedules up to 80 trains in a peak hour and provides connections between passenger services, coupling and splitting of rolling stock for intercity and local services coming from/going to Rotterdam, the Hague or Amersfoort, as well as re-use of rolling stock for commuter services towards Utrecht Overvecht and Culemborg. For each train in the timetable, we consider the possibility of rerouting trains to nearby platforms. This flexibility results in a total amount of 228 alternative train routes.

Table 15.2: Description of the Utrecht instances.

Number of delayed trains	Maximum entrance delay	Number of blocked trains	Time horizon length
1 / 3 / 5	100 / 300 / 900	0 (100%) / 1 (93%) / 2 (91%)	15 / 30 / 60

Table 15.2 describes the disturbances for the Utrecht dispatching area. Also in this case, three instances are generated for each value of columns 2, 3 and 4, yielding a total of 81 entrance delay configurations.

15.5.2 Dynamic traffic management strategies

Table 15.3 shows the effects of implementing various RDTM principles in combination with the ROMA system. Column 1 reports the reordering algorithms ("FCFS" is the First Come First Served rule and "BB" is the Branch and Bound algorithm of [6]) while Column 2 reports the rerouting strategy ("Default" means that ROMA selects for each train the default route, "TS" means that the route is chosen with the tabu search of [2]). Column 3 indicates whether departure flexibility is used or not by the ROMA system. Precisely, we consider one minute of flexible departure time for all trains at their scheduled stops.

Each row of Table 15.3 reports the average results on the 81 instances of the tables 15.1 and 15.2 for the two dispatching areas. Columns 4 and 5 show, respectively, the maximum and average consecutive delays for the Schiphol instances. The last two columns present the delays for the Utrecht instances. In both cases, delays are expressed in seconds and computed for all trains at their scheduled stops and at their exit from the network.

We next comment on the delay impact of applying different combinations of the RDTM principles for the two dispatching areas under study.

As for the reordering algorithms, the BB algorithm provides by far better results with respect to the FCFS dispatching rule since the former algorithm chooses the train orders on the basis of global information on the delay propagation. Comparing the average results of all instances, the delay reduction is more evident for the Schiphol dispatching area (36% and 12% in terms of, respectively, maximum and average consecutive delays) than for the Utrecht dispatching area (9% and 6% in terms of, respectively, maximum and average consecutive delays). This is likely due to the fact that the Schiphol timetable is more dense of trains that run with short time headways.

As for the rerouting strategies, the TS algorithm exhibits a better capacity to keep delays small with respect to the default routes for both networks, even if rerouting is only allowed as alternative platforming of trains at their scheduled stops. Comparing the average results for all instances, the delay reduction is more evident at the Utrecht dispatching area (28% in terms of average consecutive delays) rather than

Table 15.3: Average results on various configurations of the ROMA system.

Dynamic traffic management strategies			Schiphol dispatching area		Utrecht dispatching area	
Reordering algorithm	Rerouting strategy	Departure flexibility	Max delay (sec)	Avg delay (sec)	Max delay (sec)	Avg delay (sec)
FCFS	Default	No	473	68.9	252	7.0
FCFS	TS	No	420	58.6	222	5.1
FCFS	Default	Yes	466	65.2	181	3.6
FCFS	TS	Yes	386	54.9	159	2.4
BB	Default	No	313	63.4	221	6.6
BB	TS	No	255	46.2	213	4.8
BB	Default	Yes	307	63.4	154	3.2
BB	TS	Yes	249	45.6	151	2.4

at the Schiphol dispatching area (21% in terms of average consecutive delays) since Utrecht Central Station presents a larger number of alternative platforms.

As for the departure flexibility, its introduction in the Utrecht dispatching area allows to halve the average consecutive delays. On the other hand, the results in the Schiphol dispatching area are by far less effective (3% in terms of average consecutive delays). The different performance is probably due to the longer dwell times at Utrecht Central Station, leading to more possibilities to improve train punctuality by means of flexible departure times.

We have shown that each individual RDTM principle is an effective measure in managing train traffic, limiting the propagation of train delays. However, the best results are achieved when using all the three RDTM principles. For both networks, the worst configuration is the basic FCFS rule (see the first row of Table 15.3) while the best configuration is the one using the BB algorithm, the TS algorithm and flexible departures (see the last row of Table 15.3). The gaps between these two configurations in terms of average consecutive delays are 34% for the Schiphol dispatching area and 66% for the Utrecht dispatching area.

15.5.3 Effects of increasing perturbations

Figure 15.14 presents the average results obtained for six types of entrance perturbation scenarios. Three entrance perturbation scenarios (1, 2 and 3) are tested for the Schiphol dispatching area: 1 train delayed by 100 seconds (scenario 1), 3 trains delayed by 300 seconds each (scenario 2), 5 trains delayed by 600 seconds each (scenario 3). Similarly, three types of entrance perturbation scenarios (4, 5 and 6) are tested for the Utrecht dispatching area: 1 train delayed by 100 seconds (scenario 4), 3 trains delayed by 300 seconds each (scenario 5), 5 trains delayed by 900 seconds each (scenario 6). For each of type of scenario, the average results on 9 instances are reported, in terms of average consecutive delays, for the basic FCFS rule and the best RDTM configuration.

For all tested scenarios and for both the dispatching areas, the gap between the basic FCFS and the best RDTM is evident. While the results obtained from the best RDTM present an average consecutive delay slightly increasing with the number of late trains, the behavior of the basis FCFS is more erratic since this is a local decision rule and its output is less predictable.

15.5.4 Effects of increasing disruptions

Figure 15.15 presents the average results obtained for six types of track blockage scenarios. Three scenarios (1, 2 and 3) are tested for the Schiphol dispatching area: all tracks available (scenario 1), 18% tracks blocked (scenario 2), 33% tracks blocked (scenario 3). Similarly, three scenarios (4, 5 and 6) are tested for the Utrecht dispatching area: all tracks available (scenario 4), 7% tracks blocked (scenario 5), 9% tracks blocked (scenario 6). For each of type of scenario, the average results on 27 instances are reported, in terms of average consecutive delays, for the basic FCFS rule and the best RDTM configuration.

In general, there is a considerable gap between the FCFS rule and the best RDTM configuration. For the Schiphol instances, the track blockage increases the average consecutive delays for both configurations of the ROMA system while their gap remains quite similar. For the Utrecht instances, scenario 5 has a stronger impact on delays compared to scenario 4, while scenario 6 has a limited impact compared to scenario 5 for both the system configurations, even if the best RDTM configuration presents a much smaller increase of delays.

15.5.5 Effects of increasing time horizons

Figure 15.16 presents the average results for six scenarios obtained by varying the time horizon and the test site. Three time horizon scenarios (1, 2 and 3) are tested for the Schiphol dispatching area: 15 minutes with 19 running trains (scenario 1), 30 minutes with 31 running trains (scenario 2), 60 minutes with 54 running trains (scenario 3). Similarly, three scenarios (4, 5 and 6) are tested for the Utrecht dispatching area: 15 minutes with 22 running trains (scenario 4), 30 minutes with 39 running trains (scenario 5), 60 minutes with 80 running trains (scenario 6). For each scenario, the average results on 27 instances are reported, in terms of average consecutive delays, for the basic FCFS rule and the best RDTM configuration.

The results obtained for different time horizons also show a considerable gap between the two system configurations. For the Schiphol dispatching area, enlarging the time horizon of traffic prediction results in larger average consecutive delays. The very small time reserves in the Schiphol timetable cause a propagation of train delays in the network, even if the best RDTM configuration is able to strongly limit this effect. For the Utrecht dispatching area, there is a different trend of the average consecutive delays. Their values decrease appreciably, for both the system configu-

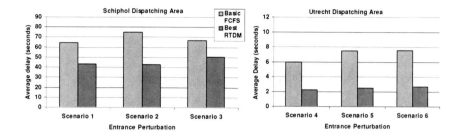

Figure 15.14: Comparison of the basic FCFS rule with the best RDTM configuration (BB, TS and flexible departures) for six types of entrance perturbation scenarios.

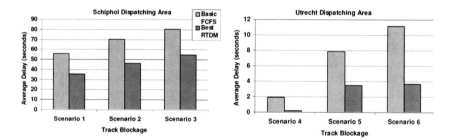

Figure 15.15: Comparison of the basic FCFS rule with the best RDTM configuration (BB, TS and flexible departures) for six types of track blockage scenarios.

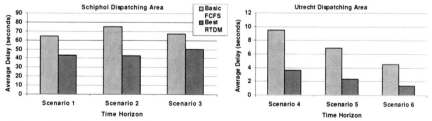

Figure 15.16: Comparison of the basic FCFS rule with the best RDTM configuration (BB, TS and flexible departures) for six types of time horizon scenarios.

rations, when the time horizon of traffic prediction is enlarged. This is mainly due to the larger amount of time reserves included in the Utrecht timetable with respect to the Schiphol one.

15.6 Conclusions and future research

This chapter presents the performance of different configurations of the ROMA system and various RDTM strategies. The results show the effectiveness of using advanced optimization algorithms with respect to simple and local dispatching procedures. ROMA can be applied to compute efficient dispatching solutions for any given rail infrastructure and timetable, also when the timetable is not conflict-free. This fact enables its usage when managing dense traffic in complex railway networks and under severe traffic disturbances, such as when emergency timetables are required and traffic controllers need support to solve conflicts.

As for the impact of railway dynamic traffic management principles (e.g., flexible departure times at scheduled stops, train reordering and rerouting alternatives), our computational results demonstrate that all the proposed principles may lead to interesting improvements. These benefits are the largest when the principles are used in combination with advanced traffic management algorithms.

Future research should address the integration of the proposed system into a larger framework, enabling to cope with several dispatching areas. To this end, it is important to address the decomposition of large problems into smaller problems to be solved by local dispatching systems, and their coordination may ensure globally viable and effective solutions for the whole rail network.

Acknowledgements We thank the Dutch infrastructure manager ProRail (specially R. Hemelrijk, D. Middelkoop, and L. Lodder) for providing the instances. This work is partially supported by the programs "Towards Reliable Mobility" of the Transport Research Centre Delft and by the Italian Ministry of Research, Grant number RBIP06BZW8, project FIRB "Advanced tracking system in intermodal freight transportation".

References

1. A. Caprara, L. G. Kroon, M. Monaci, M. Peeters, and P. Toth. Passenger railway optimization. In C. Barnhart and G. Laporte, editors, *Handbooks in Operations Research and Management Science*, volume 14, pages 129–187. 2006.
2. F. Corman, A. D'Ariano, D. Pacciarelli, and M. Pranzo. A tabu search algorithm for rerouting trains during rail operations. Technical Report RT-DIA-127-2008, Dipartimento di Informatica e Automazione, Università degli Studi Roma Tre, Rome, Italy, 2008.
3. W. Daamen, R. M. P. Goverde, and I. A. Hansen. Non-discriminatory automatic and distinct registration of primary and secondary train delays. In I.A. Hansen, A. Radtke, J. Pachl, and E. Wendler, editors, *Proceedings of the 2nd International Seminar on Railway Operations Modelling and Analysis*, Hannover, Germany, 2007.

4. A. D'Ariano. *Improving real-time train dispatching: Models, algorithms and applications.* PhD thesis, Delft University of Technology, Delft, The Netherlands, 2008.

5. A. D'Ariano, F. Corman, D. Pacciarelli, and M. Pranzo. Reordering and local rerouting strategies to manage train traffic in real-time. *Transportation Science*, 42(4):405–419, 2008.

6. A. D'Ariano, D. Pacciarelli, and M. Pranzo. A branch and bound algorithm for scheduling trains in a railway network. *European Journal of Operational Research*, 183(2):643–657, 2007.

7. A. D'Ariano, D. Pacciarelli, and M. Pranzo. Assessment of flexible timetables in real-time traffic management of a railway bottleneck. *Transportation Research Part C*, 16(2):232–245, 2008.

8. E. Goddard. Overview of signalling and train control systems. In *The 9th Institution of Engineering and Technology Professional Development Course on Electric Traction Systems*, pages 336–350, Manchester, UK, 2006.

9. R. M. P. Goverde. *Punctuality of Railway Operations and Timetable Stability Analysis.* PhD thesis, Delft University of Technology, Delft, The Netherlands, 2005.

10. R. M. P. Goverde and I. A. Hansen. TNV-prepare: Analysis of dutch railway operations based on train detection data. In J. Allan, C. A. Brebbia, R. J. Hill, G. Sciutto, and S. Sone, editors, *Computers in Railways VII*, pages 779–788. WIT Press, Southampton, UK, 2000.

11. S. Hailes. Modern telecommunications systems for train control. In *The 11th Institution of Engineering and Technology Professional Development Course on Railway Signalling and Control Systems*, pages 185–192, Manchester, UK, 2006.

12. I. A. Hansen and J. Pachl. *Railway Timetable and Traffic: Analysis, Modelling and Simulation.* Eurailpress, Hamburg, Germany, 2008.

13. R. Hemelrijk, J. Kruijer, and D. K. de Vries. Schiphol tunnel 2007. Description of the situation, 2003. Internal report.

14. A. Kauppi, J.Wikström, B. Sandblad, and A. W. Andersson. Future train traffic control: control by re-planning. *Cognition, Technology & Work*, 8(1):50–56, 2006.

15. A. Mascis and D. Pacciarelli. Job shop scheduling with blocking and no-wait constraints. *European Journal of Operational Research*, 143(3):498–517, 2002.

16. M. Mastrolilli and L. M. Gambardella. Effective neighborhood functions for the flexible job shop problem. *Journal of Scheduling*, 3(1):3–20, 2000.

17. A. D. Middelkoop and R. Hemelrijk. Exploring the effects of dynamic traffic management. In *Proceedings of Dagstuhl Seminar no. 04261 on "Algorithmic Methods for Railway Optimization"*, Schloss Dagstuhl Wadern, Germany, 2004.

18. A. D. Middelkoop and L. Loeve. Simulation of traffic management with FRISO. In J. Allan, C. A. Brebbia, A. F. Rumsey, G. Sciutto, S. Sone, and C. J. Goodman, editors, *Computers in Railways X*, pages 501–509. WIT Press, Southampton, UK, 2006.

19. G. Neil. On board train control and monitoring systems. In *The 9th Institution of Engineering and Technology Professional Development Course on Electric Traction Systems*, pages 211–241, Manchester, UK, 2006.

20. L. Nie and I. A. Hansen. System analysis of train operations and track occupancy at railway stations. *European Journal of Transport and Infrastructure Research*, 5(1):31–54, 2005.

21. NS, ProRail, Railion, and V&W. Benutten en bouwen: Het plan van de spoorsector, 2003. In Dutch.

22. J. Pachl. *Railway Operation and Control.* VTD Rail Publishing, Mountlake Terrace, Washington, 2002.

23. A. J. D. Santos, A. R. Soares, F. M. De Almeida Redondo, and N. B. Carvalho. Tracking trains via radio frequency systems. *IEEE Transactions on Intelligent Transportation Systems*, 6(2):244–258, 2005.

24. A. A. M. Schaafsma. *Dynamisch Railverkeersmanagement; besturingsconcept voor railverkeer op basis van het Lagenmodel Verkeer en Vervoer.* PhD thesis, Delft University of Technology, Delft, The Netherlands, 2001.

25. A. A. M. Schaafsma. Dynamic traffic management – innovative solution for the schiphol bottleneck 2007. In I. A. Hansen, F. M. Dekking, R. M. P. Goverde, B. Heidergott, and L. E. Meester, editors, *Proceedings of the 1st International Seminar on Railway Operations Modelling and Analysis*, Delft, The Netherlands, 2005.

26. A. A. M. Schaafsma and M. M. G. P. Bartholomeus. Dynamic traffic management in the schiphol bottleneck. In I. A. Hansen, A. Radtke, J. Pachl, and E. Wendler, editors, *Proceedings of the 2nd International Seminar on Railway Operations Modelling and Analysis*, Hannover, Germany, 2007.

27. A. A. M. Schaafsma and V. A. Weeda. Operation-driven scheduling approach for fast, frequent and reliable railway services. In E. Wendler, U. Weidmann, M. Luethi, J. Rodriguez, S. Ricci, and L. Kroon, editors, *Proceedings of the 3rd International Seminar on Railway Operations Modelling and Analysis*, Zürich, Switzerland, 2009.

28. J. Van den Top. Dynamic traffic management: planning with uncertainty to offer certainty. In I. A. Hansen, F. M. Dekking, R. M. P. Goverde, B. Heidergott, and L. E. Meester, editors, *Proceedings of the 1st International Seminar on Railway Operations Modelling and Analysis*, Delft, The Netherlands, 2005.

29. M. J. C. M. Vromans. *Reliability of Railway Systems*. PhD thesis, Delft University of Technology, Delft, The Netherlands, 2005.

30. E. Wendler. The scheduled waiting time on railway lines. *Transportation Research Part B*, 41(2):148–158, 2007.

31. T. A. White. The development and use of dynamic traffic management simulations in North America. In I. A. Hansen, A. Radtke, J. Pachl, and E. Wendler, editors, *Proceedings of the 2nd International Seminar on Railway Operations Modelling and Analysis*, Hannover, Germany, 2007.

Part IV
Water Infrastructures

Chapter 16
Flood Regulation by Means of Model Predictive Control

T. Barjas Blanco, P. Willems, P-K. Chiang, K. Cauwenberghs, B. De Moor, and
J. Berlamont

Abstract In this chapter flooding regulation of the river Demer is discussed. The
Demer is a river located in Belgium. In the past the river was the victim of several
serious flooding events. Therefore, the local water administration provided the river
with flood reservoirs and hydraulical structures in order to be able to better manage
the water flows in the Demer basin. Though this measures have significantly reduced
the floods in the basin, the recent floods in 1998 and 2002 showed that this was not
enough. In order to improve this situation a pilot project is started with as main
goal to regulate the Demer with a model predictive controller. In this chapter the
results of this project are discussed. First a simplified model of the Demer basin is
derived based on the reservoir model. The model is calibrated and validated using
historical data obtained from the local water administration. On the one hand the
resulting model is accurate enough to capture the most important dynamics of the
river; on the other hand the model is fast enough to be used in a real-time setting.
Afterwards, the focus will be shifted to the model predictive controller. The use
of the model predictive controller will be justified by comparing it to other control
strategies used in practice for flood regulation. Then, the more technical details of
the model predictive controller will be discussed in more detail. Finally the chapter
will be concluded by historical simulations in which the model predictive controller
is compared with the current control strategy used by the local water administration.

T. Barjas Blanco, B. De Moor
Katholieke Universiteit Leuven, Department of Electrotechnical Engineering, Leuven, Belgium,
e-mail: tonibarjasmartinez@gmail.com

P. Willems, P-K. Chiang, J. Berlamont
Katholieke Universiteit Leuven, Department of Civil Engineering, Hydraulics Division, Leuven,
Belgium, e-mail: patrick.willems@bwk.kuleuven.be

K. Cauwenberghs
Flemish Environment Agency (VMM), Division Water, Brussels, Belgium

R.R. Negenborn et al. (eds.), *Intelligent Infrastructures*, Intelligent Systems, Control and
Automation: Science and Engineering 42, DOI 10.1007/978-90-481-3598-1_16,
© Springer Science+Business Media B.V. 2010

16.1 Introduction

Flooding of rivers are worldwide the cause of great economic losses. The areas located in the surroundings of the Belgian rivers are no exception to this. This study focuses on the Demer river, a large river located in Belgium where severe floodings have occurred in the past during periods of heavy rainfall. In order to prevent these flooding events the local water administration (VMM water authority) installed hydraulic structures (with movable gates) in order to influence the discharges and water levels in the water systems. Extra storage capacity for periods of heavy rainfall was provided by means of flood control reservoirs. Hydraulic structures to control the flow from and into the reservoirs were also installed. The hydraulical structures are controlled by an advanced version of a standard three position controller (see Section 16.3). Though all these measures have reduced the amount and frequency of floodings in the Demer basin, past events have shown that improvement is still necessary.

In Table 16.1 a summary of the latest flooding events in the Demer basin is presented. From the table it can be seen that floodings have lead to big economical losses in the Demer basin, especially the flooding from 1998. Simulations based on a complex finite element model (Infoworks) have shown that these flooding events could have been less severe and even completely avoided if the gates would have been controlled in a different way. Therefore, the local water administration has set its target to find a new control strategy in order to control the gates. The study presented in this chapter is an attempt to improve the capacity management of the available storage volume in the Demer basin by determining a better control strategy. The proposed control strategy is a centralized version of a nonlinear model predictive controller.

The structure of this chapter is as follows. Section 16.2 presents the model used in this work. Section 16.3 presents the general idea behind control theory and gives a short overview about control techniques used in practice for the regulation of water systems. Also a nonlinear model predictive control (NMPC) scheme is presented. This NMPC scheme is applied on the model from Section 16.2 and its results are presented in Section 16.4. Finally in Section 16.5 some guidelines and ideas are presented for future research.

Table 16.1: Damage report of the latest floodings in the Demer basin.

Period	Estimated flood volume (km^3)	Cost in €
Dec 1993–Jan 1994	23.5	47 000
Jan 1994–Feb 1995	22.9	11 000
Sep 1998	32.6	16 169 000

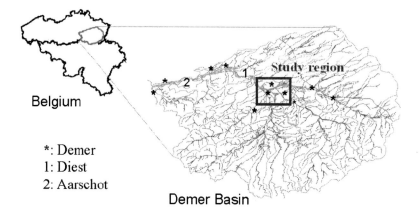

Belgium

*: Demer
1: Diest
2: Aarschot

Demer Basin

Figure 16.1: Study region in the river Demer basin in Belgium.

16.2 Hydrodynamic water system modeling

16.2.1 Introduction

A Model Predictive Control (MPC) application has been tested for the river Demer case in Belgium (Figure 16.1) in order to increase the efficiency of flood control by means of two flood control reservoirs installed upstream of the densely populated cities of Diest and Aarschot. Figure 16.2 gives a schematic representation of the water system structure (river reaches and hydraulic control structures) in the region surrounding the two flood control reservoirs, which are called 'Schulensmeer' and 'Webbekom'.

The MPC controller being developed is based on two types of hydrodynamic river models. The first model has been developed during earlier studies for the VMM water authority and is of the full hydrodynamic type. The full hydrodynamic equations of de Saint-Venant (momentum and continuity equations; [4]) have been implemented in that model for the river Demer as well as for its main tributaries. This has been done for the entire Demer basin (this means also for the areas far up- and downstream of the area of interest for this study) (see river reaches indicated by stars in Figure 16.1). The full hydrodynamic model has been implemented in the InfoWorks-RS software of Wallingford Software in the UK [19]. It is based on river bed cross-sectional data approximately every 50 meters along the modeled rivers, river bed roughness information and geometric data on all hydraulic structures (weirs, culverts, flow and water level control structures) and bridges along the course of all these rivers. The hydraulic control is based on the current fix regulations rules, which uniquely depend on up- and/or downstream water levels and discharges.

The InfoWorks-RS model has been recently extended with a real-time flood forecasting module, such that real-time flood predictions can be made (based on rainfall forecasts) with a time interval of 15 minutes and a prediction horizon of 2 days

(depending on the time horizon of the rainfall forecasts). This real-time flood forecasting model has been implemented in the FloodWorks software (extension of InfoWorks; [18]) and is currently fully operational. The forecasting results are linked
to a warning alert and alarm system, sending messages to local fire brigades and
crisis intervention authorities, but which can be consulted as well on-line by the
public.

A next step in further advancing the flood control, also in view of the adaptations
required due to the potential negative impacts of climate change, is the set up of a
flood control system. This text describes the development steps taken so far, where
an MPC-based flood control prototype algorithm has been developed and tested.

The MPC-algorithm could not directly make use of the full hydrodynamic InfoWorks or FloodWorks model, because of the long calculation times of this model.
One simulation of the entire model (to be done once in 15 minutes in the current
real-time flood forecasting setting) takes only a little less than 15 minutes. The
MPC-algorithm, however, requires multiple iterations during each prediction step.

With the aim to reduce the model calculation time, a simplified model has been
developed. This second model is of a conceptual type. It simplifies the hydrodynamic river flow process by lumping the processes in space, and by limiting the
study area to the region affected by the flood control. Lumping of the processes
in space is done by modeling of the water levels, not every 50 meters as the full
hydrodynamic model does, but only at the relevant locations. These are the locations up- and downstream of the hydraulic regulation structures, to be controlled by
the MPC-controller, and the locations along the Demer where potential flooding is
induced, to be limited by the controller. Depending on these locations, the river is
subdivided in reaches, in which water continuity is modeled (in a spatially lumped
way per reach) based on reservoir-type of models.

16.2.2 Conceptual model structure

Figure 16.2 shows the scheme of the model components for the study area around
the two flood control reservoirs 'Schulensmeer' and 'Webbekom' in the river Demer
basin. This area receives rainfall-runoff inflow via the tributary rivers Mangelbeek,
Herk, Gete, Velpe, Zwartebeek, Zwartewater and Begijnenbeek. By means of the
hydraulic regulating structures A and K7, the local water engineers can anticipate on
future flood risks. Through closing gate K7 and opening gate A, the Schulensmeer
reservoir is being filled, the downstream Demer flow reduced, and consequently the
flood risk of the cities Diest and Aarschot downstream of the study area of the reservoirs reduced. After the flood period, the Schulensmeer reservoir (which consists of
different reservoir compartments) can be emptied through the hydraulic regulating
structures D and E. The second reservoir 'Webbekom' is regulated in a similar way
by means of the hydraulic structures K18, K19, K7 at the Leugebeek river, K24*
and K30. Figure 16.3 gives an overview of the structure of the conceptual model
developed for the study area. In this scheme the river reaches are represented by
means of lines with positive flow in the direction of the arrows, the hydraulic reg-

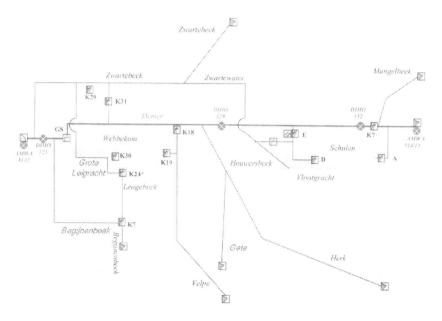

Figure 16.2: Scheme of the river network in the study area, the flood control reservoirs 'Webbekom' and 'Schulensmeer' at the village Schulen, and the hydraulic regulating structures [7].

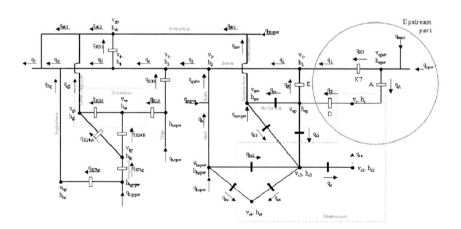

Figure 16.3: Scheme of the conceptual model for the study area of Figure 16.2 (dots for the calculation nodes, i.e., river or reservoir storage elements; lines for the river reaches; open rectangles for the hydraulic regulating structures; closed rectangles for the fixed spills or weirs).

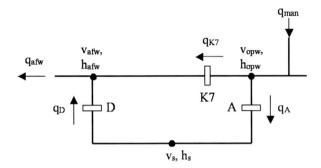

Figure 16.4: Schematic representation of the 'Schulensmeer' reduced area, together with the conceptual model variables.

Figure 16.5: Photo of the river Demer and the 'Schulensmeer' flood control reservoir in the background, together with the locations of the main water level and discharge variables.

ulating structures by means of the full rectangles, the fixed spills or overflows by open rectangles, and the model units where water storage (in the reservoir compartments or along river reaches) and water levels are simulated by nodes. The symbol 'q' denotes discharges, 'h' water levels, 'v' storage volumes, and 'k' controllable gate crest levels. The water levels and volumes are the model variables describing the state of the water system in the MPC controller. The gate crest levels are in the inputs in the MPC controller, the upstream (rainfall-runoff) discharges the disturbances of the MPC controller.

In order to test the MPC controller in the first phase of the project, a 'reduced area' around the Schulensmeer reservoir has been considered. A separate reduced conceptual model has been developed for this area. The scheme of this conceptual model for the reduced area is shown in Figure 16.4. Next to the reduction in area, also some simplifications to the physical reality have been implemented, in order to enable disconnection of the Schulensmeer area from the more downstream areas. Figure 16.5 shows photos of the region upstream of the Schulensmeer reservoir and shows some upstream model components and model variables.

16.2.3 Conceptual model building and calibration process

The conceptual model-structure has been identified and its parameters calibrated based on simulation results with the full hydrodynamic InfoWorks model. In order to do so, a combination of different modeling concepts has been applied. The flood control reservoirs have been modeled in the simplest way as storage nodes (vs, vs2, vs3, vs4, and vw), maintaining water continuity. Also the large water storage areas, which in the conceptual model are modeled in a combined way with their drainage canals, are schematized as storage nodes. This is the case for the nodes 'vvg' (draining the 'Schulensmeer' reservoir) and 'vgl' (draining the 'Webbekom' reservoir). For each of these storage nodes, a continuous relationship has been implemented between the storage volume and the mean water level (which is assumed static at each time step). These relationships have been calibrated based on the Digital Elevation Model underlying the IWRS model. An example of such calibration result is shown in Figure 16.6 for the Webbekom flood control reservoir. The water levels are shown as levels above the 'TAW' standard reference level for Belgium.

The static level assumption is valid for the flood control reservoirs (thus for the water levels hs, hs2, hs3, hs4 and hw). This is, however, not the case for the water levels 'hvg' and 'hgl' of the drainage areas. These water levels indeed vary a lot along the course of the drainage canals. The water level considered in the conceptual model then is taken at one specific location along the drainage canal. Water levels at other locations along the drainage canal are only included in the conceptual model when they are really needed (e.g., because they control hydraulic structures). They are assumed related to the primary levels considered ('hvg' and 'hgl' in this case). In case no unique relationship exists with these primary levels, the drainage canal (together with the flood zones on its banks) is modeled as a long regular river reach, as described next.

Long regular river reaches have been modeled in two different ways: (a) schematized by means of a serial connection of reservoirs, and (b) by means of the water surface profile concept (considering the water level differences from down- to upstream along the river). In approach (a), water levels are modeled in a step-wise way from up- to downstream, while approach (b) starts from the most downstream locations along the reach and considers water level changed to upstream. Approach (a) guarantees water continuity, but has the disadvantage that more downstream reservoirs in the serial connection do not necessarily have lower water levels. Outflow corrections are then needed to avoid this type of anomalies. Approach (b) does not pose this problem, because water level differences are always taken positive from down- to upstream.

In approach (b) the water level differences are depending on the discharge in the reach and the downstream water level. Primarily the following relation is tested:

$$h = h_{afw} + a \, (q^2 / (h_{afw} - h_{afw,0})^2)^b, \qquad (16.1)$$

where $h - h_{afw}$ is the water level difference to be modeled depending on the discharge q and the downstream water level h_{afw}, and where $h_{afw,0}$ is the bed level at

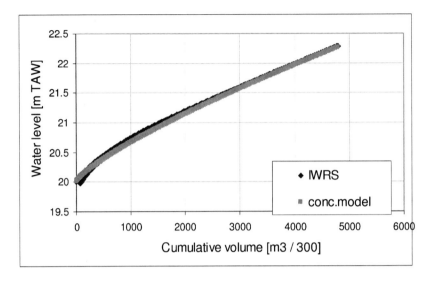

Figure 16.6: Calibration result for the storage volume-mean water level relation of the 'Webbekom' flood control reservoir.

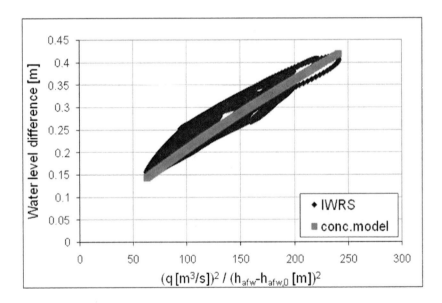

Figure 16.7: Calibration result applying approach (b) for the Demer river reach between the conceptual storage nodes h2 and h3.

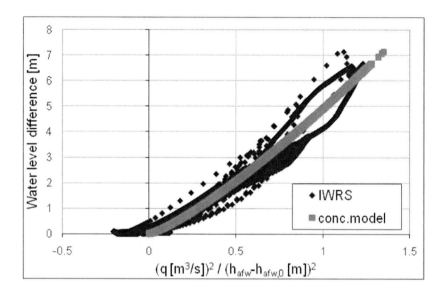

Figure 16.8: Calibration result applying approach (b) for the Grote Leigracht river between the conceptual storage nodes hgl and hzb.

the downstream location. This equation is based on the assumption of the equation of Manning [4]:

$$q = \frac{1.49}{n} A R^{\frac{2}{3}} S^{\frac{1}{2}},\qquad(16.2)$$

where the discharge q depends on cross-section area A, the hydraulic radius R, the friction slope S and the Manning coefficient n. Under the uniform flow approximation, the friction slope S equals the river bed slope, which under the normal depth assumption furthermore equals the water surface slope $h - h_{afw}$. For wide river sections, the hydraulic radius becomes approximately equal to the river bed width and thus independent on the water level. For rectangular sections, the cross-section area becomes linearly proportional to the water depth. Under these assumptions, the water level difference becomes proportional to the ratio of the squared discharge and the squared water depth:

$$q^2 / (h_{afw} - h_{afw,0})^2,\qquad(16.3)$$

which leads to equation (1). Figure 16.7 gives an example of the calibration of equation (16.1) to conceptually model the water level differences between h2 en h3 along the Demer. Figure 16.8 shows another example of the same approach but for the Grote Leigracht river (water level differences between hgl en hzb).

Approach (a), on the other hand, makes use of reservoir models , where the outflow discharge (flow to downstream) depends on the reservoir storage volume and

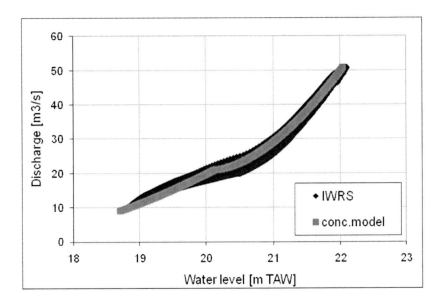

Figure 16.9: Calibration result applying approach (a) for the Demer storage node describing h3. A unique relation is shown between the water level h3, which depends on storage volume v3, and the outflow discharge (q3) of this storage.

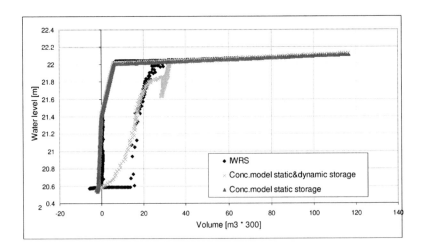

Figure 16.10: Calibration result applying approach (a) for the Leugebeek describing hlg. The hysteresis effect in the storage-volume relation has been modeled separating the total storage in static and dynamic storages.

(potentially also) on the upstream inflow discharges. The relationships between outflow discharge and storage volume and (potentially) inflow discharge, are to be calibrated to simulation results with the full hydrodynamic model. These relationships are linear in the most parsimonious case, but often of non-linear nature. Any non-linear relation can, however, be approximated by a set of piece-wise linear relations. In this study, the reservoir-model relations are identified in the following step-wise approach. First the relationship between water storage in the river reach and the outflow discharge is studied based on the IWRS results. When this relation is unique, a linear, non-linear, piece-wise linear or non-linear function is calibrated to that relation. Figure 16.9 shows for the same Demer reach as in Figure 16.6 the relation between the water level h3 (depending on the storage volume v3) and the downstream discharge. A unique relation is found, indicating that the reservoir concept is applicable, thus that both approach (a) and approach (b) could be implemented for this case. In Figure 16.9, the relation is not linear but a piecewise relation could be calibrated: a linear relation up to water level 20.2 m TAW, and a power relation for the higher water levels.

When the relation is not unique, but shows 'hysteresis effects' (see example in Figure 16.10), a more complex calibration method has to be followed. As shown in [14] based on hydrodynamic sewer system applications, hysteresis effects are to be explained by different storage-outflow relationships in the decreasing and increasing flanks of the flow hydrographs. These differences are to be explained by the differences in 'dynamic storage' in the system, and thus by the inflow discharges. In this case, the reservoir model is advanced separating the total storage in the river reach in static and dynamic storage parts. The static storage is identified as the lowest storage for a given outflow discharge (thus during the decreasing flanks of the flow hydrographs), while the dynamic storage is the difference between the total storage in the reach and the static storage identified. By plotting this dynamic storage at each time step versus the inflow discharge in the reach, and calibrating the identified relation by means of a linear, non-linear, piece-wise linear or non-linear model, a first reservoir submodel becomes ready: the submodel to predict at each time step the dynamic storage based on the inflow discharge. When this dynamic storage is subtracted from the total storage (at each time step to be calculated considering water continuity) static storage predictions are obtained. Analyzing the relationship between this static storage and the outflow discharge, and fitting a function to this relation, completes the reservoir model-structure.

The 'hysteresis' phenomenon was seen for reaches of the 'Leugebeek' river (node 'lg') (see Figure 16.10). This is a river reach where part of the reach has water levels that are upstream controlled (steep hydraulic slope) and another part downstream controlled (mild hydraulic slope).

Both approaches (a) and (b) model the water levels in a 'lumped' way, which means that only 1 water level (at one single location along the reach) is considered. When water levels at more locations are required along the reach (e.g., because of their importance in controlling hydraulic structures, or in controlling water levels in upstream river reaches, through back-water effects), relations are sought between the levels at these locations and the primary modelled water level. Because of its

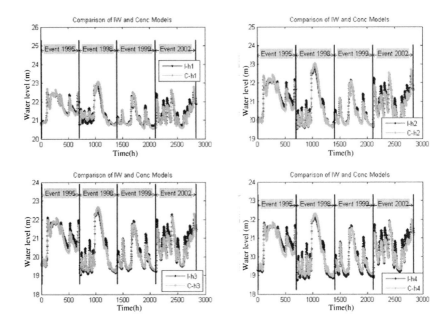

Figure 16.11: Comparison of the InfoWorks-RS (I) and conceptual model (C) results for the discharges and water levels at selected nodes; for the historical floods of 1995, 1998, 1999–2000 and 2002.

stability and mathematical elegance, preference has been given to approach (b). The hydraulic structures in the conceptual model are implemented in a way identical to the IWRS-model; the same hydraulic model equations are considered.

In the conceptual model, model equations are solved based on finite differences between successive time steps. Multiple iterations per time step are avoided in order to have the model calculation times as limited as possible. A time step of 5 minutes is considered, which is coarse in comparison with the spatial resolution of the model (average distance between the calculation nodes). For this reason, the first version of the conceptual model was strongly affected by instabilities. Flow delay terms therefore have been added to the model, making use of the linear reservoir model equation:

$$q_{out}(t) = \exp(-\frac{1}{k})\, q_{out}(t-1) + (1-\exp(-\frac{1}{k}))\, q_{in}, \qquad (16.4)$$

where q_{in} is the original flow (before delay) and q_{out} the flow after application of the delay equation.

16.2.4 Conceptual model validation

The identification of the conceptual model-structure and the calibration of its parameters, based on the methodology described above, has been based on the simu-

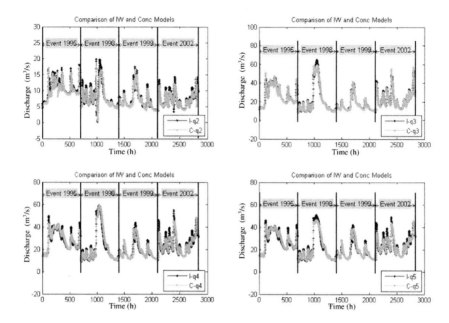

Figure 16.12: Comparison of the InfoWorks-RS (I) and conceptual model (C) results for the discharges and water levels at selected nodes; for the historical floods of 1995, 1998, 1999-2000 and 2002.

lation results with the full hydrodynamic model for two historical high flow or flood events: the flood events of September 1998 and January 2002. Validation was done based on the full set of flood events during the past 15 years; being the flood events of 1995, 1998, 1999 – 2000 and 2002. Figure 16.11–16.13 show some validation results, comparing the water level and discharge simulation results of the conceptual model with the one of the InfoWorks-RS (IW) models. Both models have a time step of 5 minutes. The model output results shown in Figures 16.11–16.13 are aggregated at the hourly time step.

Next to the comparison with the full hydrodynamic model results for historical flood events, it also has been checked whether the conceptual model still performs well for conditions (hydraulic structure regulations or gate levels) which did go beyond the range of conditions during the historical flood events. Also the model has been checked for discontinuities and stability.

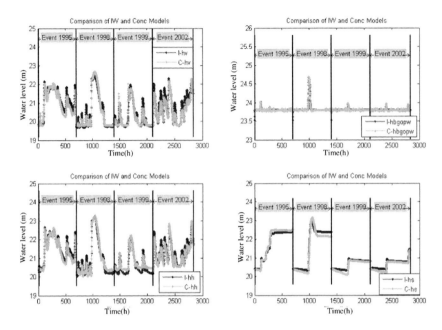

Figure 16.13: Comparison of the InfoWorks-RS (I) and conceptual model (C) results for the discharges and water levels at selected nodes; for the historical floods of 1995, 1998, 1999-2000 and 2002.

16.3 Flood regulation

16.3.1 Classical feedback and feedforward control

In Figure 16.14 a diagram of a classical feedback and feedforward controller for water system management is presented. The set-point represents the desired water level. The objective of the controller is to steer the water levels as close as possible to the desired set-points. In order to achieve this at each sampling time the different water levels are measured. This measured water level is then compared to the set-point and the deviation is passed to the feedback controller. The feedback controller will then determine a control action that tries to remove this deviation. In water system management the feedback controller is typically a proportional integral controller (PI) [12]. The feedforward controller uses measurements of the disturbance (in this case rainfall) and an inverse model in order to counteract the effect the measured disturbance will have on the water level. In the ideal case this would lead to perfect set-point regulation. In practice, however, there are uncertainties on the models and the measured disturbances which make perfect set-point regulation with a feedforward controller impossible. In order to cancel out these uncertainties a feedforward controller is always combined with a feedback controller [16].

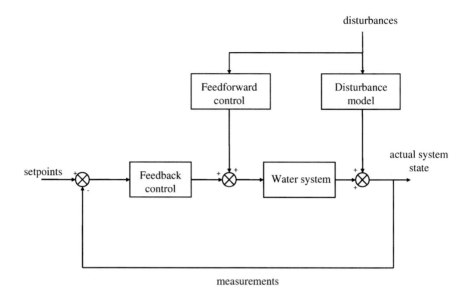

Figure 16.14: A diagram of a classical feedback and feedforward controller for water system management [15].

16.3.2 Three-position controller

Three-position control corresponds to a control mode that is based on some very simplistic rules and typically has the purpose to steer a water level to a desired set-point. A three-position controller achieves this by reacting on each deviation from the set-point by adjusting the gate position for a predetermined amount of time. A standard three-position controller consists of the following rules:

- The water level lies between desired upper and lower limit – no corrective action. The set-point lies between the upper limit and lower limit.

- The water level exceeds the upper limit – the gate is lowered in order to lower the water level.

- The water level is lower than the lower limit – the gate is raised in order to raise the water level.

Note that this controller is an example of a feedback controller. The local water administration responsible for the water management of the Demer basin uses three-position controllers for the control of the gates. However, their controllers are more advanced than the standard three-position controller. During normal operation the purpose is to steer the water levels to their set-points. This is done by standard three-position controllers. During periods of heavy rainfall the focus shifts in prevention of flooding. Therefore, the standard rules are replaced by new rules. These new rules determine the gate position based on the measurement of water levels in the water system. These rules are formulated by the local water administration based on many years of experience in regulating the Demer basin and can therefore be considered to be expert knowledge. These type of controllers have the advantage that the movement of the gate is limited which is positive with respect to wear and tear of the gates . Another advantage is its straightforward implementation.

16.3.3 Model predictive control

Management of modern water systems requires more advanced control methods than the classical feedback and feedforward. The first reason is that the classical methods are not capable of dealing with constraints on the water levels, neither can they cope with physical limitations on the gates. The second reason is that these methods are not capable to deal with future rain predictions. The third reason is that these methods are only suitable for linear systems.

A control strategy that is better suited for the management of water systems is model predictive control (MPC) [3, 13]. MPC is a control strategy that can deal with constraints on the system as well as with future rain predictions. Besides that MPC can deal with linear as well as with nonlinear models. In literature several studies can be found in which automatic control techniques are used to control a water system. There are also several studies available in which MPC is used to control water systems [11, 15–17]. These works however have as main goal to control the different water levels to some desired target value and not to prevent flooding. In these applications it is usual sufficient to linearize the system around the desired steady state in order to obtain good results. However, when trying to avoid/reduce flooding linearization around the steady state is not sufficient. During periods of heavy rainfall the complete nonlinear dynamics of the system are excited so it is very important that the MPC can deal with nonlinear model behavior. In the sequel the nonlinear MPC scheme applied in this research will be discussed in detail.

16.3.4 Components of MPC

MPC is a model based controller. These types of controllers emerge from the chemical industry where the fabricated products are produced very closely to the limits of the quality specifications [8, 13]. Therefore, it is important that the applied control strategy is capable to work as close as possible to the limits of the constraints on the process. The same observations hold for flooding regulation. In order to avoid flooding it is important that the controller pushes the water levels as close as possible to the constraints if necessary. If the controller is not capable of doing this, this can lead to flooding of some water levels that could have been avoided. MPC is a control strategy that is capable of doing this. MPC contains the following components:

- **Internal model**: The internal model describes the physical behavior of the controlled water system and is used by MPC in order to predict the future water levels within a certain time window, also called prediction horizon. The more accurate the model, the more accurate the predictions by which the performance of the controller increases. However, a trade-off must be made between accuracy of the model and its complexity. A very accurate model can come with the expense of an unacceptable high calculation time which makes it impossible to use it in a MPC framework.

- **Objective function**: This function captures the objective of the controller. The objective function penalizes deviations from desired reference values. More important water levels can be given more importance by increasing the corresponding penalization. Typically, movements of the gates are also penalized in order to avoid excessive wear of the gates.

- **Constraints**: MPC is a control strategy that differentiates itself from other control strategies by its capability to deal with constraints. In water systems two types of constraints can be distinguished. The first one concerns physical limitations of the gates. There are constraints on the maximum and minimal gate height as well as on the maximum gate movement. These constraints are hard constraints and should always be satisfied. The second type of constraints concerns limitations on the water levels. In order to avoid flooding it is important each water level stays under its corresponding flood level, the level at which flooding occurs. These constraints are soft constraints in the sense that they may be violated if the rainfall is too excessive to avoid flooding. Also note that the internal model can also be seen as a constraint on the system.

- **Optimization**: MPC performs an optimization in which it tries to optimize the objective function taking the constraints into account. This optimization leads to an input sequence that optimizes the objective function within the given prediction horizon. In this chapter the optimization will be a nonlinear program because of the nonlinear system model.

- **Receding horizon**: The first input of the optimal input sequence is applied to the system. Then, in the next sampling time the time window is shifted with 1 time step and based on the new state of the system a new optimization procedure is performed. This strategy provides robustness against uncertainties coming from modeling errors, rain prediction errors and measurement noise.

- **Measurements and state estimation**: In order to be able to make future predictions with the internal model it is necessary to know the current state of the system. This can be obtained by measurements. However, in practice it is not possible to measure the full state of the system. Therefore, based on the measurement of a subset of all the states and the internal model the full state of the system is estimated by a state estimator.

16.3.5 Mathematical formulation

In this work the internal model is assumed to be a discrete nonlinear dynamic state space model defined by

$$x_{k+1} = f(x_k, u_k, d_k)$$
$$y_k = Cx_k$$
(16.5)

with $x_k \in \mathbb{R}^{n_x}$ and $u_k \in \mathbb{R}^{n_u}$ vectors representing respectively the state, the input and the disturbances (rainfall) of the system at time step k and $y_k \in \mathbb{R}^{n_y}$ a vector containing the outputs of the system at time step k. The state x_k consists of the water levels, the discharges and the node volumes of the water system. The output y_k consists of the water levels of the system as the goal of MPC is to steer the water levels. The input u_k consists of the level of the different gates.

MPC tries to steer the future outputs to the reference value y_r while trying to keep the inputs as close as possible to the reference input u_r without violating constraints on the inputs and the states. MPC achieves this by optimizing on-line at each time step an optimal control problem. In the case of the nonlinear system (16.5) the online optimal control problem is a nonlinear programming problem defined as:

$$\min_{\substack{x_{k+1},\ldots,x_{k+N_p} \\ y_{k+1},\ldots,y_{k+N_p} \\ u_k,\ldots,u_{k+N_c-1}}} \sum_{i=1}^{N_p} (y_{k+i} - y_r)^T Q(y_{k+i} - y_r)$$

$$+ \sum_{i=0}^{N_c-1} (u_{k+i} - u_r)^T R(u_{k+i} - u_r)$$
(16.6)

subject to the following constraints for $i = 1, \ldots, N_p, j = 0, \ldots, N_c - 1$:

$$x_{k+i+1} = f(x_{k+i}, u_{k+i}, d_{k+i})$$
(16.7)

$$y_{k+i} = Cx_{k+i}$$
(16.8)

$$u_{k+i} = u_{k+N_c-1}, \text{for } i \geq N_c$$
(16.9)

$$\underline{y} \leq y_{k+i} \leq \bar{y}$$
(16.10)

$$\underline{u} \leq u_{k+j} \leq \bar{u}$$
(16.11)

$$\left\| u_{k+j+1} - u_{k+j} \right\|_\infty \leq \Delta_{max}$$
(16.12)

with N_p the prediction horizon and N_c the control horizon. The prediction horizon is usually bigger than the control horizon. The inputs after the control horizon are all equal to u_{k+N_c-1}. This optimal control problem is a nonlinear program due to the nonlinear equality constraint (16.7) representing the internal model of the water system. In order to solve this nonlinear program the nonlinear constraint (16.7) is replaced by a linear-time varying system that is obtained by doing a linearization around the states obtained by applying the optimal input sequence

$$\{u(k|k-1), u(k+1|k-1), \ldots, u(k+N_c-2|k-1), u(k+N_c-2|k-1)\} \quad (16.13)$$

obtained from the optimal control problem solved at the previous time step $k-1$. Note that for time step $k+N_c-1$ no future input can be obtained from the previous time step which is the reason this input is chosen equal to that of time step $k+N_c-2$.

This input sequence can be seen as a first guess for the solution of the optimal control problem at the current time step k. For notational convenience in the sequel this input sequence will be referred to as $\left\{ u_k^0, u_{k+1}^0, \ldots, u_{k+N_c-2}^0, u_{k+N_c-1}^0 \right\}$. At time step k a simulation of the nonlinear model (16.5) is performed by applying this input sequence and taking the future rain predictions $\left\{ d_k, d_{k+1}, \ldots, d_{k+N_p-1} \right\}$ into account. This results into a sequence of states defined by $\left\{ x_k^0, x_{k+1}^0, x_{k+2}^0, \ldots, x_{k+N_p}^0 \right\}$. Linearizations around these states give raise to the following linear time-varying system:

$$\left(x_{k+1} - x_{k+1}^0 \right) = \left(\frac{\partial f}{\partial x} \right)_k \left(x_k - x_k^0 \right) + \left(\frac{\partial f}{\partial u} \right)_k \left(u_k - u_k^0 \right). \tag{16.14}$$

This can be re-written as following linear time-varying system

$$\begin{aligned} x_{k+1} &= A_k x_k + B_k u_k + z_k \\ y_k &= C x_k \end{aligned} \tag{16.15}$$

with

$$A_k(i,j) = \frac{\partial f_i(k)}{\partial x_j(k)} \tag{16.16}$$

$$B_k(i,l) = \frac{\partial f_i(k)}{\partial u_l(k)} \tag{16.17}$$

$$z_k = x_{k+1}^0 - A_k x_k^0 - B_k u_k^0. \tag{16.18}$$

The partial derivatives in (16.16) and (16.17) are calculated numerically by means of finite differences. The linear time-varying system (16.15) is an approximation of the nonlinear model and is only valid within a certain region around the linearized states. In optimization this region is called a trust region. Taking this into account within the trust region the following optimization program gives an approximate optimal control sequence for the nonlinear model (16.5)

Optimization 1

$$\min_{\substack{x_{k+1}, \ldots, x_{k+N_p} \\ y_{k+1}, \ldots, y_{k+N_p} \\ u_k, \ldots, u_{k+N_c-1}}} \sum_{i=1}^{N_p} (y_{k+i} - y_r)^T Q (y_{k+i} - y_r)$$

$$+ \sum_{i=0}^{N_c-1} (u_{k+i} - u_r)^T R (u_{k+i} - u_r) \tag{16.19}$$

subject to the following constraints for $i = 1, \ldots, N_p, j = 0, \ldots, N_c - 1$:

$$x_{k+i+1} = A_{k+i} x_{k+i} + B_{k+i} u_{k+i} + z_{k+i} \tag{16.20}$$

$$y_{k+i} = C x_{k+i} \tag{16.21}$$

$$u_{k+i} = u_{k+N_c-1}, \text{ for } i \geq N_c \tag{16.22}$$

$$\underline{y} \leq y_{k+i} \leq \bar{y} \tag{16.23}$$

$$\underline{u} \leq u_{k+j} \leq \bar{u} \tag{16.24}$$

$$\left\| u_{k+j+1} - u_{k+j} \right\|_\infty \leq \Delta_{max} \tag{16.25}$$

$$\left\| u_{k+j} - u_{k+j}^0 \right\|_\infty \leq \Delta_T \tag{16.26}$$

with (16.26) the constraints defining the trust region for which the linear time-varying approximation is valid. Note that the optimization program is now a quadratic program. After solving optimization (1) the obtained optimal control sequence can be used to perform a new simulation leading to a new sequence of future states around which a new linearization can be made and the optimization can be redone. This sequence can be repeated until convergence is obtained. The following algorithm summarizes this optimization procedure:

Algorithm 1 *1. At time step k initialize the inputs $\left\{ u_k^0, u_{k+1}^0, \ldots, u_{N_c-1}^0 \right\}$ with (16.13).*

2. Perform a simulation with the input sequence $\left\{ u_k^0, u_{k+1}^0, \ldots, u_{N_c-1}^0 \right\}$ and obtain the future states $\left\{ x_k^0, \ldots, x_{k+N_p}^0 \right\}$.

3. Perform a linearization around these future states and solve optimization (1).

4. If the difference (based on the ∞-norm) of the new optimal input sequence $\left\{ u_k^*, u_{k+1}^*, \ldots, u_{N_c-1}^* \right\}$ with $\left\{ u_k^0, u_{k+1}^0, \ldots, u_{N_c+1}^0 \right\}$ is smaller than a prespecified tolerance then stop the algorithm, else replace $\left\{ u_k^0, u_{k+1}^0, \ldots, u_{N_c-1}^0 \right\}$ with $\left\{ u_k^*, u_{k+1}^*, \ldots, u_{N_c-1}^* \right\}$ and go to step 2.

So, after the solution of a sequence of quadratic programs algorithm 1 returns a local optimal solution to the initial nonlinear optimal control problem. Then the first input u_k^* of this input sequence is applied to the real system. At the next sampling time measurements on the system are performed and the current state of the system is estimated by means of a state estimator. However, in the sequel it will be assumed that the state of the real system and the internal model are exactly the same, so state estimation will not be considered. The addition of a state estimator is considered as future research.

Remark 16.1. In this work no use is made of multi-start optimization.

16.3.6 Constraint and cost function strategy

As pointed out earlier there are two types of constraints, namely hard and soft constraints. Hard constraints are constraints that can never be violated and are typically physical limitations on the gate movements. On the other hand, soft constraints are constraints that can only be violated in case no solution can be obtained that satisfies them all. In water systems the upper limits on the water levels are considered as soft constraints. During periods of heavy rainfall the situation can occur where it is impossible to keep each water level under its corresponding flood level meaning that the constraints are too stringent. In order to obtain a feasible solution some of the soft constraints need to be relaxed or omitted from the optimization problem. In order to do this a strategy needs to be defined that determines which soft constraints need to be relaxed. Simultaneously there also needs to be a strategy for the cost function as the objective is to steer the water levels corresponding to the relaxed soft constraints back into the feasible region.

Another reason to incorporate a variable constraint and cost function strategy is that the objective of MPC depends on the state of the water system. More specifically, during periods of light rainfall the main goal of MPC is to steer the water levels to a specified reference level. However, during periods of heavy rainfall steering the water levels to reference levels becomes less important as the focus shifts to flood prevention. This change in objectives has as consequence that the constraints as well as the objective function of MPC needs to change during operation. In the remainder of this subsection the chosen constraint and cost function strategy is outlined in more detail.

At first, during normal regulation the objective of MPC is to steer the water levels to the reference levels that are provided by the local water administration. It is important the physical limitations on the gates are satisfied. The constraints on the water levels do not play an important role here as the rainfall is too low to cause floodings. Besides this it is also important that during this period of time the water reservoirs are emptied and reach the lowest level that is allowed. Any unnecessary volume inside the reservoirs can lead to unnecessary floodings in the future. Therefore during normal regulation the cost function is adjusted to steer the most important water levels to their reference levels and to empty the reservoirs as much as possible.

During periods of heavier rainfall the expert knowledge of the operators controlling the gates in the field is incorporated in the MPC. At first sight this approach might seem restrictive as MPC is not allowed to use all possible degrees of freedom that are available. However, this is a necessary step in order for the MPC to avoid flooding properly. The reason for this is that rainfall events causing floods can last more than five days. This means that the prediction horizon N_p of MPC should be taken longer than five days in order to be able to properly avoid flooding. But the prediction horizon cannot be taken that long for the following reasons:

- The optimal control problem to be solved at each time step would be untractable as the number of unknown variables would be too large.

- Rain predictions are not accurate enough to predict so much in advance. Typically rain predictions are only reliable to predict 2 days ahead.

Therefore it is necessary to incorporate some predefined rules containing expert knowledge in order to prevent flooding. Within these rules it is the task of MPC to find the optimal control strategy. This expert knowledge contains the following rules:

1. For each water level a guard level is defined by the local water administration. As long as the water levels do not violate their corresponding guard level, it is not allowed to fill the water reservoirs. The reason behind this is to keep the reservoirs as empty as possible for as long as possible. Only when MPC detects that a guard level will get violated within the prediction horizon it is allowed to use some of the storage capacity from the reservoirs to keep the water levels beneath their guard level. This can be achieved by adding the guard levels as constraints of the optimization problem and putting a high value in the cost function for deviations of the reservoir levels from their corresponding reference level. This will ensure the water reservoirs only get filled if this is absolutely necessary to avoid the violation of the guard level constraints.

2. Water reservoirs can be used to avoid violation of the guard levels. However, it is not allowed to use the complete storage volume available in the reservoirs. For each reservoir a safety limit is defined. Once this safety limit is reached the reservoirs may not be used anymore and the guard levels will be violated. The reason for this is to ensure there is enough storage volume available in case of further future rainfall. Therefore, in the optimization the guard levels are replaced by their flood level and constraints are added to ensure the reservoir levels stay beneath their safety level.

3. If it continues raining until the point where MPC cannot keep the water levels beneath their flood level within the prediction window, it is allowed to further fill the reservoirs until the water level in the reservoirs reach their corresponding flood level. In order to achieve this the safety limits are replaced in the optimization by the flood levels. By doing this MPC can use all the available storage volume in the reservoirs in order to keep all water levels beneath their flood level.

4. If the rainfall is really excessive then flooding is unavoidable and some or all water levels will violate their corresponding flood levels and the optimization proposed in step 3 will be infeasible. In order to be able to get a useful solution from MPC soft constraints need to be removed from the optimization until a feasible solution is obtained. If a soft constraint on a specific water level is removed, the weight in the cost function corresponding to that water level is increased. This is done in order to minimize the soft constraint violation.

Note that the removal of soft constraints outlined in step 4 is based on a priority based strategy. First, different sets with different priorities are defined. Soft constraints that are more important than others are categorized into sets with higher priorities. Soft constraints of equal importance are categorized into the same priority set. If the MPC defined in step 3 turns out to be infeasible all the soft constraints belonging in the set with the lowest priority are removed from the optimization and in the cost function the weights of the corresponding water levels are increased. If the optimization still turns out to be infeasible the soft constraints from the second lowest priority set are removed and the corresponding weights are adjusted in the cost function. This procedure is repeated until a feasible solution is obtained.

16.3.7 Uncontrollable modes

In river control (irrigation channels) the discharges over the gates are typically chosen as control inputs. As a result MPC returns a sequence of future optimal discharge sequences. This sequence is passed as set-points to local controllers that try to follow their set-point as close as possible. The advantage of this approach is that the underlying system dynamics turn linear. In literature many examples can be found where this approach has been applied successfully [2, 5, 15]. All these examples concern cases where the rainfall is relatively small and where the goal is to steer the water levels to a given reference level. In this work however the main focus is flooding prevention. In this case taking the discharges over the gates as inputs might cause problems:

- It is difficult to take constraints on the gates like upper/lower limits and maximal gate movements into account as a mapping should be taken into account that maps the discharge to the gate level. This is not a straightforward operation as the discharge does not only depend on the gate level, but also on the water level up- and downstream the gate. Moreover, the water levels are strongly time-dependent which makes it even more difficult.

- The relation between the discharge over a gate depends on the gate level and the water level up- and downstream the gate. This dependence is nonlinear and by taking the discharges as inputs this nonlinearity is neglected, which can be problematic during heavy rainfall events.

- The discharge over a gate depends on the water level up- and downstream the gate. But the water levels at their turn also depend on the discharge over the gate. Therefore, if the discharges over the gates are taken as control inputs, it is not sure that the computed discharges can be achieved.

Because of these reasons in this work the gate levels are chosen as control inputs. A difficulty that arises when taking the gate levels as control inputs is that the gates can become locally uncontrollable. This uncontrollability is related to the discharge equations, system equations determining the discharge over a gate. These equations consist of different modes. In some of these modes the discharge equation is independant of the gate level. If a gate is situated in such a mode the gate becomes locally uncontrollable. Because MPC relies on local linear models this means that the controller is not able to control the gate anymore as it thinks that the gate has no influence on the water system.

In Figure 16.15 two uncontrollable modes are depicted. Gate $G1$ is closed and therefore the discharge over $G1$ is equal to zero. This is the case for as long as the gate is closed which means that the discharge over a closed gate is independent of the gate level. During normal operation the gates controlling the flow to the flood reservoirs are typically in this mode. This can decrease the performance of the controller because controllability is only recovered whenever the water level upstream of the closed gate raises above the gate level. At the other hand gate $G2$ lies much lower than its up- and downstream water level. Therefore gate $G2$ does

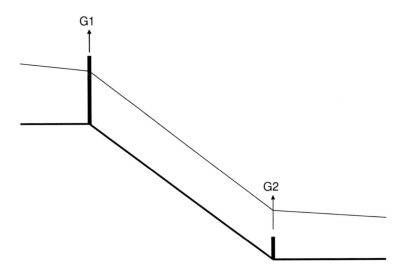

Figure 16.15: Two examples of gates that are in an uncontrollable mode.

not influence the discharge going from upstream to downstream which means the gate is also uncontrollable. Gate $G2$ can only recover from this uncontrollability if one of the 2 water levels decreases to a level lower than the gate level, or if both water levels decrease to a level at which the gate starts to have an influence again on the discharge.

Losing the controllability of a gate can significantly decrease the performance of the controller. Therefore, the controller must try to steer each uncontrollable gate back into a controllable mode. As explained previously, at each time step a simulation is done in order to approximate the nonlinear system (16.5) by a linear time-varying system (16.15). At this point a check can be done to determine for each gate whether it will get uncontrollable within the time-window. This can be done by inspecting the columns of the B_k matrix in system equation (16.15). If a certain column p of B_k consists of only zeros, then the corresponding input $u_k(p)$ is uncontrollable at time step k. Now assume that the water levels up -and downstream of input $u_k(p)$ are equal to h_u and h_d respectively. In order to steer that gate back into a controllable mode its corresponding value for the reference input u_r at time step k is modified into a value that lies between h_u and h_d. Because the controller 'thinks' the gate does not have an effect on the water levels at that specific time step, it will try to steer the gate to the new reference input and recover its controllability.

In Figure 16.16 a schematic overview is given of the complete nonlinear MPC scheme used in this study.

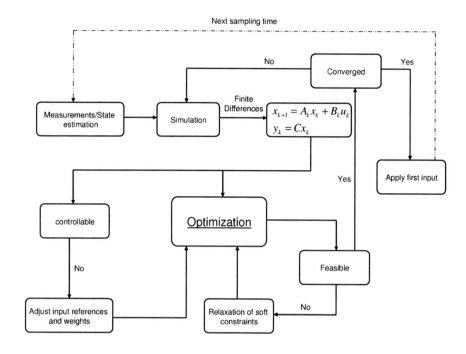

Figure 16.16: Schematic overview of the presented nonlinear NMPC scheme.

16.4 Experimental results

As mentioned in Section 16.2 this study was divided in two parts. In a first phase a 'reduced area' around the Schulensmeer reservoir has been considered and a separate conceptual model was developed for this. In a second phase a much larger region of the Demer basin was considered containing the two flood control reservoirs 'Webbekom' and 'Schulensmeer'. MPC was tested in both phases and compared with the current three-position controllers of the local water administration. In the following these results are presented and discussed.

16.4.1 Experimental results on the 'reduced area' around Schulensmeer

The 'reduced area' considered here is depicted in Figure 16.4. There are 3 water levels that need to be kept underneath their corresponding flood level. In order to achieve this MPC needs to control the 3 gates that are available. The control objectives can be summarized as follows:

- Steer h_{opw} as close as possible to 21.5 m.
- Gates cannot move at a rate faster than 0.1 m/hour.

- The upper and lower limit on the gates are the same for the 3 gates and are equal to 23 m and 20 m respectively.
- The guard level of h_{opw} is equal to 23 m; the safety limit for the reservoir level h_s is 23 m.
- The flood levels of the 3 water levels are:

   – $h_{opw} \leq 23.2$ m
   – $h_s \leq 23.2$ m
   – $h_{afw} \leq 22.75$ m

In Figure 16.17 and Figure 16.18 simulations are depicted with a three-position controller and a MPC controller respectively . Both simulations are based on the same flooding event of 1998. From the figures it can be seen that with MPC the water level h_{opw} lies much closer to the reference level of 21.5 m. Besides this one can notice that with MPC all water levels stay beneath their flood level. Only at hour 800 there is a small flooding of h_{afw}. However, the simulation with the three-position controller shows severe floodings of all the water levels especially around hour 800. The MPC clearly outperforms the three-position controller. (Note: These results were published in [1]).

16.4.2 Experimental results on the large area

The large area considered here corresponds to the area presented in Figure 16.3. The objectives of the controller are summarized in the following:

- Steer h_{opw} and hbg_{opw} as close as possible to respectively 21.5 m and 23.8 m during normal operation.
- Gates cannot move at a rate faster than 0.1 m/hour.
- For each gate an upper and a lower limit is taken into consideration.
- The guard levels for the reservoirs h_s and h_w are equal to 23 m and 22 m.
- The safety limit for h_{opw} is equal to 23 m.
- The flooding levels for the most important water levels are :

   – $h_{opw} \leq 23.2m$
   – $h_s \leq 23.2m$
   – $h_2 \leq 22.737m$
   – $hbg_{opw} \leq 24.84m$
   – $h_w \leq 22.4m$
   – $h_{afw} \leq 20.46m$
   – $h_{gl} \leq 22m$

Simulations were done with historical rain data from 1998. In Figure 16.19 a simulation with the three-position controller is presented. In Figure 16.20 the same simulation is performed but with the NMPC scheme displayed in Figure 16.16. In both figures the different flooding levels are indicated as horizontal lines. Comparing both results leads to following conclusions:

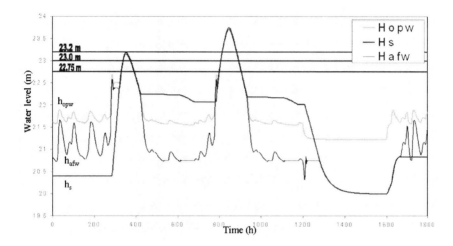

Figure 16.17: Regulation of the reduced area conceptual model with a three-position controller.

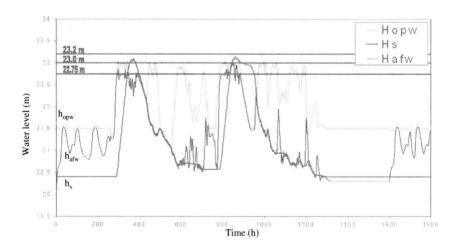

Figure 16.18: Regulation of the reduced area conceptual model with MPC.

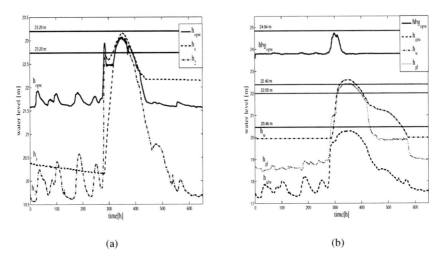

(a)   (b)

Figure 16.19: Regulation with three-position controller.

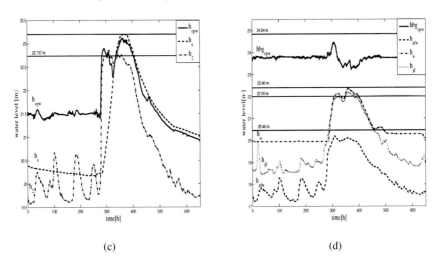

(c)   (d)

Figure 16.20: Regulation with NMPC controller.

- The NMPC controller manages to steer h_{opw} much closer to its set-point. During periods of small rainfall events NMPC regulates hbg_{opw} with a performance similar to that of the three-position controller.

- The three-position controller severely violates the flood level constraints for water levels h_w, h_{gl}, and h_2. The NMPC controller violates the flood level constraints of water levels h_2 and h_{gl}. However, the violations are very small and only last for a very short amount of time. Also note that the peaks of the violated water level are much lower for NMPC than for the three-position controller. The conclusion is that the NMPC controller significantly reduces the amount of flooding.

- It can be seen clearly that NMPC manages to steer the water levels to their limits in order to use the available capacity as much as possible. This can be seen by inspecting how the water levels h_w and h_s are pushed to their flood level constraint.

The simulations in the large area again show that NMPC outperforms the existing three-position controller.

16.5 Conclusions and future research

The main purpose of this study was to prove that NMPC can be used for real-time flood regulation and that it works better than the current three-position controller. The results in this chapter show that MPC is capable to work closer to the limits of the constraints which resulted into improved performance with respect to flooding regulation. However, there are still some issues that should be considered before implementing NMPC in the field:

- **State estimation**: In this chapter it was assumed that the full state of the system was known. In practice this is never the case. In practice only the most important water levels are measured; discharges are not measured. Therefore, it is necessary to perform a state estimation at each time step. The Kalman filter is the standard choice for state estimation of linear systems when measurements are noisy and process disturbances unknown. Often additional insights about the process can be obtained in the form of inequalities. In this case better results are obtained if the estimation problem is formulated as a quadratic program. These kind of state estimators are called moving horizon estimators (MHE) [9]. MHE can be seen as the dual of MPC. Similar to MPC, MHE uses a finite-time window of past measurements. Based on these measurements the estimation problem is optimized and an optimal state estimation is performed. MHE for nonlinear models also exists and is called nonlinear MHE (NMHE) [10]. Typically NMHE outperforms Kalman filter approaches when the underlying process models are nonlinear. In case of flooding regulation the underlying models are highly nonlinear and therefore NMHE seems the most suitable option for the state estimation.

- **Robustness**: MPC makes use of an internal model in order to make future predictions of the output. An inherent property of models is that their predictions are uncertain by definition. Models are a simplification of the real world and as a consequence the predictions made with a model are never going to be followed exactly in the real world. This means that each output returned by the model is subject to uncertainty. In case of water systems an extra uncertainty on the predictions made by MPC comes from the uncertainty of the rain predictions. These uncertainties can deteriorate the on-line performance of MPC if they are not considered in the optimal control problem. Therefore future research could consider to quantify these uncertainties. In a next step these uncertainties can be taken into consideration into the optimization problem. This can be done by extending the multiple model configuration of [15] to flooding regulation.

- **Consensus between different water authorities**: In this study the main focus was to avoid flooding in the Demer basin. However, each action taken in the Demer basin will also affect water systems that lie downstream of the Demer basin. At their turn actions taken in water systems downstream of the Demer basin can have an effect on the water levels of the water system in the river Demer basin. What must be avoided is that actions taken in order to avoid flooding in one basin leads to severe floodings in other basins. Therefore, there is a need of cooperation and coordination between the different local water management bodies. A possible way to achieve this is by using a multi-agent model predictive controller [6].

Acknowledgements The research project is funded by the Division Operational Water Management of the Flemish Environment Agency (VMM) in Belgium. Also the full hydrodynamic InfoWorks-RS model of the river Demer basin, as well as the hydrometric calibration data, were provided by this regional water authority.

References

1. T. Barjas Blanco, P. Willems, B. De Moor, and J. Berlamont. Flooding prevention of the Demer using model predictive control. In *Proceedings of the 17th IFAC World Congress*, pages 3629–3634, July 2008.
2. E. Bautista, A. J. Clemmens, and R. J. Strand. Salt river project canal automation pilot project: Simulation tests. *Journal of Irrigation and Drainage Engineering*, 132(2):143–152, March/April 2006.
3. E. F. Camacho and C. Bordons. *Model Predictive Control*. Springer, London, UK, 1999.
4. V. T. Chow, D. R., Maidment, and L. W. Mays. *Applied Hydrology*. McGraw-Hill International Editions, New York, New York, 1988.
5. M. Marinaki and M. Papageorgiou. *Optimal Real-Time Control of Sewer Networks*. Springer-Verlag, London, UK, 2005.
6. R. R. Negenborn. *Multi-Agent Model Predictive Control with Applications to Power Networks*. PhD thesis, Delft University of Technology, Delft, The Netherlands, 2007.

7. OBM-Demer. Operational river basin model Demer – Technical documentation v.2.0. Technical report, Flemish Environment Agency, Aalst, Belgium, 2003. In Dutch.

8. S. Qin and T. Badgwell. A survey if industrial model predictive control technology. *Control Engineering Practice*, 11(7):733–764, July 2003.

9. C. V. Rao, J. B. Rawlings, and J. H. Lee. Constrained linear state estimation – A moving horizon approach. *Automatica*, 37(10):1619–1628, October 2001.

10. C. V. Rao, J. B. Rawlings, and D. Mayne. Constrained state estimation for nonlinear discrete-time systems: Stability and moving horizon approximations. *IEEE Transactions on Automatic Control*, 48(2):246–258, February 2003.

11. J. Rodellar, M. Gomez, and L. Bonet. Control method for on-demand operation of open-channel flow. *Journal of Irrigation and Drainage Engineering*, 119(2):225–241, March/April 1993.

12. D. C. Rogers and J. Goussard. Canal control algorithms currently in use. *Journal of Irrigation and Drainage Engineering*, 124(1):11–15, January/February 1998.

13. J. A. Rossiter. *Model-Based Predictive Control: A Practical Approach*. CRC Press, 2000.

14. G. Vaes, P. Willems, and J. Berlamont. The use of reservoir models for the assessment of the input from combined sewer overflows into river models. In *Proceedings of the 9th International Conference on Urban Drainage*, September 2002. 16pp.

15. P. J. van Overloop. *Model Predictive Control of Open Water Systems*. PhD thesis, Delft University of Technology, Delft, The Netherlands, 2006.

16. P. J. van Overloop, A. Mursi-Batt, K. J. van Heeringen, and R. A. H. A. Thabet. Real-time-control of water quantity and quality in a re-use of drainage water project. In *Proceedings of the 1st Asian Regional ICID Conference*, September 2001.

17. P. J. van Overloop, S. Weijs, and S. Dijkstra. Multiple model predictive control on a drainage canal system. *Control Engineering Practice*, 16(5):531–540, May 2008.

18. Wallingford Software & Halcrow (UK). FloodWorks, vesion 9.5, 2008.

19. Wallingford Software & Halcrow (UK). InfoWorks-RS, version 8.5, 2008.

Chapter 17
Predictive Control for National Water Flow Optimization in The Netherlands

P.J. van Overloop, R.R. Negenborn, B. De Schutter, and N.C. van de Giesen

Abstract The river delta in The Netherlands consists of interconnected rivers and large water bodies. Structures, such as large sluices and pumps, are available to control the local water levels and flows. The national water board is responsible for the management of the system. Its main management objectives are: protection against overtopping of dikes due to high river flows and high sea tides, supply of water during dry periods, and navigation. The system is, due to its size, divided into several subsystems that are managed by separate regional divisions of the national water board. Due to changes in local land-use, local climate, and the need for energy savings, the currently existing control systems have to be upgraded from local manual control schemes to regional model predictive control (MPC) schemes. In principle, the national objectives for the total delta require a centralized control approach integrating all regional MPC schemes. However, such centralized control is on the one hand not feasible, due to computational limitations, and on the other hand unwanted, due to the existing regional structure of the organization of the national water board. In this chapter the application of MPC is discussed for both individual regional control and coordinated national control. Results of a local MPC scheme applied to the actual water system of the North Sea Canal/Amsterdam-Rhine Canal are presented and a framework for coordination between several distributed MPC schemes is proposed.

P.J. van Overloop, N.C. van de Giesen
Delft University of Technology, Department of Water Management, Delft, The Netherlands,
e-mail: {P.J.A.T.M.vanOverloop,N.C.vandeGiesen}@tudelft.nl

R.R. Negenborn
Delft University of Technology, Delft Center for Systems and Control, Delft, The Netherlands,
e-mail: r.r.negenborn@tudelft.nl

B. De Schutter
Delft University of Technology, Delft Center for Systems and Control & Department of Marine and Transport Technology, Delft, The Netherlands, e-mail: b@deschutter.info

R.R. Negenborn et al. (eds.), *Intelligent Infrastructures*, Intelligent Systems, Control and Automation: Science and Engineering 42, DOI 10.1007/978-90-481-3598-1_17,
© Springer Science+Business Media B.V. 2010

17.1 Introduction

17.1.1 Water infrastructures

Water is the most vital element in human life. It is used for drinking, agriculture, navigation, recreation, energy production, etc. For these reasons, people tend to live close to water systems and therefore run an increased risk of getting flooded. In The Netherlands, improper management of water systems has led to higher damage than necessary as is clearly illustrated by the flooding of polders in February 1998, the shutting down of power plants in June 2003 (due to the limited availability of cooling water), and the yearly high mortality rate of fish due to algae bloom and low oxygen levels in the water.

To manage the human interaction with water systems, societies have formed organizations that were made responsible for managing certain tasks on particular water systems. This has resulted in a complex system of responsibilities that is not governed by the behavior of the water infrastructures themselves, but by the existing societal and organizational structures. These structures are hereby divided at a *spatial* level and at a *working field* level:

- At a spatial level the management of large rivers is divided into several parts. These large rivers almost always run through various countries. The management of the river in each country is an important national issue in which the inflows from and the outflows to the other counties are considered as given boundary conditions.
- A division by working field is apparent from the separated departments that manage a water system with their own isolated objectives. Water boards usually have one department that is responsible for the management of the water quantity variables and processes, such as water availability and flood protection, and another department that is responsible for water quality variables and processes, such as salinity control and water treatment. In reality, these variables and processes are all part of the same water network and therefore interact.

The spatial and working field division of water management is generally considered undesirable, but difficult to change. Many studies have been carried out on trans-boundary water management of rivers and the potential of integrated water management of water quantity and quality for canal systems [14, 15]. These studies have resulted in the formation of international agreements on river inflows and outflows at a national level and agreements on target values for water quantity and water quality variables, which are used by the different departments. The agreements are updated once every couple of years, but it is evident that the dynamic behavior of water systems requires coordination at a much higher frequency, e.g., daily or even hourly. The effects of climate change only add to this need: It is expected that precipitation will intensify on the one hand, while on the other hand periods of drought will last longer [5, 9]. In order to still guarantee safety and availability of water, more flexible agreements, that are updated continuously and take into account the limitations and possibilities of the infrastructure, have to be implemented.

Figure 17.1: The water system of The Netherlands with disturbances (High sea tides, Precipitation, Inflow from upstream rivers), objectives (Ecology, Water for agriculture, Drinking water, Navigation, Energy) and control structures (Controllable structures at the sea-side and in the river, storage).

17.1.2 Water system of The Netherlands

Figure 17.1 presents the main rivers and lakes of the Dutch water system and a summary of the objectives, major control structures, and disturbances. In the East, the River Rhine enters The Netherlands at Lobith and in the South the Meuse River enters at Borgharen. Their combined flow varies over the year from 1000 to 10000 m^3/s. These rivers run from the South and the East to the sea in the North and the West. To protect the country from water excesses, in the last century, the main part of the Dutch estuary has been closed off from the sea by large dams and controllable gates and pumps. This has resulted in large reservoirs in the West and the North of the country. Downstream of these reservoirs, the fluctuating sea tide is present. Under normal conditions this sea tide fluctuates between -1 and +1 m. However,

(a) The Neder-Rhine at Driel.   (b) The Lower-Delta in Haringvliet.

Figure 17.2: Controllable structures.

during storms, the water level can reach up to +4 m. During such extreme events, there is an excess of water that has to be prevented from flowing into the western and northern parts of the country, which lie below the mean sea level. However, in the summer time there is frequently a water deficit and water from the reservoirs has to be used as efficiently as possible.

Below a more detailed outline is given on the desired behavior of the Dutch water system, the actuators that can be manipulated in order to get as close as possible to this desired behavior, and the disturbances that complicate this.

17.1.2.1 Objectives

There are a large number of different objectives with respect to water quantity and quality [5]. Water levels in the river have to be controlled in order to prevent inundation and flooding. This has the highest priority during periods with high river flows. It should also be ensured that sufficient drinking water is available for consumption. This has the highest priority during dry periods. Moreover, ships should be able to transport goods over the rivers. Blockage of this transport function has to be kept at a minimum. Also, energy savings of pump stations should be maximized, and water with a sufficiently cool temperature should be provided to energy plants. Fish should be able to swim up rivers again, and salt/fresh water transitions should be controlled to ensure a good water quality. Furthermore, increased seepage from saline ground water should be counteracted.

17.1.2.2 Actuators

Several control options are available to manipulate the flows and water levels in The Netherlands. Water can be pumped into and taken out from the large reservoirs that can store fresh water or serve as temporary storage during high flows. Controllable structures can direct water in certain directions depending on regional high flow problems, water shortages, or temperature issues. Furthermore, controllable structures can also protect against high sea water levels, control salt/fresh water transitions, and regulate water levels in the downstream parts of rivers. Figure 17.2(a) presents a typical river structure, while Figure 17.2(b) is a controllable structure that blocks the sea from entering the inland [20, 21].

17.1.2.3 Disturbances

The main disturbances that influence the large open water reaches and lakes of The Netherlands are the inflow of the rivers from Belgium and Germany, high sea tides that block the drainage capacity precipitation, and the water demand for agriculture, drinking water, and ecology. Predictions of these disturbances are currently available for up to 4 days ahead with sufficient accuracy based on measurements of river flows in the two upstream lying counties and forecasts of precipitation and snow melt [6]. The quality of these predictions is expected to improve over the coming decade. It is expected that future prediction systems can provide forecasts with a prediction horizon of up to 10 days.

17.1.2.4 Complexity of the control problem

Whereas originally the main objective with respect to water systems was safety, as illustrated above, nowadays many different additional objectives need to be taken into account as well. Consequently, when considering the complete river delta as a single system, control of this system involves solving a large-scale, multi-objective, constrained control problem. This overall control problem cannot be solved by optimizing local control actions alone. Novel control approaches have to be developed in order to be able to coordinate locally optimizing control actions.

17.1.3 Automatic control of water systems

Over the last decades, an evolution of automatic control applications in water systems has taken place. Concerning the day-to-day operation, the first attempts to implement automatic control were made by civil engineers and were based purely on feedforward control. The reason for this was that although accurate models useful for inverse modeling were available, no knowledge on feedback control was available among these civil engineers. The first successful implementations were based on feedback control [12, 19], either without or with feedforward control added. These controllers were able to keep water levels close to set-points and in this way ensured the availability of water in canals and reservoirs. Later on, projects in which controllers were successfully implemented were characterized by the collaboration between civil engineers and control engineers [2, 4, 10, 13]. A next generation of controllers that was then realized was based on the use of model predictive control (MPC) [22, 25]. This generation was able to take constraints into account. Currently, the integration of various objectives on different variables (water quantity and water quality) by means of MPC is being investigated to further improve performance.

Based on the present knowledge, the control problems for most water systems (rivers, canal networks, sewer systems, irrigation canals, and reservoirs) can be formalized into simplified models, objective functions, and constraints. These small-scale control problems can be solved with MPC. However, solving larger-scale control problems is not feasible for two reasons:

1. For larger-scale systems different parts of the system are owned by different organizational structures, that are not willing to give up their autonomy.

2. Even if all organizational structures would be willing to give up their autonomy, solving the optimization problem involved in MPC for the resulting large-scale system would require too much computational time.

Therefore, in the future, distributed MPC will have to be used in order to solve multiple local MPC problems in a coordinated way, such that overall optimal performance is obtained. Solving the control problems of the separate water systems, while having them negotiate on the interactions with other systems (managed by other organizations), is also in line with the present manner in which large-scale water systems are being managed.

17.1.4 Outline

In this chapter, we present the state of the art in physical water infrastructure control using MPC and propose a framework for coordination between individual local MPC controllers. The chapter is organized as follows. In Section 17.2, the modeling and MPC control of open-water systems is introduced. Section 17.3 focuses on the application of MPC for control of a single physical discharge station in The Netherlands. A framework for achieving coordination among MPC controllers of multiple subsystems is proposed in Section 17.4. Section 17.5 concludes the chapter and provides directions for future research.

17.2 Modeling and MPC control of open water systems

The Dutch river delta can be considered as a system of interconnected open-water canals, channels, reservoirs, and control structures. Below we describe how these components of the infrastructure are modeled, both for use in detailed simulation studies and for use in controller design. Hereby, the models used for controller design are simplified models derived from the detailed models used for simulations. In this way, solving the optimization problems involved in MPC has lower computational requirements.

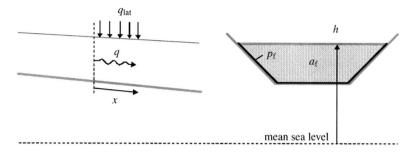

Figure 17.3: Open-water canal variables and parameters.

17.2.1 Open canals and reservoirs

17.2.1.1 Open canals

The flows and water levels (water quantity variables) in an open canal can be described by the Saint-Venant equations [3]. These nonlinear hyperbolic partial differential equations consist of a mass balance and a momentum balance. The mass balance ensures the conservation of water volume, while the momentum balance is a summation of the descriptions for the inertia, advection, gravitational force, and friction force:

$$\frac{\partial q}{\partial x} + \frac{\partial a_f}{\partial t} = q_{lat} \tag{17.1}$$

$$\frac{\partial q}{\partial t} + \frac{\partial}{\partial x}\left(\frac{q^2}{a_f}\right) + g a_f \frac{\partial h}{\partial x} + \frac{g q |q|}{c^2 r_f a_f} = 0, \tag{17.2}$$

where q represents the flow (m³/s), t is the time (s), x is the distance (m), a_f is the wetted area (i.e., the cross sectional area that is wet) of the flow (m²), q_{lat} is the lateral inflow per unit length (m²/s), $g = 9.81$ m/s² is the gravitational acceleration, h is the water level (mMSL, i.e., meters above the Mean Sea Level (MSL)), c is the Chézy friction coefficient (m$^{1/2}$/s), and r_f is the hydraulic radius (m), calculated as $r_f = a_f/p_f$, where p_f is the wetted perimeter (m) (i.e., the perimeter of the cross sectional area that is wet). Figure 17.3 gives a schematic representation of a typical open canal with its parameters.

To use the formulas in a numerical model of a canal reach, the partial differential equations are discretized in time (Δt) and space (Δx). In case these discretized formulas are simulated, the model results in time series solutions of water levels and flows at discrete locations along the reach. Also, the time series are discrete solutions in time.

17.2.1.2 Reservoirs

Reservoirs, such as lakes are modeled in a different way. Here, only the mass balance, as given in (17.1), is applied as an ordinary differential equation. The water

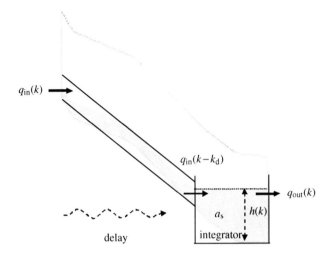

Figure 17.4: Integrator delay model of a canal reach.

level h (m) is calculated as a function of the inflows and the outflows:

$$\frac{dh(t)}{dt} = \frac{q_{in}(t) - q_{out}(t)}{a_s},$$ (17.3)

where q_{in} represents the sum of the inflows in the lake (m^3/s), q_{out} the sum of the outflows (m^3/s), and a_s the surface area (m^2).

17.2.1.3 Simplified models for controller design

For model-based controller design, the non-linear partial differential equations are usually transformed into simplified, linearized, and discrete-time models. A popular model for the water quantity variables is the integrator delay model [19]. In this model, a discrete time step index k is considered, and a discretized pure delay is placed in series with a discretized integrator. Figure 17.4 presents this model. The integrator delay model is given by the following equation:

$$h(k) = \frac{q_{in}(k - k_d)\Delta t}{a_s} - \frac{q_{out}(k)\Delta t}{a_s},$$ (17.4)

where k is the time step index and k_d is the number of delay steps.

This model functions properly for long canal reaches. For shorter reaches, resonance waves can occur that reflect on the boundaries of the reach. These need to be filtered out by means of low-pass filtering, before an integrator delay model can be fitted on the reach. The parameters can be derived from simple step tests or by application of system identification techniques [11]. Alternatively, a higher-order model can be used that does represent these waves [23].

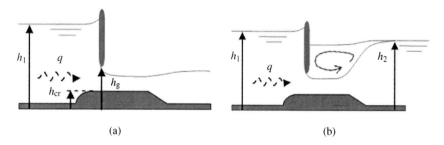

Figure 17.5: Free-flowing (a) and submerged undershot gate (b).

17.2.2 Control structures

Between canal reaches, river reaches, lakes, and the sea, controllable structures are present. By adjusting the setting of these structures, the flows between the water elements can be manipulated. The most common structures in a river delta are pumps and undershot gates.

17.2.2.1 Pumps

Pumps can be modeled in a straightforward way by imposing a flow at a certain value. In case of frequency-driven pumps this value can be between zero and the maximum capacity of the pump. Many pumps can only be switched off or set to run at maximum pump capacity. In general, pump flows are usually only slightly influenced by the surrounding water levels. The maximum pump capacity reduces though, when the difference between the upstream and the downstream water level increases. In that case, with the same amount of energy being brought into the pump, the water needs to be lifted higher and consequently the flow decreases.

17.2.2.2 Undershot gates

Undershot gates have a gate that is put into the water from the top down. The water flows under the gate. The stream lines of the upper part of the flow, just before the gate, bend down to pass the gate opening, causing the actual flow opening to be contracted. Usually, the factor μ_g by which the flow opening is contracted when compared to the gate opening is 0.63 [1].

The flow through the undershot gate can be free or submerged. In general, an undershot gate is free flowing when the downstream water level is lower than the gate height, i.e., the bottom of the gate. It is submerged, when the downstream water level is higher than the gate height. Figure 17.5 presents a free-flowing and a submerged undershot gate.

The flow through a free-flowing and a submerged undershot gate are given by:

$$q(k) = c_g w_g \mu_g (h_g(k) - h_{cr}) \sqrt{2g \left(h_1(k) - h_{cr} + \mu_g (h_g(k) - h_{cr}) \right)} \qquad (17.5)$$

and

$$q(k) = c_g w_g \mu_g (h_g(k) - h_{cr}) \sqrt{2g\,(h_1(k) - h_2(k))}, \qquad (17.6)$$

respectively, where q represents the flow through the structure (m^3/s), c_g is a calibration coefficient, w_g is the width of the gate (m), μ_g is the contraction coefficient, h_1 is the upstream water level (mMSL), h_2 is the downstream water level (mMSL), h_g is the gate height (mMSL), h_{cr} is the crest level (mMSL), g is the gravitational acceleration (m/s^2), and k is the discrete time index.

As the natural feedback mechanism between the increase in the upstream water level and the increase in the flow is described by a square root, the flow is not very sensitive to upstream water level fluctuations. Consequently, without adjusting the gate height, an undershot gate is not well suited to control this water level. However, by changing the gate height, the flow can be set precisely, allowing a well-controlled water level.

17.2.2.3 Simplified model for undershot gates

As a simplified model for the flow through a free-flowing undershot gate, the following model can be employed by linearizing (17.5) around a particular $h_1(k)$ and $h_g(k)$:

$$
\begin{aligned}
q(k+1) = q(k) + & \frac{g c_g w_g \mu_g \left(h_g(k) - h_{cr} \right)}{\sqrt{2g \left(h_1(k) - \left(h_{cr} + \mu_g \left(h_g(k) - h_{cr} \right) \right) \right)}} \Delta h_1(k) \\
& + \left(c_g w_g \mu_g \sqrt{2g \left(h_1(k) - \left(h_{cr} + \mu_g \left(h_g(k) - h_{cr} \right) \right) \right)} \right. \\
& \left. - \frac{g c_g w_g \mu_g^2 \left(h_g(k) - h_{cr} \right)}{\sqrt{2g \left(h_1(k) - \left(h_{cr} + \mu_g \left(h_g(k) - h_{cr} \right) \right) \right)}} \right) \Delta h_g(k), \qquad (17.7)
\end{aligned}
$$

where Δh_1 is the change in upstream water level (m) and Δh_g the change in gate height (m).

Considering the flow through a submerged undershot gate, the following simplified model can be obtained by linearizing (17.6) around a particular $h_1(k)$, $h_1(k)$, and $h_g(k)$:

$$
\begin{aligned}
q(k+1) = q(k) + & \frac{g c_g w_g \mu_g \left(h_g(k) - h_{cr} \right)}{\sqrt{2g(h_1(k) - h_2(k))}} \Delta h_1(k) - \frac{g c_g w_g \mu_g \left(h_g(k) - h_{cr} \right)}{\sqrt{2g(h_1(k) - h_2(k))}} \Delta h_2(k) \\
& + c_g w_g \mu_g \sqrt{2g(h_1(k) - h_2(k))} \Delta h_g(k), \qquad (17.8)
\end{aligned}
$$

where Δh_2 is the change in downstream water level (m).

17.2.2.4 Principle of MPC for water systems

MPC is a model-based control methodology meant for operational on-line control. At each decision step control actions are decided upon by solving an optimization problem. In this optimization problem an objective function that represents the control goals is minimized over a certain prediction horizon. Dynamics of the system to be controlled, operational constraints, and forecasts on, e.g., expected precipitation are hereby taken into account. The actions obtained are implemented until the next decision step, at which a new optimization is instantiated.

As the optimization has to run in real-time, it has to be fast and it has to result in a feasible solution. Therefore, typically the prediction models used are linearized and simplified models of reality, and the objective function that is optimized is usually formulated as a quadratic function. The constraints are formulated in such a way that, together with the quadratic objective function and the linear prediction model, the optimization problem is a convex optimization problem. Fast and reliable solvers are available for such quadratic programming problems.

For control of open-water systems, a typical objective function that is minimized over a finite prediction horizon is the following function, which represents the objectives for a water system with one water element controlled by a gate (via control variable Δh_g) and a pump (via control variable q_p):

$$J = \sum_{l=0}^{N-1} \left(w_e \left(e(k+1+l) \right)^2 + w_{\Delta h_g} \left(\Delta h_g(k+l) \right)^2 + w_{q_p} \left(q_p(k+l) \right)^2 \right), \qquad (17.9)$$

where J is the objective function or performance criterion, Δh_g is the change in the gate position (m), q_p is the pump flow (m^3/s), N is the length of the prediction horizon, e is the error between the value of a water level variable and the target value of this variable, w_e is the penalty on the error, $w_{\Delta h_g}$ is the penalty on the change in the gate position, and w_{q_p} is the penalty on the pump flow. This objective function encodes the control objectives of minimizing deviations from target values weighted against minimizing control effort and energy. It is straightforward to extend this objective function to multiple water elements and actuators.

For open-water systems, the physical and operational constraints are usually time-varying limitations on variables. Physical constraints, such as minimum and maximum pump flows, and minimum and maximum gate positions, are implemented as hard constraints. Operational constraints, such as maxima of water levels for safety, minima on water levels for navigation, minima on water flows for agriculture, drinking water, and ecology, are implemented as soft constraints. Slack variables, representing the amount of violation of such constraints are added to the objective function with a penalty term.

The problem of minimizing the quadratic objective function, including the slack variables for the soft constraints, subject to a linearized prediction model and linear constraints is then a quadratic programming problem with linear constraints. Many efficient solvers for such problems are available, e.g., quadprog, CPLEX, and GAMS.

17.3 MPC for the North Sea Canal and Amsterdam-Rhine Canal

Since August 2008, an MPC controller has been in operation to support the operators of the discharge complex at IJmuiden, from where the North Sea Canal and the Amsterdam-Rhine Canal is operated. Here, the precipitation of a catchment area of approximately 2300 km^2 drains into the sea. The main water ways that transport the water are the Amsterdam-Rhine Canal that continues into the North Sea Canal, as illustrated in Figure 17.6. At the end of the North Sea Canal the water can be discharged by seven gates during low tide and by six pumps when the sea water level is higher than the water level in the canal. Currently, the installed pump capacity of 260 m^3/s is the largest in Europe. Below, we detail how MPC is used there to optimize the operation of the discharge complex every hour of the day.

17.3.1 Setup of the MPC scheme

17.3.1.1 Objectives

The main objectives in the area of IJmuiden are navigation, minimum energy consumption, and to a smaller extent safety. The conflict in the operational management of the canal levels in this system is that navigation and safety require the water levels to remain close to the target level of -0.40 m with respect to mean sea level, whereas for achieving minimal energy consumption the water levels should fluctuate: During high sea water levels water should be stored in the canal, such that it can be discharged for free during low sea water levels. The objective function J is therefore defined as:

$$J = \sum_{l=0}^{N-1} \left(w_e \left(e(k+1+l)\right)^2 + w_{\Delta q_g} \left(\Delta q_g(k+l)\right)^2 + \sum_{i=1}^{6} w_{q_{p,i}} \left(q_{p,i}(k+l)\right)^2 \right), \quad (17.10)$$

where $e(k) = h_{ref}(k) - h(k)$ is the difference (m) between the target water level h_{ref} (mMSL) and the water level of the canal h (mMSL), N is the number of steps in the prediction horizon, Δq_i is the change in the discharge of all gates (m^3/s), q_i is the discharge of pump i (m^3/s), w_e is the penalty on the water level deviation, $w_{\Delta q_g}$ is the penalty on the change in the discharge of all gates, w_{q_i} is the penalty on the discharge of pump i (for $i = 1,\ldots,6$), and k is the time step index. As the energy consumption of the pumps is directly linked to the pump flow, minimizing the pump flow results in minimization of the energy consumption.

17.3.1.2 Prediction model for the canals

The North Sea Canal and the Amsterdam-Rhine Canal are wide and the prediction model can therefore be modeled as the prediction model for a single reservoir:

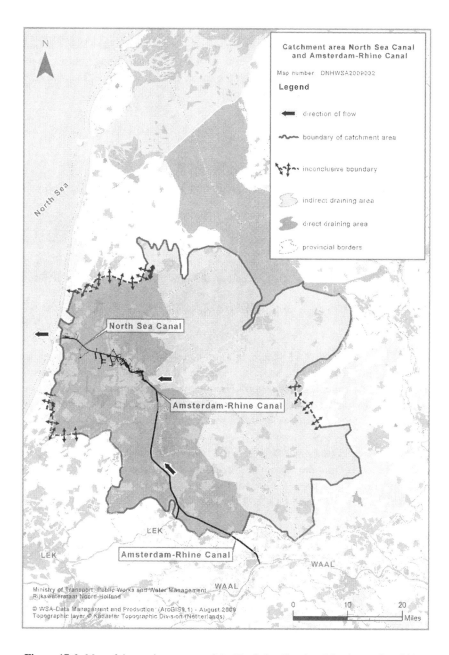

Figure 17.6: Map of the catchment area of the North Sea Canal and the Amsterdam-Rhine Canal (Illustration courtesy of Ministry of Transport, Public Works and Water Management, Rijkswaterstaat Noord-Holland).

$$h(k+1+l) = h(k+l) + \frac{q_d(k+l)\Delta T}{a_s} - \frac{q_g(k+l)\Delta T}{a_s} - \frac{q_p(k+l)\Delta T}{a_s}, \quad (17.11)$$

for $l = 0,\ldots,N-1$, where h is the water level (m) of the North Sea Canal, q_d is the disturbance inflow (m³/s), ΔT is the time step (s), a_s is the storage (surface) area (m²), q_g is the sum of the flows through the seven gates (m³/s), q_p is the sum of the flows through all six pumps (m³/s). The time step ΔT is 600 s.

17.3.1.3 Constraints

Soft constraints are applied over the prediction horizon to impose the following limitation for navigation and safety:

$$-0.55 \le h(k+1+l) \le -0.30, \quad (17.12)$$

for $l = 0,\ldots,N-1$. Time-varying hard constraints over the prediction horizon are imposed on the flow through all gates and on the flow through the individual pumps:

$$0 \le q_g(k+l) \le q_{g,\max}(k+l) \quad (17.13)$$

$$0 \le q_{p,i}(k+l) \le q_{p,i,\max}(k+l), \quad (17.14)$$

for $i = 1,\ldots,6, l = 0,\ldots,N-1$. The maximum gate flow $q_{g,\max}(k+l)$ is derived from (17.8), using predictions of the sea water level and the canal water level, and assuming that all gates are completely open. The maximum pump capacity $q_{p,i,\max}(k+l)$ is set to 0 when the water level of the canal is higher than the sea water level; it is set to the maximum capacity as specified by the pump manufacturer otherwise.

The predicted disturbance flow is calculated by means of an auto-regressive model that uses as input the measured precipitation and evaporation over the past 15 days and the predicted precipitation for the next 12 hours. The parameters of this model are identified for each month of the year.

17.3.1.4 The control scheme

The MPC optimization has been implemented in a decision support system. In this system a prediction horizon is considered with a length of $N = 144$ steps of 10 minutes, corresponding to 24 hours. Each hour the MPC controller performs its optimization, taking only a few minutes to complete. When the optimization has finished, a human operator checks the advice for the upcoming hour and accepts the advice by clicking on an acceptance button. Then, the advice for the next hour is automatically executed by implementing the actions determined for the first 6 steps of the prediction horizon.

17.3.2 Results

17.3.2.1 Without MPC control

Figure 17.7 shows the evolution of water levels, flows, and energy consumption two days before the MPC controller was activated. It can be observed that the operators maintain the water levels of the canal close to the target level (i.e., -0.4 m) by keeping a proper number of pumps running over a long period (from 14 to 23 hours and from 34 to 46 hours). The flow of these pumps balances the disturbance inflow. It can also be observed that the actions of the operators result in low power usage (around 1000 kW), when the sea water level is low, but a power usage that is at least 2.5 times higher (at least 2500 kW), when the sea water level is at its peak.

The worst situation appears when the canal water level is decreased below the target level by discharging water for free through the gates when the sea level is below the target level. After the low tide, the operator stops discharging water through the gates. However, when the water level becomes higher than the target level, the operator is triggered again to start discharging (at 6 hours on the first day), now using the pumps. By starting multiple pumps, the canal water level is kept close to the target level. However, the timing of the pumping action is costly as the starting the pumps coincides with the peak sea water level. In the considered scenario, the total amount of energy consumption is 79.5 kWh and the peak power usage is 4956 kW.

17.3.2.2 With MPC control

A period of two other days is selected after the MPC controller has been taken into operation. This period has approximately the same inflow volume and potential gravity (gates) discharge volume as the previously described days at which the MPC controller was not used. As can be observed in Figure 17.8, the result of the controller is that the canal water levels fluctuate slightly more, although still within the allowed limits. However, the pumps switch approximately twice as often in order to achieve a much more cost-effective usage. The energy consumption is 54.5 kWh and the peak power is 3528 kW. This results in a reduction of 34% in costs compared to the case without MPC. For other situations average cost reductions over a year in the order of 20% were computed. The average energy costs for operating the pumps over the past five years (2003 to 2007) has been almost 1 million euro per year. Hence, using MPC can result in a cost reduction of 200.000 euro per year.

17.4 Distributed MPC for control of the Dutch water system

As was shown in the previous section, when using MPC for control of an individual discharge station a significant performance improvement can be achieved. Ideally, such an MPC controller would be implemented for the complete Dutch water system. However, due to the complexity and the size of this large-scale water system,

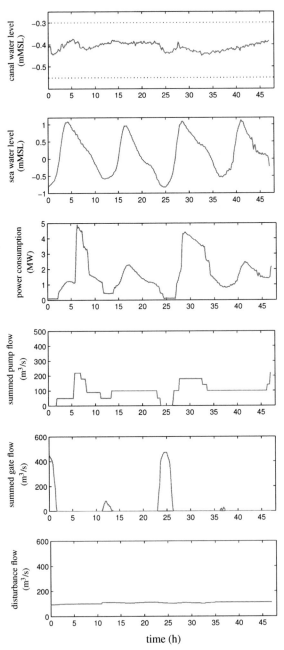

Figure 17.7: Evolution of the water levels, the flows, and the energy consumption when not using MPC.

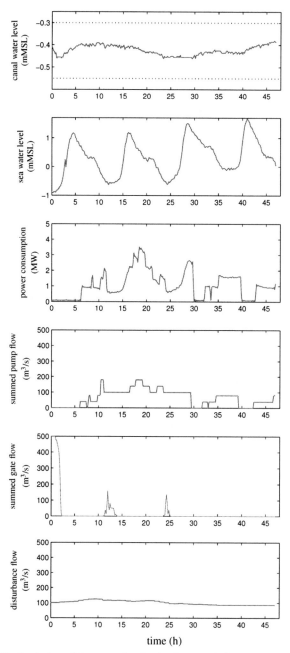

Figure 17.8: Evolution of the water levels, the flows, and the energy consumption when using MPC.

this is not feasible. Controlling such systems in a centralized way in which at a single location measurements are collected from the whole system and actions are determined for the whole system would impose a too large computational burden. To illustrate this the following estimation of the size of the Dutch water system can be given: In The Netherlands, 15000 pumps, and a multiple number of gates can be controlled. Water levels in 1000 km of rivers, 1000 different canals, and a multiple number of ditches have to be controlled. These control structures and water ways are controlled by the national water board and 26 different regional water boards.

Instead of defining an overall control problem, it should be accepted that there are multiple MPC controllers spread across the network, each controlling their own part of the network. Local control actions include activation of pumps, filling or emptying of water reservoirs, manipulating flows in certain directions of the country, closing off parts that are under threat of high sea water levels. Due to the continuing developments in information and communications technology, exchange of information between local controllers becomes practically and economically feasible, such that the local controllers have the possibility to take one another's actions into account. In order to achieve overall optimal performance, the local MPC controllers have to be designed in such a way that they account for the effects of local actions at a system-wide level using information exchange. The local controllers should thus be able to perform cooperation and negotiation with other controllers with the aim of achieving the best system-wide performance. Distributed MPC is aiming to enable this.

Similarly as in centralized MPC, the controllers in distributed MPC choose their actions at discrete control steps. The goal of each controller is to determine those actions that optimize the behavior of the overall system by minimizing costs as specified through a commonly agreed upon performance criterion that has been translated into desired water levels and flows. To make accurate predictions of the evolution of a subsystem over the prediction horizon for a given sequence of actions, each controller requires the current state of its subsystem and predictions of the values of variables that interconnect the model of its subsystem with the model of other subsystems. The predictions of the values of these so-called *interconnecting variables* are based on the information exchanged with the neighboring controllers. Usually, these interconnecting variables for water systems represent inflows and outflows between different parts of the water infrastructure.

Several authors have proposed distributed MPC strategies for control of large-scale water systems, e.g., in [7, 8, 17, 18]. These algorithms achieve cooperation among controllers in an iterative way, in which controllers perform several iterations consisting of local problem solving and communication within each control cycle. In each iteration, controllers then obtain information about the plans of neighboring controllers. This iterative process is designed to converge to local control actions that lead to overall optimal performance.

In order to employ any distributed MPC technique, first the subsystems, objectives, and constraints need to be determined. For The Netherlands this is currently being investigated. Below we first propose a division of the water system of The Netherlands into subsystems. Then, in Section 17.4.2 we illustrate the workings

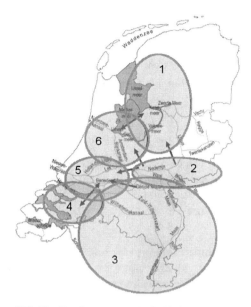

Figure 17.9: The Dutch water system divided into 6 major regions.

of a previously proposed distributed MPC scheme for control of subsystems in an irrigation canal. This illustration forms an example of how the water subsystems of The Netherlands could be coordinated in the future.

17.4.1 The subsystems of the Dutch water system

Figure 17.9 illustrates our proposal for distributed MPC control of the complete Dutch water system. In the figure, 6 major water network regions are indicated. All regions together cover the major flows in The Netherlands. Each region by itself is defined in such a way that on the one hand the flow dependencies with the other regions are minimal, whereas on the other hand the flow dependencies within each region are strong. These regions are therefore already associated with separate divisions of the Dutch national water board. For each of these regions, local control objectives are formulated. To achieve these objectives an MPC controller is associated with each region. In order to be able to take into account the interaction between the different regions, the controllers can communicate in order to coordinate their actions. The 6 major regions and their control objectives are the following:

1. **Lake IJssel** is the large water reservoir in the North of The Netherlands. This reservoir should be used for the provision of drinking water and water for agriculture in the North and West. Water should also be flowing in such a way that algae bloom is reduced, encouraging a good ecology. Furthermore, lake IJssel should store water that can be used as cooling water for power plants.

2. The **Rhine River** is the largest river of The Netherlands. In addition to the provision of water for drinking, agriculture, cooling, and ecology in the West, navigation should also be possible. Hereby, safety has to be taken into account, as the Rhine River flows through densely populated areas.

3. The **Meuse River** has to provide water for agriculture and drinking in the South. Navigation and safety are two other important aspects that have to be taken into account when managing the water levels of the Meuse.

4. The **delta of Zeeland** is the second largest water reservoir of The Netherlands. Safety in the estuary has to be ensured, while water for agriculture, drinking, and algae bloom reduction should be sufficient. Moreover, also ships navigate through the Delta of Zeeland, and hence, water levels should be sufficiently high for this.

5. The **delta of Rijnmond** should provide safety in the estuary, while providing drinking water and water for agriculture. By exploiting the open connections to the sea, the ecological state can be improved.

6. The **North Sea Canal and Amsterdam-Rhine Canal** should have a sufficiently high water level to allow navigation. Pumps at the discharge station in IJmuiden should be employed taking into account their energy consumption. Water levels should not be too high, in order to ensure the safety in the area surrounding the canals.

17.4.2 Illustration of concept using control of irrigation canals

Current research efforts are addressing controlling the water flows in The Netherlands based on the division into regions as presented above. Recently, as a first proof of concept for distributed control of water systems, we have implemented a distributed MPC scheme for control of an irrigation canal consisting of 8 individual canal reaches. Cooperating control systems for irrigation districts that have inter-dependent water demand schedules can yield a better spreading of the available water towards areas that are under increased water stress. On a larger scale, less water will be wasted. An illustration of the importance of coordination between sub-systems is the avoidance of disturbance amplification in canals consisting of canal reaches in series. When the water level in separate canal reaches are controlled simultaneously with proportional integral controllers that are tuned to give a high performance, problems could occur during operation [24]. Disturbances that occur at the downstream side are amplified at each control gate further upstream. Coordination between the canal reaches is required or a global tuning procedure for all PI-controllers needs to be used that minimizes the deviations from the set-points in all reaches [24].

In our application of distributed MPC for irrigation canals, each canal reach is controlled by an MPC controller. In order to obtain the best overall performance,

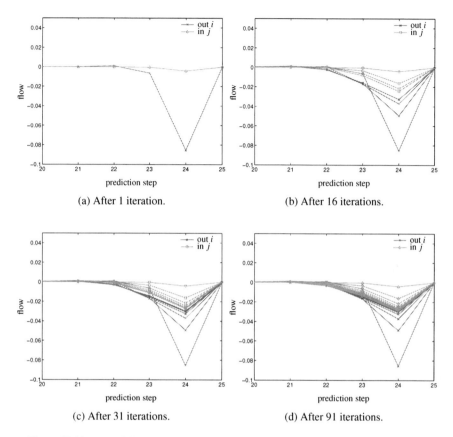

Figure 17.10: Two MPC controllers obtaining agreement over iterations on how much water should flow between two canal reaches [17].

the controllers have to reach an agreement concerning the amount of water flowing from one canal reach to the next over the full prediction horizon. At each decision step, the controllers therefore perform a number of iterations, in each of which they inform one another about desired inflows and outflows. This iterative procedure is illustrated in Figure 17.10. This figure shows at a particular decision step how the desires on the outflow from one controller become consistent with the desires on the inflow from a downstream controller over the iterations. Depending on a threshold specified in a stopping criterion the performance of the coordinated MPC scheme can be balanced with the required computational time. Using a smaller threshold can result in a performance that is less than 1% worse compared to the performance obtained by a hypothetical centralized MPC controller. However, this is at the price of a significant computational effort. With a larger threshold, the performance becomes closer to the performance of a centralized controller, although computational time requirements improve. Further details on the actual implementation are found in [16, 17].

17.5 Conclusions and future research

In this chapter, the present knowledge on control of water systems is described. Control of rivers, canals, and lakes has been discussed. Most of the control problems of individual water systems can be solved. The next step to be taken is to set objectives for multiple water systems in a larger area, in this case the water systems in The Netherlands. This requires coordination between the sub-systems in that area. A promising solution for solving this large-scale optimization challenge is to create a distributed control framework for interacting water systems that can take into account the complex dynamics of the water system on the one hand and the often conflicting objectives on the other. Future research will have to further develop this framework in order to utilize water infrastructures to their fullest potential.

Acknowledgements This research is supported by the BSIK project "Next Generation Infrastructures (NGI)", the Delft Research Center Next Generation Infrastructures, the European STREP project "Hierarchical and distributed model predictive control (HD-MPC)", contract number INFSO-ICT-223854, and the project "Multi-Agent Control of Large-Scale Hybrid Systems" (DWV. 6188) of the Dutch Technology Foundation STW.
The authors thank Peter Beuse (Division of Water and Shipping, Rijkswaterstaat Noord-Holland, Ministry of Transport, Public Works and Water Management) for carefully reading the original manuscript and for providing useful feedback.

References

1. M. G. Bos. *Discharge Measurement Structures*. International Institute for Land Reclamation and Improvement, Wageningen, The Netherlands, 1989.
2. C. M. Burt and X. Piao. Advances in PLC-based canal automation. In *Proceedings of the US-CID Conference on Benchmarking Irrigation System Performance Using Water Measurement and Water Balances*, San Luis Obispo, California, July 2002. ITRC Paper No. P02-001.
3. V. T. Chow. *Open-Channel Hydraulics*. Mc-Graw-Hill Book, New York, New York, 1959.
4. A. J. Clemmens and J. Schuurmans. Simple optimal downstream feedback canal controllers: Theory. *Journal of Irrigation and Drainage Engineering*, 130(1):26–34, January/February 2004.
5. Deltacommissie Veerman. Samen werken aan water, 2008. In Dutch.
6. Deltares. Flood Early Warning System. http://www.wlsoft.nl/soft/fews/int/index.html, 2009.
7. H. El Fawal, D. Georges, and G. Bornard. Optimal control of complex irrigation systems via decomposition-coordination and the use of augmented Lagrangian. In *Proceedings of the 1998 International Conference on Systems, Man, and Cybernetics*, pages 3874–3879, San Diego, California, 1998.
8. G. Georges. Decentralized adaptive control for a water distribution system. In *Proceedings of the 3rd IEEE Conference on Control Applications*, pages 1411–1416, Glasgow, UK, 1999.
9. Intergovernmental Panel on Climate Change. *Climate Change 2001: Impact, Adaptation and Vulnerability*. Cambridge University Press, New York, New York, 2003.
10. X. Litrico and V. Fromion. Tuning of robust distant downstream PI controllers for an irrigation canal pool. I: Theory. *Journal of Irrigation and Drainage Engineering*, 132(4):359–368, July/August 2006.

11. X. Litrico, P.O. Malaterre, J.P. Baume, P.Y. Vion, and J. Ribot-Bruno. Automatic tuning of PI controllers for an irrigation canal pool. *Journal of Irrigation and Drainage Engineering*, 133(1):27–37, January/February 2007.

12. P. O. Malaterre and J.-P. Baume. Modeling and regulation of irrigation canals: Existing applications and ongoing researches. In *Proceedings of the 1998 IEEE International Conference on Systems, Man, and Cybernetics*, pages 3850–3855, San Diego, California, October 1998.

13. I. Mareels, E. Weyer, S. K. Ooi, and D. Aughton. Modeling and control of irrigation networks: A system engineering approach. In *Proceedings of the 2nd International USCID Conference*, pages 399–411, Phoenix, Arizona, May 2003.

14. E. Mostert. River basin management in the European union; How it is done and how it should be done. *European Water Management*, 1(3):26–35, June 1998.

15. E. Mostert. Conflict and co-operation in international freshwater management: A global review. *International Journal of River Basin Management*, 1(3):1–12, 2003.

16. R. R. Negenborn, P. J. van Overloop, and B. De Schutter. Coordinated model predictive reach control for irrigation canals. In *Proceedings of the European Control Conference 2009*, Budapest, Hungary, August 2009.

17. R. R. Negenborn, P. J. van Overloop, T. Keviczky, and B. De Schutter. Distributed model predictive control for irrigation canals. *Networks and Heterogeneous Media*, 4(2):359–380, June 2009.

18. S. Sawadogo, R. M. Faye, P. O. Malaterre, and F. Mora-Camino. Decentralized predictive controller for delivery canals. In *Proceedings of the 1998 IEEE International Conference on Systems, Man, and Cybernetics*, pages 3380–3884, San Diego, California, 1998.

19. J. Schuurmans. *Control of Water Levels in Open Channels*. PhD thesis, Delft University of Technology, Delft, The Netherlands, October 1997.

20. W. ten Brinke. *De Beteugelde Rivier. Bovenrijn, Waal, Pannerdensch Kanaal, Nederrijn-Lek en IJssel in Vorm*. Uitgeverij Veen Magazines B.V., Diemen, The Netherlands, 2005. In Dutch.

21. G. P. van de Ven. *Man-made Lowlands. History of Water Management and Land Reclamation in The Netherlands*. Matrijs, Utrecht, The Netherlands, 2004.

22. P. J. van Overloop. *Model Predictive Control on Open Water Systems*. PhD thesis, Delft University of Technology, Delft, The Netherlands, June 2006.

23. P. J. van Overloop, I. J. Miltenburg, A. J. Clemmens, and R. J. Strand. Identification of pool characteristics of irrigation canals. In *Proceedings of the World Environmental and Water Resources Congress*, Honolulu, Hawaii, May 2008.

24. P. J. van Overloop, J. Schuurmans, R. Brouwer, and C. M. Burt. Multiple model optimization of PI-controllers on canals. *Journal of Irrigation and Drainage Engineering*, 131(2):190–196, March/April 2005.

25. B. T. Wahlin. *Remote Downstream Feedback Control of Branching Networks*. PhD thesis, Arizona State University, Tempe, Arizona, December 2002.

Chapter 18
Decentralized Model Predictive Control for a Cascade of River Power Plants

A. Şahin and M. Morari

Abstract River power plants interrupt the natural flow of a river and induce unde-sired fluctuations in the water level and water discharge. To prevent the adverse impacts of these fluctuations on the nature as well as on the navigation, the opera-tion of the power plants needs to be regulated to obey certain restrictions imposed by the authorities, i.e., the water levels at specific points in the river have to be kept within certain bounds and large variations of the turbine discharges need to be avoided. In this chapter we present a Model Predictive Control (MPC) scheme to manipulate the turbine discharges of the power plants located in a cascade that will satisfy the restrictions imposed by the authorities. Since a centralized MPC scheme might become computationally infeasible for large cascades, we develop a decen-tralized MPC scheme, in which the cascade is decomposed into smaller subsystems and each subsystem is controlled by a local MPC scheme. We show through simula-tions that providing a downstream communication is sufficient to prevent significant performance deterioration in decentralized MPC, which would be expected due to the lack of coordination.

A. Şahin, M. Morari
ETH Zürich, Automatic Control Laboratory, Zürich, Switzerland,
e-mail: {sahin,morari}@control.ee.ethz.ch

R.R. Negenborn et al. (eds.), *Intelligent Infrastructures*, Intelligent Systems, Control and
Automation: Science and Engineering 42, DOI 10.1007/978-90-481-3598-1_18,
© Springer Science+Business Media B.V. 2010

18.1 Introduction

River power plants generate electricity by exploiting the kinetic energy of the natural flow of the river and the potential energy due to the elevation drops. The operation of power plants creates unnatural fluctuations in the water level and discharge of the river, since the power plants interrupt the natural flow of the river by manipulating the flow through their turbines. These fluctuations are undesired, since they hinder the navigation and they have an adverse impact on the ecological equilibrium within the river and the riparian zone. Therefore, the authorities impose restrictions on the operation of the power plants to enforce that the water levels at specific locations along the river, the so-called *concession levels*, are kept within certain bounds and large fluctuations in the water discharge are avoided.

The river we consider in this work is heavily used for navigation, which is made possible by *locks* that allow ships to bypass the power plants. The operation of the locks results in a significant amount of lateral flows leaving or entering the main flow of the river, which has a strong impact on the concession levels.

The control problem for a single *river reach*, the river segment between two power plants as depicted in Figure 18.1, is formulated as follows. The concession level h_c^i, located close to the downstream power plant has to be kept within certain bounds and it represents the controlled variable. The concession level is controlled by manipulating the turbine discharge of the downstream power plant q_{out}^i. The inflow entering the reach q_{in}^i (turbine discharge of the upstream power plant) acts as a disturbance on the concession level. Note that there is a time delay for a change in q_{in}^i to appear in h_c^i, since the concession level is located at a rather long distance from the upstream power plant. The lateral flows during the lock operations $q_{lock,in}^i$ and $q_{lock,out}^i$ act as disturbances on the concession level as well. While q_{in}^i is measured, $q_{lock,in}^i$ and $q_{lock,out}^i$ are not measured. If we consider the control problem of a cascade, the controlled variables are the concession levels within each reach, the manipulated variables are the turbine discharges and the disturbances acting on the system are the inflow entering the cascade at the most upstream reach and the lateral flows during the lock operations.

Currently, a local PI controller with a feed-forward action is used at each reach to control the concession level, see Figure 18.2. The PI controller computes a control action based on the difference between the reference and the measured concession level. The turbine discharge of the upstream power plant is included in the control action with a feed-forward term in order to eliminate its effect on the concession level. The currently employed local controllers have several disadvantages. First of all, tuning the PI controller for two contradictory control objectives, minimizing both the concession level deviations and turbine discharge variations, is a demanding task. Second, the PI controllers cannot explicitly handle the constraints on the system variables, e.g., the admissible operating range of concession levels and the physical limitations on the maximum and minimum turbine discharges. Thus, the controllers have to be tuned by trial and error to satisfy the constraints, which makes

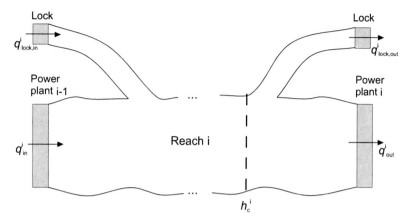

Figure 18.1: Reach description.

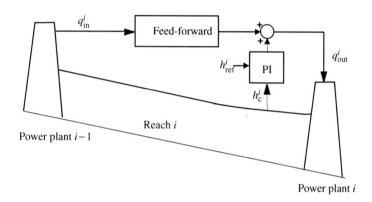

Figure 18.2: Local PI controller with feed-forward action.

the tuning of the controllers even more demanding. Even after the tuning the constraint satisfaction is not guaranteed over the full range of operating conditions.

The control of the water levels in the river reaches has been studied by several authors in the literature. In [8, 12, 21] the control of a single reservoir was considered, where a PI control was used in [21] and pole placement by state feedback method was used in [12] and [8]. In [10], a cascade of five river power plants was controlled, where a local PI controller with feed-forward action, the same as the current control scheme explained above, was used to control each reach. In none of these papers, the constraints on the system variables were taken into consideration.

A control strategy suitable for handling the constraints explicitly is Model Predictive Control (MPC). MPC is an online optimization-based control strategy, in which the control actions at each control step are calculated by solving an optimization problem. In the optimization problem a performance index is minimized subject to a set of constraints on the system variables. In [1], MPC was applied to control a single river reach, which was later implemented on a real plant. In [11] and

[23] MPC was applied to control a cascade of river power plants. Irrigation channel control has many similarities with river power plant control. A good overview of the control techniques used in irrigation channel control can be found in [20]. The control methods range from simple PI controllers [28] to H_{inf} control [6, 14] and to MPC [18, 19].

In this work we develop a decentralized MPC scheme to control a large cascade of river power plants, for which a centralized MPC scheme (controlling all the power plants from a central unit) is computationally infeasible. In decentralized MPC the cascade is decomposed into smaller subsystems and each subsystem is controlled using a local MPC scheme. While the decomposition decreases the computational effort, it deteriorates the performance, since the coordination of all the control actions is lost. In order to reduce the performance loss we use a downstream communication strategy, in which each local MPC communicates the predictions of its outflow, which is the inflow entering the downstream subsystem, to the downstream local MPC. We implement the proposed decentralized MPC scheme on an accurate simulation model of a cascade of 35 river power plants, where the length of the river is approximately 400 km. We show that the performance of decentralized MPC does not deteriorate significantly when the downstream communication is provided. We compare the performance and computation time of decentralized MPCs containing different numbers of subsystems in simulations. Based on this comparison we conclude that the decentralized MPC scheme, in which each power plant is controlled with a local MPC, is the most suitable scheme, since it has the shortest computation time and its performance is still reasonably good.

In this chapter we aim at ensuring the sustainability of the river power plant operation, but the control scheme we propose can be used for other infrastructure systems as well. Decentralized/distributed MPC schemes have already been used in the control of power networks [3, 26] and water networks [18]. The decentralized MPC scheme with downstream communication (one-way communication in general) we propose here would perform particularly well when applied to infrastructures that consist of serially connected subsystems. For instance, recently we applied the decentralized MPC with one-way communication to a cascade of irrigation channels [19] and showed that the scheme performs reasonably well.

This chapter is organized as follows. In Section 18.2, the simulation model and the control model are described. The decentralized MPC scheme is explained in detail in Section 18.3. Finally the performance of the MPC scheme in the simulations is demonstrated in Section 18.4.

18.2 Modeling

In this work mainly two models are used; a detailed simulation model, and a simple control model. The simulation model represents the river dynamics accurately and is used for assessing the performance of the controller in the simulations. The control

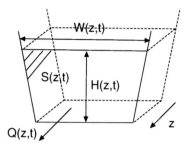

Figure 18.3: Cross section geometry.

model is a simple linear model, the so-called Integrator Delay (ID) model, and is used in the controller design.

18.2.1 Simulation model

The simulation model is developed in FLORIS [27], which is an accurate river hydraulics modeling software package. In FLORIS the river hydraulics are simulated by solving the nonlinear partial differential equations of Saint Venant numerically. The Saint Venant equations are explained in detail below.

18.2.1.1 Saint Venant equations

The Saint Venant equations are the state of the art for describing the unsteady water flow dynamics in rivers or canals [9]. They are two coupled nonlinear partial differential equations based on the assumption of one-dimensional flow and represent the conservation of mass and momentum within an infinitesimal cross section, as shown in Figure 18.3. The conservation of mass equation is given as

$$\frac{\partial Q}{\partial z} + \frac{\partial S}{\partial t} = 0, \tag{18.1}$$

and the conservation of momentum equation is

$$\frac{1}{g}\frac{\partial}{\partial t}\left(\frac{Q}{S}\right) + \frac{1}{2g}\frac{\partial}{\partial z}\left(\frac{Q^2}{S^2}\right) + \frac{\partial H}{\partial z} + I_f - I_0 = 0, \tag{18.2}$$

where t is the time (s), z is the distance along the river (m), Q is the discharge (m^3/s), H is the water level (m). g denotes the gravitational acceleration (m/s^2), S is the wetter cross section area (m^2), I_0 is the river slope (rad), and I_f is the dimensionless friction slope, see [7] for derivation of this parameter. In order to solve this pair of partial differential equations the initial conditions and the boundary conditions are needed. The boundary conditions for a river reach located between two power plants are given as,

$$Q(0,t) = q_{in}(t) \tag{18.3}$$

$$Q(L,t) = q_{out}(t), \tag{18.4}$$

where q_{in} is the inflow entering the reach and q_{out} is the outflow leaving the reach. Assuming that the initial conditions $H(z,0)$ and $Q(z,0)$ are also known, the partial differential equations (18.1) and (18.2) can be solved.

18.2.2 Control model

For the control of river reaches and irrigation channels mainly two control models are used in the literature, the linear Saint Venant model [11, 23] and the Integrator Delay (ID) model [6, 19, 22]. The linear Saint Venant model is a finite difference approximation of the linearized Saint Venant equations and it is a quite accurate control model. The disadvantages of the linear Saint Venant model, however, are that it is large and its derivation requires river topology data. In the ID model the reach is modeled as a tank with a delayed inflow. The high frequency dynamics, e.g., propagation and reflection of waves inside the reach, are neglected in the ID model. As discussed in several works [6, 22], as long as the bandwidth of the controlled system is sufficiently small, ignoring the high frequency dynamics in the control model neither deteriorates the performance nor results in instability. In [2] this argument was confirmed by performing simulations on a cascade of four power plants with two MPC schemes, using linear Saint Venant and ID models. It was shown that the performance of both MPCs was almost the same. Therefore, we use the ID model as the control model, since it is smaller, its derivation is much easier, and it requires far less river topology data.

In this section we first derive the linear model for a single reach. The model of a cascade is obtained by serially connecting the models of the reaches within the cascade.

18.2.2.1 Single reach model

In Figure 18.1 the concession level and all the flows entering and leaving a reach were shown. Let the surface area of reach i be represented by A_S^i. The time required for a change in q_{in}^i and $q_{lock,in}^i$ to result in a change in the concession level h_c^i is represented by τ^i. The ID model is described by the following equation,

$$\frac{dh_c^i}{dt} = \frac{1}{A_S^i}\left(q_{in}^i(t-\tau^i) + q_{lock,in}^i(t-\tau^i) - q_{out}^i(t) - q_{lock,out}^i(t)\right). \tag{18.5}$$

The continuous-time model (18.5) is discretized using the forward Euler approximation with a sampling time of T_s, which results in the following discrete time model

$$h_c^i(k+1) = h_c^i(k) + \frac{T_s}{A_S^i}\left(q_{in}^i(k-\delta^i) + \right.$$

$$q_{\text{lock,in}}^i(k-\delta^i) - q_{\text{out}}^i(k) - q_{\text{lock,out}}^i(k) \bigg), \qquad (18.6)$$

where δ^i is the closest integer approximation of the ratio τ^i/T_{s}.

There are two parameters to be determined in the ID model; A_{S}^i and τ^i. In [2] these parameters were determined following two approaches, system identification and computation from approximated analytical expressions using the available river topology data. By performing tests on several reaches it was shown that both approaches return very similar estimates of the parameters. Since the river topology data are already available in the database of the FLORIS model, the second approach is much more convenient. Thus, we follow the second approach here.

A_{S}^i is read directly from the available river topology data. τ^i is computed as,

$$\tau^i = \frac{L^i}{c^i + v^i}, \qquad (18.7)$$

where L^i is the length of the reach, v^i is the mean velocity of the water flow, given by

$$v^i = \frac{Q^i}{S^i}, \qquad (18.8)$$

where Q^i is the mean discharge and S^i is the mean cross sectional area. c^i is the propagation celerity of a wave, which can be approximated for shallow reaches (as in our application) by [5]

$$c^i = \sqrt{gh^i}, \qquad (18.9)$$

where g is the gravitational acceleration.

State-space representation

The state-space representation of the discrete-time ID model (18.6) is given as,

$$x^i(k+1) = A^i x^i(k) + B^i u^i(k) + G^i q_{\text{in}}^i(k) + G_{\text{f}}^i q_{\text{lock}}^i(k) \qquad (18.10)$$
$$y^i(k) = C^i x^i(k),$$

where $x^i \in \mathbb{R}^n$ is the state, $u^i \in \mathbb{R}$ is the input, $y^i \in \mathbb{R}$ is the output, given by

$$x^i(k) = \begin{bmatrix} h_{\text{c}}^i(k) \\ q_{\text{in,tot}}^i(k-\delta^i) \\ q_{\text{in,tot}}^i(k-\delta^i+1) \\ \vdots \\ q_{\text{in,tot}}^i(k-1) \end{bmatrix}, \quad u^i(k) = q_{\text{out}}^i(k), \ y^i(k) = h_{\text{c}}^i(k), \qquad (18.11)$$

with $q_{\text{in,tot}}^{i}(k) \in \mathbb{R}$ representing the total inflow entering the reach,

$$q_{\text{in,tot}}^{i}(k) = q_{\text{in}}^{i}(k) + q_{\text{lock,in}}^{i}(k). \tag{18.12}$$

$q_{\text{lock}}(k) \in \mathbb{R}^2$ is the vector of lateral flows due to the lock operations,

$$q_{\text{lock}}^{i}(k) = \begin{bmatrix} q_{\text{lock,in}}^{i}(k) \\ q_{\text{lock,out}}^{i}(k) \end{bmatrix}. \tag{18.13}$$

The system matrices in (18.10) are given as,

$$A^{i} = \begin{bmatrix} 1 & \frac{T_s}{A_S^i} & 0 & \dots & \dots & 0 \\ 0 & 0 & 1 & 0 & \dots & 0 \\ \vdots & \ddots & \ddots & \ddots & \ddots & \vdots \\ 0 & \ddots & \ddots & \ddots & \ddots & 0 \\ 0 & \ddots & \ddots & \ddots & \ddots & 1 \\ 0 & \ddots & \ddots & \ddots & \ddots & 0 \end{bmatrix}, \quad B^{i} = \begin{bmatrix} -\frac{Ts}{A_S^i} \\ 0 \\ \vdots \\ 0 \end{bmatrix}, \tag{18.14}$$

$$G^{i} = \begin{bmatrix} 0 \\ \vdots \\ 0 \\ 1 \end{bmatrix}, \quad G_{\text{f}}^{i} = \begin{bmatrix} 0 & -\frac{Ts}{A_S^i} \\ 0 & 0 \\ \vdots & \vdots \\ 0 & 0 \\ 1 & 0 \end{bmatrix}, \quad C^{i} = \begin{bmatrix} 1 & 0 & \dots & 0 \end{bmatrix}. \tag{18.15}$$

Disturbance modeling

In the current installation the lateral flows during the lock operations $q_{\text{lock}}^{i}(k)$ are not measured. Since the controller has no information about the lateral flows, in the model they are assumed to be zero $q_{\text{lock}}^{i}(k) = 0$. Thus the lateral flows act as unmeasured disturbances on the system. Moreover, there is a mismatch between the control model predictions and the real plant behavior. In order to reject the unmeasured disturbances and to compensate for the plant/model mismatch, the control model is augmented with an integrating disturbance model. We use the input disturbance modeling approach, see [17], in which an integrated disturbance state $x_{\text{d}} \in \mathbb{R}$ is added to the input variable. The augmented control model of reach i is then given as,

$$z^{i}(k+1) = \tilde{A}^{i} z^{i}(k) + \tilde{B}^{i} u^{i}(k) + \tilde{G}^{i} q_{\text{in}}^{i}(k) + \omega^{i}(k) \tag{18.16}$$

$$y^{i}(k) = \tilde{C}^{i} z^{i}(k) + \nu^{i}(k), \tag{18.17}$$

where $z^i(k) \in \mathbb{R}^{n+1}$ is the augmented state vector,

$$z^i(k) = \begin{bmatrix} x^i(k) \\ x_d^i(k) \end{bmatrix},$$ (18.18)

$\omega^i \in \mathbb{R}^{n+1}$ and $\nu^i \in \mathbb{R}$ are zero-mean white noises, named as disturbance noise and output noise, respectively, and

$$\tilde{A}^i = \begin{bmatrix} A^i & B^i \\ 0 & I \end{bmatrix}, \quad \tilde{B}^i = \begin{bmatrix} B^i \\ 0 \end{bmatrix}, \quad \tilde{G}^i = \begin{bmatrix} G^i \\ 0 \end{bmatrix}, \quad \tilde{C}^i = \begin{bmatrix} C^i & 0 \end{bmatrix}.$$ (18.19)

18.2.2.2 Cascade model

The model of a cascade containing P reaches is obtained by serially connecting the individual augmented reach models given in (18.16) via the interconnection variable $q_{in}^i(k)$,

$$q_{in}^i(k) = \begin{cases} u^{i-1}(k), & \text{if } i = 2, \dots, P \\ q_{in,c}(k) & \text{if } i = 1, \end{cases}$$ (18.20)

where $q_{in,c}(k)$ is the inflow entering the cascade at the first reach that is located most upstream.

18.3 Controller design

MPC has proven to be a powerful control technique especially in the process in-dustries due to its abilities to handle multivariable systems and incorporate con-straints in the problem formulation [16]. Unlike the conventional control strategies, it involves an online optimization algorithm. The optimization utilizes an internal control model to predict the behavior of the system. The control objectives are for-mulated in a cost function and the limitations on the system variables are expressed as constraints. At each control step, an optimal input sequence is computed by solv-ing the optimization problem over a specific time window, the so-called prediction horizon . Only the first element of this input sequence is applied to the plant. At the next control step, the time window is shifted by one step and the optimization procedure is repeated over the shifted time window. This procedure is referred to as a *Receding Horizon Policy* .

In this section we first formulate an MPC scheme for a single reach. Then, based on this formulation, we formulate a decentralized MPC scheme for a cascade of reaches.

18.3.1 MPC for a single reach

In order to predict the future behavior of the system, MPC requires the states of the system at the current time step. These states are estimated from the available measurements using a standard Kalman filter. Then the optimal control problem is solved for the estimated states. Next we explain the details of the state estimation and the optimal control problem.

18.3.1.1 State estimation

A steady-state Kalman filter [25] is designed for the augmented system (18.16) to estimate the current state at time k from the measurements of concession level $y^i(k)$, inflow $q_{in}^i(k)$, outflow $u^i(k)$. The estimation of the augmented system state $\hat{z}^i(k|k)$ at time k is given as

$$\hat{z}^i(k|k) = \hat{z}^i(k|k-1) + L^i\left(y^i(k) - \tilde{C}^i\hat{z}^i(k|k-1)\right), \tag{18.21}$$

and the one-step-ahead prediction of the augmented state $\hat{z}^i(k+1|k)$ is obtained by

$$\hat{z}^i(k+1|k) = \tilde{A}^i\hat{z}^i(k|k-1) + \tilde{B}^iu^i(k) + \tilde{G}^iq_{in}^i(k), \tag{18.22}$$

where L^i represents the optimal steady-state Kalman filter gain, which can be tuned by changing the covariance matrices of the disturbance noise ω and the output noise ν. Further information about the Kalman filter can be found in standard control textbooks, for example [25].

18.3.1.2 Optimal control problem

One of the control objectives is to minimize the turbine discharge variations $\Delta u(k)$. For this reason, in the augmented control model (18.16) $\Delta u(k)$ is used as the input variable of the system, such that

$$z^i(k+1) = \tilde{A}^iz^i(k) + \tilde{B}^iu^i(k) + \tilde{G}^iq_{in}^i(k) \tag{18.23}$$

$$u^i(k) = u^i(k-1) + \Delta u^i(k) \tag{18.24}$$

$$y^i(k) = \tilde{C}^iz^i(k). \tag{18.25}$$

Next we describe the cost function and the constraints on the system variables in detail and then we summarize the optimal control problem.

Cost function

The control objectives are minimizing the deviation of the concession level $y(k)$ from the reference value of zero and at the same time minimizing the turbine discharge variations $\Delta u(k)$. The two objectives are formulated mathematically in a cost function. For reach i a quadratic cost function V^i is defined at time t as a function of the state $z^i(t)$ and the input at previous time step $u^i(t-1)$ over a prediction horizon of N, given as

$$V^i(z^i(t), u^i(t-1)) = \sum_{k=1}^{N} \left(y^{i^T}(k)Q^i y^i(k) + s_{em}^{i^T}(k)Q_{s,em}^i s_{em}^i(k) + \right.$$

$$\left. s_{fo}^{i^T}(k)Q_{s,fo}^i s_{fo}^i(k) \right) + \sum_{k=0}^{N-1} \left(\Delta u^{i^T}(k)R^i \Delta u^i(k) \right). \quad (18.26)$$

$y^i(k)$ denotes the prediction of the concession level at time k, obtained by applying the sequence of turbine discharge variations $\Delta u^i(0), \Delta u^i(1), \ldots, \Delta u^i(k-1)$ to the system starting from the initial state $z^i(t)$ and having the turbine discharge of $u^i(t-1)$ at the previous time step. Q^i, R^i, $Q_{s,em}^i$ and $Q_{s,fo}^i$ are positive definite weighting matrices. Q^i is used to penalize the deviations of the concession level from the reference value of zero. R^i penalizes the turbine discharge variations Δu^i. $Q_{s,em}^i$ and $Q_{s,fo}^i$ penalize the slack variables s_{em}^i and s_{fo}^i that represent the violations of the soft constraints on the concession level deviations, which is explained in more detail below.

Constraints

The maximum discharge through the turbines has a physical limit, $q_{turb,max}$. The water flow above this limit is handled by the weirs, built next to the power plants. Due to economical reasons the minimum turbine discharge is limited by $q_{turb,min}$. These two limitations are included in the MPC formulation as hard constraints

$$q_{turb,min}^i \leq u^i(k) \leq q_{turb,max}^i. \quad (18.27)$$

There are restrictions on the maximum and minimum turbine discharge variations, Δu_{max}^i and Δu_{min}^i, due to the physical limitations of the equipment. These restrictions also formulated as hard constraints as well,

$$\Delta u_{min}^i \leq \Delta u^i(k) \leq \Delta u_{max}^i. \quad (18.28)$$

The concession level deviation can take values in three different regions depending on its magnitude; *preferred zone*, *emergency zone*, and *forbidden zone*. Ideally, the deviations need to be kept within the preferred zone defined by the boundaries $h_{c,min}$ and $h_{c,max}$. Deviations exceeding the preferred zone enter the emergency zone with the boundaries $h_{em,min}$ and $h_{em,max}$, where $h_{em,min} < h_{c,min}$ and $h_{em,max} > h_{c,max}$. The deviations in the emergency zone are undesired and they should be moved back to the preferred zone as soon as possible. The deviations exceeding the boundaries of the emergency zone enter the forbidden zone, which must be strictly avoided. These restrictions are formulated as soft constraints in order to avoid infeasibility of the optimization problem. The following linear inequalities with the slack variables s_{em}^i and s_{fo}^i are used,

$$h^i_{c,min} - s^i_{em}(k) \le h^i_c(k) \le h^i_{c,max} + s^i_{em}(k) \tag{18.29}$$

$$h^i_{em,min} - s^i_{fo}(k) \le h^i_c(k) \le h^i_{em,max} + s^i_{fo}(k) \tag{18.30}$$

$$0 \le s^i_{em}(k) \tag{18.31}$$

$$0 \le s^i_{fo}(k), \tag{18.32}$$

where s^i_{em} represents how much the concession level exceeds the preferred zone and it is penalized in the cost function with the weight matrix $Q^i_{s,em}$, see (18.26). Similarly, s^i_{fo} represents how much the concession level exceeds the emergency zone and it is penalized with the weight matrix $Q^i_{s,fo}$. Clearly the weight $Q^i_{s,fo}$ is selected larger than $Q^i_{s,em}$, since the forbidden zone violation is more severe.

Dead-band restriction

In order to decelerate the wear of the turbines, the number of control moves, i.e., the number of turbine discharge adjustments, needs to be small. To achieve that in practice, a dead-band restriction is applied to the control input. That is, if the absolute value of the control input Δu computed by the controller is smaller than a certain threshold Δu_{db}, it is not applied to the system,

$$\Delta u^a(k) = \begin{cases} \Delta u & \text{if } |\Delta u(k)| \ge \Delta u_{db} \\ 0 & \text{else,} \end{cases}$$

where $\Delta u^a(k)$ is the applied discharge variation and $\Delta u(k)$ is the computed one.

In order to have accurate predictions of the system, the dead-band restriction needs to be considered in the MPC formulation. The logical condition given above can be included in the MPC formulation by using the Mixed Logical Dynamical formulation [4]. The dead-band restriction on a predicted control move $\Delta u(k)$ within the horizon $k \in \{0, \ldots, N-1\}$ is modeled by the following mixed-integer inequalities using two binary variables, d_1 and d_2, [23]

$$\Delta u(k) \ge \Delta u_{db} d_1(k) + \Delta u_{min} d_2(k) \tag{18.33}$$

$$\Delta u(k) \le -\Delta u_{db} d_2(k) + \Delta u_{max} d_1(k) \tag{18.34}$$

$$d_1(k) + d_2(k) \le 1. \tag{18.35}$$

Restricting all the control moves $\Delta u(k)$ within the horizon for $k = 0, \ldots, N-1$ requires a total of $2N$ binary variables. For large horizons using $2N$ binary variables in the optimization problem might result in a large computation time (at the worst case the computation time increases exponentially with the number of binary variables). Therefore, we restrict only the first M control moves with the dead band, i.e., $k = 0, 1, \ldots, M$, for $M \le N$, and set the later control moves free. This way, the number of binary variables is reduced from $2N$ to $2M$. The intuition behind this approach is that it is not essential to determine the later control moves within the horizon precisely, due to the receding horizon policy. For smaller values of M the computation time decreases while the performance deteriorates due to the incorrect predictions. Therefore, a trade off has to be found between the computation time and the performance. By doing extensive simulation tests we determined the parameter as $M = 2$.

Optimal control problem

Having defined all the components of MPC, the control problem for reach i at time t becomes,

$$J_t^i(\hat{z}^i(t), u^i(t-1), q_{in}^i(k)) = \min_{\Delta U^i, S_{em}^i, S_{fo}^i} V^i \tag{18.36}$$

subject to

$$z^i(k+1) = \tilde{A}^i z^i(k) + \tilde{B}^i u^i(k) + \tilde{G}^i q_{in}^i(k) \tag{18.37}$$

$$u^i(k) = u^i(k-1) + \Delta u^i(k) \tag{18.38}$$

$$y^i(k) = \tilde{C}^i z^i(k) \tag{18.39}$$

$$E_1^i d^i(k) + E_2^i s^i(k) \le E_3^i \Delta u^i(k) + E_4^i z^i(k) + E_5^i \tag{18.40}$$

$$z^i(0) = \hat{z}^i(t), u^i(-1) = u^i(t-1), \tag{18.41}$$

for $k = 0, \ldots, N-1$.

where $\Delta U^{[i]}$ is the predicted sequence of the turbine discharge variations, given as

$$\Delta U^{[i]} = \left[\Delta u^{[i]}(0), \ldots, \Delta u^{[i]}(N-1) \right]. \tag{18.42}$$

Similarly,

$$S_{em}^{[i]} = \left[s_{em}^{[i]}(1), \ldots, s_{em}^{[i]}(N) \right], \tag{18.43}$$

$$S_{fo}^{[i]} = \left[s_{fo}^{[i]}(1), \ldots, s_{fo}^{[i]}(N) \right]. \tag{18.44}$$

The constraints of the system are written in a compact form in (18.40), where $d^i(k)$ contains the two binary variables $d^i(k) = [d_1^i(k), d_2^i(k)]^T$, and $s^i(k)$ contains the two slack variables $s^i(k) = [s_{em}^i(k), s_{fo}^i(k)]^T$. E_1, E_2, E_3, E_4, and E_5 are matrices with proper dimensions.

If the inflow is not measured, $q_{in}^i(k) = 0$ for $k = 0, \ldots, N-1$. If the inflow is measured as $q_{in}^i(t)$, $q_{in}(k) = q_{in}^i(t)$ for $k = 0, \ldots, N-1$. In addition to the measurement, if future predictions of the inflow $q_{in,pred}^i(k)$ for $k = 0, \ldots, N-1$ are also available, $q_{in}^i(k) = q_{in,pred}^i(k)$.

The optimal control problem in (18.36-18.41) is a Mixed-Integer Quadratic Programming problem, which is formulated in MATLAB using Yalmip [15] and solved at each time step using the commercial software package CPLEX 10.0 [13]. According to the receding horizon policy, only the first element $\Delta u^*(0)$ of the input

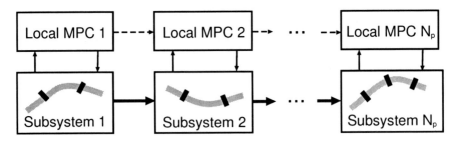

Figure 18.4: Decentralized MPC with downstream communication.

sequence is applied to the system. The optimization procedure is repeated at the next control step.

18.3.2 Decentralized MPC with downstream communication

A centralized MPC scheme, where all the power plants within the cascade are controlled by a single unit, may be computationally infeasible when applied to a large cascade. While one obvious reason for computational infeasibility is the large system size, the second reason is that the number of binary variables in the optimization problem (used for dead-band modeling) increases with the number of power plants in the cascade. Therefore, we design a decentralized MPC scheme, where the complete cascade is decomposed into N_p smaller subsystems and each subsystem is controlled by a local MPC scheme, as depicted in Figure 18.4. Note that there is no overlap between the subsystems, i.e., each power plant belongs to only one of the subsystems. The only shared variable between two subsequent subsystems is the inflow entering the downstream subsystem, which is equal to the outflow of the upstream subsystem (turbine discharge of the last power plant at the upstream subsystem).

In this work, we propose a decentralized MPC scheme with downstream communication, which has the following working principle. The local MPCs are solved sequentially starting from the most upstream reach. Each local MPC communicates the predictions of its outflow to the next local MPC. This way the downstream local MPC has a forecast of its inflow, which is useful to take preemptive actions when a disturbance enters upstreams. Providing a downstream communication between local controllers is a well-known strategy in the decentralized control of serially connected systems. In [24], it was shown for several applications that the performance of the decentralized control with downstream communication is very close to the performance of the centralized control.

18.3.2.1 Local MPC for a subsystem

Let n_j denote the number of power plants in subsystem j. P_j contains the indices of the power plants in subsystem j, such that $P_j = \{1, \dots, n_j\}$. The outflow of subsystem j is denoted as u_{out}^j, which is equal to the turbine discharge of the last power plant of subsystem j, $u_{out}^j = u^{n_j}$. The inflow entering subsystem j is denoted as u_{in}^j and it is equal to the outflow of the upstream subsystem, $u_{in}^j = u_{out}^{j-1}$, for $j > 1$.

State estimation

The states of a reach are only dependent on the inflow and the ,outflow of the reach, i.e. they are independent of the states of the subsequent reaches. Therefore, a local Kalman filter is used at each reach to estimate its states. The Kalman filter for a single reach was described in Section 18.3.1.1.

Optimal control problem

The optimal control problem for the local MPC scheme for subsystem j is:

$$J_t^j(\hat{\underline{z}}^j(t), \underline{u}^j(t-1), u_{in}^j(k)) = \min_{\underline{\Delta U}^j, \underline{S_{em}}^j, \underline{S_{fo}}^j} \sum_{i \in P_j} V^i \qquad (18.45)$$

subject to

$$z^i(k+1) = \tilde{A}^i z^i(k) + \tilde{B}^i u^i(k) + \tilde{G}^i q_{in}^i(k) \qquad (18.46)$$

$$u^i(k) = u^i(k-1) + \Delta u^i(k) \qquad (18.47)$$

$$y^i(k) = \tilde{C}^i z^i(k) \qquad (18.48)$$

$$q_{in}^i(k) = u^{i-1}(k) \qquad (18.49)$$

$$E_1^i d^i(k) + E_2^i s^i(k) \leq E_3^i \Delta u^i(k) + E_4^i z^i(k) + E_5^i \qquad (18.50)$$

$$\underline{z}^j(0) = \hat{\underline{z}}^j(t), \; \underline{u}^j(-1) = \underline{u}^j(t-1), \qquad (18.51)$$

$$u^0(k) = u_{in}^j(k) \qquad (18.52)$$

$$\text{for } k = 0, \ldots, N-1, \text{ and } i \in P_j, \qquad (18.53)$$

where $\underline{z}^j$ represents the state vector for the subsystem j, $\underline{z}^j = [z^{1^T}, \ldots, z^{n_j^T}]^T$. Similarly, $\underline{u}^j = [u^{1^T}, \ldots, u^{n_j^T}]^T$, $\underline{\Delta U}^j = [\Delta U^{1^T}, \ldots, \Delta U^{n_j^T}]^T$, $\underline{S_{em}}^j = [S_{em}^{1^T}, \ldots, S_{em}^{n_j^T}]^T$ and $\underline{S_{fo}}^j = [S_{fo}^{1^T}, \ldots, S_{fo}^{n_j^T}]^T$.

The inflow $u_{in}^j(k)$ entering subsystem j is given as,

$$u_{in}^j(k) = \begin{cases} u_{out}^{j-1}(k), & \text{if } j = 2, \ldots, N_p \\ q_{in,c}(t) & \text{if } j = 1, \end{cases} \qquad (18.54)$$

where $q_{in,c}$ is the inflow entering subsystem 1.

18.4 Simulation results

The decentralized MPC scheme is implemented on the FLORIS model of a cascade of 35 river power plants. The performance of decentralized MPC is tested in the following simulation scenario. While the system is in steady state initially, two types of disturbances act on the system simultaneously. First, the inflow $q_{in,c}$ entering the cascade at the most upstream reach has fluctuations around its initial discharge, as shown in Figure 18.5. The second disturbance is due to the lock operations, which creates lateral in- and outflows within the cascade. The locks operate about 20 times a day from 5 am to 10 pm. While the total simulation time is 100 hours, the lock operations last only for the first three days. A sample for the discharges at the first and second gates of a lock are shown in Figure 18.6. When the first gate opens it creates a lateral outflow at the upstream reach. When the second gate opens it creates a lateral inflow at the downstream reach. While the discharges at the first and second gates of a lock are the same, the discharges range from 9 to 23 m³/s at different locks (9 m³/s in Figure 18.6). While the inflow $q_{in,c}$ is measured, the lateral flows due to the lock operations are not measured in the simulations.

In the simulations several decentralized MPC schemes containing different numbers of local MPC's are tested. The same sets of constraints and controller parameters are used at the MPC formulation of each reach as given in Tables 18.1 and 18.2. The weight Q^i that penalizes the concession level deviations is selected very small so that the controller allows deviations within the preferred zone and thus, uses the

Table 18.1: Constraints on the concession levels and the dead band restriction. The same set of constraints is considered for each river reach.

Parameters	Values
Range of preferred zone [cm]	[-2,2]
Range of emergency zone [cm]	[-10,-2], [2,10]
Range of forbidden zone [cm]	$(-\infty,-10], [10,\infty)$
Dead band on turbine discharge rate [m³/s]	+/-2.5

Table 18.2: MPC parameters ($i = 1, \ldots, 35$).

Parameters	Values
Prediction horizon N	50 steps (5 hr)
Number of dead-band restricted control moves M	2
Penalty on turbine discharge variations R^i	diag(500,...,500)
Penalty on concession level deviations Q^i	diag(0.01,...,0.01)
Penalty on first slack variables $Q^i_{s,em}$	diag(2000,...,2000)
Penalty on second slack variables $Q^i_{s,fo}$	diag(10000,...,10000)
Sampling time of the linear model	6 min

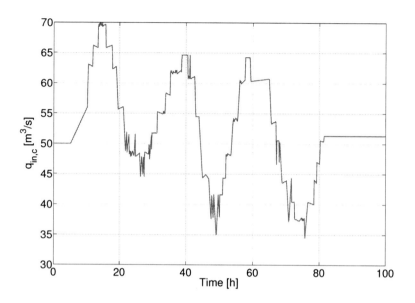

Figure 18.5: Inflow $q_{in,c}$ entering the cascade at the most upstream.

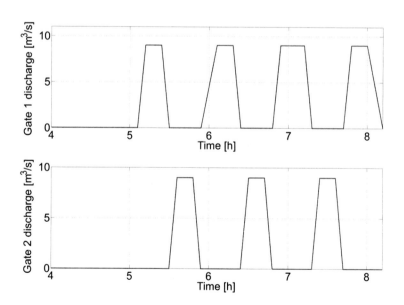

Figure 18.6: The gate discharges of a lock (shown for a limited time duration).

storage capacity of the reaches efficiently. On the other hand, the deviations exceeding the preferred zone are penalized heavily with a large $Q_{s,em}^i$. The deviations exceeding the emergency zone are penalized with a much larger weight $Q_{s,fo}^i$. The simulations are performed on a PC with a Intel(R) 2.0 GHz processor and 1 GB RAM.

18.4.1 Influence of the number of subsystems on the performance

Here we test three decentralized MPC schemes containing $N_p = 5$, $N_p = 12$, and $N_p = 35$ local MPC's. Each local MPC controls a cascade of 7, 3 and 1 reaches, respectively. The simulation results are given in Figures 18.7– 18.9. The performance of the three schemes is very similar. The deviations of the turbine discharges from the initial operating point are very close to one another for all the cases, see Figure 18.7. The high frequency oscillations in the concession levels, see Figure 18.8, are due to the lateral flows during the lock operations. The concession levels exceed the preferred zone only by small amounts and for short periods of time. The mean and maximum absolute errors of the concession levels at each plant are also very close to each other, see Figure 18.9 (Notice the scale of the mean absolute error, although $N_p = 35$ has the lowest value, its difference from the others is at most 0.3 cm, which is hardly possible to measure). In summary, the performance of decentralized MPC does not deteriorate with increasing number of subsystems and even with the most decentralized scheme $N_p = 35$, where each reach is controlled by a local MPC, it performs well.

The good performance of the scheme with $N_p = 35$ can be explained as follows. When an inflow disturbance acts on the reach, the local MPC has the measurement and the forecast of this disturbance, which is provided by the upstream local MPC. Since there is a time delay between the inflow and the concession level, the local MPC has enough time to react against this disturbance.

The computation times of the three decentralized MPC schemes are shown in Table 18.3. The decentralized MPC with $N_p = 5$ subsystems is computationally infeasible, since the maximum computation time might reach up to 16 min, while the sampling time is only 6 min. The computation time is shortest when $N_p = 35$ and in that case even the maximum computation time is much smaller than the sampling time. Since we obtain almost the same performance with the smallest computation time the decentralized MPC with $N_p = 35$ partitions is the most convenient scheme to be implemented.

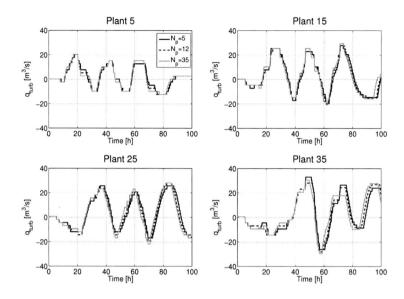

Figure 18.7: Deviations of turbine discharges from the initial operating points at the plants 5, 15, 25, and 35 for decentralized MPC with different number of subsystems (N_p).

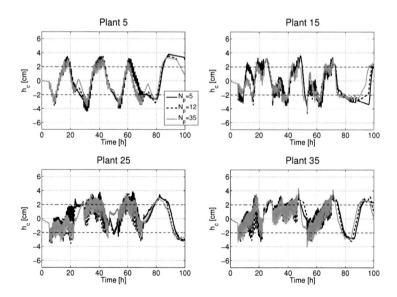

Figure 18.8: Deviations of concession levels from the zero reference value at the plants 5, 15, 25, and 35 for decentralized MPC with different number of subsystems (N_p). (Black dashed lines represent the boundaries of the preferred zone ± 2 cm.)

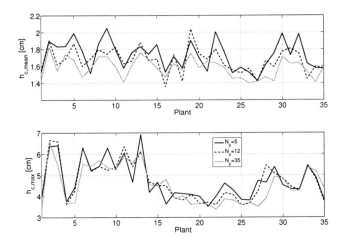

Figure 18.9: Mean absolute error ($h_{c,mean}$) and maximum absolute error ($h_{c,max}$) of concession level deviations for decentralized MPC with different numbers of subsystems (N_p).

Table 18.3: The mean and maximum computation time (the time that is required to solve the MPC optimization problems of all subsystems at each time step) of decentralized MPC with a different number of subsystems (N_p) per simulation.

N_p	Mean computation time [s]	Max. computation time [s]
5	270	1000
12	57	93
35	25	58

18.4.2 Influence of downstream communication on the performance

Here we compare the performance of decentralized MPC containing 35 local MPCs with and without downstream communication. The downstream communication strategy was explained in detail in Section 18.3.2. When there is no communication, the inflow entering the reach, i.e., the turbine discharge of the upstream power plant, is measured and this measurement is assumed to stay constant over the prediction horizon. The control performance is shown in Figures 18.10 and 18.11 (The concession level deviations as given in Figure 18.8 are not shown, since they do not contain any useful information).

The performance achieved is similar in terms of the mean and maximum absolute error of the concession levels, see Figure 18.11. However, this performance is achieved with a smaller magnitude of turbine discharges when downstream communication is used, see especially the turbine discharge variations of the 35th power plant in Figure 18.10. Therefore, downstream communication is needed for a good performance of decentralized MPC.

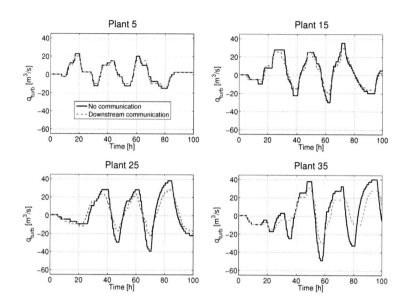

Figure 18.10: Deviations of turbine discharges from the initial operating points at the plants 5, 15, 25 and 35. Controller: Decentralized MPC scheme with 35 local MPC's.

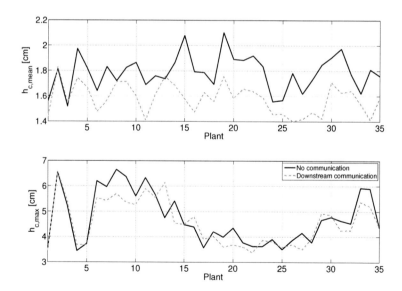

Figure 18.11: Mean absolute error ($h_{c,mean}$) and maximum absolute error ($h_{c,max}$) of concession level deviations. Controller: Decentralized MPC scheme with 35 local MPC's.

18.5 Conclusions and future research

In this work we developed a decentralized MPC scheme with downstream communication to control a large cascade of river power plants, for which a centralized MPC would be computationally infeasible. We tested the proposed scheme by implementing it on the accurate simulation model of a cascade of 35 power plants. We showed that providing the downstream communication between the local MPC's is sufficient to prevent a significant performance deterioration in decentralized MPC, which would be expected due to the lack of coordination. We compared the performance and computation time of decentralized MPCs containing different numbers of subsystems. Based on this comparison we concluded that the decentralized MPC scheme with 35 local MPCs, in which each power plant is controlled with a local MPC, is the most suitable scheme, since it has the shortest computation time and its performance is still reasonably good. As the next step before implementation on the real plant, the decentralized MPC scheme has to be tested through simulations in several hydraulic scenarios, e.g., low flow and high flow scenarios. Furthermore, its sensitivity to measurement errors needs to be tested.

References

1. T. Ackermann, D. P. Loucks, D. Schwanenberg, and M. Detering. Real time modeling for navigation and hydropower in the river model. *Journal of Water Resources Planning and Management*, pages 298–303, September/October 2000.
2. F. Althaus. Model predictive control for cascaded river power plants. Master's thesis, ETH Zürich, Zürich, Switzerland, August 2008.
3. A. G. Beccuti and M. Morari. A distributed solution approach to centralized emergency voltage control. In *Proceedings of the 2006 IEEE American Control Conference*, pages 3445–3450, Minneapolis, Minnesota, June 2006.
4. A. Bemporad and M. Morari. Control of systems integrating logic, dynamics, and constraints. *Automatica*, 35(3):407–427, March 1999.
5. G. Bollrich. *Technische Hydromechanik 1, Grundlagen*. HUSS-Medien GmbH, Berlin, Germany, 2007.
6. M. Cantoni, E. Weyer, Y. Li, S. K. Ooi, I. Mareels, and M. Ryan. Control of large scale irrigation networks. *Proceedings of the IEEE*, 95(1):75–91, January 2007.
7. J. Chapuis. *Modellierung und Neues Konzept für die Regelung von Laufwasserkraftwerken*. PhD thesis, ETH Zürich, Zürich, Switzerland, 1998.
8. G. Corriga, D. Salimbeni, S. Sanna, and G. Usai. A control method for speeding up the response of hydroelectric station power canals. *Applied Mathematical Modelling*, 12(6):627–633, December 1988.
9. B. de Saint-Venant. Théorie du mouvement non permanent des eaux, avec application aux crues des rivières et à l'introduction des marées dans leur lit. *Comptes Rendus des Séances de l'Académie des Sciences Paris*, 73:147–154, 1871.
10. D. Dumur, A. Libaux, and P. Boucher. Robust control for Basse Isere run-of-river cascaded hydro-electric power plants. In *Proceedings of the 2001 IEEE International Conference on Control Applications*, pages 1083–1088, Mexico City, Mexico, September 2001.
11. G. Glanzmann, M. von Siebenthal, T. Geyer, G. Papafotiou, and M. Morari. Supervisory water level control for cascaded river power plants. In *International Conference on Hydropower*, Stavanger, Norway, May 2005.

12. A. H. Glattfelder and L. Huser. Hydropower reservoir level control: A case study. *Automatica*, 29(5):1203–1214, September 1993.
13. ILOG. Webpage of ILOG, developer and distributor of CPLEX.
14. X. Litrico and V. Fromion. H-infinity control of an irrigation canal pool with a mixed control politics. *IEEE Transactions on Control Systems Technology*, 14(1):99–111, January 2006.
15. J. Löfberg. YALMIP: A toolbox for modeling and optimization in MATLAB. In *Proceedings of the 2004 IEEE International Symposium on Computer Aided Control Systems Design*, pages 284–289, Taipei, Taiwan, September 2004.
16. J. M. Maciejowski. *Predicitive Control with Constraints*. Prentice-Hall, Pearson Education Limited, Harlow, UK, 2002.
17. U. Mäder and M. Morari. Offset-free reference tracking for predictive controllers. In *Proceedings of the 46th IEEE Conference on Decision and Control*, pages 5252–5257, New Orleans, Los Angeles, December 2007.
18. R. R. Negenborn, P. J. van Overloop, T. Keviczky, and B. De Schutter. Distributed model predictive control for irrigation canals. *Networks and Heterogeneous Media*, 4(2):359–380, June 2009.
19. R.R. Negenborn, A. Sahin, Z. Lukszo, B. De Schutter, and M. Morari. A non-iterative cascaded predictive control approach for control of irrigation canals. In *Proceedings of the IEEE International Conference on Systems, Man, and Cybernetics*, San Antonio, Texas, October 2009.
20. J.-P. Baume P.-O. Malaterre. Modeling and regulation of irrigation canals: existing applications and ongoing researches. In *Proceedings of the International Conference on Systems, Man, and Cybernetics*, pages 3850–3855, San Diego, California, October 1998.
21. J. I. Sarasua, J. F. Ardanuy, J. I. Perez, and J. A. Sanchez. Control of a run of a river small hydro power plant. In *Proceedings of POWERENG 2007*, pages 672–677, Setubal, Portugal, April 2007.
22. J. Schuurmans, O. H. Bosgra, and R. Brouwer. Open-channel flow model approximation for controller design. *Applied Mathematical Modelling*, 19(9):525–530, September 1995.
23. C. Setz, A. Heinrich, P. Rostalski, G. Papafotiou, and M. Morari. Application of model predictive control to a cascade of river power plants. In *Proceedings of the 17th IFAC World Congress*, Seoul, South Korea, July 2008.
24. M. G. Singh. *Dynamical Hierarchical Control*. North-Holland Publishing Company, Amsterdam, The Netherlands, 1977.
25. R. F. Stenge. *Optimal Control and Estimation*. Dover Publications, New York, New York, 1994.
26. A. N. Venkat, I. A. Hiskens, J. B. Rawlings, and S. J. Wright. Distributed output feedback MPC for power system control. In *Proceedings of the 45th IEEE Conference on Decision and Control*, pages 4038–4045, San Diego, California, December 2006.
27. Versuchsanstalt für Wasserbau, Hydrologie und Glaziologie der Eidgenössischen Technischen Hochschule Zürich. Floris benutzer-handbuch, 1992.
28. E. Weyer. Decentralized PI control of an open water channel. In *Proceedings of the 15th IFAC World Congress*, Barcelona, Spain, July 2002.

Chapter 19
Enhancing the Reliability and Security of Urban Water Infrastructures through Intelligent Monitoring, Assessment, and Optimization

W. Wu and J. Gao

Abstract Urban water infrastructure systems are large, complex network systems. In delivering water to end consumers they are required to meet increasingly stringent water quality standards as well as minimum flow and pressure criteria. At the same time, water utilities are expected to become more effective and to demonstrate sustainability in both operation and profit. Water supply companies are facing growing costs related to the energy needed to meet increasing demand for water, due to leakage and water loss caused by the aging and failure of infrastructure and also due to climate changes, increased population density, etc. They also have to deal with increasing uncertainty from a range of threats including natural hazards and human-caused threats, such as climate changes and biochemical contamination. This chapter will investigate intelligent monitoring, assessment and optimization techniques that can be applied in urban water infrastructures to improve the spatial and temporal resolution of operational data from water distribution networks and address the challenge of real-time monitoring and control in large-scale complex distributed environments. We will explore techniques that can comprehensively monitor a complex, highly dynamic environment and enhance the reliability and security of urban water infrastructures through intelligent monitoring, assessment and optimization.

W. Wu
Staffordshire University, Faculty of Computing, Engineering and Technology, Stafford, UK,
e-mail: w.wu@staffs.ac.uk

J. Gao
Harbin Institute of Technology, School of Municipal and Environment Engineering, Harbin, China,
e-mail: gjl@hit.edu.cn

R.R. Negenborn et al. (eds.), *Intelligent Infrastructures*, Intelligent Systems, Control and   487
Automation: Science and Engineering 42, DOI 10.1007/978-90-481-3598-1_19,
© Springer Science+Business Media B.V. 2010

19.1 Introduction

Urban water infrastructure systems are large, complex network systems. In delivering water to end consumers they are required to meet increasingly stringent water quality standards as well as minimum flow and pressure criteria. At the same time, water utilities are expected to become more effective and to demonstrate sustainability in both operation and profit. Water supply companies are facing growing costs related to the energy needed to meet increasing demand for water, due to leakage and water loss caused by the aging and failure of infrastructure and also due to climate changes, increased population density, etc. They also have to deal with increasing uncertainty from a range of threats including natural hazards and human-caused threats, such as climate changes and biochemical contamination. Climate changes, such as rising sea level due to global warming, changing rainfall pattern, flood, severe droughts, storm surge, and strong flow are the conditions that water infrastructures must withstand. The associated risks must be considered in long-term planning. Infrastructure failures can have complex causes and catastrophic consequences; the reactive response and maintenance can be expensive. Little is known about the long-term performance of such water infrastructures. The uncertainties in water infrastructures and the importance of their security and safety to users and customers will be discussed. We will investigate a number of ways of improving capabilities for efficiently maintaining and protecting water infrastructures, such as early warning systems, damage detection, and decision-making support.

Current water infrastructure systems rely heavily on large-scale engineering and low technology systems, which often lead to inefficient operations and increased costs. The new development of hardware and monitoring technologies offers an increase in the ability to monitor and manage infrastructure at low cost. Accurate monitoring, modeling, and optimization of the water infrastructure systems are essential for understanding and assessing conditions of urban water infrastructures. Currently, monitoring systems in the water supply network are widely used; the majority of urban water supply networks include both off-line and on-line systems. However, there is no strong evidence to show that the current monitoring systems are effective and efficient. Some monitoring locations are redundant, while others lack monitoring. A number of operations are not computerized and require a full support system of human operators. The information exchange among systems is not sufficient, resulting in the duplication of effort [21]. Many data resources are still distributed manually by the water companies. This chapter will investigate various monitoring techniques and their possible applications. We will explore their usefulness toward improving the spatial and temporal resolution of operational data from water distribution networks and toward addressing the challenge of achieving comprehensive real-time monitoring and control of large-scale, complex, highly dynamic distributed environments.

Optimization of pressure and water quality monitoring in a water distribution network is essential for the ability of water companies to provide a safe and reliable water supply to customers [27]. This chapter will study the optimization of the off-line and on-line monitoring points used to protect the safety of a water distribution system. Through analysis of pressure pertinence and its quantified criteria of nodes in water supply networks, a mathematical model has been established for the optimal placement of pressure monitoring [10]. The elitist genetic algorithm (GA) method was adopted for solving this model. The model is evaluated by solving an example network and improves the accuracy and feasibility of this optimal model. It will be used to enhance water distribution system security against the threat of contaminants in emergency situations.

Another important issue in urban water infrastructure is drinking water quality. Monitoring and comprehensive evaluation of water quality in water supply networks will be discussed. The theory of fuzzy comprehensive evaluation is adopted. This theory is based on microcosmic models in the water supply network. The water quality indicators, by which the states of water quality are expressed, are selected as the input vector for the comprehensive evaluation model. The Self-Organizing Feature Map (SOM) and K-Means arithmetic are adopted in the model [28]. Since there are some problems with using the single indicator grade evaluation method, a comprehensive water quality evaluation method and an objective water quality evaluation model of water supply networks are developed. Instead of the single indicator evaluation, both the single water quality monitoring data and the overall data are considered in the model using SOM, which reflects the water quality distribution in the water supply network objectively. These approaches have been applied and evaluated in several case studies and in the actual water distribution network in China.

Effective decision making is underpinned by the ability to sense the full range of data available within the environment. A number of spatially distributed sensing points can continuously gather data from the water infrastructure. The data from the sensor network will be analyzed. Then, based on their relevance to different models and services, the data will be filtered and communicated in order to support local decision-making about modeling and calibration and to guide safety operations. It can further be used to produce an integrated and informed urban water infrastructure information management system, which can forecast, control and estimate uncertainty in water infrastructure systems and further support high level executives in the water companies to improve their intelligent decision-making capabilities and guide the urban water supply system toward operating securely and safely.

19.2 Water distribution system

This section discusses how both the physical objects that constitute a water distribution system and its operational parameters are modeled. An overview is provided of the basic principle used by a water distribution system model in simulating hydraulic and water quality transport behavior.

19.2.1 Water supply system

A water distribution system is a pressurized piping network of interconnected pipes, reservoirs, storage tanks, pumps, and regulating valves, all of which work together to provide adequate pressure, adequate supply and good water quality throughout the system. The main purpose of a water distribution system is to meet demand for potable water for living. It also provides water for fire protection.

Water demand

Flow, head, and pressure are basic hydraulic parameters for water distribution system analysis. The accurate estimation and forecast of customer water demand are essential for the precision of a water distribution system model. The optimal control of water distribution system is also dependent upon the perfect forecasting of water demand.

Demand pattern

A demand pattern is a function relating water use to the time of day. Patterns allow the user to apply automatic time-variable changes within the system. Different categories of users, such as residential, hospital, school, or industrial customers, will typically be assigned different patterns to accurately reflect their particular demand variations.

Pumps

Pumps are components that impart energy to a fluid, thereby raising its hydraulic head. They are an important part of modeling head change. The principal output parameters of pumps are flow and head gain. Flow through a pump is unidirectional, and water distribution systems will not allow a pump to operate outside the range of its pump curve. A water distribution system can also compute the energy consumption and cost of a pump. Each pump can be assigned an efficiency curve and schedule of energy prices.

Valves

Valves are links that control the pressure or flow at a specific point in a water distribution system. There are different types of valves for various functions, such as Pressure Reducing Valve (PRV), Pressure Sustaining Valve (PSV), Pressure Breaker Valve (PBV), Flow Control Valve (FCV), etc.

Some urban water distribution infrastructures are currently facing a significant number of problems. Pipes, valves, and pumps are deteriorating. As these infrastructures were built years ago, the data is fragmented, making it difficult for utility companies to use and control them. Modeling the water distribution system is especially essential in these situations for providing good services.

19.2.2 Water distribution network models

A water distribution network model is a collection of links connected to nodes. The links represent pipes, pumps, and control valves. The nodes represent junctions, tanks and reservoirs. Hydraulic and water quality simulation models are the two main types of water distribution system models. These types of models have emerged as useful tools for water distribution system planning, design, analysis, and rehabilitation. The scenarios of water distribution systems are unpredictable, because the patterns of water supply and consumer demand change every day.

19.2.2.1 Hydraulic simulation models

A hydraulic simulation model computes junction heads and pipe flows for a fixed set of reservoir levels, tank levels, and water demand over a fixed time period. The critical problem of hydraulic modeling is how to accurately represent the hydraulic scenarios by using monitoring data and how to estimate the operation states in order to be able to meet the water demand for consumers. The reservoir levels and junction demand are updated according to their prescribed time patterns, while tank levels are updated using the current flow solution. The hydraulic modeling in water distribution systems follows two laws, the laws of continuity and energy conservation. The difference in energy at any two points connected in a network is equal to the energy gains from pumps and energy losses in pipes and fittings that occur in the path between them. Particularly interesting are paths between reservoirs or tanks (where the difference in head is known), or paths around loops since there the changes in energy must have a zero sum.

Hydraulic simulation models are mainly applied under an optimal schedule, optimal rehabilitation, and scientific management. The water supply network simulation models can be classified into macroscopic and microscopic models. A microscopic model simulates more correctly the networks, static state and dynamic information, including figures, attributes, parameters, and state information. However, due to the longer simulation time of a microscopic distribution model, a macroscopic hydraulic network model is used more often in creating optimal control models for large-scale water distribution systems. But using a macroscopic model to make an optimal control model is complicated, and its reliability is poor. Therefore, an initial study of optimal control models with a simplified model has been made. This could overcome defects of optimal control models with either the macroscopic hydraulic model or microscopic model.

19.2.2.2 Water quality simulation models

Water network modeling in conjunction with water quality modeling plays an important role in explaining histories and patterns of water quality in water supply systems. The critical problem is how to estimate consumer exposure to contaminants, including locating the possibly affected areas and the dose of exposure. One approach using Monte Carlo computer simulations for water quality monitoring in mass transport through a pipe network is discussed in [20].

The water quality model has to trace, analyze and visualize the impact of contaminants on consumers in the event of a biochemical threat in the water distribution system. A Lagrangian time-based approach tracks the fate of discrete parcels of water as they move along pipes and mix together at junctions between fixed-length time steps. These water quality time steps are typically much shorter than the hydraulic time steps (e.g., minutes rather than hours) in order to accommodate the short times of travel that can occur within pipes [23].

The water quality model can track the growth or decay of a substance due to reaction as it travels through a distribution system. In order to do this it needs to know the rate at which the substance reacts and how this rate might depend on the substance concentration. Reactions can occur both within the bulk flow and with material along the pipe wall.

Reactions occurring in the bulk flow are modeled with n order kinetics. Bulk reaction coefficients usually increase with increasing temperature. The rate of water quality reactions occurring at or near the pipe wall can be considered to be dependent on the concentration in the bulk flow. The wall reaction coefficient can depend on temperature and can also be correlated to pipe age and material. Accurate hydraulic and water quality modeling in water distribution systems rely on the water demand of users and the water supply pattern. Forecasting water demand of users is thus fundamental for the water supply system.

19.2.3 Water demand forecast

The daily water demand forecast model is an important factor affecting the precision of macroscopic and microscopic hydraulic models. The combined forecast model was initially established by using daily water demand data of a reference day in the regression forecast model.

The methods of forecasting urban water demand are usually classified into two types of analysis: time series and regression. Factors influencing water demand variation between adjoining days are very relevant and are utilized in the time series analysis method, but known factors such as temperature are not taken into account. On the contrary, known factors are included in the regression analysis model, but some uncertain factors cannot be taken into account since previous observed water demand data are not incorporated in the model. In addition, the precision of regression models is mostly dependent upon the accuracy of the weather forecast. Hence, the predicted results of traditional methods are not very efficient. The combined forecast model was initiated to predict daily water demand.

After the daily water demand of a reference day was brought into the regression model, the combined model was established. This new model can overcome defects of neglecting known influencing factors in time series analysis and neglecting uncertain factors in regression analysis. At the same time, the problem of the regression model's precision being mostly dependent on weather forecast can be relaxed. The

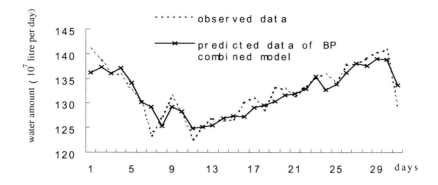

Figure 19.1: Predicted and observed values of water demand in Tianjin in July 1999.

combined model was shown as below:

$$Q_d = A_0 + A_1 Q_p + A_2 \mathrm{sgn}(T_{\max} - T_{p\max})|T_{\max} - T_{p\max}|^{A_3}$$
$$+ A_4 \mathrm{sgn}(T_{\min} - T_{p\min})|T_{\min} - T_{p\min}|^{A_5}$$
$$+ A_6 \mathrm{sgn}(H - H_p)|H - H_p|^{A_7} + Y, \qquad (19.1)$$

where Q_d is the observed value of the diurnal water demand of the predicted day; Q_p is the observed value of the diurnal water demand of the reference day (usually the previous day of the predicted day); $\mathrm{sgn}(\cdot)$ is the sign function, $A_0, A_1, A_2, A_3, A_4,$ $A_5,$ and A_6 are regression coefficients; $T_{\min}$ and $T_{\max}$ are the lowest and highest temperature of predicted day, respectively; $T_{p\min}$ and $T_{p\max}$ are the lowest and highest temperature of predicted day, respectively; H is the influence degree of holidays on the predicted day, the range of the value is [0,3]; H_p is the influence degree of holidays on the referenced day, the range of the value is [0,3]; and Y is the water amount influenced by uncertain factors that are not described in the combined model.

The regression coefficients were estimated by using the GAUSS-NEWTON method or the BP nervous network. The combined model was applied to predict the water demand in a city. Except for the first self-correlation coefficient, other coefficients are within the range of a 95% confidence region. This predicted result of the combined model was pleasing.

The result of the BP nervous network is slightly better than that of the GAUSS-NEWTON method. This showed that cooperation of some factors slightly influence the water demand fluctuation trend. In order to eliminate the influence of uncertain factors not included in the combined model, the auto-regression model of error series between observed and predicted data was built. The recursion least square method was used to estimate their parameters. This can further improve the precision of the prediction. The average relative error of the combined model with the GAUSS-NEWTON method is 1.204%, and with the BP nervous network it is 1.147%. The results of the BP combined model are shown in Figure 19.1.

19.3 Monitoring and modeling in water distribution systems

19.3.1 Data acquisition and management using GPS and GIS

Designing, establishing, maintaining, or rehabilitating a water supply system depends on information or data collected by surveyors. Establishing a geographic information system (GIS) of a water supply system requires even greater support from surveyors. However, because of the size and complexity of water networks associated with extensive urban areas, it is difficult for traditional surveying practice to meet today's engineering requirements. The Global Positioning System (GPS) technology gives us a new alternative for collecting information on water networks [12]. Collected data from water networks using GPS can achieve centimeter accuracy and save costs, time, and manpower as compared with traditional surveying methods.

A GIS for water networks is a database of pipes and accessories associated with the water distribution system. Based on a water network GIS, a real time dynamic model has been established for analyzing scenarios and for optimization and scheduling of a water distribution system. Hydraulic and water quality parameters are included in the real time dynamic model.

19.3.2 Monitoring and modeling in a water distribution system

19.3.2.1 Hydraulic pressure intelligent monitoring

Optimal placement of pressure monitors for supervisory control and data acquisition (SCADA) in water supply networks is of great importance for the accuracy of the information collected from monitors. For a typical system, there is no guarantee that a global optimum for pressure monitoring can be achieved. However, using optimization, it should be possible to find better locations than those that would be selected randomly.

The purpose of building a pressure monitoring system is to obtain pressure information in particular areas of the water distribution network, and to gain an overview of the pressure situation for the entire network. In order to minimize the number of monitors used, the selected monitoring nodes must supervise other nodes as much as possible. From this concept, we can abstract the optimal placement problem as a combined optimization problem. One constraint is the number of monitors, and the objective is to maximize the monitoring area of the placement scheme. An important aspect of building monitoring models is defining a criterion to determine if the monitor node can supervise other nodes – in other words, to determine which nodes can be delegated to one monitoring node.

The pressure correlation between two nodes is implemented by the influence of the upstream node acting on the downstream node. Thus, the essential and sufficient condition of having a pressure correlation between two nodes is that there are flow paths between these two nodes. With this precondition, the quantification method of pressure pertinence is the node head difference between these two nodes. Therefore, if two nodes A and B satisfy the two conditions below, we can say that the node head information of these two nodes can be delegated to each other:

1. there are flow paths between node A and node B;

2. $|H_a - H_b| < h$.

With the definition of pressure correlation and its quantification method, the mathematical model of optimal placement of the pressure monitor can be established. The objective is to maximize the supervised nodes. This means using limited pressure monitoring information to represent other nodes' pressure status as much as possible. The constraint is the number of monitoring nodes. In general, this constraint is a number specified in advance. The optimal model can therefore be expressed as follows:

$$X_j = \begin{cases} 1 & \text{if node } j \text{ is selected as a pressure monitor} \\ 0 & \text{otherwise} \end{cases} \tag{19.2}$$

$$T = A_1 \cup A_2 \cup \cdots \cup A_N \tag{19.3}$$

$$\text{Objective: } \max C_T \tag{19.4}$$

$$\text{Constraints: } \sum_{j=1}^{M} X_j = N, \tag{19.5}$$

where X_j denotes whether node j was selected as a pressure monitoring location; A is a set that contains all nodes which satisfy the pressure correlation conditions as monitoring nodes; T is the union set of all A sets, and C_T is the number of elements in collection T.

This model describes a combined optimization problem that can be solved by simply exhausting all possible schemes and selecting the best one. In fact, however, this is difficult to achieve, especially when the network has a large number of nodes. Several strategies were used to solve this problem, including the heuristic method, simulated annealing, and genetic algorithms. The genetic algorithm proved to be a highly efficient method [19, 26]. The elitist GA (EGA) [4] method aims to ensure that the best individual of a population obtains offspring within the following generation. EGA was chosen here to solve the problem of optimal placement of pressure monitors in our model.

19.3.2.2 Case study

A water supply network [22] with 79 nodes is used as an example. Six pressure monitors have been placed into this network. The optimal placement problem of this case network was solved by EGA and GA separately. Average fitness values and maximum fitness values of chromosomes in different generations solved by EGA are shown in Table 19.1. By comparing the results between the two methods, it is found that elitist operation plays a distinctive role in the evolution process.

Table 19.1: Average and maximum evaluation values of chromosomes in difference generations.

Generations	Average evaluation values	Max evaluation values	Code of the best chromosome
1	0.59324	0.74683	10 11 12 42 49 52
20	0.75063	0.81012	12 15 40 42 52 61
35	0.72911	0.89873	2 5 12 34 39 52
50	0.83291	0.87341	2 12 15 39 52 74
100	0.82067	0.87341	2 5 12 39 52 74

Figure 19.2 shows the curves of the results of EGA, which has an obvious excessive trend; the best individual obtained by this method has a much better fitness value than the result obtained by GA. The two average fitness curves show a more obvious evolving trend than the maximum fitness curves. In the EGA results, the average fitness curve shows a high contingency speed and better evolution space than the curves obtained by GA. This phenomenon shows that the removal of the best individuals in the elitist operation reduces the chance of populations degenerating due to the unsuccessful genes brought in by the mutation option, and at same time accelerates the contingency speed. Elitist operation also plays a role in the maximum fitness curve. In the GA results, the maximum fitness curve shows a nearly random behavior in evolution. In contrast, the best individuals of generations with EGA show a relatively steady behavior and have an obvious evolution trend.

With optimal placement of pressure monitors, we can also obtain information on the supervising area of each pressure monitor as shown in Figure 19.3, which is very important when the supervising system has been established and brought into use.

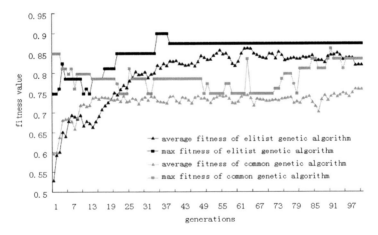

Figure 19.2: Average and maximum fitness values of each generation in the evolution process of the two methods.

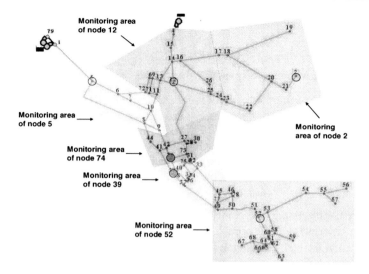

Figure 19.3: The monitoring area of each pressure monitor.

With this supervising information, the monitoring can be conducted more logically and efficiently.

The problem of optimal placement of pressure monitors of water supply networks can be abstracted as a combined optimization problem. The constraint is the number of monitors, and the objective is to maximize the monitoring area by means of an optimum placement scheme. Although truly optimal monitor locations are unlikely to be justified by the optimization model alone due to the randomness of the node elevations and pressure sensitive consumers or other special evaluation criteria, the solutions produced by the optimization will serve as a good basis for fine-tuning the final monitor locations.

19.3.2.3 Water quality monitoring

The complex water distribution network and broad area coverage are vulnerable to a range of threats, including natural hazards and human-caused threats. It is therefore essential for water utilities to provide for monitoring, predicting, and assessing of these threats and to provide an effective early detection system to address each type of threat.

For potential harmful contaminants to reach the consumer taps, the contaminants must appear within the water source or within the pipe. Harmful contaminants can be injected into the water supply via any connection of the water distribution system, such as pumps. Pumps allow the inflow of contaminants because they override the system pressure. On the other hand, backflow is obstacle. It is thus unclear as to whether the contaminants will stay in the system and reach consumer taps. Therefore, the detection and tracking of contaminants in the water supply must be performed by integrating information about the hydraulic and water distribution networks [27].

Identification of water quality monitoring points

The functions of monitoring points in the water distribution system vary in practical engineering. The following three types of monitoring points were selected:

1. Water quality monitoring points at the output of water treatment plants, which not only can directly measure the water quality to the water distribution system, but also provide the guidance for water production and the adjustment of the chlorine dosage.

2. Water quality monitoring points at areas of the water distribution networks that easily deteriorate under normal circumstances. These points are normally distributed at the end of regional, branched networks, the boundaries of the water supply, and at areas where pipe corrosion is serious. At water supply boundaries, water may be mixed from several water supplies, but these boundaries are change over time according to water hydraulic scenarios. Monitoring points are necessary in some serious pipe corrosion regions, where the water quality impact from these pipes will pass to the downstream nodes.

3. Water quality representative points, where these points should be a major proportion of the selected monitoring points.

Monitoring items

Turbidity is most commonly used as an indicator; it is a direct reflection of water quality. Chlorine is another important indicator to ensure water safety. To ensure the water quality, maintaining a certain level of residual chlorine in the water distribution system can reduce the possibility of microbial contamination [8]. On-line monitoring of residual chlorine as a water quality monitoring method is an essential supplement.

Optimization of water quality monitoring

The method of a water age index of nodes is adopted for water quality on-line monitoring, which integrates the water age of nodes with the attributes of the pipes. On the basis of the calculated water age index, the pipe attributes are converted to water ages according to the effect of pipe attributes on water quality. The water age of a node gained by hydraulic calculation combined with the converted water age from the pipe effect can be expressed as the water quality of the node. Water quality on-line monitoring points are selected by assessing the interrelationship between these nodes.

The principle of optimizing on-line monitoring using a water age index can be expressed as follows: if node i and j have absolutely the same water age index, then they also have the same or nearly the same water quality. That is to say that node

j can stand for node i. Given a standard for monitoring CC (criteria concentration), the nodes with the greatest volume of coverage will be the monitoring points.

If we select $CC = 0.6$ as the selection criteria, the procedure for the selection of monitoring points is as given below:

- First, $L = (l(i))$, for $i = 1, 2, \ldots, n$, where L is the number of selection points, and $l(i)$ is defined as:

$$l(i) = \begin{cases} 1 & \text{choose node } i \text{ as monitoring point} \\ 0 & \text{do not choose node } i \text{ as monitoring point.} \end{cases} \tag{19.6}$$

- The monitored water flow Q can be obtained as follows:

$$Q = \sum_{i=1}^{n} DC(i) = \sum_{i=1}^{n} \left\{ l(i) \sum_{j=1}^{n} d(j) R[CC](i, j) \right\}, \tag{19.7}$$

where $DC(i)$ is the demand coverage in node i, $d(j)$ is the flow in node j, and

$$R[CC] = \begin{cases} 1 & \text{for } CC \leq T_\lambda(i, j) \leq 1/CC \\ 0 & \text{otherwise,} \end{cases} \tag{19.8}$$

where $T_\lambda(i, j) = T_{\lambda_j}/T_{\lambda_i}$, for $i = 1, 2, \ldots, n$, $j = 1, 2, \ldots, n$, and $R[CC]$ is the matrix of monitoring coverage for the water age index of nodes; T_{λ_j} is the water age index of node j, T_{λ_i} is the water age index of node i; $T_\lambda(i, j)$ is the ratio of the water age of node i and the water age of node j. This ratio of water age indexes of the two nodes clearly represents the correlation of the water quality of the two nodes.

- Because the consumption of water changes according to different water supply scenarios, so the locations of selected monitoring nodes are different when the scenarios change. Therefore, a weight factor C is introduced for different scenarios. Assuming there are k scenarios of the water distribution in a cycle, the consumption of water is described as D_i, water supplying time is defined as t_i, then

$$C = [C_1, C_2, \ldots, C_k], \tag{19.9}$$

where

$$C_i = D_i t_i / \sum_{i=1}^{k} D_i t_i. \tag{19.10}$$

- Finally, the objective function of optimized monitoring locations and the constraints are given as:

Figure 19.4: The monitoring allocations of the Tianjin water distribution system (the blue lines are the main supply routes, the blue squares are the water treatment plants, and the pink circles are water quality on-line monitoring points).

$$\max Q_m = \sum_{i=1}^{k} \alpha_i C_i Q_i \qquad (19.11)$$

$$\text{subject to } \sum_{i=1}^{n} l(i) \leq N, \qquad (19.12)$$

where Q_m is the maximum water coverage flow, Q_i is the flow of node i, $l(i)$ are selected monitoring points, k is the number of scenarios, and N is the number of water nodes. Based on this equation, the selected monitoring points, which have maximum demand water coverage, can be obtained; these selected monitoring points are the optimized points.

Case study

An application was undertaken in a real water distribution system. The water supply network extends 4,320 km. The pipe material is mainly cast iron and ductile iron. It has 12 main pipes, 6860 pipes, 5397 nodes, and 1464 loops.

The selection of on-line water quality monitoring points aims to create a water quality model and to realize real time, remote monitoring of the water quality in the water distribution system. The water quality model was based on the selected monitoring points and densities.. Thirty-five on-line monitoring points were set up in the water distribution system as shown in Figure 19.4. Data transmission methods such as industrial network and CDMA (Code Division Multiple Access) were adopted for the long-range wireless data transmission mode. A HACH 1720 series Turbidity Monitoring Instrument was selected as an on-line monitoring instrument. The residual chlorine measuring equipment HACH CL17 was adopted in the monitoring principles as it meets the internationally recognized standards for monitoring instruments.

19.4 Intelligent assessment in water distribution systems

19.4.1 Hydraulic accident assessment

The possible consequences of hydraulic accidents have two aspects: one is the water leakage at the point of the accident and the state of the surrounding environment; the other is the influence of accidents on the water distribution network. The evaluation of the first aspect reflects the severity of the accident itself, and the evaluation of the second aspect reflects the state of various affected parameters when the accident occurred. The seriousness of the accident can be estimated by making a comprehensive evaluation of these two aspects. However, these two aspects are complex and interact with each other, making it difficult to build an evaluation model. Considering the characteristics of water distribution network incidents, a fuzzy comprehensive evaluation model has been adopted to evaluate the hydraulic incidents [21].

19.4.1.1 Fuzzy comprehensive evaluation for hydraulic accidents

The process of fuzzy comprehensive evaluation involves creating a set of factors, weight factor sets, an optional set, a single evaluation, and a comprehensive evaluation. Based on the evaluation of the factors, different methods (the maximum degree, the weighted average, and the fuzzy distribution method) are used to determine the evaluation results for a specific factor. The flowchart of the fuzzy comprehensive evaluation for water network hydraulic accidents is shown in Figure 19.5.

19.4.1.2 A practical example analysis of hydraulic water network accidents

The accident simulation was performed in two areas (Xin Yang Road and An He Street) in Harbin. The accident location is a cast iron pipe with a 700 mm diameter, which was laid in 1977. According to the accident analysis, a depth-first search was used to find the valves that should be closed.

When the accident occurred, the pipe in the suspension region was not taken into account for the hydraulic simulation due to the closure of the valve. This would result in a change to the overall water network topology. At the same time, the flow of

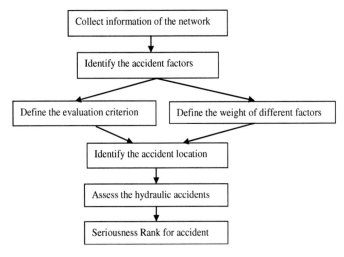

Figure 19.5: Assessment of the accident consequence.

the suspension regional nodes was zero. In comparing hydraulic calculation results of a water distribution network under normal conditions and accidental conditions, the results below can be shown.

An evaluation of the consequences of the water supply network accidents is shown in Table 19.2. The highest evaluation level was applied to the water network accident in Harbin; the level of seriousness of the hydraulic accident is 3, meaning an average level.

Table 19.2: Assessment factors degree.

First factors grade level	Secondary factors grade level	Evaluation level
Water leak and surrounding environment U_1	Water leak u_1	higher
	Surrounding environment u_2	higher affluence
	Road traffic u_3	higher
Pipe hydraulic circumstances U_2	Valve closure number u_4	less
	Cut pipe number u_5	less
	Customers density u_6	more
	Big customers number u_7	none
	Low head number u_8	more
	Node flow u_9	average
	Hydraulic gradient change u_{10}	average
Pipe water security circumstances U_3	Node age change u_{11}	average
	Pipe velocity change u_{12}	higher
	Water direction change u_{13}	average

19.4.2 Water quality assessment in a water distribution system

With the development of a water quality model, the water quality evaluation system for the water supply network has been greatly improved [13, 29]. R.K. Horton published the first indicator of water quality evaluation systems and proposed water quality evaluation methods. A series of evaluation methods for a complex system is proposed, such as the grey system theory and the matter-element analysis theory. The values of the evaluation index for real water supply networks are generally continuous real numbers. It is difficult to express the relation among the indexes, and the evaluation results may be incompatible. The comprehensive water quality category is determined according to the worst single indicator. This evaluation is too extreme, too one-sided. The results are greatly affected by subjective factors and less so by objective factors. Self-organizing features such as map neural networks (SOM) [24] and a K-means clustering algorithm [16] are used for the urban water supply network for cluster analysis and the comprehensive water quality evaluation system [28].

19.4.2.1 The selection of a water quality evaluation index for water supply networks

To create a scientific and rational clustering evaluation system for water supply networks, several indicators of water quality at the nodes and in the pipe were selected for the clustering evaluation. The research shows that there is some interrelation among the disinfectants, disinfection by-products, organic matter and bacteria. The main indicators of residual chlorine – THMs, TOC, BDOC, water age, TAAs, UV-254, demand, and pressure – are easy to monitor over the long term. Compared with other indicators, these indicators fluctuate more noticeably with the changing state of the water supply network. The indicator value for some nodes or pipes cannot meet water quality standards, so water quality deteriorates in some districts.

19.4.2.2 Water quality clustering evaluation system

The water quality clustering analysis of the water supply network is developed here. The SOM algorithm uses the batch training method, which includes the rough and fine-tuning stage. The trained network topology represents the final state of nodes of the water supply network after the process of learning and training. According to the weighted value of the trained network, the distance between the adjacent neurons is calculated. The U-matrix of the SOM network is generated in Figure 19.6(a). The plans of the variable for each indicator are received by reversing the standardization process of the variable. Each neuron in the figure contains more than one node on which the water quality is similar. The water supply network is divided into different regional types using the U-matrix, and then cluster analysis is performed using the K-means algorithm. The number of clusters is automatically selected by the Davies-Boudle index DBI, which is calculated using the following formula:

$$DBI = \frac{\sum A}{B}, \tag{19.13}$$

where A is the distance between the elements within the group; B is the distance between the groups. When the DBI is at its minimum, it is the best clustering result. The number of classes is selected. According to the weighted value of the trained SOM neurons, the state of the water quality in the nodes is classified by the K-means algorithm, and the DBI is selected as the comprehensive indicator.

19.4.2.3 Application of the water quality clustering evaluation system

A classic network example [22] is selected for the water quality clustering evaluation analysis. The input vector of the SOM algorithm is composed of the water quality and hydraulic indicator data. The hierarchical water quality indicators are t residual chlorine, THMs, TOC, BDOC, water age, TAAs and UV-254, and the hydraulic indicators are demand and pressure. After the temporary training of the typical structure, the average quality error (qe) and topology error (te) are shown in Table 19.3. The plane of the water age is shown in Figure 19.6, and the structure of the planes for the other indicators is similar.

The water quality at the nodes is classified according to the comprehensive DBI indicator. The result is shown in Figure 19.7. The average value of the water quality indicators of each class after the classification is shown in Table 19.4.

A water quality comprehensive evaluation model is then built based on the SOM and K-means algorithm. The water quality at the nodes is eventually divided into eight categories. According to the clustering results, the average value of the water quality indicators is obtained, and the water quality gradually deteriorates. It is shown that the SOM neural network has great significance for the water quality classification of the water supply network.

Table 19.3: The mean quality error (qe) and topology error (te).

Network	qe	te
7×8	5.9880	0
9×6	6.1066	0
10×5	6.0795	0.0109
11×4	6.3525	0.0326

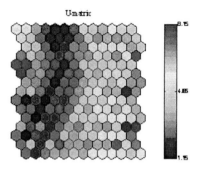

(a) The U-matrix.

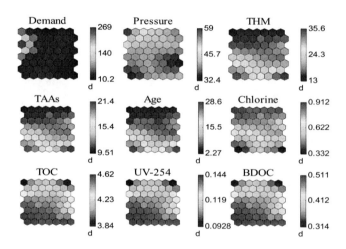

(b) The component plane figure.

Figure 19.6: The U-matrix and the component plane figure.

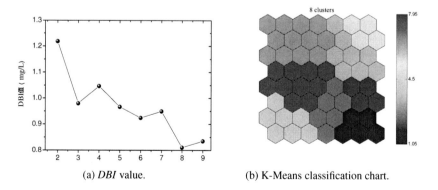

(a) *DBI* value.   (b) K-Means classification chart.

Figure 19.7: The result of comprehensive evaluation.

Table 19.4: The water quality index average value of each class node.

No.	residual chlorine (mg/L)	Water age (hr)	THM (μg/L)	TOC (mg/L)	BDOC (mg/L)	TAAs (μg/L)	UV-254 (cm^{-1})
1	0.81	6.10	16.50	4.08	0.28	7.83	0.104
2	0.74	8.99	19.42	4.10	0.31	8.48	0.105
3	0.71	9.55	20.39	4.13	0.30	8.73	0.110
4	0.67	12.14	22.36	4.16	0.32	9.29	0.114
5	0.57	14.79	25.36	4.23	0.35	11.69	0.118
6	0.56	21.25	26.92	4.23	0.36	12.57	0.118
7	0.52	25.17	29.06	4.25	0.39	13.93	0.120
8	0.17	43.45	44.65	4.28	0.45	19.23	0.122

19.5 Multi-objective optimal pump scheduling

Typically, a pumping station consists of a set of pumps with varying capacities, including constant pumps and variable speed pumps, used to pump water to the entire distribution system. Optimizing the pumping schedule has proven to be a practical and highly effective method for reducing costs without making changes to the physical infrastructure of the entire system. Depending on the number of variables and objectives considered, the problem of optimizing the pump schedule may become very complex, especially for large systems.

Multi-objective evolutionary algorithms (MOEA) have been introduced into the study of the optimal pump scheduling problem [1, 18]. Barán [1] considered four objectives to be optimized within a hydraulic model of five constant-speed pumps and compared six different MOEAs. Carrijo [2] evaluated two basic objectives, the economic objective and the objective of hydraulic benefits of the water distribution systems. In addition, Savic [25] applied MOGA to optimize the pump schedule for a system consisting of four constant pumps. However, few researchers have studied real pump systems with both constant and variable-speed pumps.

19.5.1 Multi-objective optimal pump scheduling problem

Here the optimization model implemented takes three objectives into account: the minimization of operation costs including energy costs and treatment costs, the minimization of maintenance related to the number of pump switches, and the maximization of the hydraulic service level.

Operation cost (f1)

The electrical energy cost is the cost of electrical energy consumed by all pumps at the pumping station. It is not constant throughout the day, so the rate (tariff) change is taken into account. The treatment cost accounts for a certain proportion of the whole water supply system, so it is also considered.

$$f_1 = \sum_{t=1}^{T}\sum_{i=1}^{I} S_i Q_{i,t} + \sum_{t=1}^{T}\sum_{k=1}^{K} \frac{C \cdot NP_{k,t} QP_{k,t} HP_{k,t}}{\eta_{k,t}} SP_{k,t}, \qquad (19.14)$$

where S_i is the treatment cost of pump station i per cubic meter, $Q_{i,t}$ is the discharge of pump station i at time interval t, I represents the total number of pump stations, T represents the total time steps of operational time horizon. Moreover, $SP_{k,t}$ is the electricity tariff at time interval t, $NP_{k,t}$ is the on-off state of pump k at time t, K is the total number of pumps in all pump stations, C is the unit conversion factor, $HP_{k,t}$ is the hydraulic head of pump k at time t, $QP_{k,t}$ is the flow rate of pump k, and $\eta_{k,t}$ is the efficiency of pump k at time t.

Maintenance cost (f2)

Pump maintenance costs are mainly attributed to wear and tear of pumps caused by frequently switching them on and off. However, a safe assumption is that maintenance costs increase with the number of pump switches. Therefore, a surrogate objective, the number of pump switches, is considered to be representative of pump maintenance costs. The total number of pump switches is computed simply by adding the number of pump switches at every time interval:

$$f_2 = \sum_{j=1}^{J}\sum_{k=1}^{K} \left(\max\left\{ 0; NP_{k,j} - NP_{k,(j-1)} \right\} \right), \qquad (19.15)$$

where $NP_{k,j}$ is the on-off state of pump k at time j, and $NP_{k,j-1}$ is the on-off state of pump k at time $j-1$.

Hydraulic service level (f3)

Sufficient water pressure within the system should be maintained. Conversely, excessive water pressure should be avoided as this increases the incidence of burst pipes and the amount of water lost from the system as a result of leakage. Based on Carrijo [2], to consider the benefit of meeting the required pressures at the demand nodes, an index called the pressure adequacy benefit $\Psi(i,t)$, was adopted using the following equation; in this study, the nodes are pressure monitoring nodes, which can represent the entire water network's pressure level.

$$\Psi_{(i,t)} = \begin{cases} \left(\dfrac{P_{a(i,t)} - P_{\min}}{P_{e(i,t)} - P_{\min}} \right)^{1/2} & \text{if } P_{\min} \le P_{a(i,t)} \le P_{e(i,t)} \\ \Psi_{(i,t)} = 0 & \text{otherwise,} \end{cases} \qquad (19.16)$$

where $P_{e(i,t)}$ is the maximum expected pressure for node i at time interval t, $P_{a(i,t)}$ is the actual nodal pressure, and $P_{\min}$ is the minimum pressure at node i. Thus, the objective function of the hydraulic service level to be maximized is expressed as:

$$f_3 = \sum_{t=1}^{24} \sum_{i=1}^{NN} \Psi(i,t), \tag{19.17}$$

where NN is the number of pressure control points.

With each of the three objectives defined, the multi-objective pump scheduling problem can be stated as follows:

$$\min(f_1, f_2) \text{ and } \max f_3. \tag{19.18}$$

Operational constraints

- **Limitation on pump station discharge**: This chischarge is limited by the pumps' capacity and the treatment capacity of the treatment plant at time interval t.

$$Q_{p,t} \leq Q_{\max p,t}, \tag{19.19}$$

where $Q_{p,t}$ is the actual discharge of pump station p at time interval t, $Q_{\max p,t}$ denotes the maximum discharge capacity of pump station p at time interval t.

- **Pressure control point constraint**: The pressure at control points must often maintain a minimum pressure level to ensure adequate water service and a maximum pressure level to reduce water leakage or bursts within a system:

$$P_{\min(i,t)} \leq P_{c(i,t)} \leq P_{\max(i,t)} \qquad \text{for } i = 1, 2, \ldots, NN, \tag{19.20}$$

where $P_{\min(i,t)}$ is the minimum control pressure at point i, $P_{\max(i,t)}$ is the maximum control pressure at point i, and NN is the number of control pressure points.

- **Pump efficiency constraint**: A pump should run in the designed high-efficiency interval, which is often required to be no smaller than a minimum efficiency:

$$\eta_{i\min} \leq \eta_i \tag{19.21}$$

where η_i is actual pump efficiency, and $\eta_{i\min}$ is the minimum efficiency required during operation.

- **Pump speed ratio constraint**: The pump speed ratio is defined as the ratio of the actual operating speed divided by its normal speed. Limited by the current and voltage of the power distribution, the pump installed with a variable-frequency drive is allowed to vary its speed within the range of minimum to maximum speed:

$$n_{i\min} \leq n_i \leq n_{i\max}, \tag{19.22}$$

where $n_{i\min}$ is the minimum speed ratio of pump i, and $n_{i\max}$ is the maximum speed ration of pump i.

- **Tank constraints**: A tank in a water distribution system must be operated within a minimum and a maximum allowable water level. The bounds on the water levels can be expressed as:

$$T_{k\min} \leq T_k(t) \leq T_{k\max}, \tag{19.23}$$

where $T_k(t)$ is the water level of tank k at time t; $T_{k\min}$ represents the minimum water level required at tank k, and $T_{k\max}$ denotes the maximum water level allowed at tank k. To ensure hydraulic periodicity for the next operating period, the tanks must be refilled to a prescribed water level at the end of a scheduling period, given as:

$$\left| T_k^{\text{final}} - T_k^{\text{start}} \right| \leq \Delta v_k, \tag{19.24}$$

where Δv_k denotes the tolerance of the final water level for tank k; in most cases, the water level of the tank's beginning and ending levels should be the same.

19.5.2 Solution methodology

The multiple objectives established require us to find a balance of good, trade-off solutions with respect to all objectives. In our multi-objective pump scheduling problem, each operation policy is a solution which represents an objective vector formed by the operation costs, the number of pump switches, and the hydraulic service level of that operation policy. The multiple objective evolutionary algorithm has been adopted to find or at least approximate the optimal Pareto set of operation policies.

19.5.3 Case study

The proposed solution technique has been applied to a real water network in Maanshan, shown in Figure 19.8. It comprises three water treatment plants, three pump stations, seven fixed pumps (1 standby), six variable-speed pumps, and one pressure control point at node 667 (minimum control pressure 25m and maximum control pressure 32 m). This network has an average daily demand of 220,000 (m^3/d). The running state of the water distribution system is monitored by a well-built SCADA system, and the pressure data form 34 monitoring sites located in the network can be obtained every 15 minutes. A hydraulic model of the water networks with pipes over DN200 mm has been built, containing 923 nodes and 1,052 pipes, and this model is precise enough to be used in solving the optimal scheduling problem. The electricity rate is uniform and there are no tanks or reservoirs in the system. The detailed data on the pump station and the pumps are shown in Table 19.5.

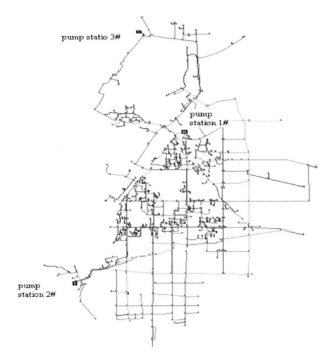

Figure 19.8: Layout of the water distribution network.

Table 19.5: Pump station setup.

PS	PNo.	Type	TC	EC	c/v
	1#	16SA—9J			c
	2#	16SA—9J			c
1#	3#	16SA—9J	0.3	0.66	v
	4#	12SH—9A			v
	5#	14SA—10B			c
	2#	24SA—10B			c
	3#	28SA—10J			c
2#	4#	28SA—10J	0.22	0.66	c
	5#	28SA—10JC			v
	6#	28SA—10JC			v
	1#	350S44			c
3#	2#	6000S47E	0.21	0.66	v
	3#	6000S47E			v

(PS: pump station, PNo: pump number; TC: treatment cost (Yuan/m^3);
EC: electricity cost (Yuan/kWh) c:constant; v:variable)

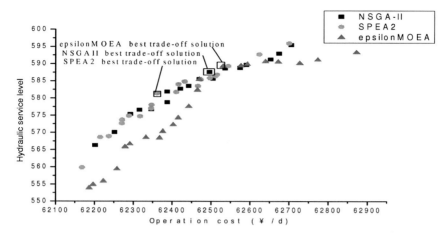

Figure 19.9: Comparison of Pareto fronts.

Table 19.6: Best trade-off solution.

Objective	NSGA-II	SPEA2	epsilon-MOEA
Operation cost	62495.8	62361.1	62531.1
Hydraulic service level	587.6	581.4	589.4
Total switch numbers	5	5	5

An accurate, reliable demand predictor for a particular water system is essential, since the pump-station optimization and related controls will be based primarily upon satisfying these predicted consumer demands for the duration of the operational control period. Data were obtained through a statistical study of the community's water demand in recent years. The scheduling time steps are not fixed, and can thus be determined by the decision makers according to the predicted water demand curve. In this study, an optimization period of 24 hours is divided into nine time intervals (0:00–5:00; 5:00–6:00; 6:00–7:00; 7:00–13:00; 13:00–18:00; 18:00–21:00; 21:00–22:00; 22:00–23:00; 23:00–24:00).

Three well known MOEAs (NSGA-II [5], epsilon-MOEA, SPEA2 [30]) combined with a repair mechanism are applied to solving the pump scheduling problem. Figure 19.9 presents a final Pareto front identified by the multi-objective evolutionary algorithm (NSGA-II, SPEA2, epsilon-MOEA) in the objective space with solutions; the final Pareto solutions of the three MOEAs have the same total number of switches, and the hydraulic service level increases with the growth of operation costs.

Once having the Pareto optimal set, decision makers need to choose the best trade-off solution from the Pareto set. Table 19.6 represents the best trade-off solutions of the three evolutionary algorithms found by the proposed approach without considering the decision maker's preference. The best trade-off solutions are also shown in Figure 19.9; SPEA2 has the minimum operation cost and epsilon-MOEA has the maximum hydraulic service level.

19.6 Water distribution system rehabilitation

19.6.1 Introduction

Water network rehabilitation is a complex problem, and many facets should be considered in its solution. Water network rehabilitation is a discrete variables, nonlinear, multi-objective optimization problem. The concepts of building decision models for this problem are varied, the main methods being general rehabilitation guides, prioritization models and optimization models [7].

Through the development of optimal theory and modeling technology, using a more comprehensive and detailed optimal model to solve the rehabilitation problem becomes possible. Optimization techniques consider the interaction of each main with the system as a whole. They allow both the performance and cost of the rehabilitated system to play a role in the formulation of the rehabilitation. The objectives and constraints vary in different optimization models, such as the network performance objective with a cost constraint or the minimization of the cost objective with a performance constraint [14, 15, 17]. These optimization techniques can also be used to consider network rehabilitation as a multi-objective problem. This approach allows a trade-off between system performance and the cost of rehabilitation. However, such techniques require large numbers of trial evaluations to obtain near-global optimal solutions. By using models of the water network system and new optimization methods (GA, particle swarm optimization (PSO), non-dominated sorting Genetic Algorithm (NSGA), NSGA-II), an optimization model of rehabilitation becomes possible.

19.6.2 Optimal rehabilitation model

The popular optimization model for the rehabilitation of water supply networks uses a minimization of rehabilitation costs and energy costs per year in the investment period as an objective, and hydraulic performance of the network as a constraint. With this concept, the general form of the rehabilitation optimization model can be expressed by the equations below [9].

It is a one-objective model with a minimization of construction costs and energy costs as an objective, and network performances and available diameters as constraints. In conventional solving methods, constraints will transform parts of the objective function by a weighting or ε-constraint method [6]. The goal of obtaining the optimal objective value with a limitation of constraints is thus achieved. This transformation can be used in resolving optimization models for water supply network rehabilitation. Unfortunately, the solution obtained by this process depends largely on the values assigned to the weighting factors used or the design of the

ε-constraint function. This approach does not provide a dense spread of the Pareto points. Therefore, the best way to solve the rehabilitation problem is to abstract the problem as the multi-objective functions given below, and to solve these functions with a true multi-objective oriented algorithm.

The concept of alteration of the optimization model for rehabilitation regards the constraints that transformed parts of the objective function with a weighting method or a ε-constraint function as an isolated objective function. Therefore, the constraints which represent the performance of the water supply network should be expressed as an objective function.

The objective function of the pipe velocity is given by

$$\min W_1 = \sum_{i=1}^{N} (P_i const + EVV_i), \tag{19.25}$$

where $P_i \in \{0, 1\}$ shows whether the velocity of pipe i meets the velocity limitation; 0 means satisfied, 1 means not satisfied; $const = 10$ and EVV_i refers to the part of the velocity of pipe i that exceeds or is less than the limitation.

The objective function of node pressure is expressed as follows:

$$\min W_2 = \sum_{j=1}^{J} (J_j const + EPV_j), \tag{19.26}$$

where $J_j \in \{0, 1\}$ shows whether the pressure of node j meets the pressure limitation; 0 means satisfied, 1 means not satisfied; $const = 10$ and EPV_j refers to the part of the pressure of node j that exceeds or is less than the limitation.

The new multi-objective optimization model can thus be expressed by the following objective functions:

$$\min W_1 = \left(P + \frac{m}{100}\right) \sum_{i \in N} (a + bD_i^{\alpha}) L_i + 3.58 \sum_{i=1}^{3} (r_i E_i T_i) \sum_{j \in N_s} \frac{H_{i,j} Q_{i,j}}{\eta_{i,j}} \tag{19.27}$$

$$\min W_2 = \sum_{i=1}^{N} (P_i const + EVV) \tag{19.28}$$

$$\min W_3 = \sum_{j=1}^{J} (J_j const + EPV) \tag{19.29}$$

and the following constraints:

$$Q_i - \sum_{j \in V_i} q_{ij} = 0 \qquad \text{for } i = 1, 2, \ldots, n \tag{19.30}$$

$$\sum (h_{ij})_1 = 0 \tag{19.31}$$

$$\sum (h_{ij})_2 = 0 \qquad\qquad\qquad (19.32)$$

$$\vdots$$

$$\sum (h_{ij})_l = 0 \qquad\qquad\qquad (19.33)$$

$$d_i \in D = \{D_1, D_2, \dots, D_z\}. \qquad\qquad (19.34)$$

19.6.3 Solution

Of the several methods (the goal attainment method, the ε-constraint method, and versions of NSGA) available to solve multi-objective optimization problems [3], NSGA-II [5] was used here to obtain the Pareto set. A penalty function is commonly used for handling constraints. However, tuning the penalty parameter appearing in the penalty function is very time consuming and is normally based on trial and error. Unless tuned properly, one may be completely misdirected in the search space. The NSGA-II based constraint-handling technique negates this problem.

For the optimization model of rehabilitation, the purpose of the content of artificial inducement mutation (AIM) is to make the selected diameters of rehabilitated pipes meet the velocity constraint until the solutions converge to the feasible field [11].

The multi-objective optimal model of water supply network rehabilitation was solved with NSGA-II and AIM. In fact, there are several different concepts involved in building multi-objective optimal models for water supply network rehabilitation. NSGA-II and AIM can still work in solving other multi-objective optimal models that build in other concepts. By introducing AIM, the population is directed to the feasible solutions rapidly, and the search for the best solution is performed in the feasible field. In short, the convergence of algorithms has greatly improved and can offer better and more feasible solutions.

19.7 Conclusions and future research

This chapter has discussed the problems and challenges of urban water infrastructures. It began with the fundamentals of hydraulic and water quality monitoring and modeling, the water demand forecasting of water distribution networks, the development of optimal placement systems of monitoring stations for hydraulic and water quality and the creation of a comprehensive model for a water distribution network. Reliability analysis of water distribution networks and intelligent assessment for hydraulic and water quality using fuzzy and intelligent agents were investigated and practically applied. Operation of multiple-objective optimal scheduling pump stations in the water distribution network was explored for energy efficiency. The multi-objective optimal rehabilitation of the water distribution network model and solutions were developed for the need of rapidly increasing demand, maintenance, and improving services. By using various intelligent techniques and new technology, it provides the feasible solutions, references and decision-making tools needed by companies in the water industry. It makes water infrastructure operation more intelligent and resilient to global climate changes and increasing population. It will enhance the security and reliability of urban water infrastructures. Future research focuses on further developing and assessing the approach proposed in this chapter.

Acknowledgements The authors are grateful for the support received from the China-UK science network, the Royal society UK, the National Science Foundation of China, and other collaboration projects with water companies.

References

1. B. Barán, C. von Lücken, and A. Sotelo. Multi-objective pump scheduling optimization using evolutionary strategies. *Advances in Engineering Software*, 36(1):39–47, 2005.
2. I. B. Carrijo, L. F. R. Reis, G. A. Walters, and D. Savic. Operational optimization of wds based on multiobjective genetic algorithm and operational extraction rules using data mining. In *Proceedings of the World Water and Environmental Resources Congress*, 2004.
3. V. Chankong and Y.Y. Haimes. *Multi-objective Decision Making Theory and Methodology*. Elsevier, New York, New York, 1983.
4. L. Davis. *Handbook of genetic algorithms*. Van Nostrand-Reinhold, New York, New York, 1991.
5. K. Deb, A. Pratap, S. Agarwal, and T. Meyarivan. Fast and elitist multiobjective genetic algorithms: NSGA-II. *IEEE Transactions on Evolutionary Computation*, 6(2):182–197, 2002.
6. L. Duan, J. B. Yang, and M. P. Biswal. Quantitative parametric connections between methods for generating noninferior solutions in multiobjective optimization. *European Journal of Operational Research*, 117(1):84–99, 1999.
7. M. O. Engelhardt, P. J. Skipworth, D. A. Savic, A. J. Saul, and G.A. Walters. Rehabilitation strategies for water distribution networks: A literature review with a UK perspective. *Urban Water*, 2:153–170, 2000.
8. N. B. Hallam, J. R. West, C. F. Forster, and J. Simms. The potential for biofilm growth in water distribution systems. *Water Research*, 35(17):4063–4071, 2001.
9. Z. Hong-bin. *Water network system theories and analysis*. China Architecture & Building Press, Beijing, China, 2003.
10. X. Jin, J. L. Gao, and W. Y. Wu. Optimal placement of pressure monitors in water supply network with elitist genetic algorithm. *International Journal of Modelling, Identification and Control*, 4(2):68–75, September 2008.
11. X. Jin, J. Zang, J. L. Gao, and W. Y. Wu. Multi-objective optimization of water supply network rehabilitation with non-dominated sorting genetic algorithm–II. *Journal of Zhejiang University SCIENCE A*, April 2008.
12. G. Jin-liang and J. Xi. Applications of GPS technology in water distribution system. In *Proceedings of the Pipelines 2004 International Conference*, San Diego, California, 2004.
13. C. D. Jones, A. B. Smith, and E. F. Roberts. Water quality status measured by node age in distribution network. *Water & Wastewater*, 28(15):36–38, 2002.
14. J. H. Kim and L. W. Mays. Optimal rehabilitation model for water distribution systems. *Journal of Water Resources Planning and Management*, 120(5):674–691, 1994.
15. Y. Kleiner, B. J. Adams, and J. S. Rogers. Selection and scheduling of rehabilitation alternatives for water distribution systems. *Water Resouces Research*, 34(8):2053–2061, 1998.
16. R. J. Kuo, H. S. Wang, T.-L. Hu, and S. H. Chou. Application of ant k-means on clustering analysis. *Computers and Mathematics with Applications*, 50:1709–1724, 2005.

17. K. E. Lansey, C. Basnet, L. W. Mays, and J. Woodburn. Optimal maintenance scheduling for water distribution systems. *Civil Engineering Systems*, 9:211–226, 1992.
18. M. López-Ibáñez, T. D. Prasad, and B. Paechter. Multi-objective optimization of the pump scheduling problem using SPEA2. In *Proceedings of the IEEE Congress on Evolutionary Computation*, pages 435–442, September 2005.
19. K. Makawa and T. Furuhashi. Cooperation and evolution of scheduling system with genetic algorithms. Technical report, Department of Information Electronics, Nagoya University, Nagoya, Japan, 1996.
20. K. A. Nilsson, S. G. Buchberger, and R. M. Clark. Simulating exposures to deliberate intrusion into water distribution systems. *Journal of Water Resource Planning and Management*, 131(3):228–236, June 2005.
21. Z. Qian, J. L. Gao, W. Wu, X. Q. Hou, Y. B. Wu, L. Tao, and N. Ran. Harbin water supply network monitoring and evaluation. In *Proceedings of the 2007 IEEE International Conference of Networking, Sensing an Control*, pages 345–349, London, UK, April 2007.
22. L. A. Rossman. EPANET Users Manual, 1993.
23. L. A. Rossman and P. F. Boulos. Numerical methods for modeling water quality in distribution systems: A comparison. *Journal of Water Resource Planning and Management*, 122(2):137–146, 1996.
24. J. Vesanto and E. Alhoniemi. Clustering of the self-organizing map. *IEEE Transactions on Neural Networks*, 11(3):586–600, 2000.
25. D. A. Savic G. A. Walters and M. Schwab. Multiobjective genetic algorithms for pump scheduling in water supply. In *Proceedings of the AISB International Workshop on Evolutionary Computing*, pages 227–236, Berlin, Germany, April 1997.
26. L. D. Whiley, T. Starkweather, and D. A. Fuquay. Scheduling problems and traveling salesman. In *Proceedings of the 3rd International Conference on Genetic Algorithms*, pages 133–140, 1989.
27. W. Wu, J. L. Gao, M. Zhao, Z. Qian, X. Q. Hou, and Y. Han. Assessing and optimizing online monitoring for securing the water distribution system. In *Proceedings of the 2007 IEEE International Conference of Networking, Sensing an Control*, pages 350–355, London, UK, April 2007.
28. W. Y. Wu, K. Chang, J. Gao, M. Zhang, N. Li, and Y. Yuan. Research on water quality comprehensive evaluation of water supply network using SOM. In *Proceedings of the 2007 IEEE International Conference of Networking, Sensing an Control*, Okayama, Japan, March 2009.
29. W. Y. Wu, B. Ulanicki, B. Coulbeck, and H. Zhao. A system approach to water quality modeling and calibration. In *Water software systems: Theory and Applications*, volume 1, pages 275–285. 2001.
30. E. Zitzler, L. Marco, and L. Thiele. SPEA2: Improving the strength pareto evolutionary algorithm. Technical Report TIK-Report 103, Computer Engineering and Networks Laboratory, ETH Zürich, Zürich, Switzerland, 2001.

Chapter 20
Long-term Sustainable Use of Water in Infrastructure Design

R. van der Brugge and J. van Eijndhoven

Abstract The key question addressed in this chapter is: how we can design and use infrastructure sustainably? This means that we do not seek to optimize the current infrastructural systems, but that we deal with the question of the role of infrastructure in long-term societal development and how we can deal with infrastructure in a more sustainable way. In this chapter we take the three-layers approach of Dutch spatial planning as a starting point: underground, infrastructural networks and the occupation layer. We look into the history of the city of Rotterdam in order to generate insight into the co-evolutionary process of these three layers and how periods of transition and lock-in alternate. We contend that sustainable design of infrastructure involves two qualities: first, it integrates the three layers in a good fashion, and, secondly, the design is flexible so it can adapt to changing social and environmental circumstances. We can conclude the following: infrastructure has a dual role. Existing infrastructure can induce lock-in, however, new infrastructures can give an impulse to transition. The infrastructure is not the trigger for transition, but a result of a city vision that directs the design of infrastructure. Infrastructure can thus be used to transform the occupation layer and to continue that purpose it should be flexible as social and environmental conditions change over time.

R. van der Brugge, J. van Eijndhoven
Erasmus University Rotterdam, Drift, Dutch Research Institute for Transitions, Rotterdam, The Netherlands, e-mail: {VanderBrugge,VanEijndhoven}@fsw.eur.nl

R.R. Negenborn et al. (eds.), *Intelligent Infrastructures*, Intelligent Systems, Control and Automation: Science and Engineering 42, DOI 10.1007/978-90-481-3598-1_20, © Springer Science+Business Media B.V. 2010

20.1 Introduction

In this chapter we focus on designing sustainable physical infrastructure, by taking stock of the principle of the Dutch three-layers approach in spatial planning: the underground, the infrastructure, and the occupation layer. We adopt a long-term perspective as to see what the role of infrastructure is in long-term social development. The perspective of this chapter thus differs somewhat from the other chapters of this book providing a 'larger picture'.

Historically, the way a region is being occupied, was strongly dependent on the underground and on the infrastructure. In a country like The Netherlands, which is a river delta, water is an important structural element, which can be seen as part of both the underground as well as the infrastructural network. The 16th century shows a tendency to substitute water for land, since the land is deemed more valuable as an underground for the occupational layer than the water. Large land reclamations turned lakes into dry polder systems. Starting in the 19th century there is even a tendency to perceive the underground as a 'blank slate' for the occupational layer. The desired underground is created by bringing in soil from elsewhere and prepared for infrastructure and settlements despite the characteristics of the original underground. Developing technology seemed to endlessly enhance the possibilities. But now, we are experiencing that this way of operating might not be the most sustainable route, and that it would be more sustainable to re-establish a more intimate connection between the three layers. However, changing the infrastructure is problematic, since infrastructure has a life span of decades and is characterized by considerable sunk cost.

The key question addressed in this chapter is: how can we design and use infrastructure sustainably? This means that we do not seek to optimize the current infrastructural systems, but that we deal with the question of the role of infrastructure in long-term societal development and how we can deal with infrastructure in a more sustainable way. To answer this we should improve our understanding of the dynamic interplay between the three layers in order to find leverages for management. To this end, we introduce the concepts of co-evolution, lock-in, transition, and resilience, that enable us to describe the dynamic interplay between the layers.

Based on these concepts, we will derive two main criteria on a conceptual level for sustainable use of infrastructure. We will further illustrate what these criteria mean by looking into the history of the city of Rotterdam in the Netherlands with regard to the three layers. The city of Rotterdam is an interesting case in this respect. Rotterdam started as a dike settlement around 1270 AD, and it has grown into the second largest city of the Netherlands, harboring the largest port of the world until recently. Part of its success can be explained by the interaction of the three layers and how the resulting infrastructure and settlement attracted new people and new commerce. The city increased in scale and transformed, leading to a new interaction pattern between the layers, requiring new infrastructures and so on. The city of Rotterdam thus provides the opportunity to analyze the interaction between the layers of underground, infrastructural network, and occupation. This chapter thus pursuits a new way – both conceptually as well as empirically – of looking at the

Figure 20.1: Three layers from top to bottom: a) occupation layer; b) network layer; c) underground (source figure: http://www.ruimtexmilieu.nl/).

role of infrastructure in societal change processes and how we may deal with this in a smart way.

20.2 Conceptual framework

In order to provide a conceptual basis for what we mean by dealing intelligently with infrastructure, we use the following five concepts: the three-layer approach, co-evolution, transition, lock-in and resilience. Each concept is outlined below, and together they form a framework from which we will derive two main criteria for sustainable use of infrastructure.

20.2.1 The three layer approach

The first concept is the three-layer approach which is used in spatial planning in the Netherlands. The White paper on Spatial planning of the Dutch government introduced the so-called 'three-layer approach': underground, the network layer, and the occupation layer [14] (Figure 20.1).

The *underground* is the system of water, soil and the living creatures in the soil. Processes in the underground are part of cycles (water, energy, various biochemical cycles) operating at a local, regional, and sometimes even global scale. On a macro level the processes in the underground look slow, certainly in comparison to events in the other two layers. However, the underground also harbors important relationships to ecosystems and the development of ecosystems.

The condition of the underground also depends on geological and climatic conditions. The Netherlands is part of a river delta area, where a number of large European rivers flow into the North sea. Long-term processes like climate change and changes in the height of the ground level (e.g., compaction of the peaty soils, when they are dry) have a strong influence on the underground, and especially on the operation of the water system. In The Netherlands water also plays an important role in the formation of the soil by sedimentation and removal of soil by water from rivers, and the coastal formation of dunes and sandy stretches by the tidal movements of the sea.

The *network layer* consists of physical infrastructures that canalize mobility and transport, like roads and railroads. But there are also connections that are not directly visible, like ICTs, and air and water routes. The physical infrastructure is the whole of roads, railroads, water connections, harbors, airports, hubs, and underground piping.

The *occupation layer* is the layer of human settlement and it is about the way in which people settle and use the underground and the infrastructures. Cultural preferences and habits and especially policy have a large influence on the way the three layers are being shaped [14].

To a large part the underground may be conceived of as given, for instance, while The Netherlands is a flat delta, Switzerland is a mountain range. Evidently, the underground is of major importance with respect to the way in which the other layers can be filled in. The Netherlands is a special case with respect to the fact that the underground has been manipulated and changed drastically by the Dutch: in earlier decades domestic lakes have been turned into polders and rivers and brooks have been canalized.

In general, the infrastructural networks are strongly coupled to the underground and to the occupation layer. The network layer is more or less the connecting medium to the underground and the occupation layer. In principle the occupation layer is the most dynamic and changing layer, however any area has an historical path dependence.

20.2.2 Co-evolution

The second concept is addressing this issue of interaction between the three layers. The three layers evolve over time and so does their interaction. Each layer has its own typical mechanisms of change, but the dynamics of change for each layer have typical time scales: the time frame of change of the occupation layer is 20–40 years; infrastructure changes over 20–80 years and changes in the underground may take more than 100 years.

The interaction between the three layers can be understood as a process of co-evolution: change at one layer triggers change at the other layers. Changes at the two other layers then influence the first layer. Hence, the trajectory through time is the result of a co-evolution between the three layers. Co-evolution refers to a specific kind of interaction, namely a selective interaction: the development at one layer enables and constrains – and therefore selects – developments at the other levels.

20.2.3 Lock-in and Transition

History shows that such evolutionary trajectories are non-linear. There are alternating periods of slow change and rapid transition. The third and fourth concepts describe these patterns of slow and rapid change. Slow change is the result of constraints between the layers. Each layer develops and as the area develops, for instance the growth in volume of a city, as a result of the constraints the interaction between the layers remains the same. This pattern is referred to as lock-in.

Transitional periods can be defined as: "a long-term continuous process of societal change during which the structure of society, or a subsystem of society, fundamentally changes" [11]. A transition refers to a specific kind of change, namely the kind of change we tend to depict as 'structural', 'fundamental', or 'transformative' in contrast to 'incremental change' or 'optimization' during a lock-in. In our three layer approach a transition is the result of the layers reinforcing the changes at each level.

Periods of transition may occur as a result of a number of causes. The first cause is when the slowly changing underground is no longer able to support the network and occupation layer, for instance as a result of sediment deposition or subsidence of soil. Secondly, it can also be triggered by the discovery of a new technology or infrastructure, such as railroad or ICT triggered transitions in history. In the third place, it can also be triggered by a disaster which destroys parts of the infrastructure and occupation layer, which opens up new opportunities for development. In the fourth place, old infrastructure may be in need of renovation, which also brings opportunities for renewal.

20.2.4 Resilience

The final concept is that of resilience [4, 5, 15]. Resilience as a concept is mainly used in the ecological literature, but is becoming increasingly central to the literature on combined human-environment systems [8, 9] or combined social-ecological systems. The three layer concept as such is a description of a social-ecological system. The resilience of a social-ecological system is currently understood as (1) the amount of disturbance a social-ecological system can absorb, (2) the degree to which a social-ecological system is capable of reorganization, (3) the degree to which a social-ecological system can increase the capacity for adaptation through learning [4].

What does this mean? In non-resilient systems, disasters, such as fires, floods, earthquakes, or stock market crashes lead to collapse of the system. In contrast, in highly resilient systems, disturbances may create the opportunities for renewal [1]. Social-ecological systems in a particular state may thus be more or less vulnerable to disturbances. In the three layer approach, the relations between the three layers can range from very flexible to very rigid. When the relations between the layers are flexible, the variability at one layer does not have direct consequences for the other layers. When the relations between the layers are very rigid, then the layers constrain variability at the other layers. However, if variability occurs, for instance through a disturbance, then the variability jumps onto the other layers as well, initiating a cascade of changes and the possible start of a transition. During the transitional period the relation between the layers loosen up and become more flexible. The system eventually stabilizes as the result of the new relations between the layers becoming more rigid again and the system shifts into a new lock-in pattern.

20.2.5 Criteria for sustainable use of infrastructure

The question can be posed whether some ways of relating the three layers can be seen as more sustainable than others. Based on the five concepts above, we take sustainable use of infrastructure to be subject to the following two criteria:

1. Infrastructure integrates the underground and the occupation layer. All other things being equal, a spatial social-ecological system will function more sustainable if the three layers are well interconnected.

2. The infrastructure should be designed in such a way that the relationships between the three layers are flexible and do not become too rigid. The system is less prone to collapse as a result of external disturbances of sorts, or the changing underground or upheavals in the occupation layer. Infrastructural design should be easily adaptable to the requirements of the underground and the occupation layer.

In order to improve our understanding of sustainable infrastructure and what this means in practice, we will investigate the relations between the layers and their dynamic interplay over time. With the above concepts in mind, we will look into the history of the city of Rotterdam and generate insight into the co-evolutionary process of the three layers. We adopt a long-term perspective in order to identify a number of periods of lock-in and transition.

20.3 Water as a central element in the history of The Netherlands

Before we will describe the case of Rotterdam we will outline two broad transitions in the way the three layers interacted in The Netherlands. In [6], different phases in the history of Dutch cities with respect to the relation between water and the occupation layer are discerned. In these consecutive periods a typical relationship can be found between the underground, the network layer and the occupation layer, shaped by the technical options and the social circumstances. Till about the year 1000 AD in the low lying parts of the Netherlands people settled on higher stretches of land where they would not normally be inundated by the sea or the rivers. The rules of the water determined where settlements arose. In this first period large settlements did not yet develop in the lower lying parts of The Netherlands.

This was no longer the case after 1000 AD as a transition took off towards a new way of relating to the underground. People started to intervene in the natural ways of the water system, by defending themselves against flooding through dikes and re-directing water flows where they did not interfere with other activities, e.g., by building culverts. In the beginning such measures were taken individually, but soon the scale of measures required cooperation. The network of waterways was a good means for the transport of goods and it offered ways to defend oneself against enemies. We can conclude that city development in the low lying parts of the Netherlands became possible once cooperative means of managing the water situation became available The same means could be used for transport and commerce. A clear example of co-evolution.

In the 19th century another transition took place: water as transport infrastructure started to give way to roads and railroads, reducing the importance of water as part of the network layer. As a consequence a new evolutionary path was taken, during which the water became less useful as a functional infrastructure and less visible. By the end of the 20th century, however, the Netherlands found itself in a lock-in, in which the lower lying parts were losing their resilience rapidly with respect to climate variability and climate change. Currently, a transition is going on in which the relationships between the layers are being changed by transforming the water infrastructure in the network layer and to give the water its proper place. The city of Rotterdam exemplifies these general developments.

20.4 Rotterdam, a water-related history

The history of Rotterdam is intimately connected with water. It can be seen as a example of how underground, networks and occupation interact and can lead to transitions into a relatively sustainable situation as well as situations of lock-in.

The growth of the city till the 17th century could be absorbed within a triangle formed by two branches of the river Rotte (now Coolsingel and Goudsesingel) [10]. After the 80-year war with the Spanish, strong growth enlarged the need for harbors and city development. Like in many other cities, city growth started to spread beyond the original dikes along the river Rotte. New polders were developed in the neighboring areas. An important new development was also the range of investments done in the areas outside the dike, directly along the river Meuse, what was called the 'Watertown': the part of the city on the riverside of the Meuse [7].

The expansion of the market and commerce was followed by expansion of housing needs in the city triangle. At the end of the 17th century most areas in the triangle were in full use. In the beginning of the 17th century, new wharfs had been designated to settle around a specific location in the "Scheepsmakershaven-Boompjes" quarter, which would locate about 30 of them. However, more than half of those

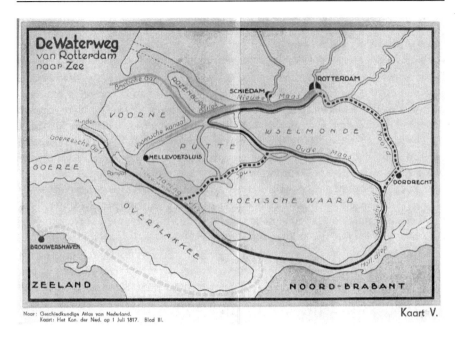

Figure 20.2: The water infrastructure from Rotterdam to the sea around 1850. Large ships had to take the black lined route (taken from [3]).

were never realized [12] turning it into an opportunity to transform this stretch of land into a luxury settlement area. In this way, rich citizens would settle in town [12]. The plan of course did not stop all citizens from moving out of town, but it was indeed an example of a sustainable coupling between the three layers. The characteristics of this area were attractive to wealthy citizens, and through the attractive surroundings, supported in a positive way the inner city development. In parallel, in the inner city more room was being created for market places, shops and extra stories on existing houses.

From the end of the 17th century until well into the 19th century the city of Rotterdam had a steady growth. The infrastructural network and the occupational layer slowly expanded. However, the dynamic interplay between the underground, infrastructural network, and the occupation layer did not change fundamentally. Therefore, we may speak of a period of lock-in. Not until the second half of the 19th century, there occurred some interesting changes in the patterns of co-evolution of underground, infrastructure, and settlement.

With respect to the underground the process of sediment deposition of silt kept going on, leading to ongoing growth of new land. Some former islands grew to become attached to the mainland. The scale of this process threatened the accessibility of the Rotterdam harbor.

The river Meuse near Brielle (The Brielse Maas) had silted up over time and so a canal was dug along the former island of Voorne as a new passage. This route was not able to handle the ships that were becoming larger and larger (see Figure 20.2) and there was a considerable risk that Rotterdam would join the earlier fate of the Flemish cities Bruges and Ghent. This was countered by excavating a new channel called the "the New Waterway" (Nieuwe Waterweg) from Rotterdam to Hoek van Holland at the sea. Initially, not enough attention was paid to the ongoing silting and so the New Waterway quickly silted up. It could only be recovered by redesigning it, keeping into mind the movement of the sandy waters [13]. This was accomplished by deepening and broadening the waterway and adding levees perpendicular to it. In this way the silt would settle further toward the sea, and not inhibit the water traffic. The levees opened up new opportunities for the settlement in the occupation layer. The levees became the basis for new harbor developments around 1900.

With respect to transport infrastructure, in the second half of the 19th century a rapid expansion of land based transport infrastructures took place. The availability of machine traction lay at the basis of the expansion. Rotterdam was initially reluctant to develop land based traffic infrastructure because of its efficient water based infrastructure. However, the threat that Rotterdam would be left out of a development that certainly would go on, finally made Rotterdam shift its stance. Also as a result of the ongoing sedimentation, the island Feijenoord in the river Meuse had become part of the southern bank of the river. The city planned to expand the city and to develop this area. The development of the North-South rail connection from Rotterdam to Dordrecht co-evolved with the settlement on Feijenoord. With the expansion of land based transport, the infrastructural function of the waterways became less important. Symptomatic for the reducing importance of the water network as transport infrastructure was that the rail connection was built along the trace of the original Binnenrotte, the place where the city of Rotterdam was first founded. Now that water in the underground did no longer fulfill a function as infrastructure for transport, it went literally out of sight.

The third development in this period is strongly related to changes in the occupation layer. Cities were becoming more densely populated; between 1840 and 1860 the population of Rotterdam grew with more than 20% [2]. As a result more people made use of surface water as a source for drinking water, but at the same time as a sewer. Especially in the relatively new polder city the discharge of the water became a problem. In the beginning of the 19th century the discharge of watery sewerage and faeces went via separate routes. Watery waste went through the waterways and solid waste was collected and used for generating fertilizer [2].

Around the middle of the 19th century cholera outbreaks were frequent in the Netherlands, and Rotterdam was one of the cities were these outbreaks started. The outbreaks in 1848–1849 and 1853 were important factors in rethinking the relationship between the city and water. The Rotterdam city council was one of the first in the Netherlands to consider the option of a flushing system. This development finally led to the development of a drinking water system and a sewerage system decoupled from the surface water.

This development started around 1840 with a plan by the city architect Rose to build new waterways in order to enhance the possibilities for inland shipping west of the city, and to enhance the water level by digging a new canal, fill up a number of smaller waterways and to start building a sewerage system for rainwater and

Table 20.1: Use of drinking water in liters per person [2].

Year	Amount of water (l)
1875	19
1880	72
1885	149
1890	196

household sewage. The plan also involved the building of a drinking water system whereby inhabitants would get drinking water from the river Meuse upstream. The plan was not accepted because it was considered way to costly. The handling of the water system re-entered the agenda only after the first two cholera epidemics in 1848–49 and 1853.

Already starting in 1851 the city started filling up canals against stench and mis-use. From this time on a large number of different plans were developed, amongst which a new plan by Rose and Zocher, with a relatively strong emphasis on enhancing water quality and the quality of the living environment. Other people developed various ideas to build new systems to keep waste interesting for re-use (fosse mobile, liernur).

In 1863 the municipality decided in favour of flushing the sewage (instead of re-using), but without considering all of the consequences. Only part of the city had a sewerage system, most of the waste was flushed in the canals. The building up of a sewerage system took off quickly in the years after, and many canals were digged up or covered. Digging up the canals necessitated a drinking water system, because many city dwellers were still used to drink water from the canals. A drinking water system had been part of some of the plans already in 1859, but initially it was decided that this should be taken up by private enterprise. Entrepreneurs, however, were not interested enough. One thought the amount of drinking water to be bought would not be large enough to earn back the investments.

When in 1866 a new cholera epidemic hit the city, the situation changed once again. The city decided to start its own drinking water system. In 1874 two basins were opened in the east of the city (upstream) to harbor water from the river. People could get tapwater in their houses or go to general water points in the city to buy water. Although it took some time before people started to trust the quality of the water, the use of drinking water spread quickly, as illustrated in Table 20.1 [2]. As the sewerage system got expanded the visibility of the water in the city got reduced.

Overall, the growing importance of land-based transport infrastructure during the decades starting from 1900 and as a result the water in the city lost its function for transport. The harbor started to move westward, together with harbor related industry. The web of small scale water infrastructure narrowed down to large waterways. Water in the city was becoming even less visible, as former canals and rivulets were reclaimed for land, in most cases related to developing land based infrastructure, e.g., roads or railroads. In these ways, the water system in the city got literally decoupled physically from the underground.

Until the end of the 20th century this decoupling of underground on the one hand and infrastructure and the occupation layer on the other continued and so we may speak of a period of lock-in. Even the bombing of the city of Rotterdam on May 14, 1940, during the second world war strengthened this lock-in. As a result of the decoupling and the one-sided attention to economic development the city lost much of its attraction, to rich citizens as well as in the role of a market place. Optimization of the harbor function in the economic sense and adaptation to the conditions posed by finding the economic optimum between speed and cost has had a positive result for the economic function of the harbor, but the attractiveness of the city had been severely diminished.

Recent changes in the vision on water in Rotterdam imply a new transitional period. Now that the possibility of climate change and the associated threat of sea level rise have become a perspective to be taken into account, there is a need to accommodate an extra amount of surface water within the city borders. In the current situation there is insufficient room for water retention, so effort is done to design new kinds of functional infrastructures that are simultaneously able to store water, like green roofs and water retention squares.

In addition, the urban water management plans suggest to once more bring back the water to the city and make it visible again, arguing that surface water creates beautiful sites and makes the city attractive, also for the highly educated people. Urban surface water adds quality to the city and may contribute to solving some of the urban problems, such as neighborhood deterioration. The new water discourse of developing a new and large water infrastructure network is currently starting to being implemented. What they do in fact is using the network layer as a means to improve the overall quality of the occupational layer.

20.5 Conclusions and future research

So what are the lessons with regard to intelligent use of infrastructure? From the history of Rotterdam and the interaction with the way water and the more general infrastructure have been interwoven in its history we can conclude the following:

- Infrastructure has a dual role. Existing infrastructure can impose constraints on the occupation layer, stimulating lock-in and inhibiting transition. However, new infrastructures can give an impulse to transition in the occupation layer.

- Infrastructures are not the trigger, but a result of a vision that directs development.

- Although infrastructure can be viewed as serving the occupation layer, you can also put it to use to transform the occupation layer.

- In that case infrastructure can be taken as a means to enable quality enhancement in the area.

- Infrastructure should be designed in such a way it can be easily adapted to meet the changing demands of the other layers over time.

What are the lessons we can draw from this history with respect to sustainable water infrastructure design? First of all, water has specific characteristics. It is part of a worldwide cycle, strongly interconnected with other earth systems, like weather/climate and movements in the underground. In local developments these interdependencies are often not taken into account because they seem so slow, but such neglect may lead to severe consequences. The New Waterway could have easily become a big disaster, because the sedimentation of the rivers had been neglected. The ever larger technological possibilities led to the belief that water works were only dependent upon this knowledge [6]. The occupation layer and transport infrastructures were developed in ways that seemed to be more and more independent upon the underground, not only in newly developed areas, but also in old inner cities. The canals and rivulets that had had an economic as well as a water regulating function, lost their economic importance, and were therefore superseded by the land based transport infrastructure. The function of the water infrastructure in regulating the water household looked less important because there were other technological possibilities.

At the moment it is becoming clear that these technological solutions do not guarantee a development that can be sustained over a longer period of time. Variations in the weather and change of climate show that the reduction in the possibilities to store water reduces the resilience of the water system. The objective with respect to the amount of water that the system should be able to store can not be reached within the space that has been left for water retention in the city as it is now. The rediscovery of the ecosystem basis of the water system prompts a rethinking of the starting points for city development in such a way that the resilience of the system is enlarged. This rethinking implies a transition to a different regime. Future research will address how to make this transition most efficiently.

References

1. W. N. Adger. Vulnerability. *Global Environmental Change*, 16(3):268–281, 2006.
2. H. Buiter. *Riool, rails en asfalt, 80 jaar straatrumoer in vier Nederlandse steden.* Walburg Pers, Zutphen, The Netherlands, 2005.
3. J. M. Droogendijk and J. S. Verburg. *Langs Rotte, Maas en Schie, Schetsen uit de geschiedenis van Rotterdam.* Wolters, Groningen, The Netherlands, 1911.
4. C. Folke. Resilience: The emergence of a perspective for social-ecological systems analysis. *Global environmental change*, 16(3):253–267, August 2006.
5. C. S. Holling. Understanding the complexity of economic, ecological and social systems. *Ecosystems*, 4(5):390–405, 2001.

6. F. Hooimeijer. Stedenbouw in een waterrijke traditie. In *Ontwerpen met water, Essays over de rijke traditie van ontwerpen met water in Nederland*. VROM, The Hague, The Netherlands, 2007.

7. P. Van de Laar. *Stad van Formaat, Geschiedenis van Rotterdam in de negentiende en twintigste eeuw*. Waanders, Zwolle, The Netherlands, 2000.

8. E. Ostrom. *Governing the Commons: The Evolution of Institutions for Collective Action*. Cambridge University Press, New York, New York, 1990.

9. E. Ostrom. Crossing the great divide: Coproduction, synergy and development. *World Development*, 24(6):1073–1087, June 1996.

10. L. J. Rogier. *Uit de geschiedenis der Nederlandsche Gouwen, Aan de monden van de Rotte en Schie*. Gregoriushuis, Utrecht, The Netherlands, 1933.

11. J. Rotmans, R. Kemp, and M. van Asselt. More evolution than revolution, transition management in public policy. *Foresight*, 3(1):15–31, 2001.

12. A. Van der Schoor. *Stad in Aanwas, Geschiedenis van Rotterdam tot 1800*. Waanders, Zwolle, The Netherlands, 1999.

13. G. P. Van de Ven. *Leefbaar Laagland, Geschiedenis van de waterbeheersing en landaanwinning in Nederland*. Matrijs, Utrecht, The Netherlands, 2003.

14. VROM. Nota Ruimte (White paper on spatial development), 2004.

15. B. Walker, C. S. Holling, S. R. Carpenter, and A. Kinzig. Resilience, adaptability and transformability in social-ecological systems. *Ecology and Society*, 9(2), 2004.

9 789048 135974